# Robotic Exploration of the Solar

## Part 3: Wows and Woes 1997–2003

Paolo Ulivi with David M. Harland

# Robotic Exploration of the Solar System

**Part 3: Wows and Woes 1997–2003**

 Springer

Published in association with
**Praxis Publishing**
Chichester, UK

Dr Paolo Ulivi
Toulouse
France

Dr David M. Harland
Space Historian
Kelvinbridge
Glasgow
UK

SPRINGER–PRAXIS BOOKS IN SPACE EXPLORATION

ISBN 978-0-387-09627-8     ISBN 978-0-387-09628-5 (eBook)
DOI 10.1007/978-0-387-09628-5
Springer New York Heidelberg Dordrecht London

Library of Congress Control Number: 2012944403

Cover design: Jim Wilkie
Project copy editor: David M. Harland
Typesetting: BookEns, Royston, Herts., UK

Printed on acid-free paper

Springer is a part of Springer Science+Business Media (www.springer.com)

*To Irene*
*and to the memory of my number one fan,*
*grandmother Myriam*

# Contents

# Foreword

I was both honored and delighted to be invited to write the foreword to this third volume in the outstanding series *Robotic Exploration of the Solar System*, written by Paolo Ulivi and edited by David Harland. With my lifelong passion for the exploration of space and my great fortune to be able to contribute to it in my work at the Jet Propulsion Laboratory, I already considered the first two volumes to be an essential resource for anyone who wishes to gain insight into these extraordinary interplanetary adventures. If for no other reason then, being asked to provide these introductory remarks would give me the earliest possible opportunity to read the eagerly awaited next installment, and just as I hoped, this is an excellent addition to the series.

This volume covers another remarkable period in humankind's use of robots to extend our reach into the solar system. The story is bracketed by extensive chronicles of two of the most impressive missions ever undertaken. It begins with Cassini. Despite its difficult development history, easily forgotten now that it has been operating so smoothly in orbit around Saturn for nearly eight years, the mission has delivered spectacularly on its many ambitious goals to reveal the nature of the rings, the diverse moons, and the planet itself. As with all the projects described, the book contains a wealth of fascinating details on the ever-resourceful solutions in development and operations to engineering challenges plus the scientific rewards that accrue. As we follow the journey through the solar system and the intensive exploration of the Saturn system, it is as if we are reading the robot's own diary.

The book concludes with Spirit and Opportunity, the Mars Exploration Rovers. The extensive description of the itinerary of these two explorers on the surface of Mars is presented with the same richness as for Cassini, blending what the probes did with what we learned as a result. A preceding account of the many Mars missions that never made it much beyond the stage of intriguing ideas or tantalizing technology demonstrations on Earth reminds us that those that do make it to space represent only a fraction of the creative effort devoted to investigating the solar system.

In between these missions, the book documents all the others, each with its unique constellation of objectives and constraints, problems and solutions, rewards and disappointments. Different as they are, they all have in common great ambition. Sometimes the goal is to undertake a large, comprehensive mission, such as Cassini, and sometimes it is to find ways to conduct a worthwhile and productive mission on a low budget, as Deep Space 1 (DS1), Stardust, Hayabusa, and others did.

During the period that this book spans, I devoted more than seven years to DS1. I began at the first meeting that led to the formulation of the New Millennium Program – when it was only a vague concept for testing advanced, high-risk technologies on dedicated interplanetary missions so that subsequent projects could rely on them without the cost and risk of being the first users – and ended by approving transmission of the commands to conclude DS1's operations in 2001. I greatly enjoyed reading about our experiences on that project, learning not only how to incorporate these new and exotic technologies into a spacecraft but also how to use them in flying the mission. Following the successful conclusion of the 11-month mission to provide a thorough assessment of the systems, we adopted a very different goal. With no further technology objectives, and no longer part of the New Millennium Program, DS1 was devoted to comet science. Early in this two-year expedition, the failure of the sole source of three-axis attitude knowledge, a star tracker using conventional technology, presented us with one of the most daunting obstacles faced by an operations team. Despite everyone's initial impression that such a loss was catastrophic and that the spacecraft should be retired because it had already exceeded the objectives of its primary mission, my team undertook a long and exceedingly difficult rescue effort. Our successful restoration of full attitude control enabled us to acquire NASA's first close-up images and other data from the nucleus of a comet. The account here returned me to the thrill of that adventure.

Although Russia did not conduct any deep-space flights during this time, the exploration of the solar system was not limited to the United States. In addition to international collaborations on most projects, impressive and exciting missions led by Europe and Japan are a fundamental part of the story. I followed all of the missions at the time with great interest and hope for success, whether my own contribution was major, minor, or nonexistent, and in every case I was fascinated with the details presented in these pages. The facts are well researched and documented with extensive citations, making this reference book the best place to begin understanding not only what occurred (and, in many cases, what did not occur), but also how and why.

As I continue my own efforts to reveal more of the solar system's secrets as well as my deep interest in following all other such activities, I also look forward to future volumes in this series.

*Marc D. Rayman*
*JPL, California*
*April 2012*

# Author's preface

The second part of Robotic Exploration of the Solar System ended with launches in 1996, but discussed missions in flight at that time through to their completion. This third part covers missions launched during a short 7-year period between 1997 and 2003, featuring two of the most productive planetary missions ever: Cassini and the twin Mars Exploration Rovers. It was a hectic period, with frequent launches of one "flagship" and many low-cost missions, but a period also marked by frequent failures which cast doubt on the low-cost approach and eventually led to it being essentially abandoned. Notably absent in this period was the former Soviet Union, a key player in the previous parts of this series, but it contributed to the successes of other nations by providing scientific instruments, valuable know-how as well as "launch services". Also noteworthy in this period is the number of low-cost sample return mission that were flown and, with mixed success, provided samples from comets, asteroids, and from our star.

The two most successful missions described in this part are still functioning and producing data at the time of writing. Unfortunately, this means that their histories are incomplete and many of the results from their investigations, especially the most recent ones, are either not yet available or must be regarded as highly preliminary.

In this as well as in the next part of the series, I made some arbitrary choices of which missions to include. In particular, I left out all of the solar observers and other spacecraft stationed at the Lagrangian points of the Sun-Earth system, except for Genesis because it was a sample return mission and because it was actually part of a series of small planetary missions. And I included the Spitzer space telescope because it was placed in a space probe-like solar orbit.

*Paolo Ulivi*
*Toulouse, France*
*April 2012*

# Acknowledgments

As usual, there are many people that I must thank. First, I must thank my family for their support and help. I found invaluable support from the library of the aerospace engineering department of Milan Politecnico, as well as from members of the Internet forums in which I participate. Special thanks go to all of those who provided documentation, information, and images for this volume, including Giovanni Adamoli, Bruno Bertotti, Peter R. Bond, Andrea Carusi, Philip Corneille, Dwayne Day, Brian Harvey, Colin Pillinger, David S. F. Portree, the late Patrick Roger-Ravily, Ingo Richter, Anne Marie Schipper, John T. Steinberg, Paolo Tortora, Vu Trong Thu, and Makoto Yoshikawa. I apologize if I have inadvertently left out anyone. I also thank all of my friends for their support in this time of big changes in my professional life, which is also reflected in the four-year hiatus between this and the previous part of the series. In addition to all of those already mentioned in the first volumes, I must add Simone, Laura, Domenico, Rita, Annalisa, Teresa, Mario, Carlo, Maxent, Vijay, Benjamin, Christophe, Romain, Flora, Alexandra, Damien, Cyril, Boussad and all of my work colleagues from all over the world that I had the privilege of working with in Metz, Lorraine, and in Toulouse.

I must thank David M. Harland for his support in reviewing and expanding the subject, and Clive Horwood at Praxis as well as the new interfaces at Springer for their help and support, notably Maury Solomon. I must thank Marc D. Rayman for the Foreword, for proof reading, and first hand comments on the section on the Deep Space 1 mission.

Although I have managed to identify the copyright holders of most of the drawings and photographs, in those cases where this has not been possible and I deemed an image to be important to the story, I have used it and attributed as full a credit as possible; I apologize for any inconvenience this may create.

# 7

# The last flagship

## BACK TO SATURN

As the Pioneer 11 spacecraft was en route to Saturn and the twin Mariner Jupiter–Saturn (later Voyager) missions were under development, the United States started to study possible orbital missions to follow up this fast reconnaissance of the ringed planet.

Some early proposals of a mission intended to explore Saturn, its system and its intriguing moon Titan were made at NASA Ames Research Center in 1973, reusing technology then under development for the Pioneer Venus and Jupiter Orbiter with Probe (later Galileo) missions. In 1975, the Space Science Board of the National Research Council recommended the development of an orbiter to undertake an in-depth exploration of Saturn and its system of moons and rings.

At about this same time, the Martin Marietta Corporation was contracted by Ames to examine in detail concepts and architectures for the exploration of Titan. Of course, Titan, the solar system's second largest moon after Jupiter's Ganymede, had until then been accessible only to terrestrial telescopes, and little was known of its atmosphere except for its very existence and the presence of methane, identified spectroscopically in 1943, and very likely other complex hydrocarbons. As such, it was evident that Titan represented an environment similar to that of the primitive Earth.[1]

A number of possible missions were studied, including an adaptive penetrator that would autonomously 'decide' whether to use a parachute to slow its fall or to use a retrorocket for deceleration, depending on the density of the atmosphere that it encountered. Another possibility was an orbiter and lander that would first conduct orbital studies and then, after atmospheric braking, make a soft landing.

Fantastic ideas were proposed, including a hot-air balloon, taking oxidizer with it all the way to Titan to burn ambient methane. Titan balloons were also the focus of some European studies. In 1978 the French scientist Jacques Blamont proposed a solar-heated air balloon for Titan. Blamont was also behind the joint French–Soviet Eos balloon for Venus.[2]

1

The most conventional idea from the study, was that of a dedicated orbiter and a Viking-like lander with a payload similar to that of the Martian probes.[3] A Saturn orbiter and a Titan lander mission was also one of the studies revealed by JPL in 1976 as part of the 'Purple Pigeons' program of high-visibility planetary missions.[4] The earliest launch opportunity identified for a Shuttle and a four-stage departure rocket was 1985, with spacecraft arrival at Saturn in 1990. So little was known of Titan that the design of the lander would be difficult. However, it was expected that a soft landing would be assisted by the relatively small mass of the moon and by the atmosphere, which was then believed to be between 20 and 100 per cent as dense as that of Earth. It was already hypothesized that Titan might have standing bodies of liquid hydrocarbons on its surface, so the lander would probably require to be made buoyant.[5]

An in-depth exploration of Saturn and Titan became scientifically compelling in the wake of the Voyager flybys of 1980 and 1981, due in part to what they showed but also for what they were unable to reveal. In particular, the close flyby of Titan by Voyager 1, which was one of the major drivers of the mission, did not show any surface detail on what proved to be a haze-enshrouded body. Just like Venus, Titan appeared to Voyager's cameras as a featureless globe, except for some evanescent detail in the atmosphere. To find out what was hidden beneath the haze, a mapping radar similar to that which was under development for the Venus Orbiting Imaging Radar (later Venus Radar Mapper) would be required. By this time, NASA studies were leaning toward a Galileo-like mission to Saturn, possibly equipped with a pair of atmospheric probes: one for Saturn, the other for Titan. At about the same time, the Scientific Application International Corporation undertook a 6-month study of the exploration of Saturn and Titan. The results were presented at a workshop in the summer of 1983. It envisaged a mission involving a number of vehicles, including two Titan orbiters, a Titan flyby bus, an atmospheric probe and penetrator, a rocket to sound the upper atmosphere, three small balloons and either a large balloon or a blimp. The Galileo-like orbiters would be aerocaptured in Titan's atmosphere.[6] The aerocapture maneuver seems to have been first proposed by the French astronautics pioneer Robert Esnault-Pelterie in 1929. It would involve flying the vehicle deep in the atmosphere of the target upon arrival in order to shed most of its kinetic energy whilst consuming very little propellant (essentially, only for attitude control). On reaching apoapsis in the capture orbit, a small propulsive maneuver would raise the periapsis out of the atmosphere. Whilst aerocapture has been proposed for a number of missions over the years, it has yet to be practically demonstrated.

Also during the early 1980s, Daniel Gautier of the Meudon observatory in Paris suggested that France and the United States could develop a joint Saturn orbiter in much the same manner as West Germany and the US were developing the Galileo Jupiter Orbiter and Probe. As such a joint mission would be extremely expensive, Gautier and Wing-Huan Ip of the Max Planck Institute in Germany merged their independent ideas for a Saturn mission and submitted a joint proposal involving 27 other researchers to the European Space Agency's 1982 call for scientific missions. If approved, it was expected that such a project would then provide the basis for a joint US–European effort. Links between the European and American studies were

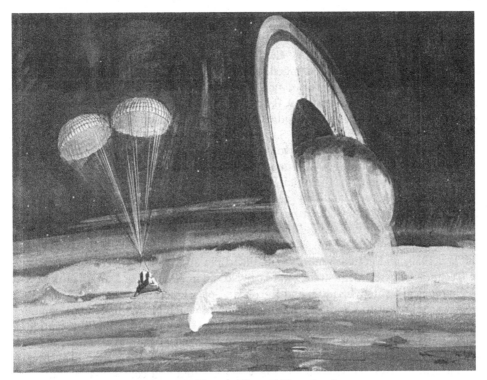

The 1976 "Purple Pigeon" Titan lander.

developed in the US by Tobias Owen of the University of Hawaii. In exploiting the expertise of each partner, as then perceived, it was envisaged that ESA would build the orbiter and NASA would build the Titan probe. Never having built a vehicle for atmospheric entry, ESA had extremely limited expertise in this field. The proposal was named Cassini in honor of the seventeenth century Italian–French astronomer Giovanni Domenico Cassini, who devoted much of his work to studying Saturn, its moons and rings, in the process discovering the moons Iapetus, Rhea, Dione and Tethys, and the fact that the ring contained a narrow gap in its span, now known as the Cassini division.

The early 1980s were also the time of the Solar System Exploration Committee report that would shape planetary exploration in the US for at least the next decade. As related in the previous volume, the committee recommended four core missions for implementation: the Venus Radar Mapper, the Mars Geoscience/Climatology Orbiter, the Comet Rendezvous/Asteroid Flyby (CRAF), and a Saturn orbiter, Titan Probe and Radar Mapper mission.[7] In parallel with the Solar System Exploration Committee study, the European Space Foundation and the US National Academy of Sciences set up a Joint Working Group to explore the possibility of US–European cooperation in planetary missions. This group recommended a Saturn orbiter with a mission concept and scientific objectives which essentially replicated those of the Cassini proposal. Initially, to save costs the

Saturn orbiter was to be a spare Galileo spacecraft whose capabilities, whilst not yet demonstrated in space, were expected to be exceptional. However, in a change of roles, the spacecraft provided by the US would deliver a short-lived lightweight atmospheric probe for Titan supplied by the Europeans. The orbiter already had a fixture for a probe, a system for relaying the probe's data, and a large-capacity data recorder for scientific observations. It could be modified to operate a Pioneer Venus-style radar mapper. The report of the Joint Working Group envisaged a launch in February 1990 and arrival at Saturn exactly eight years later. During its interplanetary cruise, the craft would undertake flyby inspections of asteroids (830) Petropolitana and (250) Bettina, and possibly others.[8]

However, the report of the Solar System Exploration Committee made the reuse of the Galileo spacecraft unlikely, as it favored the adoption of a standardized, low-cost Mariner Mark II spacecraft bus. The first such mission was to be CRAF, and Cassini soon became the second prospective mission of the series. In principle, the decision helped to reduce the cost of the mission, but it also slipped its approval to the late 1980s and actually subordinated it to the approval of CRAF.[9]

Technology assessment studies were carried out at JPL and in Europe at the ESA Technological Center (ESTEC) and European Space Operation Center (ESOC). But ESA was reluctant to sign up to a joint mission to which NASA had yet to commit, and, of course, the recent unilateral cancellation of the US part of the International Solar-Polar Mission made ESA all the more skeptical. Cassini and another proposal were selected for assessment in 1983. Other candidate missions were added after a second call in 1985. These included the Vesta asteroid and comet mission, and the CAESAR (Comet Atmosphere Encounter and Sample Return) comet coma sample-return. Cassini was always rated above the other projects, but remained contingent on a commitment from NASA.[10]

In 1986 the US National Research Council rated an intensive study of Saturn and its system as the top priority for the exploration of the outer planets.

JPL concluded its preliminary 'Phase A' study for Cassini in 1987. Launch was to be on a Shuttle in May 1994, and the Centaur G-prime stage would inject it into a 3-year orbit which had its aphelion in the asteroid belt. There, the spacecraft would perform a deep-space burn to lower its perihelion and set up an Earth flyby in June 1997. This close pass would give it the energy required to head out to Saturn, with arrival in 2002. A number of flyby opportunities were identified for the first circuit of the Sun and for the subsequent cruise to Saturn. These included small objects, intermediate ones, and 115-km asteroid (24) Themis. In line with NASA policy, the targets of opportunity during the cruise would continue a 'staged' exploration of the minor planets.

The baseline mission design called for a 4-year exploration of the Saturn system that featured several dozen encounters with Titan and a number with the other icy satellites, in addition to distant orbits in the magnetospheric tail and polar orbits for particles and fields studies of the auroras and to perform high scientific value radio-occultations as the orbiter passed behind the rings. With the exception of the polar orbits and of shorter revolution periods caused by the smaller mass of Saturn, the mission design was similar to the 'mini tour' of the Jovian system planned for the Galileo orbiter. The European battery-powered Titan probe would be released some

The Mariner Mark II Cassini, as envisaged in the late 1980s.

10 days before the orbiter reached periapsis at the conclusion of its capture orbit, on a trajectory to intercept Titan. It would penetrate the atmosphere at an altitude of about 500 km, traveling at a relative speed of about 6 km/s. The ambient pressure in the upper atmosphere would be extremely small but would rapidly increase, along with the density, and after a few hundred kilometers the probe would have canceled its horizontal speed. At an altitude of about 175 km the probe would have slowed to subsonic speed, the heat shield, having completed its function, would release and a parachute would open for a slow descent. On Earth, this environment would still be considered as space, but on Titan, owing to the low gravity, there would be enough air. Instruments would start to study the atmosphere's structure and composition. Monitoring the Doppler shift of the transmission from the lander during its descent would provide (as in the case of the Galileo atmospheric probe) a measurement of the speed and dynamics of winds. With the rapid increase of density and pressure, the parachute would be released and replaced by a smaller one (or alternatively the probe could free-fall) so as to reach the surface within a reasonable time. The entire descent would take about 3 hours. It was hoped that the probe would survive impact at about 4 m/s and undertake a brief surface mission. As there was no way to know what the surface environment would be like, the probe would have to be designed for a variety of conditions, including splashing down in a hydrocarbon ocean. Since power and data-rate constraints prohibited providing the probe with a direct link to Earth, the orbiter would require an antenna specifically to receive this transmission for relay to Earth. The surface mission would end when the orbiter, passing Titan at an altitude of 1,000 km, set below the probe's horizon.[11,12]

## SELECTION AND 'DESCOPING'

The selection process for Cassini was a long and complicated one, with the initial call for proposals published by ESA in 1982 leading to its selection 6 years later for implementation in the 1990s and arrival in the 2000s. Not surprisingly, given the difficulties of US planetary exploration and space science during the 1980s, and the effects of the Challenger accident, ESA was actually the first to commit itself to the mission, on 25 November 1988 choosing the Titan probe as the first medium-size mission of its Horizon 2000 plan, over four competitors which included the Vesta asteroid and comet mission, an orbiting radio-telescope, an ultraviolet telescope and a gamma-ray astronomy mission. By now, too, the probe had been named in honor of Christiaan Huygens, the seventeenth century Dutch astronomer who discovered Titan and first explained the mysterious varying appearance of Saturn in terms of a flat ring around the planet.[13] Not having yet received the go-ahead, NASA was engaged in a succession of 'definition studies' of the Mariner Mark II spacecraft and of its first two projected missions. In the end, to ease approval of the concept, CRAF and Cassini were proposed in 1988 as a single item in the agency's budget request, but it was not approved until November 1989 at a reduced budget.

The baseline 1994 Shuttle launch was no longer possible for reasons of schedule, and because, as a result of the Challenger disaster, the Centaur stage was no longer available for Shuttle carriage. The mission was switched to the Titan IV, which had been built by the Department of Defense to launch its heaviest payloads and was at that time the most powerful and most expensive rocket in the arsenal. Three launch opportunities were identified which required flybys of Jupiter, in addition to several of either Venus or Earth. The earliest opportunity was in December 1995, the next in April 1996, and the last in 1997. It was decided to aim for 1996. After launch, Cassini would encounter the 80-km main belt asteroid (66) Maja in March 1997 in a heliocentric orbit which would return it to the vicinity of Earth in June 1998 for a gravity-assist that would set up a 3.5 million km flyby of Jupiter in February 2000 and arrival at Saturn in October 2002.

The selection of the payload was announced in September 1990, by which time detailed engineering studies of the hardware were underway. Unlike previous joint missions, owing to their different funding and management procedures this time the two space agencies decided to pursue separate instrument selections. In particular, ESA continued its procedure of selecting a payload on behalf of its member states, which then provided the funding; and NASA continued its procedure of selecting and financing instruments and experiments itself. This management scheme worked so well that it was adopted for all future joint US–European missions.

The Cassini spacecraft would be controlled from a dedicated support area at JPL, while Huygens would have a control room at the European Space Operation Center in Darmstadt, near Munich, with all commands to the probe being passed to JPL for transmission through its Deep Space Network.

As adapted for the Cassini mission, the Mariner Mark II spacecraft was to have two scan platforms, one carrying pointing-critical remote-sensing instruments such as cameras and spectrometers, and the other with a turntable that would rotate at 1 rpm in order to provide the widest possible coverage for mass spectrometers, dust and plasma detectors, plasma-wave sensors, etc. The magnetometer would have its own dedicated boom.[14],[15]

The first change in schedule occurred in 1991, and brought the launch of Cassini forward to 1995, ahead of CRAF, which would now begin its mission in February 1996. On the new plan, Cassini would have one gravity-assist with Venus and one with Earth, then encounter the 38-km asteroid (302) Clarissa on its way to Jupiter. But this plan was short-lived, and budget cuts delayed both missions (which were now projected at a combined cost of $1.85 billion) to 1997. CRAF would launch in April and Cassini in October. There could be no further slippage in the launch date if Cassini were to benefit from a crucial gravity-assist at Jupiter. For ESA, the delay increased the overall mission cost but provided more time for engineering studies prior to initiating the construction of actual hardware. Several months later, NASA and the German space agency, its principal CRAF partner, decided to cancel that mission altogether. For NASA, this eased its financial problems and ensured that it could proceed with Cassini, whose survival was also in jeopardy, while Germany was more than happy because the recent reunification of the nation was creating a host of financial problems. The cancellation of CRAF effectively killed the Mariner Mark II concept.[16]

Faced with escalating costs, in April 1992 it was decided to save $250 million by deleting the scan platforms from Cassini and transfer all the instruments, as well as the antenna for receiving the Huygens probe's transmission, to body-fixed positions. As this would make it impossible to relay the data from the probe to Earth in real-time, the data would have to be recorded on board for later replay. To overcome the limitations of being body-mounted, three instruments were provided with actuators to enable them to rotate individually. This redesign would complicate the planning of observations and make the mission less productive, because Cassini would have to reorient itself to face its instruments toward each individual target and store the data onboard. This made a high-capacity data recorder imperative. After making its observation, Cassini would have to turn to aim its antenna at Earth to download the data, just as had the Magellan Venus radar mapper. Moreover, because the maximum angular speed of the vehicle was 18 times less than that of the intended high-precision scan platform, this would rule out tracking objects during extremely close flybys. On the other hand, this decision enabled Cassini to survive the severe financial crisis that faced NASA in general and space science in particular.

Another cost-saving measure was to delay full development of most of the flight software until after launch. While this would give "the appearance of having a large staff to maintain a spacecraft that was doing relatively little" in space, it would save money at a time when other developments were in need of investment. Moreover, it would allow the team to learn from flight experience, rather than have to deal with and correct anomalies. Of course, the two systems on board, that for attitude control and that for handling commands and data, would have to be provided with adequate capabilities for launch, cruise navigation, course correction, hardware maintenance and payload calibration. As a result of this decision no targets of opportunity were scheduled while crossing the asteroid belt, and unless specific funding was obtained there would be little if any scientific data taken during the planetary gravity-assists. In the worst case, scientific observations were planned to start only in early 2004, a mere 6 months prior to Saturn orbit insertion.

Nevertheless, Cassini was subjected to enormous budgetary pressure in 1992 and 1993. Prior to the deletion of the scan platform, the mission was projected at $1.68 billion. By November 1993, after a restructuring exercise designed to make it more "cost effective", it had been cut to $976 million. Also in 1993, an agency-to-agency agreement was signed by NASA and ASI (Agenzia Spaziale Italiana, the Italian Space Agency) and transformed the following year to a government-to-government agreement which helped NASA to offload some of the cost. As a result, the Italian contribution ended up covering the high-gain and low-gain antenna assembly, a large portion of the radio system, half of the Titan mapping radar, and the visible channel of the visible and infrared mapping spectrometer. Even so, the mission narrowly escaped cancellation in 1994. By then NASA management was stressing the adoption of the "faster, cheaper, better" concept of small missions which would yield first-rate science at the cost of only a few hundred million dollars. Cassini was derided as the obsolete approach to planetary exploration of flying large 'flagships' which involved so many instruments and experiments (and careers) that they could not afford to fail. (NASA administrator Daniel Goldin likened Cassini to Battlestar Galactica.) Cassini escaped being canceled, largely owing to its international flavor. By this time ESA had already committed some $300 million to the Huygens probe, for which hardware integration and testing had begun. A letter from ESA's Director General Jean-Marie Luton to the US Vice President Albert Gore emphasized that Europe viewed "any prospects of unilateral withdrawal from the cooperation on the part of the United States as totally unacceptable". Further, such an act "would call into question the reliability of the U.S. as a partner". He hinted at repercussions on a number of joint programs, including the International Space Station. Although the pressure to cancel the mission was defeated, the US funding was further trimmed to $755 million. As "the last of the dinosaurs", Cassini survived yet another attempt in the US Congress to kill it in 1995.

During the frequent "descoping" exercises, the orbiter finally lost the fixed relay antenna, meaning that the high-gain antenna would have to be used to receive the transmission from the Titan probe. Another cost-saving decision was to use a spare Voyager wide-angle camera optics.[17,18] In order to power Cassini, three new RTGs (Radioisotope Thermal Generators) had to be built. Owing to lack of applications, production of RTG-grade low radiation and high heat output plutonium oxide had been halted in the US, and had to be resumed especially for Cassini. If it appeared that meeting the 1997 launch window were at risk, an old Ulysses spare RTG was available as a last resort.

In the end the total cost of the mission was estimated at $3.3 billion including launch vehicle, operations and international contributions. The development and construction of the orbiter and its instruments was estimated to have cost a total of $1.422 billion, the launcher cost about $427 million, and the rest was accounted for by the RTG power system and contributions to the Deep Space Network through to September 2008. The European Space Agency contributed $500 million and Italy a further $160 million.

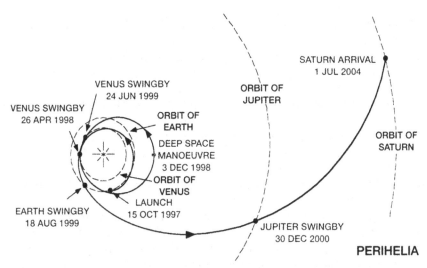

CASSINI - VVEJGA OCT 1997

# INTERPLANETARY TRAJECTORY

Cassini's circuitous journey from the Earth to Saturn via Venus and Jupiter.

As finally approved, the convoluted mission design following launch in October 1997 included two flybys of Venus in 1998 and 1999, a relatively close perihelion at 0.67 AU that would impose the highest thermal loads that the vehicle would endure during the entire duration of its mission, an Earth flyby in August 1999, a relatively distant flyby of Jupiter around the turn of the century, and then arrival at Saturn in November 2004. The baseline launcher was the standard Titan IV. The US military, however, were developing an uprated, more powerful version by replacing the two metal-cased solid-propellant boosters with lightweight carbon composite ones, and if this development were successful NASA intended to use it. It would enable the spacecraft to carry more propellant and would bring forward the arrival date to July 2004. The tankage Cassini inherited from Mariner Mark II had been sized for the greater propulsion requirements of the CRAF mission, so the additional propellant required no changes. This would also facilitate an orbital tour with almost twice the number of satellite encounters.

## THE FLAGSHIP

The Cassini spacecraft had a mass at launch of about 5,650 kg (slightly different masses have been cited), with over 3 tonnes of propellant and the 320-kg Huygens probe. The orbiter had a dry mass of about 2,125 kg and was the heaviest US-made unmanned spacecraft yet to be dispatched beyond Earth orbit. Only the Fobos and Mars 8 spacecraft sent by the Soviet Union and Russia respectively were heavier, weighing in excess of 6 tonnes.

The orbiter was a 6.8-meter-tall stack consisting of a lower equipment module that held the main engines and RTGs; a long propulsion module with tankage and plumbing; a 12-sided upper avionics bus and equipment module of an architecture developed by JPL in the early 1960s for the Mariner-R missions sent to Venus; and finally the high-gain antenna. Attached to the stack were the remote-sensing pallet, the particles and fields pallet, and the Huygens probe with its support hardware and release mechanisms. Other pallets were attached to the upper module, including the radar bay and the radio and plasma-wave antennas and instruments. A long boom projecting from the upper equipment bay held the magnetometer. The three RTGs mounted at the bottom of the stack were to provide 816 W of power at the start of the mission, declining to 641 W at its nominal end. This power would be distributed as 30 V direct current to all users.

The propulsion system was provided by Lockheed Martin. A pair of redundant 1960's-era main engines were installed at the bottom of the stack and had separate plumbing and gimbals capable of directing their 445-N thrust through 25 degrees in each of two axes. This was the first time that redundant main engines were used on a deep-space mission, and the spacecraft could swap between them without ground intervention in as little as 10 minutes. The engines burned self-igniting monomethyl hydrazine and nitrogen tetroxide stored in tanks that comprised half of the length of the spacecraft. Their nozzles had been coated with a thin layer of refractory ceramic that was particularly fragile, and thus vulnerable to strikes by micrometeoroid dust. A small crater in the ceramic could lead to the wall of the thruster burning through, and the loss of the engine. A protective cover was therefore installed on the engines that could be rolled on and off many times to conduct firings, and be jettisoned if ever it failed to open. The engine was extensively tested at White Sands in New Mexico, with a full 200-minute burn clearing the way for its use on the mission. A separate propulsion system with a helium-pressurized tank for 132 kg of hydrazine supplied sixteen 0.1-N and eight 1-N thrusters in two redundant sets. These were in clusters around the lower equipment module to provide thrust in directions parallel to the principal axis and perpendicular to it, and were for small velocity corrections and for attitude control. The attitude control system was responsible for stabilizing the spacecraft, pointing the antenna at Earth for communications, and pointing the instruments at their targets during encounters and data-collection sessions. Attitude would usually be controlled by a trio of reaction wheels plus a spare, and provide a highly stable platform. However, fast slews during satellite flybys and radar passes of Titan were to be performed using the less accurate but more powerful thrusters. Attitude determination would use three redundant inertial platforms, each equipped with four solid-state gyroscopes, plus wide-field star trackers capable of tracking up to five stars simultaneously. Coarse attitude determination during safing events and other emergencies would be by Sun sensors. A longitudinal-axis accelerometer was included in the attitude control system to determine when to shut down the engine during burns.

The upper equipment module contained electronics and avionics. In particular, the command and data subsystem was to receive commands from Earth, verify and process them, and then supply them to the scientific payload and other systems. It would also receive data and process, format, encode and prepare it for transmission to Earth. A number of technological improvements were included in the electronics bay: solid-state switches were used in place of relays and fuses, and two redundant solid-state data recorders were carried instead of reel-to-reel tape recorders to store up to 1.8 Gbits of data each. The 4-meter-diameter high-gain antenna was provided by ASI, and was the largest non-deployable antenna for space applications. It was made of carbon composite on an aluminum honeycomb and, in spite of its size, was a mere 100 kg in mass. The design of the antenna was particularly difficult, since it was required not only to work as an antenna at −200°C at Saturn, but also to act as a sunshield in the early part of the mission during which its temperature would reach 180°C. One low-gain antenna was co-located with the high-gain antenna feed, and a second was installed beneath the structure that supported the Huygens probe. These were to be used at times when the main antenna could not be pointed at Earth; i.e. mainly during the early phases of the interplanetary cruise, when nearby Earth, and when the antenna itself was being used as a sunshade. An S-band receiver was also mounted on the high-gain feed to receive the transmission from Huygens during its atmospheric descent. The mission was to employ four distinct bands for telemetry, requiring a complex selective reflector at the feed of the antenna. Of the four, the Ka-band system had particular stability requirements because it was to be used for high-precision radio experiments, including attempts to detect gravitational waves. The X-band system would be used during most of the mission and provide 20 W of power through redundant amplifiers. When the 70-meter dishes of the Deep Space Network were used as receivers on Earth the X-band could provide a peak data rate of 116 kilobits per second, but if the 34-meter dishes were used the rate was just 36 kilobits per second. Thermal control would be mostly passive, with black and gold insulating blankets covering most of the vehicle's body. The electronics bays would radiate heat to space when required by way of automatically actuated metal louvers. Electrical and radioisotope heaters were used to protect thrusters and engines from the intense cold of space. As mentioned above, while in the inner solar system the high-gain antenna would serve as a sunshade to prevent the remainder of the stack from overheating.

The scientific payload comprised a suite of 12 instruments having a total mass of 362 kg. With around 300 members, the scientific team to analyze the data was one of the largest yet assembled for a planetary mission. Cassini was to determine the composition and properties of Saturn, its moons and rings; determine the processes operating in the atmospheres of Saturn and Titan, and on the surfaces of the moons and rings; investigate the magnetosphere of the planet and its interactions with the solar wind, rings and satellites; map the surface of Titan at wavelengths from radar to ultraviolet; study the other satellites and the icy particles in the Saturnian system; study Saturn's radio and plasma-wave emissions; assess the biological potential of Titan; and seek evidence of gravitational waves passing through the solar system.

The camera or 'Imaging Science Subsystem' (ISS) incorporated two boresighted optics. The f/3.5, 200-mm-focal-length wide-angle camera was a Voyager spare. The

Assembly technicians in the back give a sense of the size of the Cassini spacecraft.

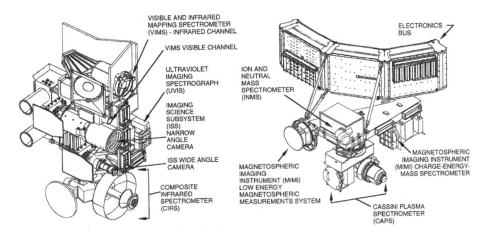

VISIBLE AND INFRARED
MAPPING SPECTROMETER
(VIMS) - INFRARED CHANNEL

VIMS VISIBLE CHANNEL

ULTRAVIOLET
IMAGING
SPECTROGRAPH
(UVIS)

ION AND
NEUTRAL
MASS
SPECTROMETER
(INMS)

ELECTRONICS
BUS

IMAGING
SCIENCE
SUBSYSTEM
(ISS)
NARROW
ANGLE
CAMERA

ISS WIDE ANGLE
CAMERA

MAGNETOSPHERIC
IMAGING
INSTRUMENT (MIMI)
LOW ENERGY
MAGNETOSPHERIC
MEASUREMENTS SYSTEM

MAGNETOSPHERIC
IMAGING INSTRUMENT
(MIMI) CHARGE-ENERGY-
MASS SPECTROMETER

COMPOSITE
INFRARED
SPECTROMETER
(CIRS)

CASSINI PLASMA
SPECTROMETER
(CAPS)

The location of Cassini's remote sensing (left) and particles and fields instruments (right).

f/10.5, 2,000-mm narrow-angle camera used Ritchey-Chrétien optics specially created for Cassini. Unlike Voyager, both cameras used 1,024 × 1,024-pixel 12-bit CCD (Charge-Coupled Device) imagers instead of the now obsolete and technically inferior vidicon sensors, and were able to sense wavelengths that were inaccessible to vidicons. Each camera had a pair of filter wheels of a design inherited from the Hubble Space Telescope. Each wheel on the wide-angle camera had 9 filter slots and each wheel of the narrow-angle camera had 12 slots, giving a total of 18 and 24 filters respectively. Each wheel had one neutral filter, in addition to blue, green, red and infrared filters and specialized methane, methane frost, water frost, ammonia frost and hydrogen-alpha filters. The narrow-angle camera also had ultraviolet and polarizing filters. Cassini was not initially expected to be able to discern the surface of Titan, but in the early 1990s it was realized that the moon's atmosphere ought to be relatively transparent at particular infrared wavelengths. Once this was verified by ground-based telescopes and by reanalysis of Voyager data, a special filter was developed to enable Cassini to see through the layers of haze that formed the upper atmosphere. On a 1,000-km flyby of Titan the narrow-angle camera would have a theoretical spatial resolution of just 6 meters, but in practice the haze would scatter the light and reduce the resolution by several orders of magnitude.

The other satellites were to be observed with a resolution of at least 1 km for the largest ones and 10 km for the smallest. However, targeted flybys were to obtain images with resolutions of 1 to 10 meters. It was considered particularly important to make at least one close flyby of Iapetus, the 'ying-yang' moon with the largest variation in albedo, or reflectance, in the entire solar system. Its dark hemisphere reflected just 3 per cent of the light that it received from the Sun, as black as coal, and similar to a cometary nucleus. Its bright hemisphere reflected typically 35 per cent of the incident light, and some patches reflected up to 60 per cent. The origin of this dichotomy, which was discovered by Giovanni Domenico Cassini himself, was unknown and had not been revealed by the Voyagers. In particular, it was not possible to tell for sure whether it was endogenic, i.e. dark or bright material that had seeped out from the interior, or exogenic, as if one hemisphere of the moon had been 'painted' by material originating from elsewhere. Enceladus was to be another focus of attention. Voyager 1 did not have a good view of this moon but Voyager 2 established it to be a small, extremely reflective object having a surface resembling fresh snow or water ice. Moreover, Enceladus had old cratered terrain juxtaposed to young craterless plains which looked as if they might have been produced within the last 100 million years by some kind of internal activity. Even the older terrain showed extensive modification. This satellite should receive some internal heating from gravitational tides, being in orbital resonance with the larger moon Dione, but that was not expected to be sufficient to cause volcanism in the manner of Jupiter's satellite Io. At best, scientists expected to find activity in the form of geysers. The smaller moons were to be targeted by Cassini to improve on the observations made by the Voyagers. Cassini was also expected to provide confirmation of some of the tiny moonlets suspected from Voyager data, and to discover new ones. In addition, navigation images were to be used to refine the ephemerides of all the satellites.

Extensive observations were to be made of the ring system, in particular on the first and final orbits of the primary mission, when Cassini would not be in the plane of the rings. A search was to be conducted for more moonlets orbiting within gaps, and for braids and kinks along the thread-like F ring. Particularly good observations would be possible at the time of arrival, when Cassini would fly only a few thousand kilometers directly above the rings. After raising its periapsis, Cassini would no longer approach the rings so closely. The CCD cameras were sufficiently sensitive to make detailed observations of the tenuous outer G and E rings.

Since a single 12-bit raw image from Cassini could include up to 1.6 megabytes of data, compression algorithms and 8-bit reduction would be used to yield a more manageable amount of data with varying degrees of information loss. The primary mission was expected to return a total of about 50,000 images of Saturn, its moons and rings, representing an enormous leap forward in both quantity and quality over the Voyagers.

Two spectrometers formed the composite infrared spectrometer, covering near-infrared and far-infrared wavelengths. It inherited components from several similar instruments, including the high-sensitivity spectrometer developed for a Jupiter–Uranus Voyager follow-on in the 1970s that was never flown, and Mars Observer's infrared radiometer. This instrument would be particularly well suited to studying Titan's atmosphere, measuring pressure and temperature profiles and identifying its constituents. Voyager's infrared instrument had only been able to detect some of the complex chemicals, but laboratory and analytical studies predicted there would be other molecules, most of which the new instrument was capable of detecting. In addition, the results should provide insight into the photochemical processes going on in the highest levels of the moon's atmosphere.

The visual and infrared mapping spectrometer was a joint US, Italian and French instrument to provide information on the composition of atmospheres and surfaces of moons by obtaining multispectral images in which every pixel contained a high-resolution spectrum of the corresponding spot, with 96 channels in the visible range and 256 in the infrared. In a sense, it produced 3-dimensional images consisting of a 2-dimensional 64 × 64-pixel image and an additional spectral dimension. For this reason, the images were called 'cubes'. Separate infrared and visible channels were fed light by a 230-mm Ritchey-Chrétien and by a Shafer telescope respectively.

A twin-channel ultraviolet imaging spectrograph was part of ESA's contribution and was to image, measure and analyze emissions present in the far-ultraviolet and extreme-ultraviolet. It was equipped to determine the hydrogen-to-deuterium ratio in Saturn's atmosphere, which ought not to have changed since the formation of the planet. It also had operating modes that would allow it to track solar and stellar occultations by the atmospheres of Saturn and Titan and by the rings.

Six instruments were dedicated to particles and fields. The magnetometer was a derivative of the Ulysses instrument and combined a 3-axis flux-gate and a helium magnetometer, the former mounted inboard and the latter outboard on an 11-meter boom. Compared to the instrument on Ulysses, the Cassini one would have more modern electronics and the capability to accurately measure an intense field such as that generated by Saturn. This instrument would combine scientists from the United Kingdom, the United States and Germany. Pioneer 11 and the Voyagers had shown Saturn's magnetic field to be "remarkable in its unremarkableness". Whereas other magnetic planets had tilted dipole axes, dipoles displaced from the planet's rotation axis, or axes changing in polarity, Saturn's dipole field was almost perfectly aligned with the planet's rotation axis – although it did appear to be displaced along it. The magnetometer would measure higher order components of the field, determine the extent to which it differed from a simple dipole, and detect changes relative to the previous flyby measurements. It was also to study in detail the interactions of Titan with Saturn's magnetosphere, and of the magnetosphere with the solar wind. As for Titan itself, Voyager 1's close flyby had failed to detect any measurable magnetic field.

The radio and plasma-wave experiment had a dipole and a monopole antenna (each 10 meters long and mutually perpendicular) to detect electric fields, a search-coil magnetometer, and a Langmuir probe mounted at the end of a 1.5-meter boom. The instrument was to detect, amongst other things, electrostatic discharges. These were discovered by the Voyagers' planetary radio-astronomy package and initially interpreted as discharges within the rings.[19] However, investigations in the 1980s established that they were due to lightning in the planet's atmosphere. The package was also to address the rotation period of Saturn. In the absence of solid surfaces to observe, the rotational periods of Jupiter, Uranus and Neptune had been determined by timing the modulation period of the magnetic field, which was presumed to be synchronized with the rotation of the core. This had been possible because the axes of these fields were not closely aligned with the rotation axes of their planets. In the case of Saturn, however, that modulation was too small to measure and the period had been determined by assuming that periodicities in kilometric-wavelength radio emissions detected by radio-astronomy packages of the Voyagers matched the core rotation. But it was hard to understand why such radio emissions were modulated at all in the presence of a magnetic field whose axis was co-aligned with the rotation axis. Also, atmospheric scientists were skeptical because the rotation determined in this manner implied that there would be very fast, almost supersonic winds near the equator. A package of three instruments was to characterize plasma processes in the planet's magnetosphere and radiation belts, take measurements in the plasma torus that occupies the orbit of Titan, and measure ions escaping from the icy satellites. It was mounted on an actuator that could swivel though 208 degrees of azimuth, and comprised an ion mass spectrometer, a narrow-angle ion beam spectrometer and an electron spectrometer to enable it to measure the composition, density, velocity and temperature of ions. A triple-sensor magnetospheric instrument consisted of a low-energy magnetospheric sensor, a charge-energy-mass spectrometer, and an ion and neutral atom 'camera'. This was to image the plasma environment around Saturn and remotely determine the charge and composition of the ions. An ion and neutral mass spectrometer would directly determine the chemical, elemental and isotopic composition of ions in the planet's magnetosphere and rings, as well as in the upper atmosphere and ionosphere of Titan during close flybys.

The dust analyzer would be a million times more sensitive than the detector on Pioneer 11, the only dust instrument previously operated in the Saturnian system. It inherited techniques and components from Vega, Giotto, Galileo and Ulysses, and consisted of two separate instruments: a dust analyzer proper to measure the mass, charge, speed, direction of arrival and composition of dust particles; and a high-rate detector which was optimized for the very dusty environment such as at ring plane crossings, but could record only fluxes. Like the previous dust instruments, this one would be managed by the German Max Planck Institute, with the high-rate detector being provided by the University of Chicago.

The radar was a multimode instrument that could be used as a synthetic-aperture radar, an altimeter, a scatterometer or a radiometer. It used five illuminators at the feed of the high-gain antenna to produce circular adjacent spots in the cross-track direction. A narrow spot was located at the nadir of the spacecraft, bounded by two to the left and two to the right, altogether extending several degrees from the axis of the antenna. Depending on the altitude, this arrangement could illuminate a strip on the surface of Titan in the direction of travel varying between 120 and 460 km in width, with a resolution along the track as fine as 350 meters and varying between 420 meters and 2.7 km across the track. This was about three times poorer than the capability of Magellan at Venus. In the Ku-band the radar produced a peak output power of 63 W. In altimeter mode the central beam was used to obtain swaths with a vertical resolution of about 50 meters over a spot area some 25 km in diameter. Scatterometer mode would be used at large distances from Titan to characterize the nature and texture of different types of terrain by the way in which they reflected radio waves. Radiometric mode would measure radiation from an object in a purely passive manner. After calibration and spacecraft slewing, a typical radar session for Titan would begin 70 minutes out with scatterometry, radiometry and altimetry runs as the spacecraft closed in from 22,500 km to 4,000 km, then radar imaging would begin about 16 minutes out and continue through closest approach. High-resolution radiometric data would be collected passively while the instrument was acquiring imaging data. Approximately 1 per cent of Titan's surface would be 'seen' on each

Cassini mated to Huygens (on the left). Before launch, the two spacecraft would be wrapped in golden thermal blankets.

flyby, with the swaths from the primary mission adding up to about 20 per cent of the surface. Since none of the other instruments were boresighted with the radar, it would not be possible to simultaneously obtain remote-sensing data to complement the radar swaths. Until Cassini began to unveil it, Titan was the largest unexplored surface of the solar system.

The radio-science experiment had several objectives. The most obvious was to use flybys to determine the mass of each satellite. The Voyager flybys had provided fairly crude results for most of them, and a better determination would enable more detailed models of their interiors to be built. Flybys of Titan could also measure the amount of deformation due to Saturnian tides, to obtain insight into the strength and elastic properties of its icy crust. In addition, the experiment would use radio-occultations to probe atmospheres, ionospheres and the rings. The radio system was also to conduct two fundamental physics experiments. When Cassini was in opposition to Earth after the Jupiter flyby, it would try to detect gravitational waves by measuring any minute periodic displacement between itself and Earth that might occur. And in 2002 a solar conjunction would be exploited for a more precise determination of the gravitational bending of electromagnetic waves in the vicinity of a mass such as the Sun in order to further measure the accuracy of Einstein's General Theory of Relativity.

The power available to Cassini would not be sufficient to run all of its systems and instruments simultaneously, so scientific sequences would have to be carefully crafted with this constraint in mind, favoring radar, imaging, or particles and fields observations. The typical daily activity in orbit around Saturn would include about 16 hours of collecting remote-sensing data followed by 8 hours of data playback to Earth. While downloading data, the high-gain antenna would be aimed at Earth and the spacecraft left to slowly roll around the antenna's axis as the particles and fields instruments swept the ambient environment.[20,21,22,23,24,25,26,27,28,29,30]

## HUYGENS: ESA'S MASTERPIECE

Huygens was probably the most heavily instrumented planetary atmospheric probe yet flown. It was built in Cannes by the prime contractor Aerospatiale (now Thales Alenia Space) in one of the oldest aerospace facilities in Europe, using components and systems provided by all member states of ESA and some from the US. Indeed, over 40 companies and institutes participated in the development.

The initial concept was for Huygens to be built inside a conical heat shield with a deployable skirt decelerator that would be discarded after the probe had slowed to subsonic speed, with ejectable covers for the instruments. The heat shield would be made of beryllium. Moreover, onboard artificial intelligence was to ensure that the descent trajectory was adapted to the parameters of the atmosphere as measured in-situ. However, the design was revised and simplified, with the probe proper being cocooned between the heat shield and a jettisonable aftbody.

In its final design Huygens had a mass of 320 kg, plus 30 kg of support systems installed on Cassini; the descent probe represented 200.5 kg. The 79-kg heat shield

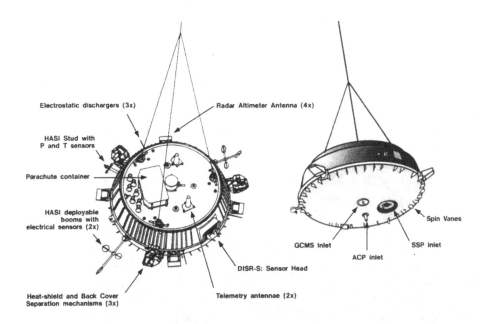

The Huygens descent module. (ESA)

was a 120-degree sphere-cone, 2.7 meters in diameter. It was the most advanced entry vehicle yet developed in Europe, and drew on the technologies developed for the canceled Hermes space shuttle. It consisted of a carbon composite honeycomb structure to which were bonded tiles of an ablative material incorporating a felt of silica fibers buried in a matrix of phenolic resin. A 'milder' thermal protection was sprayed directly on the shield's rear structure. From the little data that was available (mostly from the flyby and radio-occultation by Voyager 1) an engineering model of Titan's atmosphere was created to study the entry aerodynamics, and to assist in the design of the heat shield and parachute. Assumptions had to be made about the composition of the atmosphere. In particular, argon was believed to comprise a few per cent but was undetectable spectroscopically. It could substantially influence the hypersonic aerodynamics and heat balance of the probe. The American engineers who designed the Viking entry 'aeroshell' had faced a similar problem in the form of the flawed measurements of argon in the Martian atmosphere by Mars 6.[31] In the case of Titan a 21 per cent argon content was initially assumed, but this was later reduced to a less conservative 6 per cent. An aluminum shell coated with spray-on silicon elastomer foam formed the back cover that completed the thermal protection of the probe. The descent module was a stubby truncated cone with two aluminum honeycomb platforms supporting internal instruments and systems, plus the fixtures for parachutes and external appendages. The platforms were joined by an aluminum shell that was extensively stiffened to prevent it from buckling when contracting in the extremely cold atmosphere. Titanium struts connected the main platform with the outer shell. The probe was completed by a 1.25-meter-radius dome which faced

Huygens' "Special Model 2" used for drop tests. The rectangular box was the main parachute container.

forward during the descent. A hole on the top platform was to equalize the internal and ambient pressures during the descent.

Huygens was to spin for stability and to provide its instruments with 360-degree horizontal coverage. To ensure this, 36 spin vanes were mounted on the fore dome and accelerometers would measure the axial deceleration and rate of spin. A swivel link decoupled the spinning descent module from the parachute lines. Four square antennas were mounted on the periphery for use by a pair of radar altimeters which operated at slightly different frequencies. Information on the rate of descent and the spin would be fed to all instruments which required it. Three short rods extended at 120-degree intervals as preferred conductors for any lightning. Extensive testing was carried out to characterize Huygens' susceptibility to lightning, and protection against this threat. This was to prevent a repeat of the experience with the Pioneer Venus probes, all of which suffered failed instruments at altitudes below about 12 km, most likely as a result of electrical discharges. This precaution was taken even though the Voyager 1 flyby had shown no evidence of lightning.[32,33,34]

While attached to Cassini, the temperature of Huygens would be controlled by a 5-cm-thick foam layer of thermal insulation, a multilayer blanket of foil insulation, and 35 internal Radioisotope Thermal Heaters (RTH). A white-painted window on the forward face of the heat shield would allow excess heat to escape during flight. An umbilical link from the orbiter supplied power to the probe for periodic health checks and instrument calibrations, but the instruments were not to be used during

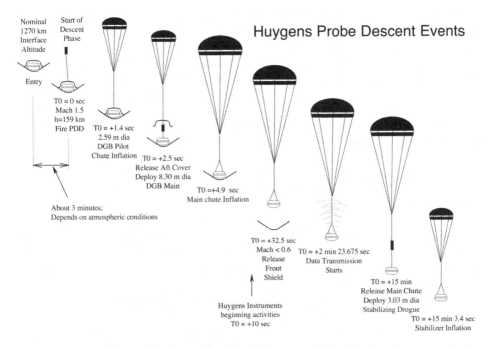

Huygens Probe Descent Events

Nominal | Start of
1270 km | Descent
Interface | Phase
Altitude

Entry

T0 = 0 sec
Mach 1.5
h=159 km
Fire PDD

T0 = +1.4 sec
2.59 m dia
DGB Pilot
Chute Inflation

T0 = +2.5 sec
Release Aft Cover
Deploy 8.30 m dia
DGB Main

T0 =+4.9 sec
Main chute Inflation

About 3 minutes;
Depends on atmospheric conditions

T0 = +32.5 sec
Mach < 0.6
Release
Front
Shield

T0 = +2 min 23.675 sec
Data Transmission
Starts

Huygens Instruments
beginning activities
T0 = +10 sec

T0 = +15 min
Release Main Chute
Deploy 3.03 m dia
Stabilizing Drogue

T0 = +15 min 3.4 sec
Stabilizer Inflation

The Huygens descent profile.

the cruise. The link could also be used to upload commands to the probe, revise its descent sequence, and set the wake-up timer shortly prior to release. The umbilical would be severed at deployment. Since direct intervention from Earth during entry and landing would be impossible because the distance between Saturn and Earth introduced a time delay of several hours, Huygens would effectively be on its own, performing a series of programmed commands and responding to a variety of inputs using onboard control and decision-making systems. As it had no trajectory control capability of its own, the entry trajectory would be entirely reliant on targeting by its carrier. A triple-redundant wake-up timer and a gravity switch were to detect atmospheric deceleration and initiate the descent sequence. After its release from the orbiter, the probe would draw power from five lithium sulfur dioxide batteries whose total capacity of 1,800 Wh would ensure a mission duration of 153 minutes with a comfortable margin; the descent was expected to last no more than 2.5 hours, and the final few minutes were expected to enable the probe to provide some data from the surface. The orbiter was programmed to continue listening for more than 3 hours. A pair of redundant 12-W transmitters on the descent module were to send data at 8 kilobits per second through independent helical wide-beam antennas to the orbiter, which would store the data on its redundant recorders. Barring the unlikely total failure of either the transmitters or the recorders, this was designed to ensure that the data would be secured for relay to Earth. The total dataset was expected to be about 175 megabits.

Although it was not designed primarily to survive a landing, it was realized early on that Huygens would probably withstand the modest impact speed. It was shown that after splashing into liquid the probe would remain buoyant for many minutes with the camera above the 'waterline' and instruments mounted on its underside in direct contact with the fluid. It would probably survive landing on a solid surface unless the material was very hard, in which case the instruments would be crushed. The complete descent sequence was tested by releasing a model, including the heat shield, from a stratospheric balloon in northern Sweden in 1995. This replicated as much as possible Titan's atmosphere in terms of Reynolds and Mach numbers, and was a complete success.[35] It also confirmed that Huygens was likely to survive the landing, because even though it made contact at a vertical speed twice that intended for the real probe the only damage was a bent antenna.[36]

Huygens was equipped to determine the composition of Titan's atmosphere, its structure, temperature and winds. In addition to studying the chemical composition and physical properties of aerosols and clouds, it would analyse the material at the surface. Moreover, it was to investigate the upper atmosphere and ionosphere. This comprehensive analysis required that the probe make a lengthy and stable parachute descent through the atmosphere.

Voyager 1 had revealed Titan to be completely enshrouded by tenuous layers of haze hundreds of kilometers thick, in total sufficient to completely hide the surface from view at visual wavelengths. The atmosphere was primarily molecular nitrogen with a surface pressure of about 1,500 hPa. For such a small body to have a surface pressure 50 per cent greater than that at terrestrial sea level meant its envelope must contain an order of magnitude more gas than is present in Earth's atmosphere. The large amount of inert nitrogen made this the only atmosphere in the solar system (other than Earth) to be nitrogen-dominated. The second most abundant constituent was methane, constituting a few per cent. At ground level the temperature was estimated to be a frigid −180°C, but the pressure was still sufficient to allow methane and ethane to remain liquid and form clouds and rain, as well as rivers, lakes and seas. At high altitudes, nitrogen and methane molecules would be dissociated by solar ultraviolet and by the energetic particles in the Saturnian magnetosphere, later to recombine as hydrocarbons such as acetylene, propane and hydrogen cyanide. The latter was particularly interesting because adenine, one of the DNA nucleotides, is a polymer of hydrogen cyanide. Life as we know it, however, was unlikely to arise on Titan owing to the very low temperature. The complex molecules formed in the upper atmosphere would rain or snow to the surface. By computing the quantity of hydrocarbons produced over the history of the solar system, seas kilometers deep could exist, as could blankets of organic snow kilometers thick. The crust of the moon itself was believed to consist mainly of water ice or an 'anti-freeze' mix of water and ammonia. The presence of methane was in itself a sort of mystery, because the molecule would be destroyed by solar ultraviolet over a relatively short timescale of 10 to 20 million years, so it must be being replenished. This could be lakes or oceans of liquid hydrocarbons, or possibly cryovolcanism – the extrusion of a methane-rich cryogenic fluid from the moon's interior.[37]

Six instruments with a total of 39 sensors were part of the 48-kg payload of the descent module, including two provided by US researchers.

A gas chromatograph and mass spectrometer had its heated air inlet located on the front dome of the probe, and could operate either as a simple mass spectrometer or by breaking down the sample prior to analysis in the gas chromatograph column. It was also equipped with reservoirs to store atmospheric samples collected at high altitude for analysis later in the descent. The heated inlet to this instrument would make it useful after landing for vaporizing surficial material. Furthermore, the mass spectrometer was connected by way of a temperature-controlled duct to an aerosol collector and pyrolyzer. This instrument had a separate sampler that was to be used twice during the descent: once from the top of the atmosphere down to an altitude of 40 km, and again from 23 down to 17 km. It essentially consisted of a filter on which the aerosols suspended in the atmosphere could be collected, and an oven to vaporize the sample through three temperature increment steps. The gases from the oven would then be passed into the mass spectrometer. These two instruments were expected to yield the most interesting results relating to the chemistry of Titan. In particular, they were sufficiently sensitive to detect small concentrations of organic molecules and determine whether these had undergone polymerization to create the precursor macromolecules of biochemistry, such as amino acids, tholines (amino acid precursor polymers) and nucleotides.

A complex multi-sensor optical instrument was to take images and make spectral observations at ultraviolet, visible and infrared wavelengths. It would measure the upward and downward balance of the atmosphere and the degree to which sunlight was being scattered, and image clouds with a side-looking horizon sensor. The most amazing part of the instrument, at least for the public, were the three cameras with different image and pixel sizes which looked downward and to the side. They used a clever design consisting of a single CCD and fiber optics. Also accommodated on portions of the same CCD were the solar scattering experiment and other spectral and photometric observations. The instrument was developed by Lockheed Martin for the University of Arizona. The camera was to operate during the descent and hopefully show the probe drifting across the surface. Twelve sets of three images at different azimuths would provide a full panorama from the horizon almost to the nadir. At an altitude of several hundred meters a 20-W lamp was to be switched on in order to collect the spectral reflectance data that would yield information on the surface composition. In effect the lamp would fill in the parts of the solar spectrum that had been filtered out by the atmosphere. Images would be compressed prior to being sent to Cassini using an algorithm similar to the JPEG (Joint Photographic Expert Group) standard. This would cause some loss of information but the priority was to transmit as many pictures as possible in the short time available. The camera was protected by a transparent cover that would be discarded after the heat shield had been released.[38]

Another multi-sensor instrument consisted of a sensitive 3-axis accelerometer mounted near the center of mass of the probe, fine and coarse temperature sensors, a pressure sensor, and an array of electrical field sensors to detect lightning and to measure the electric properties of the atmosphere and possibly also the surface. The temperature and pressure sensors were mounted on a fixed boom. Two deployable booms held the electrical sensors at a distance from the probe. A microphone was also mounted on one of the booms. As a late addition, the instrument would be able to process echoes from the radar altimeter and obtain additional information on the properties of the surface. The accelerometer would be the only instrument to collect data during all phases of the entry, descent and landing. Its data would be stored on board for transmission to the orbiter together with the real-time data.

The Doppler wind experiment consisted of a pair of ultra-stable atomic rubidium oscillators, one on Huygens and the other on Cassini, to measure the Doppler shift from either of the probe's transmitter signals from which information on the speed of winds on Titan could be extracted. The matching high-precision oscillator on the orbiter was installed on only one of the two telemetry receivers, that for channel A. As a result of Voyager and Earth-based observations, scientists expected Titan to have super-rotating winds similar to those of Venus. By offsetting the antenna from the rotation axis the signal would also provide data on the direction and rate of the descent module's spin, and on the pendulum and swinging motions suspended from its parachute.

Only the surface science package was primarily designed to study the properties of the surface. This instrument consisted of several sensors that were optimized for landing on liquid, but would provide valuable information on a solid surface. Two of these sensors were to measure the dynamics of the impact: an accelerometer to record decelerations from 0 to 200 g, and a 55-mm-long pole on the base that was equipped with force transducers to perform penetrometry measurements for several milliseconds. Between them, the accelerometer and penetrometer could characterize the hardness of the surface on a scale ranging from soft unpacked snow to hard ice. If the landing occurred on liquid, the atmospheric accelerometer would also record wave motions. A tilt-meter would measure either the angle at which the probe came to rest or the oscillations of waves. This instrument would also provide backup data on the spin rate and pendulum motions during the parachute descent.

There was no instrument capable of directly measuring the composition of any surface liquid, so this would have to be inferred by indirect methods. One such was a measurement of the refractive index of the fluid using an optical refractometer, regarded as a fairly sensitive indicator of the composition. A hot-wire thermometer would simultaneously measure the temperature and the thermal conductivity of the fluid. The time taken for an acoustic signal to cross a 10-cm gap would measure the speed of sound in the atmosphere or any fluid, again placing constraints on their composition (measuring, in fact, their mean molecular mass). Another experiment was to detect acoustic echoes from the surface in the last few hundred meters of the descent. In effect it was a sonar, and the nature and local roughness of the surface could be inferred from such echoes. In the case of a splashdown, it would attempt to measure the depth of the fluid. A capacitance sensor was to perform a series of electric measurements in the atmosphere and on the ground. The final instrument of the surface suite was a density meter for use in the atmosphere or in a fluid.[39]

No landing site requirements were imposed other than it must be in sunlight for imaging and other observations, and in line of sight of both the orbiter and of Earth.

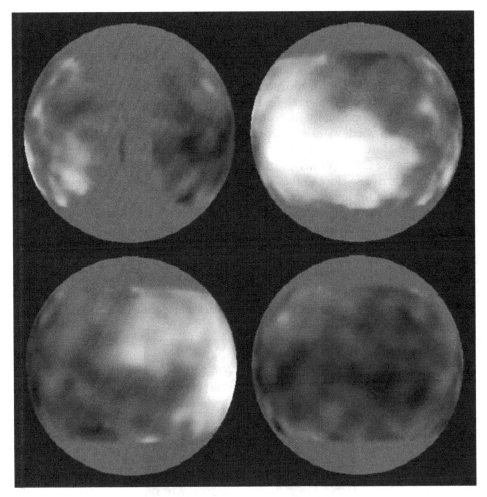

Four hemispheric views of Titan from 1994 Hubble Space Telescope data. The bright 'continent' would later be officially named Xanadu. (University of Arizona Lunar and Planetary Laboratory; STScI)

Owing to the absence of data on the nature of the surface, the target was decided by the illumination and trajectory constraints. The entry ellipse was 200 km wide by 1,200 km long with its center at 18.4°N, 200°E. After the target had been selected, observations by the Hubble Space Telescope and large terrestrial telescopes in the 'infrared window' showed it to be located within a bright 'continent' that was later dubbed Xanadu.[40],[41],[42],[43],[44]

## LAUNCHING THE GIANT

The launch window for Cassini opened on 6 October 1997 and ran to 4 November, with less favorable opportunities until 15 November that would progressively delay the mission's arrival at Saturn to December 2004, then to July 2005, and finally to December 2005. There was a great desire to achieve the launch window, as this was the final favorable opportunity for using a gravity-assist at Jupiter to reach Saturn. Secondary launch windows existed from late November 1997 to mid-January 1998 and another in March and April 1999, but in neither case would Jupiter be available and the alternative Earth–Venus–Earth–Earth–Saturn cruise would last so long that Cassini would arrive so close to the 2009 Saturnian equinox that the rings would be poorly illuminated for imagery and science.

This would be the first time that the powerful Titan IV was used for a planetary mission and, given the desire for smaller and less expensive missions, also the last. At several hundred million dollars each, the use of a Titan for a budget-constrained scientific mission would simply be unrealistic. In fact, the availability of the more powerful version of the rocket had allowed JPL to advance the arrival to July 2004, a date that had been chosen because it would enable Cassini to make a close flyby (some 3 weeks prior to orbit insertion) of the dark outer satellite Phoebe, of which

The Titan IV used to launch Cassini and Huygens on the launch pad. This was the only time the most powerful (and most expensive) US unmanned rocket of the 1980s and 1990s would be used for a planetary launch.

very little was known even after the Voyagers. Orbiting at 215 Rs (Saturnian radii) or 13 million km in a retrograde manner that hinted it was a captured body, Phoebe would otherwise be beyond the reach of the spacecraft during its orbital tour of the system. After the loss of a Titan IV in 1993, NASA briefly considered abandoning this rocket and instead using a complex, risky and expensive scenario in which two Shuttles would separately carry the spacecraft and the escape stage into low orbit to be mated by astronauts, but upon reflection stuck to the original plan.

Meanwhile, experts in celestial mechanics were busily designing the Saturnian orbital tour. The tools and techniques developed for Galileo at Jupiter were reused. Whereas Jupiter had four large moons that could be utilized to shape the orbit of the vehicle, the only 'mission enabler' for Cassini was Titan, almost the same size as Ganymede. Owing to its mass being fifty times that of any other Saturnian satellite, Titan would provide the principal means of shaping the spacecraft's orbit. A single flyby at the minimum distance allowed by its atmosphere would change Cassini's speed relative to Saturn by 840 m/s, which was far more than could be delivered by its propulsion system. Over the 4-year primary mission, Titan could deliver a total velocity change of the order of 33 km/s – more than the combined effects of Venus, Earth and Jupiter during the interplanetary cruise. Cassini could not approach closer than 950 km from Titan because that would involve flying through the outer fringes of its atmosphere, which was sufficiently dense to cause loss of attitude control; this was dubbed the 'tumble density'. Once the risk had been better assessed, this limit would be raised or lowered. Scientists were interested in inspecting the inner icy satellites, particularly Enceladus. These were to be encountered essentially on the way from one Titan flyby to the next. As was the case for Galileo there would be 'targeted' and 'non-targeted' encounters. The non-targeted flybys would not require specific orbit-shaping maneuvers and would usually be relatively distant. Again as with Galileo, tracking and imaging of the satellites against the background of stars would provide a means of navigating between encounters, with the position of the orbiter in the Saturnian system being known to within about 10 km. Ground-based telescopes had collected accurate positions of the Saturnian satellites to enable their ephemerides to be refined, and when Earth and the Sun passed through the plane of the rings at the equinox of 1995 most of the moonlets discovered by the Voyagers were recovered to refine their orbits.

By 1998 tradeoffs between project scientists and mission designers reduced the candidate tours from the eighteen initially proposed to just three which belonged to two 'families'. Since the first 1.2 years of the primary mission would be the same in each case, this period had been finalized by the time of launch. The remaining part of the tour would be decided no sooner than the 1999 Earth flyby. The options were compared for the science they would deliver, for the most efficient inclusion of icy satellite flybys, and for compliance with operational constraints and other tradeoffs. The 'ideal' tour had various scientific and operational constraints. For example, no targeted flybys could be scheduled within the 18-day periods centered on Saturn's solar conjunctions, when communications with Earth would be erratic. There would be one such period about every year. The tour was to include low-inclination orbits for ring occultations, and high-inclination ones for polar atmosphere occultations and ring imaging; orbits with a distant apoapsis opposite to the direction of the Sun for magnetosphere tail studies, and ones with the apoapsis up-Sun in order to allow full-disk imaging of the planet. And ring occultation orbits had to occur as early as possible in the tour because the ring plane, as viewed from Earth, would appear to close as Saturn approached its equinox – a geometry that would render occultations useless.

The primary driver for each Titan flyby would be the imparted orbit perturbation rather than science, although of course the large number of encounters would allow a number of different scientific observations and viewing geometries. For example, a significant fraction of the surface would be mapped by the radar, and high-latitude passes would map its polar regions as well as adjusting the inclination of Cassini's orbital plane.

At least three flybys of Enceladus and several more of the other satellites were a 'must have' for every satellite tour, and each tour had to offer non-targeted flybys at distances under 100,000 km for most of the icy moons. Hyperion and Iapetus were particularly difficult targets to squeeze in without compromising the scientific objectives of the rest of the tour, and in the end they could only be reached by low-inclination orbits. Iapetus, in particular, was a difficult target owing to its distance from Saturn and relatively steep orbital inclination of 15 degrees. And of course, a useful flyby would have to facilitate imaging of the terrain at the transition between the dark and bright hemispheres.

The most efficient tour design proved to start off with equatorial orbits, then tail excursions, up-Sun apoapsis orbits, and finally high-inclination orbits. Each phase would last about one year. To implement this strategy a maneuver known as the 'pi-transfer', or 180 degree transfer, would be required between the second and third phases. After encountering Titan inbound to Saturn, Cassini would then encounter it again outbound, with the second encounter causing the periapsis of the vehicle's orbit to rotate a full 180 degrees from over the night-side of the planet to over the day-side and thus produce an orbital geometry with new opportunities for observing Saturn.[45,46,47]

Assembly of Cassini started at JPL in early 1996 and that of Huygens in Europe in March. On being completed in September, the orbiter was subjected to a series of thermal and other grueling tests. It was flown to Cape Canaveral on 21 April 1997 for final tests, integration with its RTGs and mating with Huygens. Huygens itself was delivered in early May. A CD was mounted beneath the remote-sensing pallet on Cassini bearing the signatures of over 600,000 individuals, including those of Giovanni Domenico Cassini and Christiaan Huygens, which had been scanned from their manuscripts. Huygens had a similar CD with over 100,000 messages.

On 23 February the uprated Titan IVB launcher was tested for the first time in flight, successfully delivering a military 'early warning' satellite into orbit. Cassini would be its second payload. A number of last-minute glitches were encountered. First the rocket developed some small leaks. Then the conditioners that circulated cool air within the aerodynamic shroud were set too high and damaged Huygens' thermal insulation. The probe had to be removed, disassembled and cleaned. This slipped the first launch opportunity to no earlier than 13 October. The White House granted permission to launch the nuclear-powered spacecraft on 3 October. There had been much public debate on the subject, as on previous missions. Although the arguments of those opposing the launch were often ill-informed and at times risible the debate showed that the use of plutonium in space was, and is likely to remain, a sensitive issue. All efforts must be made in the future by the space agencies to make the public, and in particular environmental groups, aware of the real risks involved. When the rocket and its payload were finally ready, the launch had to be postponed twice due to winds at high altitude that could have fatally damaged the structure of the rocket.

Cassini lifted off from Pad 40 on 15 October 1997 at 8:43 UTC in the pre-dawn hours at Cape Canaveral. Early in its ascent, the rocket took an azimuth that would prevent the RTGs from falling in Africa in the event of a mishap. Eleven minutes after liftoff, the Centaur upper stage achieved a low parking orbit. Nineteen minutes later the Centaur reignited over the coast of west-central Africa for an escape burn that was to last 7 minute 15 seconds. A brief loss of telemetry cause some anxious moments, but all was well. The Centaur achieved the heliocentric orbit required for a flyby of Venus and then maneuvered to orient the spacecraft so that its high-gain antenna served as a sunshield. Cassini was released 43 minutes after launch, and 10 minutes later established contact with the Deep Space Network in Australia.[48]


## THE LONG CRUISE

The first checkout of Huygens was performed only 8 days after launch. Except for a low signal strength later identified as external interferences from the orbiter's high-gain antenna, the probe was healthy. Meanwhile, the spacecraft deployed the boom of its Langmuir probe and the wire antennas of its plasma-wave detector.

The launch had been so accurate in terms of both the velocity and trajectory that the trajectory correction on 9 November was just 2.7 m/s. This was a very welcome propellant saving. This time the main engine was exercised, but the second course correction on 3 March 1998 was made using the smaller thrusters. Later that month, Cassini reached its closest distance to the Sun of 0.676 AU. The second Huygens checkout was also impaired by solar noise entering Cassini's radio system through the high-gain antenna. On 24 March Cassini entered safe mode but was recovered within 36 hours. The problem occurred as flight controllers were switching between the main and backup star sensors, and it was discovered that a slight misalignment existed between the two. Although this was well within the acceptable tolerance, it made the attitude control system believe that it had lost track of its target stars. The solution was simply to increase the software tolerances.

On 26 April Cassini made its first Venus flyby, passing at a distance of 284 km over the northern hemisphere and over the night-side with a peak relative speed of 11.8 km/s. The gravity-assist boosted its velocity by 7.1 km/s and deflected its path through more than 70 degrees. This had the effect of stretching the aphelion to 1.58 AU, beyond the orbit of Mars. For several hours just after closest approach, Cassini was occulted by the bulk of Venus from the point of view of Earth. Another, shorter

North Ecliptic View

Groundtrack

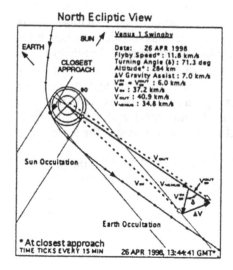

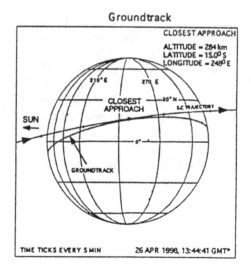

The trajectory and ground track of Cassini during the first flyby of Venus, in 1998.

Sun occultation occurred also. Most scientific instruments remained dormant during the flyby in order to save the operational and data analysis costs. But the radar did bounce radio waves off the surface of Venus. The reflected energy was measured but no imaging was attempted. Moreover, transmission of the full engineering and scientific data from the flyby was delayed until early May because the antennas of the Deep Space Network were busy with other missions. Navigation had been so precise that the course correction planned for just before the flyby, and another for 3 weeks after it, were both deemed unnecessary.[49]

On 3 December 1998 an 88-minute deep-space maneuver of 450 m/s set up the correct trajectory for the second Venus flyby, and 4 days later Cassini reached the aphelion of its stretched orbit. In late December 1998 and January 1999 the vehicle was in opposition to the Sun, enabling the high-gain antenna to communicate with Earth. This was so because at opposition Earth would pass more or less between the Sun and the spacecraft, and the antenna, serving a sunshade, was also aimed in the direction of Earth. This facilitated a 4-week instrument checkout and calibration exercise. Observations of alpha Virginis (Spica) were made by the remote-sensing instruments for calibration. On returning to the inner solar system the dust analyzer was activated; it would operate for most of the remainder of the mission. A course correction on 18 May fine-tuned the second Venus flyby. On 24 June Cassini flew by the planet, this time passing over the southern hemisphere and over the day-side at a minimum altitude of 603 km and a peak relative speed of 13.6 km/s. As before, an Earth occultation occurred during the flyby. Prior to the encounter, it was noted that the planet would cross the field of view of the remote-sensing instruments, so these were activated. The camera was commanded to take a sequence of 24 images for calibration purposes. One of these captured the evening terminator of the planet, showing some atmospheric features. A single long exposure of the night-side and a

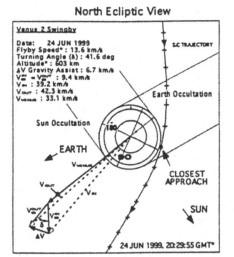

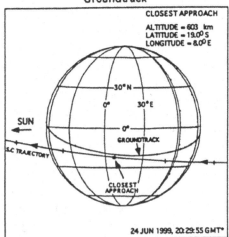

North Ecliptic View

Venus 2 Swingby

Date:         24 JUN 1999
Flyby Speed* : 13.6 km/s
Turning Angle (δ) : 41.6 deg
Altitude* : 603 km
ΔV Gravity Assist : 6.7 km/s
$V_{\infty}^{in} = V_{\infty}^{out}$ : 9.4 km/s
$V_{\infty}$ : 39.2 km/s
$V_{out}$ : 42.3 km/s
$V_{venus}$ : 33.1 km/s

Sun Occultation

EARTH

Earth Occultation

CLOSEST
APPROACH

SUN

24 JUN 1999, 20:29:55 GMT*

Groundtrack

CLOSEST APPROACH

ALTITUDE = 603  km
LATITUDE = 19.0° S
LONGITUDE = 8.0° E

SUN

S/C TRAJECTORY

CLOSEST
APPROACH

24 JUN 1999, 20:29:55 GMT*

* At closest approach
TIME TICKS EVERY 5 MIN

The trajectory and ground track of Cassini during the second flyby of Venus, in 1999.

This is the only 'clear' image of Venus taken by Cassini during its second flyby, showing what seems like waves in the atmosphere along the terminator. (NASA/JPL/Space Science Institute)

short series of calibration observations were made by the mapping spectrometer to measure the light scattered within the optics. The ultraviolet spectrograph obtained several spectra as well. For 2 hours at the time of closest approach on each Venus flyby the plasma-wave package 'listened' for high-frequency radio bursts made by lightning. This instrument was much more sensitive than its equivalent on Galileo, already used for such an experiment. A dozen 'events' were recorded, but they did not appear to be caused by lightning. Overall, therefore, the evidence suggested that either lightning was very rare, occurring once per hour or so, or was different from terrestrial lightning and was mostly undetectable at high frequencies.[50,51] This flyby increased Cassini's heliocentric velocity by 6.7 km/s, raising its perihelion to 0.717 AU and its aphelion to 2.6 AU. Well before reaching aphelion, however, a flyby of Earth would further boost the spacecraft's energy. Leaving Venus behind, Cassini continued to close on the Sun for 5 more days before making its perihelion passage on 29 June. It would never again be that near the Sun.

Four corrections between July and early August 1999 refined the aim to Earth. The possibility of Cassini inadvertently entering the atmosphere, disintegrating and releasing its plutonium-laden RTGs was minimized by robust trajectory design and precise navigation. The chances of a flyby at less than 1,000 km were estimated at one in a million. Until the week before the flyby, the spacecraft was on a trajectory which would miss Earth by thousands of kilometers. This would ensure safety from re-entry even if the spacecraft were somehow incapacitated. Three days prior to the flyby, the magnetometer boom was deployed in order to calibrate that instrument in our planet's well-known magnetosphere. The particles and fields instruments were also operating. These calibrations were assisted by satellites in Earth orbit.

On 18 August Cassini passed 1,166 km over the southeastern Pacific Ocean, not far from the zenith at Easter Island. It was then traveling at a velocity of 19 km/s relative to our planet. Around closest approach, it was in eclipse for over an hour. It departed in a direction opposite to the Sun, flying down the magnetospheric tail. As the spacecraft emerged back into sunlight it was spotted by amateur astronomers in Australia. Nine instruments were operated during the flyby, taking data for calibration and science purposes, including imaging the Moon and studying the Earth's magnetosphere. In contrast with the 'radio silence' at Venus, the plasma-wave experiment noted an almost continuous buzz of lightning bursts. Cassini and the Wind satellite jointly 'listened' for radio emissions from Jupiter. The radar was calibrated in scatterometry and radiometry modes by observing the southern hemisphere of Earth. Cassini took a sequence of images of the Moon from a range of 375,000 km in order to calibrate its two cameras. The visual-spectrum images showed the near-side with a resolution of about 2 km at best which were of no scientific interest. The infrared mapping spectrometer obtained eleven full scans and two partial scans at a resolution of 175 km. These were re-evaluated 10 years later, after the detection of the hydroxyl radical on the surface of the Moon, a tracer of the presence of water, and not only confirmed this detection but also gave a map of its distribution over a portion of the near-side which implied that in places it was present in concentrations as great as 1,000 parts per million of the lunar regolith.[52] Cassini exited the magnetosphere tail at about 60 Earth radii, but re-entered the tail

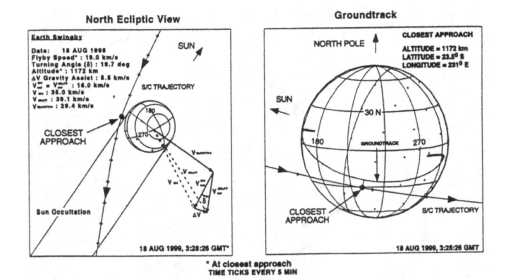

**North Ecliptic View**

Earth Swingby

Date:      18 AUG 1999
Flyby Speed* : 19.0 km/s
Turning Angle (δ) : 19.7 deg
Altitude* : 1172 km
ΔV Gravity Assist : 5.5 km/s
V∞ⁱⁿ = V∞ᵒᵘᵗ : 16.0 km/s
V∞ : 35.0 km/s
Vₒᵤₜ : 39.1 km/s
Vₑₐᵣₜₕ : 29.4 km/s

SUN

S/C TRAJECTORY

CLOSEST
APPROACH

Sun Occultation

18 AUG 1999, 3:28:26 GMT*

**Groundtrack**

CLOSEST APPROACH
ALTITUDE = 1172 km
LATITUDE = 23.5° S
LONGITUDE = 231° E

NORTH POLE

SUN

30 N

180    GROUNDTRACK    270

CLOSEST
APPROACH

S/C TRAJECTORY

18 AUG 1999, 3:28:26 GMT

* At closest approach
TIME TICKS EVERY 5 MIN

The trajectory and ground track of Cassini during its August 1999 flyby of Earth.

several times as the solar wind wafted it across the spacecraft. The magnetospheric imaging instrument investigated the tail to very large distances. The Earth flyby had the effect of increasing Cassini's heliocentric velocity by 5.5 km/s and stretching its aphelion to 7.2 AU, between the orbits of Jupiter and Saturn.

Shortly prior to the flyby, the choice of the Saturnian tour had finally been made. The primary mission would include 74 orbits around Saturn, 44 flybys of Titan at distances between 950 and 2,500 km, three targeted encounters of Enceladus, and one each of Hyperion, Dione, Rhea and Iapetus, and 30 non-targeted flybys of most of the others and many of the smaller moonlets. Unfortunately, it was not possible to squeeze in a close flyby of Mimas. Of particular interest was the first orbit after orbit insertion, because a propulsive maneuver at apoapsis would set up the first Titan encounter scheduled for 27 November 2004. Three weeks before the flyby, Cassini was to establish a collision course and release the Huygens probe. Two days later, a deflection maneuver would set up a 1,200-km flyby of Titan suitable for receiving the transmission from the probe. In the event of a problem, the release of the probe could be delayed to the second encounter with Titan on 14 January 2005. During the primary mission, the period of the orbit would vary between 7 and 155 days, the periapsis between 2.6 and 15.8 Rs, and the inclination between 0 and 75 degrees for a comprehensive program which included imaging and remote-sensing of satellites, ring imaging, magnetospheric studies, and occultations of Earth, the Sun and stars by the planet, by Titan and by the rings.

Thirteen days after the Earth flyby, an engine burn refined the course for Jupiter, and then on 25 September Cassini crossed Mars' orbit as it prepared to enter the asteroid belt. This included turning the vehicle so that the inlet for the dust analyzer was ideally oriented to collect dust. Meanwhile, the ultraviolet spectrograph started

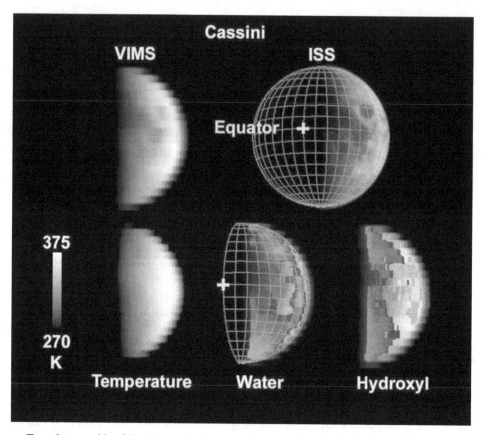

Top: the nearside of the Moon photographed by the imaging spectrometer (VIMS) and camera (ISS) during Cassini's flyby of Earth. Bottom: the temperature of the visible side of the Moon, and a map of water- and hydroxyl-bearing minerals. (NASA/JPL-Caltech/ USGS)

a program of observations of the hydrogen and helium which was penetrating the solar system from interstellar space.

On 23 January 2000 the ability of Cassini to autonomously track a moving target was tested by turning the wide-angle and narrow-angle cameras and remote-sensing package toward asteroid (2685) Masursky, at that time some 1.6 million km distant. Discovered in May 1981, this had been named after Harold Masursky, a geologist who participated in many of the early US planetary missions. The engineering test was mostly successful. Although 15 × 20 km, the asteroid could not be resolved at such a large distance. Useful scientific data was obtained nevertheless, including a high-resolution spectrum by the infrared spectrometer. Despite its association with a family of S-class bodies, Masursky did not appear to be S-class.

In February the probe-to-orbiter radio link was tested for the first time. This involved European controllers generating a signal to simulate the probe's Doppler shift, attenuation, etc. NASA transmitted this signal to Cassini, where it would be

The 'star' at the center of the frame is asteroid (2685) Masursky, 1.6 million km from Cassini. (NASA/JPL/Space Science Institute)

recorded as if it had actually originated from Huygens. The results of the test were unexpected: a major anomaly in the receiver onboard Cassini would prevent it from locking on to Huygens' signal, with the result that it would lose, or corrupt, most of the data. Further tests confirmed the problem. The bandwidth of the receiver on the spacecraft was simply too narrow to accommodate the entire range of Doppler shift that the relative motion between Cassini and Huygens would produce. This would significantly jeopardize the Huygens' science investigations. The narrow bandwidth had been an early choice made when there was to have been a dedicated antenna on the orbiter to receive the transmission from the probe, then when it was decided to delete this antenna and instead employ the high-gain antenna it became inappropriate. However, the subject was not raised on any technical review during development, nor was it identified in ground tests. In fact, end-to-end checks of the radio system in realistic conditions were not required by the project and were never undertaken. Various possible solutions were investigated, such as revising the software to take advantage of the probe's clock bias or modifying the descent by changing the parachute deployment times. One option was to delay the deployment of Huygens until after Cassini had measured the winds on Titan, so that ways could be devised to exploit the winds to reduce the peak Doppler shift.[53] While engineers on both sides of the Atlantic addressed this problem, Cassini left the asteroid belt. It was only the seventh spacecraft to fully traverse this region of space and, as had its predecessors, it emerged unscathed.

In May Cassini finally switched all its communication to the high-gain antenna, which was no longer needed as a sunshade. This enabled particles and fields as well as dust data to be collected more or less continuously from then on. The first new flight software was uploaded for use during the Jupiter flyby. This software would allow Cassini to collect first-class scientific data for the first time since launch. The dust analyzer was operating as Cassini crossed the orbit of Wild 2. This comet was to be visited 4 years later by the Stardust spacecraft carrying a similar detector. But

Cassini did not detect any dust particles. Also during May the plasma from a solar coronal mass ejection washed over Cassini, which was in solar conjunction and on the other side of the solar system from Earth.

On 14 June Cassini made a 0.6-m/s burn to refine its aim for Jupiter. The timing of this encounter was designed to establish a trajectory to Saturn that would enable Cassini to make a flyby of the mysterious moon Phoebe while inbound to Saturn, since that would serve to slow the spacecraft and reduce the propellant required for orbit insertion. While passing a relatively distant 10 million km from Jupiter in late December, Cassini offered a unique opportunity to study the magnetosphere of that planet because Galileo was still in orbit, on its second mission extension. Early in its mission Galileo had suffered a fault that severely limited its ability to transmit large volumes of data. Its high-gain antenna had lost lubrication while in storage in the 1980s, and once in space it had failed to fully deploy. Synergistic particles and fields observations were planned. The two spacecraft were to use techniques which had been developed for use by multiple-spacecraft missions in the vicinity of Earth. During closest approach, Cassini would be monitoring the solar wind upstream and outside Jupiter's magnetosphere, while Galileo would be immersed in it. Then after the encounter Cassini would move into the magnetotail as Galileo reached apoapsis just outside the magnetosphere on the dusk side, and the two spacecraft would swap roles. Scientists worried that Cassini might miss the magnetosphere altogether, but the closest point of approach was dictated by its Saturn requirements, and making observations at Jupiter was at bonus. The flyby was sufficiently distant that if the solar wind were favorable, Cassini would remain outside the Jovian magnetosphere on the inbound leg and enter the tail only later. The Cassini flyby also provided an opportunity to observe the planet and its satellites, although only at long range and over a brief period. Cassini was to make a number of coordinated observations with Galileo and terrestrial telescopes spanning the electromagnetic spectrum to monitor the planet's meteorology, lightning and auroral displays, and Io's volcanic activity, dust streams and plasma torus.[54] Although the joint mission would officially begin in October 2000, in February and May the two spacecraft were in good positions for stereoscopic observations of Jovian radio emissions.

Stability tests of the attitude control system were carried out in September in preparation of the flyby. Long exposures of star fields demonstrated that the camera could precisely stare at a target for periods as long as 32 seconds. The instruments were then checked and calibrated, and observations of Jupiter started on 1 October, by which time Cassini was 84.7 million km out. Color imaging began 3 days later. Also in October, the ultraviolet spectrograph initiated observations of the Io torus. Neutral atoms from Jupiter were first detected by the magnetospheric imaging instrument shortly after the start of observations. Ions 'picked up' by the solar wind were also detected at Cassini's location. These originated from gases issued by Io's volcanoes. Volcanic dust that had been accelerated by Jupiter's magnetic field had first been detected when Cassini was 6 months and 1 AU out. Whereas Ulysses and Galileo had recorded tens of thousands of hits in collimated jets, Cassini observed a continuous flux and only several hundred hits.[55]

By the middle of November Jupiter filled the frame of the narrow-angle camera

The shadow of Europa transiting across the disk of Jupiter on 7 December 2000. The moon itself is not in the frame. (NASA/JPL/Space Science Institute)

and mosaicking was required to cover its full disk. At one point in the inbound leg Cassini was swept by another solar coronal mass ejection. Galileo, which was then descending from apoapsis, documented how Jupiter's magnetosphere reacted to this change in the solar wind. On this and other occasions, joint observations by the two missions confirmed that those radio emissions from Jupiter's magnetosphere which were not related to Io were triggered and controlled by the solar wind and resulted in auroral processes similar to those on Earth. On another occasion in December, while the Hubble Space Telescope was observing the Jovian auroral displays in the ultraviolet, Cassini's plasma sensors and magnetometers monitored the parameters of the solar wind upstream.

On 18 December Cassini turned its attention to the largest of the irregular outer satellites, Himalia. Although the flyby range was 4.44 million km, Himalia was at that time the closest known member of the Jovian system to Cassini. At a resolution of 27 km per pixel, 186-km Himalia spanned at most 6 pixels in the narrow-angle camera. This was sufficient to resolve its irregular shape and to measure its albedo, which confirmed its resemblance to the carbonaceous asteroids and supported the hypothesis that it had been captured by Jupiter.

Meanwhile, on 15 December Cassini unexpectedly and autonomously switched

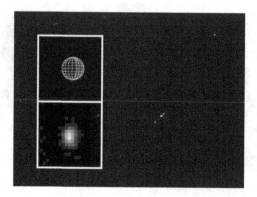

A Cassini image of Himalia (arrowed), the sixth largest moon of Jupiter. The insets show the geometry of the observation, with the half lit moon illuminated from the left, and the tiny disk of the moon enlarged 10 times. (NASA/JPL/Space Science Institute)

from reaction wheels to thrusters for attitude control. Engineers at JPL discovered this only two days later. It appeared that one of the wheels was experiencing more friction than usual, and hence drawing a higher current. To continue the scheduled program of remote-sensing would use too much hydrazine fuel, so on 19 December the decision was taken to temporarily terminate scientific observations that required precise pointing. Cassini was then oriented with the high-gain antenna facing Earth for optimum communications while the problem was investigated. Experiments that did not require an accurate orientation continued. Unfortunately, it was not possible to complete the planned Himalia observations to determine its rotation period. Also lost were a number of satellite and ring observations. Diagnostic tests showed that the wheel needed increased torque when spun up at low speed, but then rotated as it should at higher speeds, which held out the prospect that it was a one-off problem. It appeared to have been caused by lubrication that had not spread evenly when the wheel had been spinning for a protracted period at low speed. Further tests cleared Cassini to resume its observational program from 28 December, just in time for the closest approach. However, it was decided to minimize wear on the reaction wheels during the cruise from Jupiter to Saturn and use the thrusters instead.[56] Early on the same day, at 140.2 Rj, Cassini crossed the Jovian bow shock a full day earlier than expected; an event duly recorded by the plasma-wave experiment. A weakening of the solar wind had evidently allowed the magnetosphere to temporarily inflate well beyond 100 Rj. The shock front was seen to flap back and forth many times through to 3 January, but Cassini never really penetrated beyond the magnetopause into the magnetosphere proper.

At 10:05 UTC on 30 December 2000 Cassini reached its closest approach to the planet at 9.72 million km, equivalent to about 136 Rj, and well outside the orbit of Callisto. It passed Jupiter at a relative speed of 11.6 km/s on a trajectory very close to the equatorial plane. The powerful gravity bent the spacecraft's trajectory toward Saturn and the 2.22-km/s increase in its heliocentric velocity stretched the aphelion to 9.28 AU. Imaging of Jupiter's ring, satellites and atmosphere continued until 15

Ganymede over the limb of Jupiter and the Red Spot. This image was taken on 6 January 2001, less than a week after Cassini hit the minimum distance from the planet. (NASA/JPL/Space Science Institute)

January, by which time Cassini was 18 million km from the planet and in position to start imaging its night-side in search of auroras.

Movies covering full rotations of Jupiter's atmosphere from the ultraviolet to the near-infrared were taken during the inbound leg in order to monitor the evolution of atmospheric features. Near the time of closest approach the disk occupied a 3 × 3-frame mosaic. The best resolution of 58 km per pixel was less than achieved by the Voyagers and Galileo but no previous mission provided uninterrupted monitoring over such a lengthy interval. Changes in atmospheric jet speeds relative to Voyager observations were detected, in some cases amounting to several tens of meters per second. Observations of the Great Red Spot swallowing smaller convective storms suggested that the larger, longer-lasting hurricane sustained itself by absorbing the energy of smaller systems. At higher latitudes small eddies were seen to form and dissipate in days and weeks, and Cassini was even able to document the merging of the remaining white ovals which had appeared in the 1930s south of the Great Red Spot. For the first time, huge masses of 'air' rising in the dark 'belts' and absent in the bright 'zones' were observed, and there were dozens of thunderstorms. In all likelihood, the Voyager-era model of the bright zones being where air expands and cools as it ascends and the dark belts being where it descends back into the interior was too simplistic. The most impressive atmospheric observations were of a large ultraviolet-dark oval approximately 28,000 km by 18,000 km (about the size of the Great Red Spot) at about 60°S that was seen to form early in the encounter. This oval was visible only intermittently from late October, was then seen to stretch, and may have begun to dissolve by the time the spacecraft left the planet behind in late December. A similar spot had been seen by the Hubble Space Telescope in 1997. A hint at the nature of the one observed by Cassini was that its southern edge seemed to match the latitude of the strongest auroral emissions, raising the prospect that it was created by the energetic particles 'raining in' from the magnetosphere reacting with an ultraviolet-absorbing chemical – possibly methane, acetylene or some other hydrocarbon. In the north, a wavy pattern which was visible in ultraviolet images appeared to match cloud waviness seen in blue-filtered images.

Many of the science sequences were dedicated to the Galilean satellites, and in particular to Io. Over 500 images of the volcanic moon were taken during the flyby. On 30 and 31 December, volcano monitoring was undertaken from a range of 10 million km. Long exposures showed at least two vast plumes. One, 400 km high, originated from Pele in the equatorial region and was not in full sunlight. The other appeared to originate at a high northern latitude and later analysis showed its source to be Tvashtar Catena, which was on the far side of Io from Cassini at the time of the observations. Tvashtar Catena was where Galileo had photographed the iconic 'lava curtain' during its November 1999 flyby of the moon. On 1 January 2001 Io was observed while in Jupiter's shadow and simultaneously crossing the center of the plasma torus. The glow of electrons interacting with the tenuous atmosphere of the moon was recorded in visible light. Another eclipse 4 days later provided some very good spectral information on the glows. When Europa was imaged in eclipse it also displayed atmospheric glows. On 29 December and 11 January Ganymede was imaged in eclipse. Images of the smaller moons Metis and Adrastea were obtained on both the inbound and outbound legs in order to refine their orbits and to confirm that they were the main sources of ring material. Thebe was spotted in some of the images. During the approach phase, no fewer than 160 images were taken to search for small as-yet-undiscovered moonlets traveling between the orbits of Amalthea and Callisto, but none were found. Many observations targeted the ring in order to better understand the type and size of its constituent particles. These observations were an excellent rehearsal for when Cassini would study the flamboyant rings of Saturn.

Other instruments were also obtaining data. The composite infrared spectrometer profiled the abundance of minor hydrocarbons with depth and latitude in Jupiter's atmosphere, providing an interesting comparison with the Voyager data. Seasonal effects were evident between the northern autumn of 1979 and the northern summer of 2000. Despite operating at larger distances than it was designed for, even when at the time of closest approach, the energetic neutral atom imager was able to detect an unsuspected toroidal cloud of gas whose densest regions were just outside the orbit of Europa. This cloud consisted predominantly of hydrogen, oxygen and water molecules stripped from the moon's icy surface by Jovian magnetospheric particles. The satellites were also monitored by the infrared spectrometer, albeit at a very low spatial resolution. On 2 and 3 January 2001, as the vehicle receded from Jupiter, the radar was used as a radiometer by scanning the high-gain antenna across the planet and its environs to record the thermal emission from the atmosphere as well as the radio emissions from high-energy electrons in the radiation belts. At the same time, the magnetospheric imaging instrument mapped the shape and composition of the inner magnetosphere.

As it receded from Jupiter, Cassini had a view over the dark side of the planet. In the second phase of a joint study, it observed the auroral displays on the night-side while the Hubble Space Telescope did so on the day-side. Auroral emissions were seen to vary on short timescales, probably in response to changes in the solar wind pressure. For 8.5 hours on 9 January, and again for over 13 hours the following day, instruments indicated the spacecraft was flying within Jupiter's magnetospheric tail 14 million km 'downwind' of the planet. Models of the shape of the magnetosphere showed that a region of increased solar wind was causing the dusk bow shock to wash back and forth over Cassini. This was confirmed by Galileo, which was at that time climbing to apoapsis. As might be expected for a compressed magnetosphere, the Hubble Space Telescope observed a small but bright auroral oval several days later. However, no correlation was evident between the state of the solar wind and X-rays observed by the Chandra satellite in Earth orbit to be originating from the Jovian auroras.

Cassini made a correction on 28 February to refine its aim for Saturn. When it ceased imaging on 22 March 2001 it had transmitted approximately 26,000 pictures of the planet, its satellites and ring, which was comparable to a Voyager flyby. It continued to cross in and out of the magnetospheric tail shock until early March, by which time it was approximately 800 Rj beyond the planet. The Jupiter 'millennium mission' officially concluded on 31 March.[57,58,59,60,61,62,63,64,65]

## GLITCHES AND SOLUTIONS

Now on the final leg of its cruise, Cassini adopted a low-activity status. The main engine propellant lines had to be flushed for at least 5 seconds each year to prevent dirt and oxidation accumulating in the pipes and filters, so small course corrections were executed. Meanwhile, the project team were defining scientific sequences for the orbital tour consistent with the many constraints, in particular those arising from the fact that the instruments were on fixed mounts rather than on scan platforms. Of course, when any particular observation would be made could not be decided until shortly before each encounter, in order to provide the necessary flexibility and cope with the inevitable new discoveries, but the baseline plan was drawn up.[66]

Some scientific observations were made in the 3 years between the Jupiter flyby and the start of activities leading to Saturn orbit insertion. Most of these made use of the radio link over three wavelength bands and provided high-quality Doppler navigation data even when the vehicle was very close to the Sun in the sky and the radio waves passed through the plasma of the corona. This was to facilitate accurate navigation at Saturn orbit insertion, which was due to occur just a few days ahead of solar conjunction. The link was demonstrated during the conjunctions of June 2001 and June 2002. During the latter, Cassini passed within 1 degree of the center of the solar disk, which was itself half a degree in diameter.

From November 2001 to early January 2002 Cassini performed an experiment to detect gravitational waves passing though the solar system. The accurate tracking and stable orientation during this time enabled scientists to determine, among other things, the minuscule acceleration due to the thermal radiation of the RTGs, which was of the order of a few millionths of a billionth of a meter per second squared. This was a particularly interesting result because ill-modeled radiation acceleration from RTGs was considered to be the most likely cause of the 'Pioneer Anomaly'.[67]

The most interesting experiment, however, was carried out during the 2002 solar conjunction, at which time Cassini was more than 7 AU from the Sun. Two radio carriers were sent to it by a Deep Space Network antenna specifically built for the experiment, to which the radio system of the vehicle responded using all three of its radio carriers. By analyzing the returned signal it was possible to precisely measure the time delay and lengthening of the path of an electromagnetic signal on a line of sight that passed close to a massive body (in this case the Sun) due to the relativistic deformation of time-space. The results confirmed Einstein to within 10 parts per million; the best so far.[68]

A number of technical glitches were encountered and solved during this final leg of Cassini's interplanetary cruise.

In May 2001 engineers found substantial haze in images of the bright star Spica that were taken to calibrate the camera and refine knowledge of its pointing relative to the other remote-sensing instruments. Since leaving Jupiter the spacecraft had been using thrusters for attitude control so as to save the reaction wheels for later in the mission, and it seemed that thruster exhaust had found its way onto the optics. Since similar contamination problems had occurred during other missions the usual solution was adopted, which was to heat the optics to evaporate the contaminant. Of course, this would only work if the problem was indeed contamination of the optics and not a fault of the CCD detector. In spite of the contamination, on 13 July the camera was aimed at Saturn for the first time. The fuzzy image of the planet and its rings spanned only several dozen pixels and Titan was a mere dot close alongside. On 14 January 2002 steps to decontaminate the optics were started by raising their temperature by almost 100°C. Images taken in May showed a positive response and by July the quality was back to normal.

The Huygens Recovery Task Force was continuing its investigation of how best to ensure that Cassini would be able to receive the transmission from the probe. The relay system was exercised in February 2001 to better define the anomaly, and the task force finally published its conclusions on 29 July.

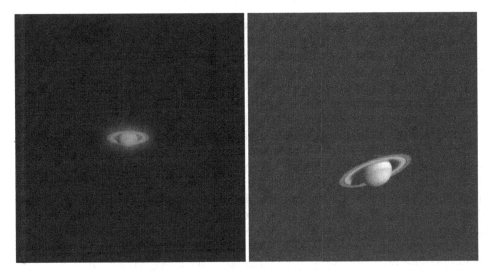

Two images of distant Saturn taken respectively on 13 July 2001 (left, the first picture of Saturn ever taken by Cassini) and on 21 October 2002 (right). Camera contamination shows heavily in the image at left. Titan is visible as a bright star in the 2002 image. (NASA/JPL/Space Science Institute).

It was recommended to add a 4-hour pre-heating phase of Huygens' transmitters prior to atmospheric entry. This warm-up had been scheduled for 45 minutes, but extending it would slightly reduce the data stream frequency and improve margins. This would not seriously drain the batteries. If only this revision were implemented, it was estimated that less than 10 per cent of the data would be received and all of the scientific goals would be lost. Inevitably, therefore, the orbital profile of Cassini would have to be extensively modified. One requirement was to release Huygens as soon as possible in the primary mission because to retain it would obscure the field of view of some of the instruments and the added mass would significantly increase the propellant consumed in maneuvering. Also, there was a desire to retain as much of the original tour as possible, since so much time and effort had been devoted to designing it.

The solution that was recommended involved drastically increasing the distance between Cassini and Titan at closest approach and reducing the delay between the orbiter and probe from 4 to 2 hours. To achieve this, the first two orbits of the tour were replaced with three shorter ones. In the original scenario Huygens would have performed its mission at the first Titan flyby in late November. During the descent the distance between Cassini and Huygens would have ranged between 77,000 and 27,000 km as the former was inbound for a 1,200 km flyby. In the revised scenario the probe would be released on 25 December and reach Titan on 14 January 2005, less than two months behind schedule. The distance between the two would range between 71,000 km at entry and 60,000 km at the nominal end of mission. The descent would occur while Cassini approached, passed through minimum distance, and began to leave the moon behind, so the relative velocity (and the Doppler shift) would start at a maximum of about 3.8 km/s, decrease to zero at closest approach and then increase to 0.8 km/s; always remaining within the tight receiver band. On the originally intended trajectory the relative velocity would have remained almost constant at 5.6 km/s. One month later Cassini would return to Titan and resume the intended orbital tour. There was a backup option to postpone the delivery one more orbit and resume the tour almost 6 months later, but this would use more propellant and would mean losing the first two Enceladus encounters.

In addition to guaranteeing that all of the expected data from Huygens would be recovered, the revised mission profile had a number of advantages. From a purely engineering consideration, high-gain antenna pointing requirements were found to be less severe than for the original plan. As regards science, the revision advanced the first Titan flyby from late November to late October 2004. A second flyby, with Huygens still attached, would occur in December. During these flybys, the strength and direction of the winds on Titan would be characterized to assess their impact on the Doppler shift of the transmission from the probe during its atmospheric descent. And of course early observations would improve the ephemeris for Titan, which in turn would facilitate more accurate high-gain antenna pointing instructions.

Observations of the other satellites were also impacted. Cassini would now make a relatively close flyby of Iapetus one week after the probe had been released. This was welcomed because it would permit observations of the hemisphere opposite to that which would be visible on the previously scheduled single targeted encounter. Another bonus was an extra Enceladus flyby in February 2005. On the minus side, the increased range of Cassini from Titan during the probe mission would decrease the strength of the received signal to one-tenth of that originally intended, and to implement the new tour would require an additional 95 m/s velocity change which was equivalent to about one-quarter of the available propellant reserve. This would be consumed by extending the orbit insertion burn in order to shorten the first orbit by 2 months. The periapsis raising burn would also need to be significantly longer. And additional propellant was needed to open up the miss distance at Titan after the probe was released. Given the ample margins provided by the excellent navigation thus far, this unexpected use of propellant was not an issue for the primary mission but it would consume one-third of that earmarked for a possible extended mission.

The landing ellipse remained near 190°W, but in order to ensure that the Sun and Earth would be visible to both Cassini and Huygens during the entire duration of the descent the target was displaced 20 degrees southward.

Other possible options were not implemented. One involved using the clock drift of the probe's transmitter. The clock had drifted significantly during the flight, and this could change the apparent Doppler shift and hence compensate in part for the problem. But how much it would compensate was difficult to predict, being due to the behavior of the hardware at low temperatures. Another proposal that was held in reserve was to insert sequences of 'zeroes' into the data stream in order to increase the signal-to-noise ratio.

Later in 2001 the attitude control system was exercised to evaluate its ability to slew Cassini sufficiently accurately to maximize the received signal from the probe. The new descent profile was simulated using Earth antennas as the transmitter, and the recovery plan was fully validated.[69],[70],[71] Meanwhile cruise data collection was continuing. Cassini refined its course on 3 April 2002, at about the same time as the radio and plasma-wave package began to detect radio emissions from Saturn, which was about 2.5 AU away. However, more technical glitches were in store and by the fall there were signs of hardware wear, with one of the reaction wheels in particular occasionally showing increased friction. It was decided to keep the balky wheel in reserve and switch over to the backup. On 21 October Cassini took more pictures of its destination, now 1.9 AU away, capturing the planet from a perspective that was impossible from Earth. Also visible but still merely a spot, was the orangeish Titan.

Activities began to surge in 2003 with three maneuvers instead of the usual one. These were to refine the encounter with Phoebe. The first was made on 1 May and the second on 10 September, this time using the attitude control system thrusters. In the meantime, new software had been uploaded to perform the orbit insertion burn, the periapsis raising burn, the Titan targeting and the Huygens relay. In particular, it incorporated autonomous decision-making capabilities for the insertion burn. A full rehearsal of this maneuver was carried out in August, and a burn of the main engine on 1 October validated the systems and algorithms that would be employed. Cassini had gone into solar conjunction in late June. Idle commands were sent from Earth as the angular distance to the Sun diminished to assess how reliably the probe was receiving them. A second General Relativity experiment was undertaken during this time, but a malfunction in the radio system made the data difficult if not impossible to interpret. Only one month later a component of the Ka-band transmission system failed. This would degrade the Doppler data obtained from the next solar opposition and during the Saturn orbiting phase of the mission.

On 22 July 2003, 1.08 AU from Saturn, bursts of radio waves were detected that were initially interpreted as emissions from lightning in Saturn's atmosphere. It was then realized that these bursts were from Jupiter. Cassini would continue to detect Jovian emissions for years, even when that planet was over 10 AU distant.[72] Other data was also collected. For a year starting in mid-2003 the ultraviolet spectrograph monitored the glow in the Lyman-alpha band caused by interstellar gas entering the solar system, and this was correlated with observations by the equivalent instrument on Voyager 1, which was then almost 100 AU from the Sun.[73]

## ARRIVAL AND FIRST DISCOVERIES

While Cassini and Huygens were being planned, built and flown, knowledge of the Saturnian system was advanced significantly. In 1988 American scientists began to use the Goldstone deep-space communication's antenna as a radar to investigate the surface of Titan, the high reflectance of which implied a 'rough' surface that was incompatible with the presence of liquid hydrocarbons. The most important finding, however, derived from a program of observations by the Hubble Space Telescope begun in 1994. Images in the near-infrared showed several dark areas and a single large bright 'continent' that straddled the equator on the leading hemisphere. This singular feature, which had already been noted by radar, was later named Xanadu. It was not possible to conclusively establish from these images whether there were 'seas' of hydrocarbons but the dark areas could be lowlands in which organics had pooled. Other infrared observations using the world's largest telescopes discovered short-lived clouds of methane at high altitudes, as well as south polar clouds. The asymmetry in the appearance of the atmosphere over the northern and southern hemispheres discovered by Voyager 1 was realized to be a seasonal phenomenon involving the transport of haze from one hemisphere to the other. The presence of liquid bodies of hydrocarbons on Titan was hotly debated. In late 2001 and 2002 the radio-telescope at Arecibo in Puerto Rico, functioning as a very low resolution radar, recorded mirror-like reflections from areas on the surface that hinted at large, flat features that could be calm lakes or seas. Infrared images and spectra, however, showed characteristics resembling those of Ganymede, a moon covered in water ice with little evidence of organic sediments. Finally, observations from Earth in 2003 and 2004 in support of Cassini failed to detect any mirror-like reflection of sunlight, thereby eliminating the prospect of a global ocean.[74,75]

Systematic observations of Saturn by Cassini started in December 2003 when it was 111 million km out. The ultraviolet spectrograph detected a cloud of atomic hydrogen at the orbital radius of the E ring on 25 December. This was seen again in February and early March 2004 before becoming asymmetric and diffusing over the next few weeks. Suggestions for the source of this cloud were impacts occurring in the E ring and ice particles released by Enceladus. The E ring was one of the most mysterious of Saturn's rings, extending from 3 to 8 Rs and encompassing the orbits of four major icy moons: Mimas, Enceladus, Tethys and Dione. It was a broad and diffuse ring, with its greatest density in the vicinity of Enceladus. In contrast to the other rings, the E ring lacked evidence of collisions and accretion, suggesting that it originated from particles released by one of these moons, most probably Enceladus. In the 1990s the Hubble Space Telescope had discovered a torus of hydroxyl ions coincident with the E ring. The inner part of the magnetosphere contained a lot of hydroxyl and oxygen, with hydrogen further out. Most if not all of these molecules and ions were suspected of originating from the dissociation of water ice particles in the E ring. Moreover, the fact that they would be lost within 100 million years indicated there was a mechanism replenishing the E ring with ice. This was similar to what the volcanoes of Io did for Jupiter's magnetosphere, although on a smaller scale.

From 10 to 30 January 2004 Cassini studied the solar wind almost continuously, measuring the magnetic field and the density and velocity of its plasma. In addition, the plasma-wave experiment recorded auroral activity on Saturn in concert with the Hubble Space Telescope, which observed auroras in the ultraviolet once every other day. Cassini was on average 78 million km from the planet and approaching it from almost directly up-Sun. Scientists were particularly lucky that on 15 and 25 January the spacecraft was swept by two shock waves in the solar wind which hit the planet 17 hours later. On these occasions the Hubble Space Telescope saw a brightening of the auroras and Cassini detected a matching increase in the radio noise from the planet. On 25 January the aurora brightened dramatically and completely filled its dawn side. The manner in which the planet's magnetosphere responded to the solar wind showed some similarities to that of Earth, some to that of Jupiter, and several unique characteristics. Unlike the Earth's magnetosphere, which was populated by solar wind plasma that 'leaked in', the magnetospheres of Jupiter and Saturn were mainly 'fed' by material shed by their satellites. But the Voyagers had revealed the Saturnian satellites to be much weaker sources of plasma than those of Jupiter (and in particular Io). As on Earth, auroras on Saturn were controlled by the solar wind. On Jupiter the auroras were connected with the Io flux tube. However, there were significant differences in how the solar wind interacted with the magnetosphere of Saturn, probably due to the planet's greater distance from the Sun.[76,77,78,79]

Also beginning in January the dust analyzer started to record bursts of impacts by microscopic particles, with an intensity that increased as Cassini approached the planet. This dust probably originated from the A ring, became electrically charged, was 'picked up' and accelerated by the planet's magnetic field to over 100 km/s and escaped into interplanetary space. Similar but stronger bursts had been recorded by many missions (including Cassini) while approaching Jupiter. The instrument found these motes to be a silicate material with relatively little water or other volatiles. In effect the particles were a free sample of the 'dirt' in the rings.[80,81]

Imaging began on 6 February. Cassini's view was of the southern hemisphere of the planet and the illuminated face of the rings. The 9th marked the start of weekly color imaging to track the knots and small clumps in the thread-like F ring and to monitor features in the atmosphere, in particular to measure wind speeds. The south polar 'collar' was plainly visible, as were delicate details in the atmosphere and the rings. During imaging sessions from late February to late March Cassini saw two storms, each 1,000 km wide, coming into contact, interacting, shearing and finally merging on 19/20 March. The latitude where this occurred, 36°S, was later dubbed 'storm alley' owing to the frequent activity. Electrostatic discharges from one such storm were recorded in May.

From an engineering point of view, over the period of one whole week in early March a final rehearsal of the Huygens' relay mission was successfully carried out.

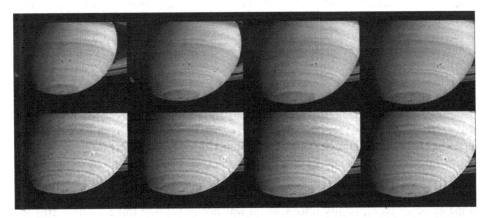

A time lapse sequence starting on 22 February 2004 and ending on 22 March, showing two storms in the southern 'storm alley' interacting and eventually merging. (NASA/ JPL/Space Science Institute)

On 22 March bursts of electrons from the bow shock were noted for the first time, a clear sign that the spacecraft was about to enter the realm of the ringed planet.

Low-resolution monitoring of Titan began in April as images taken through the methane filter started to reveal surface details, and shortly thereafter their resolution exceeded that of the Hubble Space Telescope. Atmospheric detail enabled the wind speeds to be measured for input into the Huygens' relay anomaly solution. Saturn and its atmosphere were the focus of composite infrared spectrometer observations which obtained temperature maps, and of the radar which made distant radiometric scans for calibration against data from terrestrial radio-telescopes.

The area where Cassini would cross the ring plane was being closely scrutinized by the camera and from Earth in an effort to detect any debris or other objects that might pose a risk to the spacecraft. If the plane crossing had to be moved outside the F and G rings, this would significantly degrade the tour of the satellites. By mid-May, with no hazard identified, the go-ahead was given for the final targeting maneuver. On 22 May the main engine cover was rolled back in preparation for the 34.71-m/s correction 5 days later to refine the Phoebe flyby. This served as a rehearsal for the procedure for the orbit insertion burn. There was concern about the helium pressure regulator, which had allowed too much propellant to leak into space over the years. The engine cover was left open in readiness for orbit insertion.

Between 9 February and 1 June Cassini took 800 narrow-angle frames in four sequences specifically to image faint rings, to recover known small satellites and to search for new ones. Pandora and Prometheus, the F ring 'shepherds', were spotted on 10 March, years after they were last seen; as were Janus, Epimetheus, Pan and Atlas. On the final day of the satellite search, two new ones were found in circular equatorial orbits 194,000 and 211,000 km from the planet, i.e. between the orbits of Mimas and Enceladus. Initially designated S/2004 S1 and S2 these moonlets were later named Methone and Pallene. They were respectively 3.2 and 4.4 km in size. It was also established that Pallene had been present as a streak in a single Voyager 2 image and was then dubbed 1981 S14.[82] The discovery of Methone was particularly welcome from a celestial mechanics point of view because it provided an improved estimate of the mass of Mimas, for which no close encounters, and hence no high-quality tracking data, were planned.[83]

All eyes now turned to the unique flyby of Phoebe. Although this dark, distant and retrograde moon had seemed an oddity when it was discovered in 1899, it was actually the largest of a population of small irregular moonlets which orbit at large distances from the planet. In recent years astronomers had discovered many such objects. Some, known as the Norse group, traveled in similar orbits to Phoebe which suggested that they might be fragments chipped off it by impacts.

Cassini was to pass some 2,000 km from Phoebe. The closest point of approach was initially to have been about 56,000 km but in consideration of the high interest in inspecting an object that seemed likely to be a captured Centaur asteroid a closer range made sense, especially since Cassini was not to venture this far out during its primary mission tour. An even closer flyby would have provided high-quality data on the moon's mass and gravity field, but Phoebe was believed to be shedding dust and the risk of the spacecraft being damaged was deemed unacceptable.

Ground-based observations of the moon in support of the flyby had established it to be primarily gray. Spectroscopy showed the presence of water ice, confirming a relationship with comets and the Centaur asteroids. When observations showed that Phoebe rotated in about 9 hours it was realized that this relatively rapid spin would enable Cassini to map the entire surface of the moon, albeit with a fraction of it at a reduced resolution. Controllers used navigation images of the moon to refine the observation schedule in quasi-real time. In fact, last-minute updates were needed to ensure that it would be in the field of view of the remote-sensing instruments. By 7 June Cassini had closed to within 2.5 million km of the moon and the resolution of its images already matched the best by the Voyagers. Pictures taken in the ensuing days from within 1 million km clearly showed craters and provided hints of a heavy bombardment. On 10 June images from 650,000 km out showed even more craters and a large depression near the north pole which was a huge crater possessing steep walls. Another crater-like depression straddled the southern pole.

The day of the Phoebe flyby was 11 June. Seen from a range of 140,000 km the moon sported not only dark gray terrains but also white patches so bright that they were saturated in all subsequent images. As expected, Phoebe had a non-spherical shape, making it one of the largest such bodies known in the solar system. But its self-gravitation was sufficient to have produced a roundish shape. In images taken from 32,500 km the moon almost filled the camera's field of view with a resolution of 190 meters. A heavily cratered terrain was seen, although some of the craters had a rather subdued appearance; an indication that they must be fairly old. Cassini sped by at 2,068 km with a relative speed of 6.4 km/s. The difference with respect to the plan was in large part due to uncertainties in the ephemeris of the satellite, and only in minor part due to errors in the targeting maneuver. It managed to map the entire surface over three rotations at a resolution better than 2 km. At closest approach the resolution was 13 meters. The mass of the moon was inferred from radio Doppler tracking and its shape and volume were calculated by analyzing the images. These determinations combined to give an average density that was remarkably low, only 60 per cent greater than that of water, suggesting a porous mixture of ice and rock. Nevertheless, with the exception of Titan, this density was greater than for the inner icy satellites of the Saturnian system. This was further evidence that Phoebe was a captured body.

Phoebe was revealed to be almost saturated with impact craters. There were over 130 exceeding 10 km in diameter. The largest (later named Jason) was the one near the north pole. At 100 km, it spanned half of the moon's diameter. Phoebe survived the impact but the shock probably fractured its interior and broke it into blocks that re-accreted. The northern and eastern walls of the crater showed evidence of recent bright landslides whose aprons might account for the hummocky debris on its floor. Layering was clearly visible in the walls of many craters, with a top dark layer and brighter streaks running to the floor. This implied that the interior of Phoebe was pristine material rich in water ice, with most of the surface being mantled by a layer of darker material typically several hundred meters in thickness. Giant icy boulders tens to hundreds of meters in size resided in craters, as well as on intercrater plains and hilly terrains.

The imaging spectrometer mapped the surface at a resolution of 500 meters per pixel during the flyby, measuring its composition. It was the most compositionally diverse, primitive and non-differentiated surface yet observed anywhere in the solar system. There were patches of iron-bearing minerals, water ice, bound and trapped water and carbon dioxide, nitrogen-bearing and cyanide compounds, organics and clays. The inferred composition was similar to that of other bodies in the outer solar system, such as Triton and Pluto, but was incompatible with the regular satellites of Saturn. In particular, the presence of carbon dioxide ice implied Phoebe had never warmed sufficiently for this to sublimate to space. This meant the moon formed in the colder realms of the outer solar system as a Centaur or a Kuiper Belt object, and was later captured by Saturn. The signature of water ice was confirmed from ultraviolet data. The spectrometer also sought comet-like emissions from volatiles, but did not find any.

The composite infrared spectrometer obtained global and regional temperature maps of the day-side and night-side, plus scans at high spatial resolution at the time of closest approach. It measured an average temperature of 110 K, with temperature variations around large craters and other topography. Overall, the surface reflected only 1 to 6 per cent of sunlight. The thermal inertia implied the surface to be a layer of powdered regolith that was a good insulator. Functioning in its radiometry and scatterometry modes the radar 'probed' the first few tens of centimeters of surficial material and found it to contain a substantial amount of dust.[84,85,86,87,88,89,90,91]

Cassini rapidly left Phoebe behind, and on 16 June made a very small maneuver to establish the requisite trajectory for orbit insertion. Scientists then announced an extremely puzzling observation. As previous missions had approached Saturn their plasma-wave experiments had detected kilometric-wavelength radio emissions that were widely interpreted as a 'signature' of the planet's core rotating. However, the period of the emissions detected by Cassini was 6 minutes (1 per cent) longer than in 1980 and 1981 when the Voyagers flew by. Such a slowdown had already been noticed in Ulysses radio observations between 1994 and 1997. In the absence of a known physical means of slowing the core to the degree measured over such a brief interval, the logical inference was that the radio emissions were not synchronized to the rotation of the core.[92]

By mid-June most of Titan was mapped at resolutions as fine as 35 km using the narrow-angle camera, a 10-fold improvement on imaging from Earth. Intriguingly, there were no clearly recognizable geological structures and no craters. Long-term imaging documented the formation and dissolution of clouds near the south pole.

To conclude the remote-sensing observations in the run up to the orbit insertion maneuver Cassini made a multispectral survey of the illuminated side of the rings, revealing the overall 'sandy' hue of the B ring and subtle variations in color which resulted from the different compositions of the particles in each ring. Observations on 21 June revealed two more satellites orbiting just inside and outside the core of the F ring. They were designated S/2004 S3 and S4 but were never assigned proper names because it proved difficult to secure follow-up observations to confirm their nature. It is considered likely, in fact, that these were just clumps of F ring particles that dispersed over time. Indeed, dozens of such 'disappearing moonlets' were seen in

Phoebe photographed from a distance of 32,500 km. Jason is the large crater at top. (NASA/JPL/Space Science Institute)

2004 alone. Approach-phase imaging also revealed a new faint ring, R/2004 S1, which matched the orbit of Atlas, one of the small moonlet that were discovered by the Voyagers.[93]

Eight days prior to arrival, with the entire sequence for the orbit insertion burn uploaded and rehearsed, Cassini was put into a state of minimal activity in which it would simply maintain its high-gain antenna pointed at Earth and send engineering and such scientific data as could be obtained in that orientation. At the same time continuous tracking by the Deep Space Network began. On 27 June the spacecraft first crossed the bow shock. At 49.2 Rs, this was 50 per cent further out than in the case of Pioneer 11 and the Voyagers. The bow shock washed back and forth several times until Cassini finally penetrated the magnetosphere 20 hours later at a range of 40.5 Rs.[94,95,96]

It had originally been intended that Cassini would have a steerable antenna to relay data to Earth during the orbit insertion burn, but this had been deleted to save

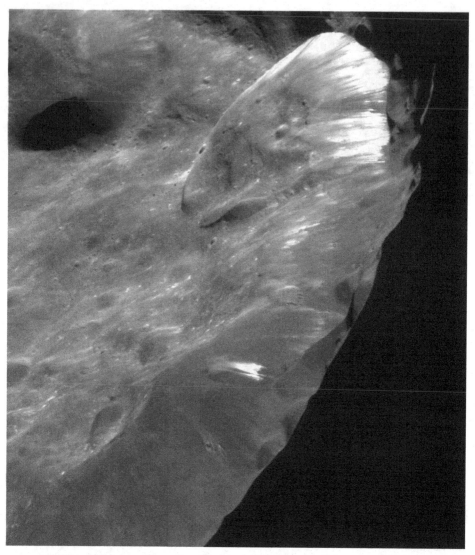

Bright material on the walls of Jason. (NASA/JPL/Space Science Institute)

money and at the time of its launch the expectation was that the spacecraft would perform this key maneuver without returning real-time telemetry. But after losing a mission to Mars in 1993 and two more in 1999 without telemetry (of which more in chapter 9) engineers sought ways to ensure a downlink during the burn. Cassini was too remote for direct intervention but in the event that it was lost telemetry would at least indicate whether the anomaly had occurred when crossing the ring plane, in which case it could be due to the environment, or during the braking burn and was probably a hardware failure. It would not be possible to use the high-gain antenna because that would be pointing well away from Earth during the burn. To aim it at

Earth would require the burn to be executed in a non-optimal attitude which would consume the propellant intended to facilitate an extended mission. Moreover, such a burn would last far longer than that planned and would prevent unique scientific observations from being made. In any case, Cassini would poll its subsystems only once every minute or so, and an instantaneous catastrophic failure would more than likely go unreported. There would also be a total of 53 minutes during which it was occulted by the A and B rings. It would be behind the dense B ring for about 30 per cent of the duration of the burn and, as Voyager 1 illustrated, even if the high-gain antenna were used the signal would be extremely attenuated or lost. And of course, Saturn would be a few days from solar conjunction at the time, and the signal would be further weakened by its passage through the corona. Sensible proposals to maintain communications without using the high-gain antenna included turning the antenna between ring-plane crossing and the start of the burn to provide a snapshot of the vehicle's health, or alternatively transmitting a powerful carrier signal using the low-gain antenna which could be tracked by receivers on Earth to monitor how the burn was slowing the spacecraft. However, unlike a high-gain signal, a low-gain signal would be attenuated by the less dense A ring as well as by the B ring.

After consulting with NASA headquarters, JPL chose to implement the second strategy since it would not preclude observations of the rings. It was also decided to schedule a brief snapshot using the high-gain antenna immediately after the burn. The start of the burn would not be at the optimal time, as it would occur 30 minutes ahead of that and shortly after the inbound ring-plane crossing. This decision, at the expense of a slight underperformance, would allow about 75 minutes after insertion for observations of the rings in a particularly favorable geometry and at a very close range that would not be repeated during the primary mission. The 'early' burn would also provide time for an 'engine switch' if this were to prove necessary. The entire maneuver was rehearsed in May.

Several minutes before midnight on 30 June telemetry was cutoff when Cassini slewed to point its high-gain antenna forward to act as a dust shield for the body of the vehicle. Even if the antenna were to be hit, small pits on its surface would have little effect on its performance. Coming from the day-side and from the south at 22 km/s, Cassini crossed the ring plane northbound at 0:47 UTC on 1 July at a range of 2.628 Rs, within kilometers of its target. Pioneer 11 and Voyager 2 had also passed through this 30,000-km gap between the F ring and the tenuous G ring, although the latter had been hit by a number of particles.[97] The particles and fields instruments were active throughout the duration of the insertion maneuver, in the hope that the magnetometer might detect higher-order components of the magnetic field. Also, the cover was jettisoned from the inlet of the mass spectrometer. Instruments duly recorded plasma created by the disintegration of ring particles striking the craft. Safely through the ring plane, the vehicle turned to point its engines along the velocity vector. The orbit insertion burn began at 1:12. Cassini slowly slewed 0.008 degrees per second in order to keep the thrust of the engine more or less parallel to the Saturn-relative velocity vector, which rotated 46 degrees during the duration of the burn. At 2:38 it reached its closest point of approach, passing a mere 19,880 km (0.33 Rs) above the cloud tops. This was several hundred kilometers nearer than for

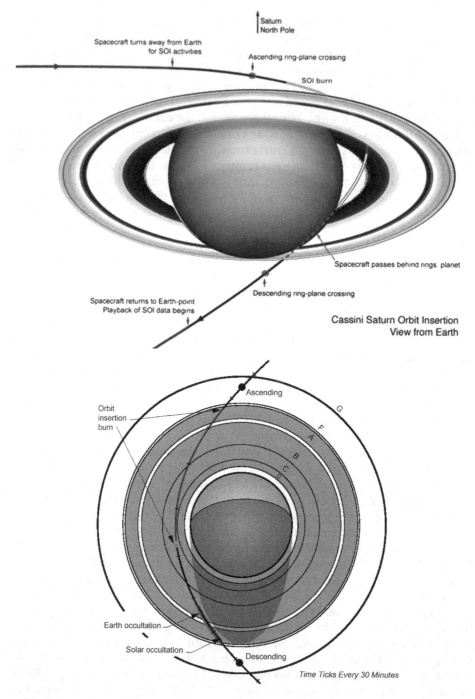

The Saturn orbit insertion maneuver seen from Earth and from above the planet (with the Sun at top).

Pioneer 11. The engine continued to fire until an accelerometer sensed that the desired reduction in speed had been achieved. The engine cover was then closed in preparation for the second ring-plane crossing.

The low-gain transmission from Cassini had been acquired intermittently during the A ring occultation, then reappeared in full strength as the spacecraft crossed the gap of the Cassini division, and was totally blocked by the B ring. It reappeared in time to show the end of the burn, at which time the Doppler shift was within 5 Hz of that predicted. The burn was about 1 minute shorter than nominal because the thrust of the engine was slightly greater than expected. It had fired for a total of 96 minutes and 24 seconds and had slowed the vehicle by 622 m/s, consuming 830 kg of propellant. The final 18 minutes had reduced the period of the orbit. Immediately after the burn there was also a 20-second burst of high-gain telemetry that provided three 10,000-bit packets of data that showed the spacecraft to be in excellent health. As a result of the burn Cassini had a speed of 30.53 km/s relative to Saturn and was in a 1.3 × 150.5 Rs orbit with a period of 116.3 days, inclined at 16.8 degrees to the equatorial plane. Some 22 years after being conceived, the vehicle had become the first artificial satellite of the ringed planet. This was the culmination of a landmark year for JPL which also saw two rovers successfully landed on Mars and a flyby of comet Wild 2 by the Stardust mission. The orbit was so close to that intended that a major trim maneuver scheduled for 3 days after insertion was canceled. Within 20 minutes of the end of the burn Cassini was already making scientific observations from a vantage point 16,000 km above the unilluminated face of the rings, ten times closer than would be possible at any other time during the primary mission.

About 45 minutes after the burn Cassini was occulted by the bulk of Saturn, as viewed from Earth, and 3 minutes later it flew into the shadow cone of the planet. The occultation and eclipse lasted just over half an hour. The spacecraft then turned to use the high-gain antenna as a dust shield for the southbound ring crossing that occurred at 4:34, at a distance of 2.632 Rs, again in the gap between the F ring and the G ring. Safely through, it turned to resume observations, this time of the illuminated face of the outer A and F rings.

About 6 hours after cutting contact at the start of the insertion sequence, Cassini resumed Earth-pointing in order to download the engineering and science data that it had collected. The two ring observation sequences yielded a total of 85 images at optical resolutions ranging between 100 and 350 meters, a level of detail previously achieved only in Voyager radio and stellar occultations. There were 43 images of the unilluminated face. This sequence began on the outer C ring, continued into the inner B ring, skipped the core of the B ring, resumed on its outer portion and then the Cassini division onto the A ring. In these pictures, dark bands were either dense regions of the ring that blocked light or the blackness of space seen through gaps. The C ring images captured the thin Maxwell gap and the narrow eccentric ring that resided within it, clearly showing a wave running through its middle. Seen at high resolution the B ring had a puzzling appearance because some parts of it showed a fine structure whilst others appeared smooth, with undulations in brightness spaced hundreds of kilometers apart. In the Cassini division there was another narrow eccentric ringlet dubbed the Huygens ring and a wavy feature representing a resonance with the small moonlet Prometheus which traveled along the thickest portion of the gap.

The most interesting images showed the A ring and were rich with details and unexpected discoveries. Two of the high-resolution pictures showed four propeller-shaped features about 5 km across which appeared to be the result of perturbations of ring particles less than 1 meter across by objects in the range 10 to 1,000 meters. The moonlet Pan was at the upper end of this size range. By extrapolating from the small area in which these four 'propellers' had been observed it was calculated that there could well be a total of 10 million such objects within the rings. Simulations showed that for a 1 km moonlet, gravity would be the principal actor, clearing a full gap around the path of its orbit. For smaller objects, the gap would be rapidly filled by frequent impacts and a propeller-shaped density enhancement would be created by the competing influences of gravity and orbital velocity, decreasing at increasing distance from Saturn. This discovery shed some light on the origin of the rings. If they were the raw material of a satellite that had failed to accrete then the particles would be expected to have suffered so many mutual collisions over the ages that by now only objects less than 1 meter in size would be left. The presence of hundred-meter-sized objects argued for the rings being the remains of a shattered satellite. It would be difficult to obtain new images of the propellers before 2006, when Cassini was to begin to increase its inclination with respect to the ring plane.[98,99] The Encke gap was found to be occupied by faint ringlets and a non-uniformly dense central ring. Density waves occurred along the gap's rim when the perturbing influence of Pan caused particles to bunch into denser streams. Similar but larger spiral density waves had already been seen by the Voyagers in the outer A ring, where they were caused by the influence of Mimas. Density waves occurred when the perturbations by moons caused the orbits of some ring particles to elongate and intersect those of other particles. The amplitude and wavelength of the wave shortened with distance from the source of the disturbance and eventually faded out. Studying such patterns would yield insight into the density of the rings, as well as place constraints on their thickness. The density of the A ring appeared to be greatest around the Encke gap, decreasing both inward and outward. The ring itself did not seem to exceed 10 to 15 meters in thickness. Density wave phenomena were not unique to planetary rings, they were also present in the arms of spiral galaxies. So knowledge gained from Saturn's rings would have far-reaching implications. The gravitational influence of an as-yet-unseen moonlet was evident in 'wispy' features along the edges of the Keeler gap, a 35-km region near the outer rim of the A ring and of the classical ring system.

After the ring-plane crossing Cassini had imaged the illuminated face of the ring system starting at the F ring and progressing inward to the Keeler and Encke gaps, capturing also the inner shepherd moon Prometheus. Images of the faint but very dynamic F ring resolved (for the first time) ropy, straw and mottled structures, as well as narrow strands on either side of the bright core. The sheet of dust interior to the core was crossed by regular 'channels', radial structures and streamers which linked Prometheus to the ring, transporting matter between the two. These images confirmed the multi-stranded nature of the F ring already suspected from Voyager

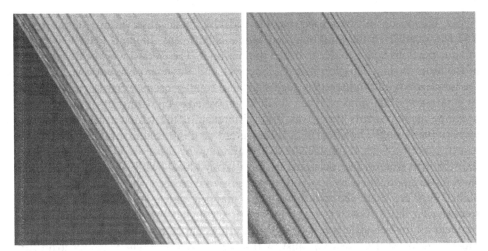

A dark-side image of the outer edge of the Encke gap and the region exterior to it (left). Density waves caused by Pan are clearly seen, as well as complex 'ropy' structures. This image covers about 180 km and the resolution is about 270 m per pixel. A dark-side image of density waves in the A ring (right). This image was taken about 135,000 km from the ring plane. (NASA/JPL/Space Science Institute)

A day-side image of the Encke gap showing its wavy inner edge. (NASA/JPL/Space Science Institute)

occultation data. The illuminated view of the inner edge of the Encke gap showed a sequence of scalloped radial waves and ropy structures downstream of Pan, which had just passed by. Such features could be caused by the gravitational influence of this moonlet if the plane of its orbit were slightly inclined to the rings, because that would cause waves of particles to 'dance' above and below the ring plane. Analysis of these structures enabled the mass and density of Pan to be measured. It proved to be extremely porous with a density only 50 per cent that of water ice.[100],[101]

The composite infrared spectrometer had also made observations of the rings. The temperature of the unilluminated face was between −200 and −160°C with the opaque portions being colder than the thinner ones. The coldest was the densest part of the outer B ring. This confirmed Pioneer 11 and Voyager observations, although the temperature differences they recorded were smaller because Saturn was close to its equinox and the rings received sunlight almost edge-on. Contrary to models and previous measurements there was little difference in the temperatures of the two faces of the A ring. The ultraviolet spectrograph and infrared mapping spectrometer investigated the rings at unprecedented resolutions for these wavelengths and found that the gaps were in fact occupied by dark rocky objects while the rest of the rings were mostly populated by icy bodies whose size increased with distance from the planet. Between them the two instruments mapped water, volatiles, rock and dust spanwise on the A, B and C rings, as well as through some of the divisions. The A ring had an ice-rich spectrum whereas other rings appeared more rocky. The F ring was found to consist mostly of dust, with little if any ice.[102] The mass spectrometer detected oxygen ions in an 'atmosphere' over the A ring.[103]

The particles and fields instruments, requiring no particular attitude while taking data, operated throughout the orbit insertion sequence. As shown by Pioneer 11 and the Voyagers, Saturn's radiation environment was relatively benign at that distance. Maps of the distribution of energetic neutral atoms obtained by the magnetospheric imaging instrument as Cassini skimmed over the rings revealed the presence of a previously unknown radiation belt in the space between the inner edge of the D ring and the body of the planet.[104] The high-resolution data from the magnetometer was corrupted, but scientists believed they could overcome this loss using other data.

About 200,000 small particles hit Cassini during the ring-plane crossings. As in the case of Voyager 2 the clouds of plasma created when the particles disintegrated on striking Cassini were detected by the plasma-wave experiment as radio bursts. A peak rate of 680 strikes per second was recorded, but any damage was borne by the high-gain antenna. Engineering data stored on board would show that Huygens had emerged unscathed.[105,106]

On 2 July, about 31 hours after orbit insertion, Cassini had the first non-targeted encounter of its primary mission, flying by Titan at a range of 339,000 km. On the revised tour this encounter was dubbed T0. Although fairly remote, it was a good opportunity for early remote-sensing observations. Being well south of the moon's orbit, the instruments had an excellent perspective on the south polar region and the southern portion of the Saturn-facing hemisphere. At that distance Titan fit into the field of view of the narrow-angle camera. Because of scattering by the atmospheric haze the resolution was significantly worse than the theoretical value. Images nicely matched the ghostly features seen in the infrared by large terrestrial telescopes, and although a variety of surface features with different shapes and appearances were in evidence scientists did not know what they represented geologically. It seemed like going back centuries, to when astronomers debated the nature of ill-defined features on Mercury or Mars. Images targeting the atmosphere clearly showed bright clouds up to 700 km across hovering 30 km above the south pole. These were also seen by the infrared mapper. The manner in which the clouds changed during the encounter provided a measure of the wind speeds. The persistence of polar clouds suggested the presence of either mountains or seas to which they could be anchored; although such features were not evident. Preliminary multispectral cubes were obtained by the visible and infrared mapping spectrometer but to save on bandwidth only a few of its spectral channels were downloaded. Spectra of the mysterious dark and bright areas were interpreted in terms of the percentage of water ice and hydrocarbons in each. It was obvious, however, that the dark areas did not represent the long-sought hydrocarbon oceans. The instrument revealed the atmosphere to glow with ionized methane over the day-side, with the maximum brightness occurring at an altitude of about 350 km. The night-side was glowing with carbon monoxide emissions. Large amounts of nitrogen were expected to have been stripped from the atmosphere over the ages but, surprisingly, the plasma spectrometer did not detect the torus that this leakage should have formed coincident with Titan's orbit around the planet. It was unclear whether this was because the rate of leakage was slower than predicted or because the ions were being removed from the magnetosphere by some mechanism. Cassini did confirm the existence of the hydrogen torus that had been inferred from Voyager observations.

One of the objectives of this first distant flyby was to collect atmospheric data to validate the models used to design the Huygens mission. Although the atmosphere proved to be slightly denser and warmer, a risk review undertaken several days later validated the model.[107,108,109] The next day, Cassini, climbing to apoapsis, was able to image Iapetus at a range of 3 million km. Solar conjunction began on 6 July. After spending about 400 hours in the planetary magnetosphere, Cassini flew back into the solar wind at high southern latitude at the end of the conjunction. It crossed the bow shock ten more times between 56 and 85 Rs until finally leaving it behind on 14 July.[110,111] The first night-side infrared maps of the planet were taken over the morning terminator from a position near apoapsis and they showed clouds and other features deep in the atmosphere that would normally be hidden by sunlit haze. Dark areas in infrared maps were opaque regions silhouetted against the thermal glow of the interior.[112]

On 23 August a 51-minute, 393-m/s burn raised Cassini's periapsis by 300,000 km in order to clear the densest parts of the ring system and establish conditions for the first close encounter with Titan. It was the third-largest engine burn of the entire mission. Since no more such burns would required, the propellant tanks were then repressurized and the helium lines shut off – all subsequent propulsive maneuvers would be made in 'blow down' mode, using the remaining gas pressure to force the propellants into the engine. The 9-million-km apoapsis 4 days later marked the start of revolution A, the first of three (designated alphabetically) that had been added to the tour in overcoming the Huygens' relay problem.

During the summer and early fall, Cassini made extensive observations of Saturn over a large part of the electromagnetic spectrum. A number of radio bursts lasting tens of milliseconds from storms in 'storm alley' were recorded, including one in September that was dubbed the 'dragon storm' for its great size. Moreover, Cassini verified that the period of the kilometric-radio emission had increased by 6 minutes since the Voyagers. Lengthy remote-sensing sequences were devoted to monitoring

A mapping spectrometer image of Titan taken during the 2 July 2004 'T0' flyby. Details of the surface are visible through the haze, as well as bright clouds over the south pole. (NASA/JPL/University of Arizona/USGS)

A view of the south pole of Saturn, taken on 28 July. (NASA/JPL/Space Science Institute)

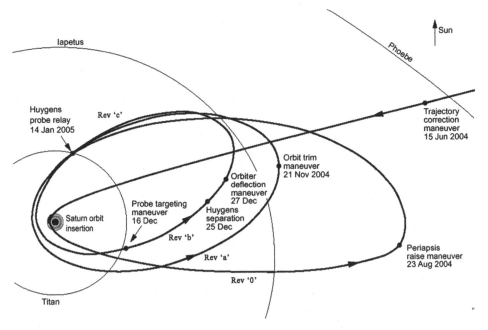

The trajectory of the first seven months of Cassini at Saturn, from the Phoebe flyby in June 2004 to the atmospheric entry of Huygens in January 2005.

the planet's atmosphere. Wind speeds in the equatorial jet stream, which had been a significant fraction of the speed of sound in 1981, were revealed to have slowed down. On the other hand, they had sped up relative to measurements by the Hubble Space Telescope in the 1990s and early 2000s. At other latitudes the wind profiles mostly matched those measured by the Voyagers. One possible explanation for this was that Cassini was viewing clouds at different depths and hence recording winds at different atmospheric levels. The atmosphere proved to be enriched in carbon relative to hydrogen. The carbon-to-hydrogen ratio was seven times greater than in the Sun and twice that of Jupiter. This was consistent with theories of the formation of the solar system which predicted an enrichment in heavier elements at increasing heliocentric distance. Infrared spectra indicated a relative depletion of phosphine (PH3, or 'swamp gas') in the southern belts. Formed at high temperatures deep in the atmosphere, phosphine was readily destroyed in the chilly upper atmosphere. Its depletion in the southern belts relative to the equatorial regions provided a means of mapping the vertical transport of chemicals in the atmospheric circulation system. In contrast, phosphine was enhanced in the south polar region.[113]

Between 6 and 7 October, Cassini's ultraviolet spectrograph monitored the star xi Ceti being occulted by Saturn's rings. In addition to recording the exceptionally sharp edges of the rings, this occultation revealed the presence of structures only a few tens of meters across. By observing many stellar occultations over the 4-year primary mission, it would be possible to study the fine structure of the ring system.

Images of Iapetus from 1.1 million km were taken on 17 October as Cassini was inbound toward Saturn and showed a string of bright dots crossing the dark leading hemisphere. These proved to be 20-km-tall peaks that rivaled the volcanoes of Mars for the distinction of being the highest mountains in the solar system. In retrospect, there was a hint of their presence in the lower resolution Voyager imagery. Cassini showed Cassini Regio on the leading hemisphere to be heavily cratered. Navigation images had been taken of Iapetus and the other satellites to refine their positions and (by the perturbations they imposed on each other) their masses. An important early result was that the mass and density of Iapetus, which were believed to have been well constrained by Pioneer 11 and Voyager measurements, were some 25 per cent greater. This seemingly innocuous result would have important repercussions. If the Huygens probe were released on schedule, the arc of its ballistic path to Titan would pass close to Iapetus and if the mass of that moon were imprecisely known then unaccounted perturbations could cause the probe to miss its very narrow entry corridor. Although scientists were confident that they would soon have a fix on the mass of the moon, three options were being considered to minimize its effect on the mission. One was to release Huygens after the flyby, with a delay of some 10 days. This would achieve the specified entry corridor but the burn to deflect Cassini onto the desired flyby path would substantially increase. A second option was to release Huygens on a later orbit but that would also affect the orbital mission by delaying its scientific observations. The third option was to release Huygens as planned but open its range from Iapetus to reduce the possible perturbation. This would disrupt the sequence already developed for the Titan encounter on 13 December and delete a Titan occultation in February, but in the end this option was chosen. As a result of more than doubling the Iapetus range to 117,000 km the altitude of the Titan flyby in December 2004 was reduced from 2,336 to 1,200 km and that of February 2005 was raised from 950 to 1,579 km.

In October observations were made in an effort to discover new satellites. Six narrow-angle frames on the 21st revealed a faint moving point of light that proved to be a moon some 5 km in size. Initially designated S/2004 S5 it was later named Polydeuces. Further imaging revealed it to occupy the trailing Lagrangian point of Dione. Although Cassini's trajectory would pass within 6,500 km of Polydeuces in February 2005 the discovery came too late to modify the science sequence to make a closer inspection.[114,115]

Inbound to Saturn for its first periapsis, Cassini was to cross the orbit of Titan on 26 October. This would be the eagerly anticipated first targeted flyby of the cloud-enshrouded moon. It was called encounter Ta because it was on revolution A of the tour. About 300 times closer than the encounter in July, it would facilitate a number of preliminary investigations.

The ion and neutral mass spectrometer made observations specifically designed to determine whether a 950-km Titan flyby would be safe. Data from this and later flybys, as well as from Huygens, indicated that the density of the upper atmosphere was greater than predicted and 950 km was judged to be too risky. Even at higher altitudes the attitude control system had to work hard to maintain the orientation of the vehicle against aerodynamic torques. The decision was taken to raise the limit from 950 km to 1,025 km for the flybys of April and September 2005. Moreover, measurements were used to validate models for Huygens' descent. The instrument produced other scientific results by detecting nitrogen, methane and some complex hydrocarbons and nitrogen-bearing molecules. It also determined the isotopic ratios of carbon and nitrogen in anticipation of more detailed measurements by Huygens, and a measurement of noble gases in support of the Huygens mission.[116]

The composite infrared spectrometer was also "at work for Huygens" collecting atmospheric temperature and density profiles. Unfortunately, much of this data was lost. The instrument detected molecules at the north pole which were not present in the southern hemisphere, hinting at some sort of seasonal effect. The results mostly confirmed the models used to design the Huygens mission.

Titan had been monitored from Earth and at long range by Cassini itself in late September and early October and the polar clouds had brightened dramatically. But as Cassini closed in the activity seemed to abate.[117] On the inbound leg, the camera took a cloud movie as well as color mosaics at increasing resolution. The rare mid-latitude clouds that had been visible until just a few days earlier were now gone. A detached layer of haze was found at an altitude of 500 km, which was more than 150 km higher than during the Voyager 1 encounter in 1980. This change was also attributed to a seasonal effect. Interspersed with observations by the remote-sensing instruments, the radar collected scatterometry data of the Xanadu 'continent'. Near closest approach the camera took high-resolution pictures of the 'coastline' where Xanadu met the darker plain to the west. Images were also obtained of the dark area that included the target for Huygens. This was also observed by the multispectral mapping spectrometer. Seen by the camera the surface of Titan appeared far more geologically active and complex than scientists had expected, with signs of wind- and even liquid-driven processes. The notable absence of impact craters confirmed that the surface must be geologically youthful and possibly active. However, these early images failed to indicate any obvious global methane reservoir. There had to be one in order to replenish the methane in the atmosphere. Linear streaks of dark material on the bright areas suggested the winds carried a fine dust rich in organics, and this observation facilitated a preliminary determination of the wind circulation regimes. Scientists would have to wait for radar observations to investigate these further. Interestingly, parallel dark and bright streaks crossed the Huygens target. A bright and dark circular feature 30 km across dubbed 'the Snail' was located within the dark Shangri-La area west of Xanadu. It was interpreted as a possible cryovolcanic dome several hundred meters tall with two flows extending to the west. The highest resolution multispectral cube covered the area of the Snail, which was later named Tortola Facula. The attitude at closest approach was controlled using the thrusters, lest atmospheric drag overwhelm the reaction wheels. The relatively poor pointing accuracy that resulted made the observations of the mapping spectrometer "highly erratic".[118,119,120] Another bright feature in Shangri-La was dubbed 'Great Britain' for its shape and later officially named Shikoku Facula.

After taking hundreds of pictures and multispectral cubes, Cassini slewed for a brief radar session. An altimetry run at mid-northern latitudes from east to west was made prior to the imaging swath covering areas not seen by the camera. It was thus not possible at this early stage of the mission to correlate the radar swath with any

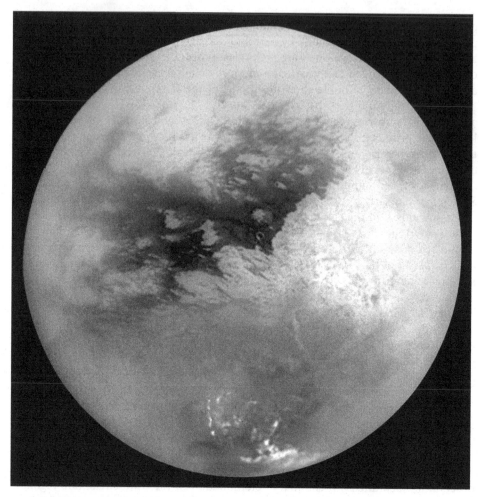

A nine-picture mosaic of Titan taken on the inbound leg to the first flyby on 26 October 2004. The bright continent Xanadu is right of center and the Huygens landing site is just left of center. Note how details at the center of the disk are more sharp than those at the limb, due to atmospheric absorption. Clouds are seen over the south pole. (NASA/JPL/ Space Science Institute)

specific topography. The radar altimeter revealed an extremely flat surface with a very small variation along the 400-km track. However, this track did not cross any albedo boundaries. The imaging swath covered a $120 \times 2,000$ km area north of the area viewed by the camera, with only a small overlap. This was caused, of course, by the need to reorient the spacecraft between camera and radar observations. The radar swept an arc of 100 degrees across the moon, covering less than 1 per cent of its surface at a best resolution of 300 meters. As usual for mapping radar imagery, bright areas corresponded to rough terrain that scattered the illuminating energy in all directions, including back to the receiver, whereas dark areas corresponded to flat

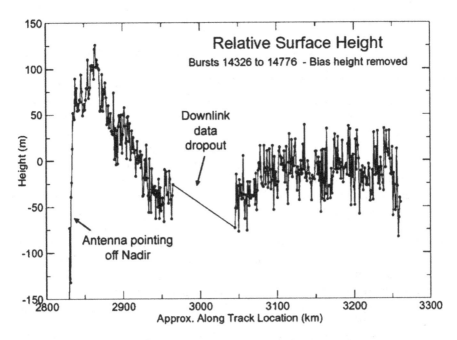

Cassini's altimetry swath of Titan during the 'Ta' flyby. Altimetry variations along the 400 km track amounted to 150 meters at maximum. (NASA/JPL)

Part of the first imaging radar swath at Titan. The image is 150 km wide and 250 km long and has a best resolution of 300 meters. Bright areas correspond to rough terrain, dark areas to smooth terrain. Note the presence of apparent flows and bright sinuous river-like features. No bona fide crater is visible in the area. (NASA/JPL)

terrain that reflected it in only one direction, most often away from the receiver. There was not necessarily a correspondence between the bright features seen in the optical imagery and the bright features seen in the radar imagery. The interpretation of radar features was also complicated by the fact that as yet no visible-light images of that area existed. One-quarter of the swath was radar-dark and the rest appeared to be 'mottled'. Lobate features resembling fluid flow fronts were visible, as well as mysterious fan patterns and linear streaks. River-like channels were ubiquitous, as were small isolated hills. Narrow sinuous features several tens of kilometers long and less than a kilometer wide could be canyons. Dark crescent-shaped spots 20 km wide and an archipelago of irregular dark patches "somewhat like the silhouette of a cat's head" could be methane lakes, their flat surfaces making them appear black. What was immediately apparent was the absence of a widespread ocean. Nor were there any unambiguous impact features. The radar swath imaged several bright and circular features whose deposits suggested they were associated with icy volcanism. A dark 180-km circular feature named Ganesa Macula was tentatively interpreted as a cryovolcanic dome somewhat similar to the 'pancake' volcanoes of Venus. Its rim was bright and probably rough terrain, but a depression some 20 km wide at its center seemed more likely to be of volcanic than impact origin. An area covering tens of thousands of square kilometers named Winia Fluctus looked like it might be a cryovolcanic flow.[121,122]

The point of closest approach was at an altitude of 1,174 km from Titan, above latitude 39°N, 26 km closer than planned as a result of uncertainties in the moon's ephemeris. During the 20 minutes the magnetometer spent within the ionosphere it confirmed the Voyager 1 observation that Titan does not have an intrinsic magnetic field. The ion and neutral mass spectrometer found neutral atoms in the outer part of Titan's exosphere being energized by particles in the Saturnian magnetosphere. These observations facilitated the development of new, less simplistic models of the exosphere and its interaction with the ambient environment.[123,124] Depending on the strength of the solar wind, Titan is sometimes inside the magnetosphere and at other times in interplanetary space. The radio and plasma-wave experiment searched for lightning in the moon's atmosphere, but the lengthy bursts it detected were from the Sun.

Continuing in towards periapsis, Cassini took the first good resolution pictures of Dione and of the trailing hemisphere of Tethys. Another clump-turned-moonlet on the inner edge of the F ring, designated S/2004 S6, was estimated to be less than 5 km in size. As observations would show, this remained intact for a year and was one of the longest-lived such objects. Detailed analysis would help to clarify the origin of many F ring structures, such as clouds of particles that appeared suddenly and then dispersed over weeks or months. One theory held that they were caused by meteoroids hitting the ring, whilst according to another they resulted from impacts between embedded objects. What was particularly interesting about S/2004 S6 was that its orbit intersected the ring core at high speed, giving credibility to the second explanation. In order to avoid cluttering the nomenclature with short-lived objects, after S/2004 S6 clumps within the F ring were given unofficial designations starting at 'C7' and usually were not widely reported in the astronomical community. At least 16 clumps had been seen by 2006, and some of them were suspected of being recoveries of S/2004 S3, S4 and S6. (Remarkably, a large number of dust-clumps-turned-moonlets had already been discovered in the vicinity of the F ring in 1995 by the Hubble Space Telescope in observations made when the rings were edge-on to the line of sight and the absence of glare enabled faint details to be seen.) A new ring some 300 km wide was discovered and designated R/2004 S2. It was between the Atlas ring R/2004 S1 and the orbit of the F ring shepherd Prometheus. It was so close to Prometheus that from time to time it could be run into by the moon. Both R/2004 S1 and S2 were more like Jupiter's tenuous ring than the other rings of the Saturnian system.[125,126,127]

Two days after the Titan flyby Cassini reached periapsis at 6.17 Rs. It was then in a 48-day orbit which, by design, was three times the period of Titan's orbit. As a result, in December Cassini would again pass Titan at the same orbital position as in October and since the axial rotation of the moon was synchronised with its orbit the view would be of the same hemisphere.[128] Imagery on the day after periapsis specifically targeted Prometheus and its interactions with the F ring. Narrow-angle camera views showed kinks, knots, strands, sheared gaps and streamers – including one bridge of dust that linked the core of the ring to Prometheus. Part of the reason for the F ring being so dynamic seemed to be the fact that it was located near where the planet's tidal forces balanced gravitational mass accretion processes, preventing the accretion of a moon. Particles within it would thus be continuously constructing and destroying structures.

The trailing hemisphere and the northern polar regions of Tethys seen on 28 October 2004. (NASA/JPL/Space Science Institute)

On reaching apoapsis at 78 Rs on 21 November Cassini started revolution B and on 13 December, inbound to Saturn, it made a 1,192-km flyby of Titan at a speed of 6.1 km/s. All the remote-sensing instruments were used on this encounter, dubbed Tb, to complement the observations made on Ta, except that on this occasion the radar was not used. A 2-day monitoring sequence showed the return of mid-latitude clouds traveling at several meters per second. The mapping spectrometer observed these clouds rising to high altitudes within half an hour and then dissipating over an hour or so, most likely as methane droplets rained out. The spectrometer, moreover, was able to detect the dissipating south polar clouds which were no longer visible to the camera. Mid-latitude clouds were also being monitored by terrestrial telescopes. Because they seemed to cluster at one longitude, one hypothesis was that they were caused by cryovolcanoes or geysers, with the vent releasing methane. This area was surveyed by radar on the successive flybys, but no volcanic feature was discovered. However, there was a mountain range that could force the prevailing airflow to rise and promote condensation of methane, thereby forming bright clouds.[129]

With the radar inoperative, it was the camera and the other optical instruments that returned most of the data. The camera surveyed the equatorial regions around Xanadu, where linear markings running east to west suggested that the surface was shaped by the action of the wind. Long curving dark features appeared to be rivers. The absence of mirror-like reflections implied that if this was their origin, then they were dry at the time of the observation. Prominent dark linear and angular features could be the result of large-scale tectonic processes. The most interesting feature seen during this flyby was a semicircular bright area southeast of Xanadu that was 550 km across. Most evident in the near-infrared it was initially dubbed 'the Smile' and later named Hotei Arcus. It was visible in long-range images taken in July just after orbit insertion, and had actually been noticed in pictures taken by telescopes in 2003.[130,131] More low-resolution images of the Huygens target were obtained. The absence of obvious changes over an interval of one and half months implied a solid surface. There was a 150-km-diameter circular structure west of the Huygens target that could be a crater with a central peak and a possible 300-km multi-ringed basin in the dark region northeast of Xanadu but once again there were no unambiguous impact craters.[132] Temperature profiles in the upper atmosphere were obtained over the entire visible hemisphere and provided further reassurance that the atmospheric models used in planning the Huygens mission were valid. The composite infrared spectrometer measured wind speeds and mapped methane and carbon monoxide. Significant differences in the abundance of more complex hydrocarbons and nitriles were recorded relative to the Voyager data, but the flybys were at different times of the local year and the variations were probably seasonal.[133] The composite infrared spectrometer scanned the atmosphere for 2 hours seeking new molecules and then switched to the surface where, intriguingly, it was not able to distinguish between the bright and dark areas. Departing, instruments were trained on the portion of the north polar regions that were illuminated by sunlight reflecting off Saturn, and also where it was winter and dark, in search of lightning and 'hot spots'.

During the Titan flyby the ultraviolet spectrograph observed occultations by two stars: lambda Scorpii over the southern hemisphere and alpha Virginis (Spica) over

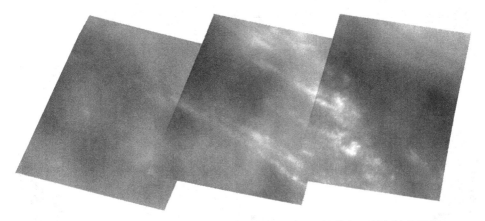

Southern mid-latitude clouds seen on Titan during the 'Tb' flyby. (NASA/JPL/Space Science Institute)

the northern hemisphere. The first observation was with the spacecraft stabilized by its reaction wheels and the second after the transition to using thrusters. As a result, Cassini was extremely stable for observing lambda Scorpii and was able to detect methane, ethylene, acetylene, ethane, diacetylene and hydrogen cyanide as the star light was filtered by the moon's atmosphere from 1,600 to 450 km of altitude. Only methane was detected below 600 km. Unfortunately, the vehicle drifted in attitude while observing Spica, impairing the quality of the results.[134] During the Ta and Tb flybys the magnetospheric imaging instrument showed that energetic neutral atoms extended out at least 40,000 km from Titan. In the absence of an intrinsic magnetic field to protect the moon, the ions in the magnetosphere could directly interact with neutral atoms in the upper atmosphere and excite them to escape velocity, creating this exosphere.[135] The December flyby deflected Cassini into an orbit which would return to Titan with a flyby range of 4,600 km in mid-January, if not corrected.

On 15 December, continuing towards periapsis, Cassini had a non-targeted flyby of Dione and viewed its anti-Saturn hemisphere from a range of 72,000 km. The camera examined the bright wispy terrain on the trailing hemisphere discovered by the Voyagers, presumed to have been 'painted' by some form of cryovolcanism. The new images revealed the wisps to be exposures of bright, fresh ice along the steep walls of scarps and cliffs in heavily cratered terrain. Later that day, Cassini made its periapsis passage at 4.76 Rs.

The last of sixteen check-ups of Huygens were carried out in mid-September and late November. In early November there was a rehearsal of the descent and relay operations. Then ESA re-validated the entire descent trajectory taking into account the various models of Titan's atmosphere that had been developed in recent years. Significant discrepancies existed in how NASA and ESA engineers computed and predicted aerodynamic heating during the descent, so the emphasis was placed on a common conservative calculation methodology. On 16 December it was decided to proceed with the deployment as planned. An 85-second, 11.9-m/s probe-targeting maneuver put the spacecraft on a collision course with Titan. On 23 December the

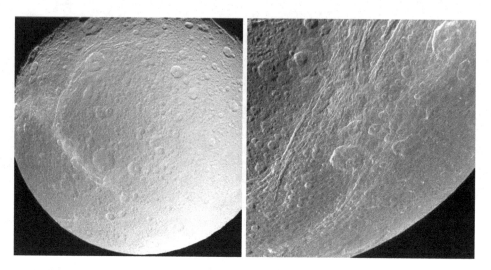

An image of the anti-Saturn hemisphere of Dione seen on 14 December 2004, resolving Voyager's enigmatic wispy terrain as parallel ridges and cracks (left). A higher resolution close-up of wispy terrain from the 14 December non-targeted flyby (right). (NASA/JPL/Space Science Institute)

entry geometry was refined. Ten seconds prior to the moment of separation on 25 December (Christmas eve at JPL) thrusters on Cassini were inhibited from firing. At 2:00 UTC springs pushed Huygens away at 33 cm/s while a helical groove gave the probe a 7.5-rpm counterclockwise spin. Sixty seconds later the attitude control system was switched back on to stabilize the orbiter. All this occurred while out of contact with Earth. When contact was re-established, telemetry clearly showed that Huygens had been successfully released and the channels to record temperatures in the probe had been disconnected. Moreover, the temperatures of parts of the orbiter previously shaded by the heat shield had changed. And the spinning probe had left a measurable 'signature' in the magnetometer data. Some 11.5 hours later Cassini took a 5 × 5-frame mosaic that showed Huygens 18 km away as a bright star with Titan and Rhea in the background in order to accurately measure its trajectory.

On 28 December Cassini made a 153-second, 23.8 m/s burn to change its path so that it would not follow the probe into Titan's atmosphere and would instead make a flyby with its closest point of approach occurring at just about the time Huygens would reach the surface.

On the last day of 2004, Cassini and its dormant probe reached apoapsis at 60 Rs and began revolution C. A few hours later they made the Iapetus flyby which had caused such consternation, passing by at a range of 123,400 km and at a relatively slow rate of 2.0 km/s.

The new images of Iapetus had a resolution of 700 meters at best, and revealed the string of dots seen 2 months earlier on Cassini Regio to be a 20 km high and 70 km wide mountainous ridge which extended at least 1,300 km or 110 degrees along the

The bright star at the top of this wide-angle image is Huygens, 18 km from Cassini and 12 hours 23 minutes after release. (NASA/JPL/Space Science Institute)

equator and gave the moon the appearance of a walnut. Craters embedded in the ridge testified to its great age. On one side of Cassini Regio it extended into a huge impact basin, and on the other side it passed through a cluster of isolated mountains onto the bright terrain. Because they were visible in Voyager 2 images poking over the limb, the latter were dubbed the Voyager Mountains. Unfortunately, the whole ridge had been missed back then. Cassini Regio had appeared almost featureless to the Voyagers but the new camera was able to see detail in the dark terrain. At least five impact basins with diameters exceeding 350 km were visible. Streaks of bright material extended from the northern polar region onto the dark Cassini Regio.

Images and spectral observations addressed the ying-yang appearance of Iapetus to find out whether the dark material was 'painted' on a bright surface or vice versa. There were hints that as the moon traveled through space it accumulated material which darkened the leading hemisphere. Streaks appeared to radiate away from the center. The brightness of steep slopes and crater walls implied that the dark material was only a thin veneer. The infrared spectrometer showed the dark material to have the physical consistency of fresh fluffy snow, rather than a viscous flow of ice. Low-resolution mapping spectrometer scans established the bright hemisphere to be water ice slightly contaminated by organic material, whilst the dark hemisphere was a mixture of organics and polymerized hydrocarbons with only a small amount of water. Spectral variations from the central point of Cassini Regio to its boundaries were observed for the first time. Although it was known that the surface of Phoebe was spectrally different from the dark hemisphere of Iapetus, a popular hypothesis was that Iapetus has swept up dark debris originating from Phoebe. Remarkably, Cassini's ultraviolet spectrograph found Phoebe and Iapetus to share some spectral absorption features – but on the bright side of Iapetus, not on the dark one!

Radio tracking of Cassini served to measure the mass of Iapetus. From this an average density only 10 per cent greater than water ice was calculated. The camera obtained long exposures of Iapetus dimly illuminated by Saturn-shine to enable its size and shape to be measured. Iapetus proved to be an ellipsoid with semi-axes of 749 × 747 × 713 km. This non-spherical satellite is one of the mysteries of modern celestial mechanics. The axial rotation of Iapetus is synchronized with the period of its orbit at about 79 days. This makes it the most distant synchronous moon in the solar system. But its shape is consistent with a spin period of 15 to 17 hours. It has been calculated that it would require twice the age of the solar system to slow the spin from 16 hours to 79 days by tidal interactions. The solution to this discrepancy would require both an interior capable of dissipating energy sufficiently for Iapetus to spin down over the age of the solar system, as well as a rigid crust to prevent the ellipsoidal shape from relaxing into a sphere. Perhaps as it spun down the stresses formed the equatorial mountain range. Another possibility was that the moon once had a moonlet of its own and a ring of debris, both formed by a large impact. Tidal interactions from the moonlet would have spun Iapetus down much more rapidly. Moreover, debris would have rained down on almost horizontal trajectories to form the ridge. As Iapetus was spinning down, the process of momentum transfer would have caused the moonlet to recede and finally escape, although it would have been retained by Saturn. This interesting theory would explain why only Iapetus sports a ridge, of all the solar system's known moons. That is, being relatively massive and far removed from the center of mass of the Saturnian system Iapetus could retain a

A distant image of Iapetus taken during the December 2004 flyby showing the dark Cassini Regio, bright polar areas, large basins, and the equatorial ridge. (NASA/JPL/ Space Science Institute)

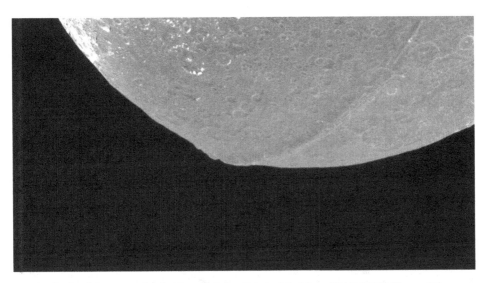

The limb of Iapetus, clearly showing the equatorial ridge. (NASA/JPL/Space Science Institute)

The bright northern polar regions of Iapetus (left). A long-exposure (hence the trailed stars) of the night side and bright hemisphere of Iapetus dimly illuminated by Saturn-shine (right). (NASA/JPL/Space Science Institute)

satellite in a quasi-stable orbit.[136] Other theories explain the ridge without addressing the synchronous spin. One argues that the interior of Iapetus was less viscous than pure ice, perhaps because large quantities of ammonia were present, and a two-cell convection pattern occurred at some time and produced the ridge on the surface.[137,138,139,140,141,142]

Leaving Iapetus and its mysteries behind, Cassini made a 'cleanup' maneuver on 3 January 2005 to refine its path for the forthcoming Titan flyby. It would return to

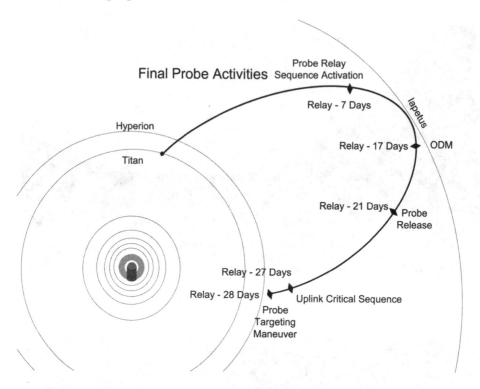

A timeline of the events leading to the atmospheric entry of Huygens. Note that the probe was released on the outbound leg of the trajectory, and the entry occurred on the inbound leg.

Iapetus only in September 2007. Four days later, the vehicle ceased non-essential activities, including science data collection, in preparation for the Huygens mission.


## HUYGENS' DESCENT

Two teams set out to track Huygens' descent from Earth. The first team, from JPL, wished to provide a backup determination of Titan's winds. A similar experiment on Galileo had gained unprecedented results impossible to obtain from the orbiter itself. The second team, European, were to use Very Long Baseline Interferometry (VLBI) to determine Huygens' position with an accuracy of just 1 km. Both teams were helped by Saturn being near opposition and at minimum distance from Earth, and by Earth being a mere 30 degrees off Huygens' antenna boresight, making real-time radio tracking possible. Because it would be difficult to detect the 12 W signal over a distance of 9 AU (1.35 billion km), a network of fifteen radio-telescopes monitored Huygens radio 'channel A', six of which were directly participating in the Doppler wind speed experiment.

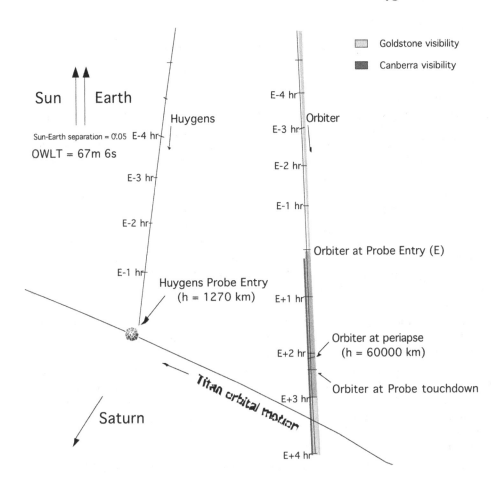

The trajectories of Cassini, Huygens and Titan during the 14 January 2005 entry.

Early on 15 January 2005 Huygens was rapidly and silently approaching Titan. At 4:41 UTC its three redundant clocks reached zero and triggered the powering on of electronics, accelerometers and transmitters. Then at about 6:51 the receivers on Cassini were switched on, configured to receive the two-channel transmission from the probe. Or so it was believed... At 1,500 km altitude the accelerometers began to detect the minuscule braking effect of the atmosphere. A few seconds before 9:06 the probe passed the 1,270-km mark which had been chosen to represent entry into the atmosphere. It was traveling at 6,022 m/s. The fact that the flight path angle of 65.2 degrees was within 0.2 degree of ideal testified to the very precise trajectory bestowed by the mothership.

It was initially planned for Cassini to use its camera to try to image the fireball of the probe's entry into Titan's atmosphere but this was canceled because it would require turning the high-gain antenna away from the moon. Several large terrestrial telescopes tried to see the fireball but with negative results because the probe was too small and the fireball too weak to be detectable at a large distance. On the other hand, they did provide a useful snapshot of the state of Titan's atmosphere at that instant.

Over about 4.5 minutes, Huygens was slowed from 6 km/s to about 400 m/s. The deceleration profile was mostly as predicted. The peak deceleration of about 13 g occurred 100 seconds after entry, at which time the temperature of the forward heat shield reached 1,700°C and the back cover 275°C. At Mach 1.5 (which was about 400 m/s) and a height of about 155 km a mortar deployed the 2.59-meter-diameter pilot parachute. This then detached the aft cover and drew out the main 8.3-meter parachute. The heat shield was released exactly 30 seconds later. Once the flow around the descent module had stabilized, the ports of the instruments were opened, their caps being jettisoned, booms were extended and measurements began. By that time the probe was well into the orange haze. At an altitude of 148 km it began to transmit to Cassini, which was passing overhead. Over an hour later the Green Bank radio-telescope on Earth detected a faint signal from Titan. The modulation of the signal could not be interpreted but the carrier wave provided confirmation that the probe had survived entry. The data recorded by Cassini would not become available for several hours.

At 146 km the gas chromatograph started to collect data. The camera snapped its first sequence of images as the probe descended from 143 to 130 km, showing only a murky sky. The mass spectrometer collected its first sample at 140 km and three more at 85, 55 and 20 km. The aerosol pyrolyzer started to collect its first sample at 130 km; the inlet was closed at 35 km. Fifteen minutes after deploying the main parachute, the probe replaced it with a 3.03-meter one in order to increase the rate of descent. The batteries would not last all the way to the surface otherwise. It was now at 111 km and falling at 5.4 m/s. Although the descent module initially swung to and fro on its harness by as much as 20 degrees, the fact that it later adopted a steady tilt of 3 degrees implied that its parachute was being swept along by a strong wind.

Several minor glitches had little impact on the mission. For some reason the Sun sensor lost sensitivity and returned valid data only when the probe was high in the atmosphere. And with the first half of the descent being rougher than expected the sensor lost track of the Sun for most of the time. Then one of the ion sources of the mass spectrometer was lost to an electric malfunction, precluding the measurement of carbon monoxide. A more serious, and as yet unexplained problem was the spin of the probe. It started at 7.5 rpm, this being the rate imparted by Cassini at the time of release, decelerated in the atmosphere and actually stopped 10 minutes into the descent, and then changed direction. As a result, by 1,000 seconds into the descent it was spinning at 10 rpm in the *opposite* sense. It then rapidly spun down to match the prediction, but clockwise rather than counterclockwise. The simple explanations such as the spin vanes having been incorrectly installed were discounted early on. It appears this anomalous behavior derived from unmodeled aerodynamic interactions between the descent module and the parachutes. The canopies had slightly less drag than expected, but were still well within acceptable performance.

More image sequences were taken between 80 and 75 km, and again between 60 and 54 km, in both cases showing only fog. At 60 km the radar altimeters were switched on. They were not expected to lock onto the surface until 45 km but while descending to this altitude they were capable of operating as meteorological radars. They did not detect any rain-laden cloud, however. At about this time several radio pulses were recorded that could have been distant lightning, but the microphone did not register any signature to suggest thunder. A controversial observation was made by the altimeter before locking onto the ground, returning echo patterns that looked like rain. It was later realized, however, that this echo was caused by thermal noise within the instrument.[143]

Another panorama was obtained at altitudes between 49 and 20 km. The surface details were still mostly hidden by haze but indistinct dark and bright regions were just discernible. This first mosaic covered an area of approximately 130 km square with the resolution increasing toward its eastern edge, indicating that the probe was traveling east-northeast at about 20 m/s. Bright regions were separated by darker ones. A bright and apparently hilly region was etched by canyons that opened onto a dark 'sea', with clouds above what looked like beaches and bays. No craters were visible.

The haze was present at lower altitudes than expected. It cleared only at 30 km, which was at least 20 km lower than predicted. Huygens again had trouble locating the Sun below the haze. This made reconstructing the orientation of aerial imagery difficult – a task that was already complicated by the fact that the descent module was not spinning as predicted. In effect, images were taken at essentially random azimuths rather than at fixed ones as intended, with the result that there were large

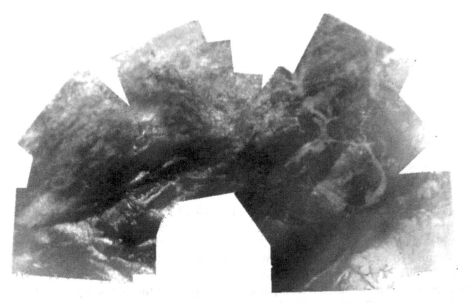

A high-altitude mosaic from Huygens images. At this altitude, only indistinct dark and bright regions are visible. (ESA)

A medium-altitude mosaic of Titan. Note the resemblance of the panorama to a terrestrial shore, complete with a coastline and rivers. (ESA)

A low-altitude mosaic. The same area is visible on the right lower portion of the previous mosaic. (ESA)

gaps in the panoramas. Halfway through the second panorama at 35 km the probe entered a region of strong turbulence. Between 25 and 20 km the second aerosol sample was collected. Below 20 km the probe penetrated a thick layer of haze that was saturated with methane. Contrary to expectations this haze persisted, to some degree, down to the surface. According to temperature profiles and weather models this stratification would be a semi-permanent state, with the drizzle layer in contact with the surface. Droplets made of a mixture of methane and nitrogen would have a very low freezing temperature. This composition was mostly verified by chemical analyses, and at 8 km the upward-looking spectrometer implied that the atmosphere had become saturated with methane.[144,145]

A medium-altitude mosaic was collected at altitudes between 17 and 8 km with a resolution of about 40 meters at the start and progressively improving. The bright terrain occupying the northern portion of the panorama was now visibly cut through by dark branched features that were several tens of meters wide and up to 10 km in length. These were readily interpreted as fluid-carved channels and riverbeds. Small circular spots at the start of some of the branches were strongly suggestive of ponds where rain runoff pooled prior to draining down the channels onto the open plain. Bright streaks could indicate either where the surface had been washed clean or an upwelling of methane ice had oozed through a fracture. The overall impression was of a rainfall drainage network. If this were true, weather models suggested that the channels would be washed by flash floods during the rainy seasons at the equinoxes of Saturn's 29-terrestrial-year orbit of the Sun. With the next equinox due in 2009, there was every chance that Cassini would be able to monitor seasonal changes on the surface of Titan.[146]

A stereoscopic analysis of the few 'paired' Huygens images found that the bright terrain that was transected by channels was around 100 meters higher than the dark terrain. It also had a more rugged topography than the dark plains. A hill seemed to rise 200 meters. The boundary between the bright landmass and the darker plains resembled a shoreline, possibly because dark hydrocarbons had pooled there with what looked like beaches, tombolos, etc. White streaks of methane fog were visible all across the mosaic. There were bright 'islands' on the dark plain, around which liquid appeared to have flowed. The scene looked remarkably like aerial pictures of Earth because similar processes appeared to be at work. One scientist joked that it seemed "Huygens had dropped into the Italian lake country".

A third, low-altitude panorama was taken as the probe descended from 7 km to 500 meters with a resolution better than 20 meters per pixel. It had ceased drifting eastward at about 6.5 km and was now moving slowly west-northwest. During the final 15 minutes on its parachute it traveled about 1 km in the direction of a bright feature that looked like an island just offshore from the bright hilly region. A ridge near the center of the mosaic was crossed by dark channels. The landing site would later be localized along the continuation of one of these channels.

Meanwhile, at 11:12 Cassini reached its closest point of approach, passing by at an altitude of 60,003 km; just 3 km greater than intended. Other than listening to Huygens and recording its own magnetometer data, the orbiter made no scientific observations whatsoever during this Tc encounter.

At low altitude, instruments on Huygens were configured to prepare for landing. At 700 meters the lamp was switched on and spectral data was collected. The dark surface did not reflect light like silicate rock, but rather as a rock-hard shell of ice. Acoustic sounding by the sonar showed a smooth surface with gentle undulations. The camera was still operating, and the last aerial picture was taken just 215 meters above the surface.

At 11:38:11 UTC Huygens hit the surface at a vertical speed of 4.54 m/s, easily surviving the 15-g impact. It came to rest at a 3-degree tilt, possibly because it was perched on a small rock. Having been carried 160 km eastward by winds during the entire descent, it landed at approximately 192.32°W, 10.25°S, just 7 km off-target. The descent from 155 km lasted 2 hours 27 minutes 50 seconds. It was longer than expected, but provided a bonus in atmospheric science data. The site on the eastern rim of Adiri, a moderately bright area west of Xanadu, was named Antilia after the mythical island lying between Europe and America to symbolize the international and continent-bridging character of the mission.

The measurements of the penetrometer at impact indicated a surface possessing the consistency of sand, packed snow, or wet clay, with perhaps a more rigid crust "suggestive of crème brûlée" as one of the scientists put it. The 'soil' was probably a mixture of ice flecks, hydrocarbon snow and liquid methane. It was also possible that the penetrator had hit a pebble and pushed it aside. The accelerometer package recorded the probe gradually moving several centimeters in the process of settling.

Images of the surface gave the impression of either a dry riverbed or a pebbly dry beach. A dark fine-grained soil was visible immediately in front of the camera, with 'rocks' resting on the level surface. Over fifty rocks were counted, ranging in size from 3 mm to 15 cm. It was unlikely that larger rocks were present in that area, as they would have shown up in the last aerial images. Of course, they were not *bona fide* rocks, but rather lumps of rock-hard water ice. Dark trails on one side of some rocks suggested the possible direction of a recent flow of surface liquid. A 'rockless' stripe cut across the panorama several meters in front of the camera. This was also visible in the aerial images. It appeared to be a creek in which the flow had been sufficient to wash away the rocks. A bright spot at the lower right of the view from the surface was illumination from the lamp on the underside of the probe. The infrared reflectance spectrum of the surface was compatible with complex organics, and clearly showed the presence of water ice. The ground had a brownish hue and the sky was probably orangeish. Light scattering in the atmosphere masked the Sun, but if it had been visible from the surface of Titan it would have been as big and as

The surface of Titan as seen by Huygens. 224 pictures of the same scene were returned after the probe landed. (ESA)

bright as a car's headlight at a range of 150 meters. The diffuse illumination at the surface was comparable to that 10 minutes after sunset on Earth.

The orientation of Huygens and of its long series of after-landing images was not known, but it was evidently not facing east because that was the direction in which the wind would have carried the parachute, and this was not visible. From the few shadows it seemed that the Sun was to the upper left, in which case the camera was facing more or less south.

The cameras took a total of 376 images during the descent and 224 after landing, or about 3 every minute. Those taken on the surface were all of the same scene and held the prospect of detecting movement. However, the lossy compression of the images made such an analysis difficult if not impossible. A bright elongated object that appeared on the bottom of some images could have been a splashing rain drop or something blown onto the lens by the wind. Or possibly the heat from the lamp on the lander was boiling chemicals out of the ground. Despite the evidence of dry riverbeds and methane-saturated haze near the ground, Huygens did not see rain or liquid on the surface. Nevertheless, the soil certainly seemed to be humid as a result of absorbing rainfall.

Cassini listened to the probe for 1 hour 12 minutes on the surface before passing below the horizon of the landing site, effectively terminating the mission. The most westerly terrestrial radio-telescope listened to the faint signal from the surface for 3 hours 13 minutes before Earth monitoring was discontinued because the orbiter was no longer in position to receive data. Such a long lifetime was unexpected, but the probe evidently had power remaining. An extrapolation from the telemetry showed that the batteries would have expired after another 17 to 20 minutes. The protracted transmission prompted concern that some instruments must have failed to switch on and draw power, but fortunately this was not the case.

Just half an hour after breaking contact with Huygens, Cassini turned to Earth and started to replay the data that it had recorded. It initially sent a 'snapshot' which indicated that it had properly received and stored data from the probe, then the full set of data was replayed twice with the Deep Space Network stations in Spain and California both recording to ensure that no data was lost. Only after it was verified that all the data had been retrieved was memory in the solid-state data recorders of the orbiter released for other use. The ESA Space Operations Center in Darmstadt, Germany, duly took receipt of 130 megabits from just over 220 minutes of science observations.

In addition to the unpredicted spin of the probe and the loss of the Sun sensor, there had been several glitches. For a while the altimeter had a fault which made it indicate half of the actual altitude. A few minutes after landing, one of the sensors in the surface science package ceased working. These were trivial in comparison to the embarrassing discovery that the command to switch on the ultrastable oscillator for telemetry channel A on Cassini had been omitted from the uploaded sequence, with the result that all the data on that channel was irrecoverably lost! Scientists had debated amongst themselves whether to send the same data over both channels or to double the total amount of data by transmitting half of it on channel A and the other half on channel B. A compromise had been reached in which the spectrometric data would be duplicated, but in order to double the number of pictures some 350 would be returned on one channel and as many on the other. In the event only the pictures on channel B were recovered. Although these were sufficient to satisfy the science objectives of that experiment, channel A was the sole source of data for the Doppler experiment to measure wind speeds, and this was lost.

The impression gained from the cumulative data was aptly described as that of a "Peter Pan world" frozen early in its development.

Accelerometers had profiled the atmosphere from 1,500 km down to the surface. At high altitudes the density and temperature profiles did not match observations by the Voyagers, in particular the Voyager 1 radio-occultation data. At 250 km the temperature reached a maximum at about –86°C. Below 200 km, in the range where Huygens made most of its observations, the profiles nicely matched predictions. At 44 km the temperature reached its minimum of just under –200°C, only 70 degrees above absolute zero. In addition, the ionosphere was analyzed and a layer with the maximum electron density was recorded at about 60 km.

The gas chromatograph and mass spectrometer yielded vital information for the development of theories about the formation of Titan and indeed of the Saturnian system as a whole. A vertical profile of the concentration of methane showed it to be much more abundant near the surface than in the stratosphere, confirming that it would condense out at low altitudes. This also proved that bright veils seen by the camera were indeed methane hazes. Methane was the primary carbon-carrying gas. Carbon dioxide would remain frozen at such low temperatures; it was estimated to represent only 0.03 per cent of the atmosphere. Since water would also be frozen, there was no water vapor in the atmosphere. The gas chromatograph detected small amounts the noble gas isotope argon-36. The heavier krypton and xenon were even scarcer; in fact, they were rarer than predicted by theories of how the solar system formed. This revealed several insights into the history of Titan and its atmosphere. First, it must have formed in relatively warm conditions, with the noble gases being replaced by molecules such as ammonia that would condense in such conditions. If this was the case, then Titan (and therefore Saturn) evidently formed in a warmer orbital position nearer to the Sun than they are now. Moreover, the relatively low abundance of argon-36 reported by this experiment implied that the nitrogen in the atmosphere was formed by the dissociation of ammonia because frozen ammonia would not trap and carry argon with it at low temperature, whereas nitrogen would. The detection of ammonia would be a major discovery because it would suggest cryovolcanism of frozen water, made more buoyant by ammonia 'anti-freeze'. An alternative theory would have nitrogen released by ammonia ices in the aftermath of high-energy impacts by comets and asteroids. The second isotope, argon-40, had been detected by Cassini during the October 2004 flyby. Since potassium could be expected to have migrated to the rocky core of the moon and argon-40 is one of the radioactive decay products of potassium-40, this would lend support to the theory that at some time in the moon's history there had been conventional volcanism. The isotopic ratio of nitrogen-14 to nitrogen-15 was also measured. The lighter isotope was relatively rare, implying that a significant fraction of the atmosphere must have escaped to space since Titan formed. In fact, these measurements suggested that it had lost between two and ten times the current atmospheric mass. The absence of similar isotopic anomalies for carbon confirmed that this must be being replenished in the form of methane. Biological processes could theoretically be involved but the ratio of carbon-12 to carbon-13 confirmed that no such reactions were occurring on the moon. Hence complex organic molecules must be the product of volcanism and photochemical reactions. To conclude the isotopic results, the atmosphere of Titan proved to be significantly enriched in deuterium relative to the average of the solar nebula.

Heating the aerosol to 600°C identified ammonia, hydrogen cyanide and several other organic molecules. Organics were probably formed in the upper atmosphere by photochemical reactions and then rained down to the surface in aerosol droplets. Complex organics were not expected to be common except in the upper atmosphere and on the ground, where they would collect. Hence it was not surprising that none were detected in the samples. No instrument detected amino-acids, despite the fact that the atmosphere was confirmed to contain all of the key ingredients required for their production.

The instruments continued to operate on the surface. Three minutes after impact, the time required for the chromatograph inlet to heat the humid soil sufficiently to release a sudden burst of gas, the concentration of methane in the air increased by about 30 per cent. No water was detected at that time, possibly because the inlet was not in direct contact with the surface. The chromatograph collected gases from the surface for 69 minutes, with the evaporation showing signs of abating only after 50 minutes. As predicted, more complex molecules such as cyanogen, benzene and carbon dioxide were detected on the surface.

Of course, the data from the instruments designed for a splashdown was not very meaningful. The temperature measured at the surface was −179°C, just 94 degrees above absolute zero, and the pressure of about 1.47 hPa was slightly less than 1.5 atmospheres. Other observations were made after landing. For example, reflections by secondary lobes from the relay antenna served to characterize the roughness of the surface for several kilometers west of the landing site.

Fortunately the Doppler wind experiment was not entirely lost. Eighteen radio-telescopes in the US, Australia, China, Japan and Europe had listened to the carrier signal from Huygens during its descent and, although the analysis was difficult, it was possible to derive wind profiles with an accuracy of 1 m/s. In the end Doppler data received on Earth met all of the mission objectives. The early results published utilized only the combined data from two radio-telescopes. The high-altitude winds were more or less in the direction of the moon's rotation. The greatest wind speed of 120 m/s occurred at an altitude of 120 km. This exceeded the rate of rotation. In fact, super-rotation had been suspected ever since the first atmospheric models were developed in the 1990s. Strong gusts were observed between 100 and 80 km, and then between 75 and 65 km the winds faded away to stagnant breezes of just a few meters per second. Wind speeds at lower altitudes were determined by combining Doppler data, pictures of the terrain beneath the probe as it drifted on its parachute, and data from the atmospheric structure instrument. In the preliminary results, data was missing for 14 km down to 5 km due to a gap in the radio-telescope coverage, but by integrating imagery data it was possible to state that below 5 km the winds did not exceed 1 m/s. That said, below 7 km the wind direction became extremely variable, probably due to a convective layer near the surface. Doppler data showed that when the landing occurred the air was stagnant and the probe was stable on its parachute. The winds at ground level could be estimated from the rate at which the probe lost heat. Although shielded by a 5-cm layer of foam insulation, Huygens lost some 400 W of heat. This cooling would require surface winds blowing at no more than 20 cm/s.[147,148,149,150,151,152,153,154,155,156,157,158]

After the successful completion of the Huygens mission an investigation was carried out to identify why the channel A receiver on Cassini had been left off. The cause proved to be a simple and embarrassing one, although for obvious reasons it received little publicity. The problem was a lack of coordination between the JPL engineers managing Cassini and the European engineers managing Huygens in a realm governed by the ITAR (International Traffics in Arms Regulations) export-control rules imposed by the US administration! Because of ITAR, in fact, JPL was discouraged from providing assistance to ESA. No systems engineer was assigned the task of coordinating the work-sharing with ESA. Hence, for example, JPL gave ESA telemetry from earlier Cassini-Huygens communication rehearsals without US engineers being able to examine and review the data. Another issue contributing to the problem was the fact that the receiving radio system was classified as company confidential by its maker, and the full technical specifications were not disclosed to its users. The problem ought to have been spotted during the March 2004 rehearsal but the overworked team missed it. However, other issues, including the fact that the pointing angle of Cassini during the rehearsal made it impossible to simulate a signal from Huygens, allowed the fact that the oscillator had not actually been switched on to remain hidden in the telemetry.[159]

## THE PRIMARY MISSION

Two days after relaying for Huygens, Cassini had non-targeted flybys of Mimas and Enceladus and then made its periapsis passage at 4.82 Rs. Particularly intense studies of Enceladus were undertaken despite the minimum range of 200,000 km. Cassini took its first pictures of the night hemisphere with a geometry that would enhance the appearance of dusty or evanescent structures. The resulting images showed a relatively bright spot over the south pole. There were no Voyager images of this region, so clarifying what this spot was would be a task for a future flyby. Images of the day-side showed for the first time the leading hemisphere that had not been visible when Voyager 2 passed by. Although low resolution owing to the range, this region seemed to be as flat and crater-free as portions of the trailing hemisphere, with curvilinear wrinkles and ridges.

On 1 February, near apoapsis, Cassini executed an 18.68-m/s burn which finally put it on course to resume the original orbital tour. While heading back in, the craft flew by Titan on 15 February. It was designated T3 because it would have been the

Reaching periapsis after delivering Huygens in the Titanian atmosphere, Cassini took this amazing view of Mimas with Saturn crossed by the shadow of the rings in the background. (NASA/JPL/Space Science Institute)

third targeted encounter with this moon on the original satellite tour. It was initially to have been a low-altitude flyby which provided a radio occultation, but in revising the Huygens release to allow for the unreliable estimates of the mass of Iapetus the range of the Titan encounter had changed. The radio-occultation team had yielded control of the trajectory to the radar team in exchange for a pass in March 2006 that would provide the desired radio-occultation. On the original plan the radar would not have made its first scans of Titan until the following September. Occurring in almost the same position relative to Saturn and its magnetosphere, the 1,579-km flyby had a geometry similar to the previous ones and was primarily for imaging and remote sensing. Both the camera and the radar would be on, the former taking mosaics of much of the leading hemisphere and the hemisphere opposite to Saturn,

This image of the night-side of Enceladus, taken on 22 January 2005, shows a bright flare over its southern polar regions. Future observations would prove it to be caused by geysers spewing water vapor. (NASA/JPL/Space Science Institute)

in particular Xanadu and Hotei Arcus, but some images would be of areas already examined by radar. The imaging spectrometer was to study areas east of the 'Snail' and northwestern Xanadu. The composite infrared spectrometer would measure the winds at high northern latitudes where a polar vortex caused air to descend to lower altitudes. Then the radar would make an altimetry run across a contact from dark to bright terrain in order to find out whether it was a topographic boundary. The radar would then make an imaging swath during the 40 minutes around closest approach. On the way out there would be further altimetry, as well as some scatterometry and radiometry measurements.

The radar imaging swath covered 1.8 million square kilometers and included the first views of Xanadu. The track ran parallel to that of the October flyby, but offset to the south. It crossed features and terrains similar to those previously imaged but a number of new features were discovered, including the definitive identification of a 450-km impact basin named Menrva and the 80-km crater Sinlap. Initially dubbed 'Circus Maximus' (which, of course, is not actually circular), Menrva had a radar-bright outer ring and an eroded inner ring. The southwestern rim seemed degraded and may have been modified by a fluid flow, with material having entered the basin in some places. A branched network of channels breached the southern wall of the crater. All these characteristics made Menrva relatively ancient. Its size and depth enabled some constraints to be placed on the evolution of the moon's icy crust. The crater Sinlap was 1,300 meters deep and had an extensive and well-defined bright ejecta blanket. It was seen at the eastern end of the swath and had been evident as a 'cold spot' in radiometry and scatterometry obtained by the radar in October. There was a circular structure 180 km across which resembled a Venusian volcanic dome. Ganesa Macula had a central collapsed caldera and radar-bright flows emanating from it for tens of kilometers. Meandering river-like channels were in evidence, as well as fan-shaped deposits downstream. Some channels were 200 km in length and appeared to flow northeast away from Menrva and Xanadu. They seemed to be dry

'Circus Maximus', as 450-km crater Menrva was originally nicknamed. (NASA/JPL)

'Cat scratches', long dark lines oriented east-to-west were later proved to be fields of longitudinal dunes whose 'sand' consisted of organics. (NASA/JPL)

at the moment, without unambiguous lakes or pools of liquid. Unlike the deeply sculpted channels seen during the first flyby and during the Huygens descent, these were shallow like desert wadis. The radar-brightness of the channels was probably because they were littered with pebbles and small 'rocks', as at the Huygens site. A large area where several channels appeared to converge was interpreted as a fluvial deposit and named Elivagar Flumina. About 20 per cent of the swath, starting just west of Menrva, was covered by thin dark linear features many tens of kilometers long and spaced several kilometers apart, all with essentially the same east-to-west orientation. Superimposed onto the terrain, these 'cat scratches' strongly resembled terrestrial linear dunes.[160]

Inbound to periapsis, Cassini had non-targeted flybys with several of the smaller satellites. Less than 3 hours after passing periapsis on 17 February, it made its first close flyby of Enceladus, dubbed E0, some 1,261 km above the equatorial region of the Saturn-facing hemisphere. The southern portion of the trailing hemisphere was imaged at a resolution of 100 meters. The surface appeared to have been shaped by extreme tectonic activity creating extensional and compressional features as well as troughs and cracks with icy exposures up to a kilometer high. There were also some features that might be glaciers or viscous cryovolcanic flows. The infrared mapping spectrometer detected only pure water ice on the surface. If other molecules were present they were in such small quantities as to be undetectable. The radar showed the surface to be coated with a layer of pure water ice tens of centimeters thick. The composite infrared spectrometer measured the thermal emissions from the night-side. The dust detector recorded hundreds of hits. The most amazing discovery was made by the magnetometer, which found that Saturn's magnetic force lines did not wrap closely around the moon but, as in the case of Titan, were displaced just as if the field were loaded with heavy ions from an atmosphere. Any atmosphere would be a very rarefied one, extending perhaps up to one moon-diameter into space. The ultraviolet spectrograph was not able to confirm this finding when it scrutinized the star lambda Scorpii being eclipsed by the equatorial region of Enceladus; there was absolutely no attenuation of the starlight at either occultation ingress or egress. The mass of the moon inferred from the 100,000-km encounter by Voyager 2 in 1981 was only approximate. Tracking of Cassini refined this measurement and revealed the density to be about 60 per cent greater than thought. This meant that Enceladus was not almost pure water ice, but probably had a rocky core constituting the inner two-thirds of its radius.

Cassini did not encounter Titan upon returning to periapsis in March. Instead it headed straight to its second targeted flyby of Enceladus. This finally restored the intended primary tour in which this encounter was designated E1. On 9 March the orbiter flew by at an altitude of 497 km with the closest point of approach over the equatorial anti-Saturn hemisphere, more or less antipodal to the first flyby. Two and a half hours later it reached periapsis. All of the remote-sensing instruments were operated, although for an 80-minute interval the vehicle faced its high-gain antenna forward as a shield against E-ring particles. The composite infrared spectrometer mapped the day-side temperature of the anti-Saturnian hemisphere in search of 'hot spots' and other thermal anomalies that might indicate geological activity. From its observations, as well as night-side data from the previous pass, the thermal inertia of the surface was measured and found to be compatible with unconsolidated snow. The thermal balance of the moon was also measured by the radar operating in its radiometry mode. The plains that had looked so smooth in the kilometer-resolution Voyager 2 pictures were revealed by Cassini to be intensely fractured and grooved by thin parallel lines several tens of meters across, and in some places by chasms hundreds of meters deep. Some of these young ridge networks appeared to coexist with cratered and hence older plains. Other instruments sought to collect material leaking from the moon in order to perform a chemical analysis. The magnetometer gained further evidence for an atmosphere; although to explain its non-detection by the ultraviolet spectrograph over the equator it had to be hypothesized that this was confined to the south pole. The persistence of this gaseous envelope (more correctly called an exosphere rather than an atmosphere) from one pass to the next implied a

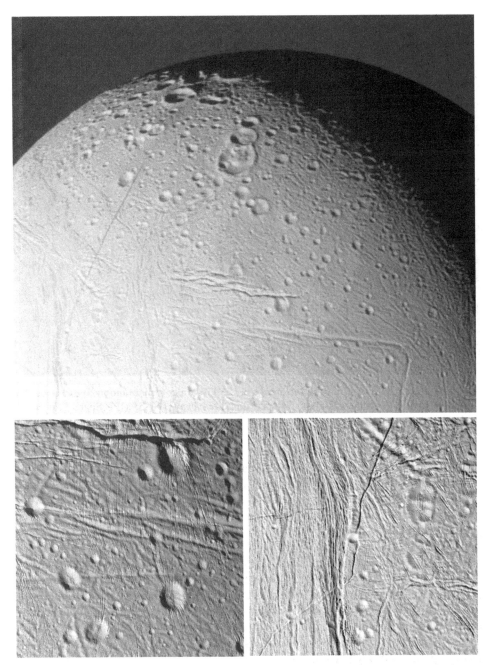

Three views of the surface of Enceladus taken during the 9 March 2005 flyby. The icy surface of the moon is pockmarked by craters that have been heavily modified and by tectonic grooves and cracks. (NASA/JPL/Space Science Institute)

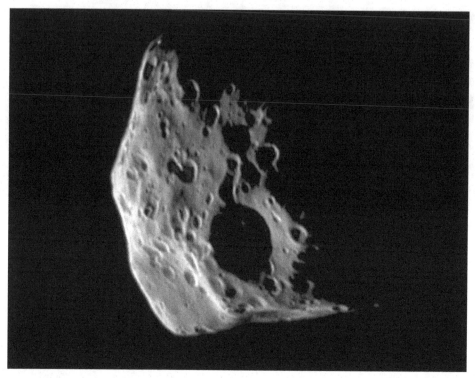

On 30 March 2005 Cassini took this image of the 116-km co-orbital moon Epimetheus. At the time, the probe was about 74,600 kilometers away. (NASA/JPL/Space Science Institute)

replenishment process, otherwise it would have been lost to space by exceeding the moon's 235-m/s escape velocity. The magnetometer detected the possible signature of an induced magnetic field. This could imply the presence of a conductive fluid beneath the surface, as well as water and water-based ions 'picked up' by Saturn's magnetic field. Finally, receding over the night-side the camera took distant images that showed an even larger bright spot over the south pole than that seen during the non-targeted flyby in January.

As Cassini climbed from periapsis, it made an 83,000-km non-targeted flyby of Tethys. Apoapsis was on 19 March. On returning to the inner system the spacecraft got good views of Epimetheus and Janus, a pair of co-orbital moons. Epimetheus was irregular and pockmarked with craters. Although heavily cratered, Janus had a distinctly spheroidal shape. In both cases the craters were softened. Spectra showed their surfaces to be primarily water ice. Their low densities implied that they were loose aggregates of icy particles.

On 31 March, two days after periapsis, Cassini made its first outbound flyby of Titan. Designated T4, this provided the first opportunity to view the Saturn-facing hemisphere. The infrared mapping spectrometer collected data on the composition of the northern hemisphere landforms, and monitored clouds and other atmospheric

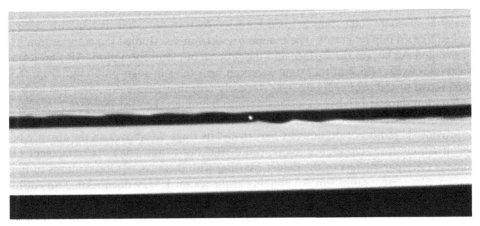

The discovery picture of the "wavemaker moon" Daphnis within the Keeler gap. (NASA/JPL/Space Science Institute)

features. A particular study was made of Hotei Arcus. As the brightest spot on the moon this was taken to be the warmest, and possibly the site of a recent impact or a cryovolcanic flow. To the infrared spectrometer and radiometer, however, it was no warmer than its surroundings. While it was possible that Hotei Arcus was a persistent cloud, it was more likely to be some kind of surface feature. The radar collected radiometry including low-resolution data of Hotei Arcus, and scattero-metry from a small patch of terrain north of the dark equatorial region of Fensal. Half an hour prior to closest approach the spacecraft reoriented itself for high-resolution 'visual truth' of the regions surveyed by the radar and altimeter during the previous flybys. The closest point of approach was at an altitude of 2,404 km passing over the dawn terminator. The particles and fields instruments operated throughout, with the radio and plasma-wave instrument attempting to detect lightning bursts.

On returning for its next periapsis Cassini inspected Pan, Mimas and Calypso, passing in particular within 82,500 km of Mimas. It also observed an occultation of epsilon Orionis by the mid-southern atmosphere of Saturn. Images taken during this periapsis covered an arc about 60 degrees in longitude along the F ring, recording multiple dark channels that spread out fan-wise from the position of Prometheus. Numerical simulations showed that channels and streamers were created whenever Prometheus penetrated the inner, dusty region of the ring.[161]

Outbound on 16 April, Cassini returned to Titan for the T5 encounter. Since only 16 days had elapsed since the previous flyby, during which the vehicle completed just over one revolution of the planet, this was the shortest period between a pair of encounters to-date. The closest approach was at an altitude of 1,027 km at a latitude of 74°N. Particles and fields data was collected at high latitudes and the interaction between the moon's plasma environment and the magnetosphere was investigated. At closest approach, priority was given to the plasma spectrometer, which sampled the ionosphere. At the same time, taking advantage of the close passage, the ion and neutral mass spectrometer made in-situ analyses, finding various complex organics at altitudes in excess of 1,200 km. High-resolution images were taken of the Saturn-facing hemisphere, including a 1,000-km-wide feature far from the equator which looked like an impact crater. The mapping spectrometer studied the composition of Fensal and Aztlan on the surface and the process of cloud formation. Also targeted was Hotei Arcus which at that time was both near the sub-solar point and, from the perspective of the orbiter flying over the early morning terminator, near the limb.

No more targeted flybys of moons were to be undertaken for the next 3 months, during which Cassini would complete five orbits. This 'window' had been designed to provide six radio-occultations by Saturn and its rings. Moreover, in the absence of perturbations from close encounters, radio tracking would enable a refinement of some gravity parameters of the planet, starting with its mass.[162] Observations of the system, of course, continued. An imaging sequence on 1 May documented an entire rotation of a putative object within the 35-km Keeler gap in the A ring in order to ascertain whether this had caused the ripples seen at the edges of the gap in images taken immediately after orbit insertion. A barely resolved 7-km object was indeed present in six pictures. This was designated S/2005 S1 and later named Daphnis. Unsurprisingly, its orbit was practically circular and coplanar with the rings.[163] Two days later, passing periapsis, Cassini was occulted by the rings from the vantage point of Earth for the first time since orbit insertion, and this was used to probe the classical A, B and C rings using radio signals at three frequencies. It was only the second time this radio-occultation experiment was attempted for Saturn's rings; the first being by Voyager 1 during its 1980 flyby. The A ring appeared to consist of centimeter-sized particles. So, too, did the C ring. The inner part of A ring and the entire B ring seemed almost devoid of small particles, but they were predominant in the outer part of the A ring and the D ring. The data resolved fine structure in the B ring, showing it to contain closely spaced ringlets and a core in which matter was four times denser than in the A ring. The densest part of the B ring appeared to be divided into five distinct bands, with the central band, some 5,000 km wide, being the densest and comprised of meter-sized bodies.[164]

The next day, 4 May, the narrow-angle camera obtained its third F ring 'movie'. The previous movies had been made in November and April. For these sequences the camera simply stared at one ansa throughout the 15-hour period of its orbit. The results established that, contrary to expectation, the strands on either side of the ring were not concentric ringlets but an enormous spiral which was wound at least three times around the planet. From dynamical simulations it was inferred that the spiral began as a localized cloud in early 2004, then spread out and wound in. The fact that the moonlet S/2004 S6 crossed the core of the ring at its intersection with the spiral hinted that this vast structure might somehow be associated with the dust clump-moonlets. The spiral could have been created by ring particles scattering off the surfaces of moonlets in high-speed impacts. Alternatively it could be the result of a large impact on the F ring, with S/2004 S6 being one of the largest fragments. The fact that this moonlet was no longer confidently detected after the end of 2005 was proof that it must indeed have been just a clump of dust. Prometheus also acted to perturb some embedded objects, which in turn perturbed the ring core and spiral. All these characteristics made the F ring a unique place in the solar system in which to study the effects of collisions over short timescales.[165,166]

At its next periapsis on 20 May, Cassini passed almost precisely between the rings and the Sun for the first time. It was a chance to observe the 'opposition spot' of sunlight reflected off ring particles. Further such observations would be possible in June.[167] On 21 May Cassini took low-resolution pictures of the fully illuminated and overexposed disk of Enceladus from a range of 102,000 km, to observe for the first time the south pole that would be the target for the next encounter. This region was crossed by four dark parallel lines, dubbed 'tiger stripes'. On the same day the spacecraft observed an occultation of delta Orionis by Saturn. Three days later there was an occultation of omicron Ceti (better known as Mira) by the outer portions of the rings, with the results showing that fine structures such as wakes made the rings more opaque at egress than at ingress.

A non-targeted flyby of Titan on 6 June at a range of 426,000 km provided the second opportunity to see the south pole of the moon. Comparison between these images and those obtained 11 months earlier provided the first evidence of surface changes. Some of the dark spots had changed shape, some had vanished, and some new ones had appeared. One possibility was that the clouds in this area the previous October (which disappeared just before Cassini made its flyby) had dumped a lot of rain. Narrow-angle images in the near-infrared showed a kidney-shaped dark object 230 × 70 km across that was an excellent candidate for a hydrocarbon lake. It was dubbed Ontario Lacus for its resemblance to the North American lake. Whatever it was, it was present in pictures taken during the distant flyby of July 2004, just after the spacecraft arrived in the system.

A ring occultation of alpha Leonis (Regulus) occurred at the next periapsis on 8 June. This time the ultraviolet spectrograph achieved a spatial resolution of a few meters and detected a 600-meter opaque object in the F ring, nicknamed 'Mitten',

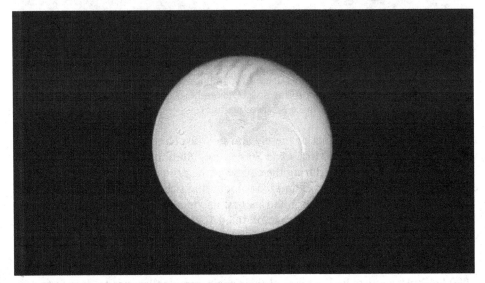

A distant image of Enceladus showing the 'tiger strips' at the south pole (at top). (NASA/JPL/Space Science Institute)

On 4 June, 2005, Cassini took this distant (1.2 million km) showing the semi-circular bright Hotei Arcus at the center of the disk. The bright spots at bottom are clouds over the south pole. (NASA/JPL/Space Science Institute)

that could be either a moonlet or merely another dust clump. Three days later there was another occultation of Mira. Over 80 such occultations would be monitored by the ultraviolet spectrograph during the course of the primary mission, many of them detecting bodies in the F ring estimated to be in the size range 9 to 27 meters.[168]

Between 9 and 11 June Cassini had its first encounter with the irregular, chaotic moon Hyperion, passing it at a range of 168,000 km. Its craters showed intriguing deposits of dark material on their floors.

A few hours before Cassini reached the periapsis of its 11th orbit on 14 July it had a very close encounter with Enceladus. The altitude for E2 was to have been about 1,000 km but scientists had convinced their engineering colleagues to drop it to 166 km. Owing to the moon's weak gravity, the flyby altitude could be modified without

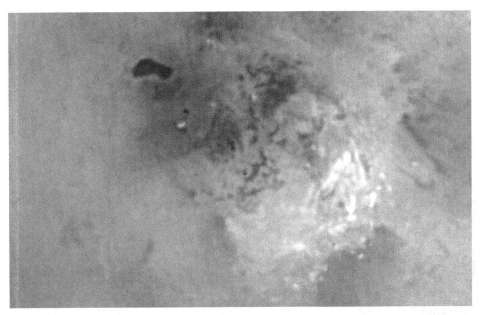

Several distant images of the south pole of Titan were combined to obtain this view, showing the dark Ontario Lacus. (NASA/JPL/Space Science Institute)

substantial influence on the departure trajectory. In this case the approach would be at a high southern latitude in order to obtain the first close-up view of the intriguing polar region, and the closest point of approach would be at 23°S over the anti-Saturn hemisphere. This was an eagerly awaited opportunity to investigate the 'localized' atmosphere, possibly even directly sampling it. However, imaging on the way in, which was essentially toward the ring plane, required extensive tests to evaluate the possible effects of dust impacts on the optics. Eight hours prior to the flyby, the remote-sensing instruments started to observe the moon from a range of 288,000 km. The polar terrain appeared to be a circular region practically devoid of craters, and hence youthful. It was crossed by the four 'tiger stripe' cracks 130 km in length, 2 km in width, and spaced 40 km apart. They were 500 meters deep and flanked by ridges 100 meters tall that ended abruptly with hook-shaped bends. The 'stripes' were later officially named Alexandria, Baghdad, Cairo and Damascus; i.e. A, B, C and D. In between them was finely grooved bright terrain that extended to about 55°S. At this latitude there was a boundary made of folded ridges, hills and valleys; probably compressional terrain. Other craterless fields and cracks extended to the equator. The few impact craters in evidence appeared deformed because they were transected by crevices, and because the icy surface had undergone an isostatic response in which the floors of the craters had domed up.

Higher resolution measurements were obtained during the last 2.5 hours of the approach, as instruments 'rode along' with the pointing selected to suit the camera. The cameras returned a score of images of the polar landscape. Wide-angle pictures had a resolution of 40 meters per pixel at best whilst narrow-angle images were ten

The 'opposition spot' of sunlight reflecting off the B ring, seen by Cassini on 26 June 2005. At the time the probe was about 478,000 kilometers from Saturn. (NASA/JPL/ Space Science Institute)

times better. The young age of the surface was confirmed by the virtual absence of craters at this fine scale too. It was littered with boulders tens to hundreds of meters in size, with little or no blanket of fine-grained ice crystals. The composite infrared spectrometer produced thermal maps of both the day-side and the night-side of the moon. Owing to its high reflectivity, the surface of Enceladus was expected to be one of the coldest places in the Saturnian system, and, indeed, on average it was a distinctly chilly 75 K, about –200°C. This was compatible with a single temperature measurement of the whole disk made by Voyager 2. A patch of terrain at the north pole was observed on the way out. This had been in continuous darkness since the 1995 equinox and was –240°C. The tiger stripes were amazingly warm, at between –159°C and –116°C. This warmth was all the more surprising since the region in which the tiger stripes were located was receiving heat from the Sun only at grazing angles. It was evident that the heat source must be internal and the rate of leakage prodigious. The mapping spectrometer found the signature of crystalline water ice. This was unexpected, because on being exposed to the space environment and to cosmic radiation ice would lose its crystalline structure and become amorphous in a transformation that would require at most several decades. Finding crystalline ice therefore meant that the terrain must be extremely young. Once again, the manner in which the magnetic field of Saturn was being bent by the moon was inconsistent with a spherical, symmetrical object. In fact, the deviation in the field first became evident at a distance of 27 Enceladus radii.

With 21 minutes to go Cassini turned in order to observe gamma Orionis, which was occulted 17 minutes later. This time the ultraviolet spectrograph targeted a star significantly brighter in the ultraviolet than lambda Scorpii in February to obtain a higher signal-to-noise ratio. The results showed clear evidence of the presence of an atmosphere. But only the ingress at a latitude of about 76°S showed the attenuation. There was none at egress over a northern latitude. The envelope was not distributed

An image of the southern hemisphere of Enceladus showing the tiger stripes. (NASA/JPL/Space Science Institute)

This smeared image was the highest resolution picture of Enceladus during the 14 July 2005 flyby. It shows the fractured, boulder-strewn southern polar terrain. (NASA/JPL/ Space Science Institute)

globally, but was concentrated at high southern latitudes. After the occultation the moon rapidly crossed the field of view of the instruments as the vehicle headed for closest approach. Three minutes later Cassini slewed back around in order to target the rapidly shrinking Enceladus. It continued to observe for almost 2 hours, initially over the day-side and then over the night-side. The particles and fields instruments operated throughout the encounter.

The solution to the puzzling observations of the polar terrain and tiger stripes, of the localized atmosphere, and of the interactions with the magnetic field of Saturn, was provided by the mass spectrometer. On previous encounters it had been facing backward and provided no particularly interesting data. This time, however, it was facing forward and it detected a strong signal from water molecules. In fact, water was present in such a large quantity that it was possible to obtain a density profile within 4,000 km of the moon and to detect a clear density peak 35 seconds prior to the moment of closest approach, at which time Cassini was at an altitude of about 250 km and almost directly over the tiger stripes. All these observations indicated a comet-style process with geysers or jets spewing directly out from the south polar region. The possibility of geysers had been discussed by scientists ever since the Voyager flybys, but this encounter was the first to provide firm evidence. Columns of gas illuminated by the Sun could explain the puzzling bright spot over the south pole that was first noted in distant images in January. Other than water, which was also observed by the ultraviolet spectrograph, the gases detected in the plume by the mass spectrometer included significant amounts of carbon dioxide, a molecule that could be either carbon monoxide or molecular nitrogen (they have the same molecular mass), and methane, hydrogen and argon. Acetylene and propane constituted a few per cent. If there was any ammonia then it did not exceed 0.5 per cent.

Starting 20 minutes before closest approach the high-rate dust detector recorded a cloud of micrometer-sized particles in the vicinity of the moon. The impact rate peaked 1 minute before closest approach, by which time Cassini was 400 km above the south polar region. Dust analyzer data were too compressed to be scientifically exploitable, but good spectra had been recorded during the February flyby for about twenty particles, and for most orbits thousands of impacts had been analyzed while crossing the E ring. The particles sampled near Enceladus matched the composition of those populating the E ring, thereby establishing that they were related. In fact, two different populations of particles were found, one corresponding to almost pure water ice grains and the other having substantial impurities in the form of organics and silica-rich minerals.[169] From all of these observations the jets appeared to be releasing 150 to 350 kg, or "about a bathtub's worth", of water every second; a flux similar to that of the Old Faithful geyser in Yellowstone National Park. The gas would expand at up to 500 m/s and readily escape the moon. Ice grains would trail along at a considerably slower velocity. Most would fall back to the surface to form snow, but about 1 per cent would escape into the E ring. It was possible that over time the jets had deposited a thin veneer of bright ice over the entire surface of the moon, making it one of the most reflective bodies of the solar system – its albedo is 99 per cent; for comparison, our own planetary satellite has an average albedo of only 13 per cent.

Although Enceladus had some characteristics of a comet, its jets were powered by an internal source of heat rather than by sunlight. The origin of this internal heat was a puzzle. Given the small size of the moon, heat from the decay of radioactive isotopes in the rocky core would not provide a temperature as high as that observed. Perhaps, as in the case of Io, the volcanic moon of Jupiter, the heat was generated by tides. Resonance with Dione, which made one revolution around Saturn in the time required for Enceladus to complete two, would prevent the orbit of Enceladus from circularizing, thereby keeping its 'heat engine' alive. This theory could explain why the tiger stripes were aligned in the direction of maximum tensile stress of the tidal bulge. Alternatively the heat could be due to small libration motions superimposed on the rotation period (technically, a spin-orbit libration resonance) amounting to a few degrees; librations larger than this were ruled out by Cassini observations. Also unexplained was why similarly sized Mimas, despite being closer to Saturn and in a more eccentric orbit that would impose greater tidal stresses, gave every appearance of being inert. In this case the heat would result from the act of rubbing together the rims of the tiger stripes as the bulge induced by Saturnian tides periodically shifted in phase with the eccentricity from one revolution to the next, in a manner similar to cycloidal ridges seen on Jupiter's moon Europa. The two sides of the crack were estimated to shift several tens of centimeters every 33-hour orbit. Moreover, during half of each orbit the sides of the cracks would be drawn apart to enable gases and vapors to emerge, and during the other half of the orbit they would seal the vent. If this theory were correct, models of the shifting tidal bulge could be used to predict when the sulci would be opening and therefore when a plume could be found there. On the other hand, to be effective in creating geysers this model would require the icy crust to be decoupled from the rocky core, for example by liquid water. The hot spots seen by the infrared spectrometer at the south pole were radiating to space an estimated 1,000 GW of heat. This suggested the presence, possibly at very shallow depths under the crust, of water near its 'triple point', co-existing as gas, liquid and solid, with boiling water forming the steam in the jets. The whole process would be greatly helped by the presence of ammonia, as an ammonia-water mix would have its triple point at a temperature close to that measured by the infrared spectrometer. The major obstacle for this model was that ammonia was not present in the geysers, at least not in the quantities required, and, moreover, it had not yet been detected on any of the icy satellites in the system.

An ammonia-water mix, plus internal heat, could create pockets of low-density material, or diapirs, that would crack the smooth plains as they moved closer to the surface. Consequently, they would modify the moment of inertia of the moon and upset its spin, making it wobble and 'roll over' into a new equilibrium position in which the blob would be at one of the poles. The rise of blobs of warm ice might be a rare event, but it would be episodic. The presence of diapirs would have 'testable' implications. First, they ought to create a mass anomaly corresponding to a volume of lower density material which should be detectable by gravity measurements. For this reason, several future flybys were earmarked for radio-science gravity studies. Moreover, if Enceladus had indeed reoriented itself to locate the low density pocket over the south pole, the rest of its surface ought not to show any difference in the cratering rates of the leading and trailing hemispheres, since they would also have reoriented during the satellite's history. And, most intriguingly, it was possible that pockets of liquid water mixed with methane and complex hydrocarbons beneath the surface could present a benign environment for low-temperature, slow-metabolism forms of life. Thus Enceladus joined Europa as one of the few objects in the solar system on which life could conceivably have developed. But some scientists have rejected a subsurface ocean, pointing out that carbon dioxide, methane and nitrogen would not dissolve in liquid water in the quantities detected in the plumes. They have suggested instead that the water was present in the form of ice, with the gases trapped in its crystalline structure. In this case the prospects for finding life would be low. If this gas-rich ice were exposed to space, the result would be an explosive decomposition in which the liberated gases would carry sufficient water to produce the plumes.[170,171,172,173,174,175,176,177,178,179,180,181,182,183,184,185,186,187,188,189]

One hour after leaving Enceladus behind, Cassini had a non-targeted flyby of the tiny moonlet Epimetheus at a range of 77,000 km.

At periapsis in early August Cassini inspected Rhea, Pan, Mimas, Titan and the south polar regions of Dione. The non-targeted flyby of Mimas at 61,000 km on 2 August provided some of the best data yet for this small, heavily cratered moon. No other major activities were assigned to this orbit, however, because the system was passing through solar conjunction. Returning to periapsis later in the month, in an orbit inclined at 22 degrees, the spacecraft took a series of long-range pictures of Hyperion to document its chaotic rotation. On the day of periapsis, 20 August, the narrow-angle camera took a sequence of 26 images to monitor the dark side of the rings as alpha Scorpii (Antares) was occulted. Four of the images showed seven propellers inward of the Encke gap, and a later sequence revealed another one a few hundred kilometers outside the gap. All the central moonlets appeared to be in the size range 10 to 100 meters.[190] The stellar occultation itself was monitored by the

An amazing image of Mimas on the background of Saturn taken during the 2 August 2005 distant flyby. (NASA/JPL/Space Science Institute)

ultraviolet spectrograph and infrared mapping spectrometer, both of which detected an increase in the opacity of the F ring about 10 km from its core and 500 metres wide, possibly a dust moonlet that was dubbed 'Pywacket'.[191]

Outbound from Saturn, Cassini had the T6 flyby of Titan on 22 August, in this case reaching 59°S. Priority was assigned to the composite infrared spectrometer to obtain a full temperature map of the atmosphere in terms of longitude, latitude and altitude. Moreover, the composition and aerosol distribution of the atmosphere were studied. Although limited by the relatively large 3,660-km minimum distance, the camera took the first high-resolution views of areas south of Xanadu. The particles and fields instruments studied the manner in which the day-side of the ionosphere interacted with the planet's magnetosphere.

One mystery was why the 'ring spokes', so evident during the Voyager flybys in 1980 and 1981, had not yet been seen in Cassini imagery. Perhaps the formation of

Spokes photographed by Cassini on 5 September 2005, for the first time since the 1981 Voyager 2 flyby. (NASA/JPL/Space Science Institute)

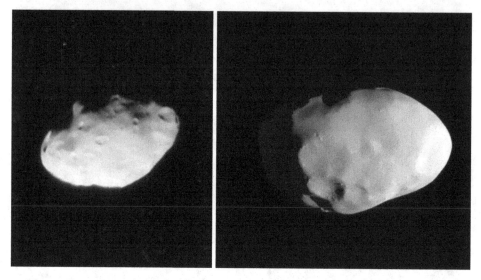

The small moon Pandora imaged on 5 September 2005. Note its subdued features and craters (left). Tethys' Lagrangian moon Telesto (right). (NASA/JPL/Space Science Institute)

spokes was a seasonal phenomenon. If so, they ought to reappear when conditions were right.[192] As Cassini reached periapsis on 5 September 2005 it imaged spokes on the unilluminated face of the B ring just about to enter the planet's shadow. But at only several thousand kilometers in length and 100 km across they were an order of magnitude smaller and fainter than those documented by the Voyagers. Plasma measurements confirmed the hypothesis that spokes formed by dust being levitated away from the rings. The process depended on the density of plasma in the inner Saturnian system, which depended on the angle between the ring plane and the Sun. This explained why spokes were so prominent in the early 1980s, were then absent, and did not reappear until that angle decreased once again, more than a year into Cassini's tour. Subsequent observations by Cassini's spectrometers proved that the spokes mainly consisted of micrometer-sized particles which could readily become electrically charged and levitated away from the ring plane.[193]

On the same day, 5 September, Cassini made a non-targeted flyby of Pandora at a range of 52,000 km. The outer shepherd of the F ring, Pandora was an elongated ellipsoidal body about 80 km long with many eroded craters and smooth intercrater plains.

On crossing the orbit of Titan on 7 September, Cassini made its T7 flyby, which was the most southerly pass to-date. It was summer in the southern hemisphere and the density of the upper atmosphere there was expected to be greater. In view of a number of unresolved issues concerning drag in the detached layers, it was decided as a precaution to raise the point of closest approach by 50 km to 1,075 km and also to use the thrusters to control the vehicle's attitude.[194] The inbound leg was entirely over the day-side, and after reaching the lowest altitude over the dawn terminator it left over the night-side. The camera and mapping spectrometer made observations during the approach but the radar had priority. It was to start with radiometry of the northern hemisphere, make a short altimetry scan, and then an imaging swath from just east of Xanadu to a high southern latitude on a track designed to investigate the presence of the lakes inferred from camera imagery. Unfortunately, a software error prevented the use of one of the solid-state data recorders. As a result, 12 minutes prior to closest approach, with the first recorder filled, the radar imaging was halted after only 8 minutes. Also lost was the altimetry scheduled for the way out.

As it turned out, the data in the first recorder covered a swath 1,970 km long and 300 km wide with a resolution of 300 meters at best. It began with Tseighi, a bright region southeast of Xanadu, and ran southeast across the Saturn-facing hemisphere through Mezzoramia, a semicircular dark area, toward the pole. At the start of the swath was a mottled terrain with faint circular features. A number of bright-rimmed circular features with dark floors were deemed too small for impact craters. In fact, bolides less than 1 km in size would break up in the dense atmosphere before they could reach the ground. Although these structures might be of cryovolcanic origin the absence of any flows was puzzling. Elsewhere, hills were arranged in a roughly semicircular pattern which suggested they were the remnants of a larger structure, perhaps an impact crater, that had been eroded and partly buried. This terrain gave way to a bright plateau dissected by channels and deep canyons that ran poleward with broad fans where debris had accumulated. In places the channels were seen to run for hundreds of kilometers and end abruptly southward of 60°S in a relatively smooth radar-dark terrain. The boundary between the radar-bright plateau and the dark terrain was scalloped like a shoreline, complete with gulfs, peninsulas, islands, etc. The radar-dark Mezzoramia could be a dry lake or sea filled with a fine 'sand' of organics. Although it was dry at the time, the area gave the appearance of being a runoff collector. It was at this point that the radar imaging swath was prematurely curtailed.[195,196] Forty minutes after the flyby the ultraviolet spectrograph observed an occultation of alpha Pegasi by Titan.

What Cassini had not yet found was any trace of liquid hydrocarbons that could have carved these rivers and filled these seas. Weather models implied that it rarely rained on Titan because evaporation was so slow at the frigid temperatures and the atmosphere would take decades to saturate sufficiently for heavy rain. On the other hand, on this and earlier flybys, the visual and infrared mapping spectrometer saw

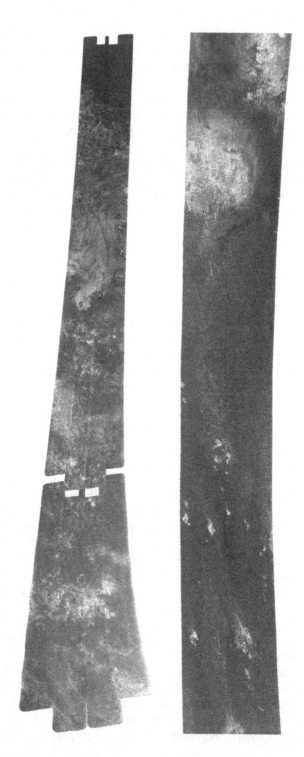

The full T7 Titan radar swath (top). The 'shoreline' of Mezzoramia is on the right side of the swath. The central and highest resolution portion of the T8 radar swath (bottom) was occupied by radar-dark parallel linear dunes. (NASA/JPL/Cassini RADAR team)

clouds in a band at 60°N that spanned all longitudes. They were at altitudes in the range 30 km to 60 km. These characteristics showed that they were made of ethane droplets.[197]

One effect of the T7 flyby was to return Cassini to the equatorial plane to enable it to perform a complex and fast-paced ballet with the satellites. After completing a little over one revolution it would encounter Tethys and Hyperion two days apart. Returning to periapsis it would pass Dione. Finally, after almost one further orbit, it would return to Titan on the inbound leg. The distances of the Tethys and Hyperion flybys were initially to have been 32,000 and 1,000 km respectively, but they were drastically reduced. This revision cost about 8 m/s of the propellant budget, but the science to be obtained from the closer encounters was expected to be well worth it. During the 3 Rs periapsis on 23 September the radar was operated as a radiometer to make a pole-to-pole image of the thermal emission from the planet's atmosphere. The spacecraft then made a 91,000-km flyby of Calypso in the trailing Lagrangian point of Tethys. It was an elongated irregular object about 25 km long, with a large crater or depression halfway along its principal axis. As seemed to be typical of the small satellites of Saturn, its surface appeared smooth.

About 5 hours after periapsis Cassini flew by Tethys at an altitude of 1,495 km. At 9 km/s, the relative velocity was about 50 per cent greater than a typical Titan flyby. Observations on the way in were of regions that had not been imaged by the Voyagers. Cassini then turned its attention to the Ithaca Chasma canyon and to the huge crater Odysseus, both of which were discovered by the Voyagers.[198] Images of Ithaca Chasma at a best resolution of 18 meters showed it to be heavily cratered, indicating a great age. Scientists were interested in the interactions between Tethys and the E ring to determine whether the moon contributed material to the ring that encompassed its orbit. The ultraviolet spectrograph observed a stellar occultation in

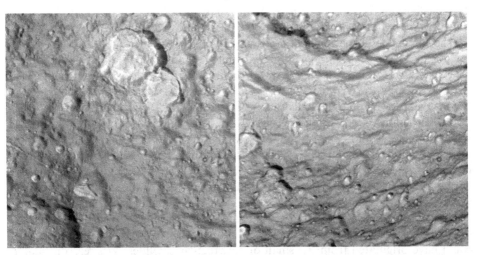

Two high resolution images of the surface of Tethys taken during the September 2005 flyby. (NASA/JPL/Space Science Institute)

search of a tenuous envelope similar to that of Enceladus but nothing was detected. Also unlike Enceladus, the planet's magnetic field wrapped tightly around Tethys. Nevertheless, radar data showed its surface to be coated with extremely clean water ice. Whilst the source for this was obvious for active Enceladus, it was not for the seemingly inert Tethys.

On 26 September, almost exactly 48 hours after Tethys, Cassini made the only targeted encounter with Hyperion of the primary mission, passing it by at a range of 479 km. Orbiting at about the same distance from Saturn as Titan, the relative speed of the encounter was also similar at 5.6 km/s. In the 1980s mathematicians realized that the irregular moon was in a chaotic rotation state, meaning that even such basic characteristics as the orientation of its spin axis and rotation period could change in an unpredictable manner over days or weeks. Observations by Cassini over the last year had confirmed that the spin axis moved at random through the body and across the sky. It was therefore not possible to say which part would be visible during the flyby.[199] Four scientific themes were pursued: namely the internal structure of the moon, its geology and composition, the possible presence of a tenuous atmosphere and volatiles on the surface, and its interactions with the planet's magnetosphere. In addition, scientists were eager to map the surface temperature distribution of a body in chaotic rotation and to seek evidence of dust in the vicinity of the moon in order to determine whether Hyperion, rather than Phoebe, was responsible for 'painting' Iapetus. Radio tracking was used to determine the mass of Hyperion and whether it was a 'rubble pile' or a solid block of ice. Determining its mass was important from a mission navigation point of view as well. Hyperion was in a resonant orbit with Titan in which it completed four revolutions in the time taken by Titan to complete three. Although the perturbations of Titan had been expected to be minor, it turned out that they were impairing the navigation accuracy of the many flybys.[200] In the end, the gravitational attraction of Hyperion changed the vehicle's speed by a mere 0.1 m/s. This put the moon's density at only 60 per cent that of water, meaning that it must be a very porous body made mostly of ice and empty space.

The camera took meter-resolution imagery of Hyperion while the spectrometers investigated the surface composition and sought signs of volatiles in the ultraviolet. The surface resembled a sponge, saturated with craters. The largest crater had been partially seen by the Voyagers as an arcuate ridge named Bond-Lassell Dorsum. It spanned a little less than the longest axis of the moon and had a central mound on its floor. The craters were elongated pits. There was dark material pooling on their remarkably deep floors and also in other topographic lows. The pits were likened to deep 'Sun cups' in which the dark deposits, absorbing heat from sunlight, gradually melted into the icy surface. Alternatively, of course, the pits could have been really deep craters caused by the low density and high porosity of the moon. The porosity was also cited to explain why the craters had no ejecta blankets, the logic being that the impactors would essentially have punched through empty space. The source of the dark material was not known, but it could be dust from Phoebe. Alternatively, it could have originated from the small outer irregular satellites. Whereas on tidally-locked Iapetus the 'paint' was on the leading hemisphere, in the case of Hyperion the chaotic rotation state would distribute it more randomly. The presence of craters

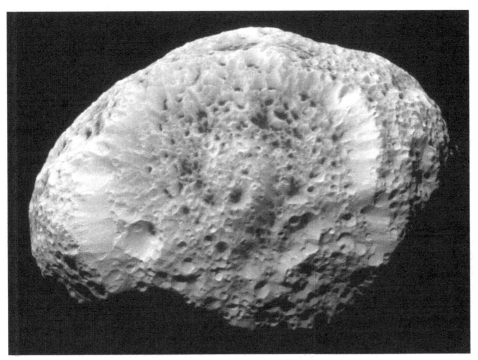

A 'full disk' image of the chaotic moon Hyperion. (NASA/JPL/Space Science Institute)

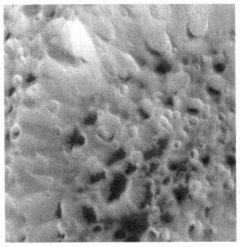

A close-up of Hyperion showing deposits of dark material and the appearance of craters peculiar to the sponge-like moon. (NASA/JPL/Space Science Institute)

superimposed on the dark material showed it to be a thin veneer. The spectrometers found exposed water ice on the rims of some craters, plus frozen carbon dioxide as well as complex organic molecules similar to those that had been found on Phoebe and on the dark hemisphere of Iapetus. To the radar, however, the surface appeared to be composed of almost pure ice only slightly dirtier than the ice of Enceladus or Tethys.[201,202,203]

Cassini reached apoapsis on 2 October and began its 16th revolution of Saturn. Nine days later, just hours before periapsis, it passed by Dione at an altitude of 499 km for the only targeted encounter with this satellite of the primary mission. While still 111,000 km away during the approach the camera obtained a mosaic of images to map regions of the anti-Saturn hemisphere that had been poorly observed by the Voyagers, or not seen at all. Around closest approach it targeted the 'wispy terrain' and saw systems of fractures of various ages. They resembled inert counterparts of the tiger stripes of Enceladus. The composite infrared spectrometer made a thermal map of the visible hemisphere in search of internal activity around the fractures and cracks but there were no 'hot spots'. Although marked by signs of ancient activity including icy volcanic flows, the surface of Dione was geologically dead. As in the case of Hyperion, Dione appeared to have received a fairly uniform veneer of dark dust. The plasma environment was studied to see how the moon interacted with the magnetosphere. The magnetometer found Dione to be surrounded by a thin cloud of ions, similar to that of Enceladus but much smaller and less dense. If the origin was the same as for Enceladus – geysers – then Dione would be spewing only just over half a kilogram of material into space per second, a full three orders of magnitude less than Enceladus. The camera looked for plumes at the moon's limb on this and subsequent occasions, but found none.

An hour and a half after passing Dione, Cassini had its closest approach of the primary mission to Telesto, a small moon orbiting in a Lagrangian relationship with Tethys. Viewed from a range of 9,550 km, Telesto was revealed to be 24 km in size and remarkably smooth, even more so than the other small satellites. A large crater was visible, and boulders marked the rims of others which were almost obliterated. All the smaller moons had a subdued appearance owing to a thick blanket of swept up dust. After imaging Telesto, Cassini took radar scatterometry of the night-side of Dione. Less than 24 hours later, a large 'cleanup' burn established the trajectory for future flybys.

The T8 flyby of Titan was on 27 October, inbound to Saturn. Cassini came in over the day-side and left over the night-side, with the lowest altitude of 1,353 km near the dawn terminator and only a few degrees north of the equator. Approaching Titan, the camera took distant cloud monitoring images as well as higher resolution pictures closer in. However, the primary instrument for this encounter was to be the radar. As usual, the pass began with radiometry and scatterometry, in this case over most of Shangri-La, a large part of Xanadu, and portions of Adiri and Dilmun (two bright areas adjacent to Shangri-La). There was then a short altimetry track running across Shangri-La that passed about 700 km northwest of the Huygens site. Cassini began taking radar imaging data 15 minutes prior to closest approach and continued for half an hour, obtaining a 5,000-km-long swath whose width ranged from 450 to

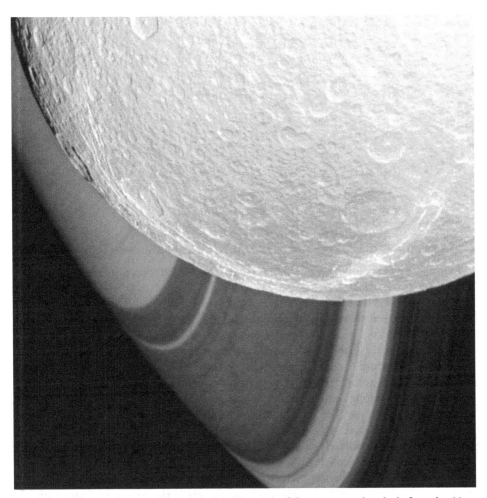

A wide-angle view of Dione on the background of Saturn seen shortly before the 11 October 2005 flyby. (NASA/JPL/Space Science Institute)

180 km. It began close to the Huygens site at a relatively high altitude and hence at modest resolution, and then proceeded westward with increasing resolution. It was initially difficult to identify the exact landing site due to the differing resolutions of the camera and the radar, and to the difficulty of matching features seen at visible wavelengths with those seen by the radar. The site was finally pinpointed using two isolated dunes that were in side-looking high-altitude panoramas taken by Huygens, some 30 km to the north. Unfortunately the channels near the landing site could not be discerned in the radar imagery, whose resolution in this area was no better than 1,000 meters.[204] Just to the west of the Huygens site were long curvilinear chains of features oriented east to west that were several hundred kilometers in length. Since these projected 'radar shadows' they were interpreted as mountain ranges, with the tallest peaks rising just 600 meters. Mountains were unexpected on Titan because it

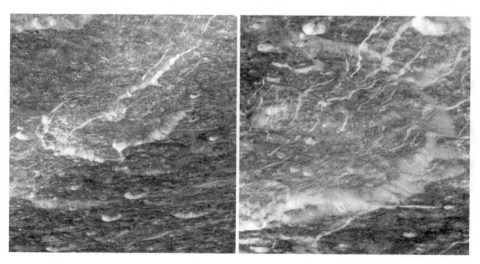

Two close-up views of 'wispy terrain' on Dione. (NASA/JPL/Space Science Institute)

had been supposed that the icy material would relax. Some kind of crustal upheaval must therefore be at work, forming mountains and sustaining them against isostatic relaxation.

About two-thirds of the swath was of the dark area Belet, and covered with 'cat scratches'. The nature of these features, and indeed of the dark equatorial areas, was established by the sometimes risky process of considering terrestrial analogs: they were vast fields of longitudinal dunes similar to those found in deserts such as the Namib, Sahara, and parts of the Gobi. Like those seen on previous occasions, these dunes were oriented east to west and the radar viewed more or less perpendicular to their long axis. This enabled their heights and slopes to be measured from the 'radar shadows' that they projected: they were up to 1,500 km long, several kilometers to 200 km across, and 100 to 150 meters tall. The presence of dunes placed significant constraints on the nature of the surface, surficial material, winds, and atmospheric processes. Whatever the nature of the 'sand' it was dark when seen by the camera, and dry because otherwise the particles would stick together and be unable to form dunes. In multispectral 'cubes', dunes showed up as water-ice-free, indicating that the grains were made mostly of small solid organic particles, or cores of ice coated with organics. Spectra showed the presence of complex organic molecules such as benzene. Liquids destroy dune fields on Earth. So the very presence of dunes in the equatorial zone of Titan would imply that this area remained mostly dry. Covering 20 per cent of the surface and representing probably half a million cubic kilometers of material, the dune fields might be the largest inventory of organics on the moon. Longitudinal dunes like those on Titan would be formed by wind-carried particles. However, this contradicted basic principles of atmospheric circulation as well as observations by Huygens. On the one hand scientists were at odds to explain wind-shaped features that required much stronger surface gusts than implied by Huygens observations, and on the other hand the dunes seemed to indicate that surface winds on the moon were blowing and transporting sand from west to east, opposite to the axial rotation. The winds were expected to be primarily powered by Saturnian tides, the heat of insolation being too weak. However, the contradiction could be resolved by supposing that the winds changed direction at equinox, as do Earth's monsoons. If so, then dunes would be produced by brief, but strong, periodic winds at equinox blowing against weak winds that prevailed for the remainder of the Saturnian year.

Other features were present in the swath, including scattered circular structures and dune-filled basins. Although they lacked well-defined rims and ejecta blankets, some of the circular features could be small impact craters. The dune fields were interrupted by a terrain that was crossed by channels. Dune fields appeared to occur almost exclusively within 30 degrees of the equator, because the radar-dark areas at higher latitudes imaged on previous flybys did not show 'cat scratches'. In fact, the evidence from the first four radar swaths was that channels and river-like features were more common at mid-latitudes, where storms and methane rainfall were more likely.[205,206,207,208,209,210] On the way out, the radar obtained altimetry across the dark region Senkyo. The altimetry tracks were only several hundred kilometers long, and recorded only modest variations in elevation.

The T8 flyby occurred just upstream of the plasma wake created by the rapidly rotating magnetosphere washing over the slowly orbiting moon, and the spacecraft made a deep pass through the zone in which ions 'piled up'. The particles and fields instruments performed observations, in particular the magnetometer.

Three days later, as Cassini crossed the orbit of Dione outbound from Saturn, the plasma spectrometer confirmed the presence of a torus of electrons associated with this moon. Tethys was suspected of being responsible for a similar torus.[211]

The T8 flyby had increased Cassini's orbital period from 18 to 28 days, and just 12 hours before periapsis on 26 November it made its only targeted encounter with Rhea of the primary mission, concluding a busy series of flybys of most of the large icy satellites. As it approached the day-side of the rough cratered moon, the camera took mapping mosaics at ranges from 540,000 km down to 54,000 km. It targeted wispy terrain on the trailing and anti-Saturn hemispheres, as well as a 75-km crater on the leading hemisphere, later named Inktomi, whose brightness suggested that it was relatively young. In general, the ridged and cracked surface resembled Dione. Imaging at closer range showed the rim of Inktomi with a best resolution of about 32 meters per pixel. One priority was radio tracking to precisely determine the mass of the moon. For this and other similar experiments, Cassini received a signal from Earth and retransmitted it on three frequencies with terrestrial antennas monitoring the Doppler effect. The closest point of approach occurred at an altitude of 504 km and at a high relative speed of 7.3 km/s. The ultraviolet spectrograph searched for a tenuous atmosphere derived from the impact of small particles. No envelope was detected either by this instrument or by the ion and neutral mass spectrometer. The composition of the surface was studied by the ultraviolet spectrograph and infrared mapping spectrometer. The radar collected scatterometry and radiometry data on the way out.

As this was the first flyby to pass through the magnetospheric wake downstream of an icy moon the particles and fields instruments played a key role. In particular,

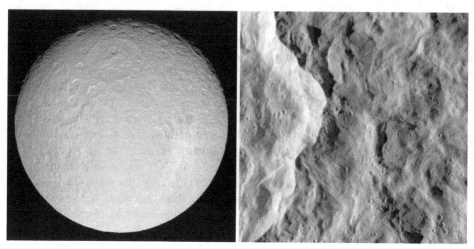

A full disk image of the heavily cratered moon Rhea (left). A close-up of the surface of Rhea, taken during the November 2005 flyby (right). (NASA/JPL/Space Science Institute)

they were to test whether Rhea was a source of matter and plasma. Approaching the moon, the plasma spectrometer and magnetospheric imager observed a decrease of magnetospheric electrons, followed by three steep drops in the electron flux before the spacecraft crossed the 'shadow' downstream of Rhea. While receding from the moon, this sequence occurred in reverse. No anomalous gas or dust envelope was detected to account for such dimming, so it seemed that Rhea must have a series of rings and a disk of debris that extended out to an altitude of 5,000 km. Such material would be gravitationally bound to the moon. The three dips seemed to suggest the presence of at least as many thin ringlets or ring arcs in the equatorial plane composed of sparse meter-sized boulders that would be difficult to discern even using forward-scattered sunlight. The existence of very thin rings and arcs might indicate the presence of minuscule shepherding moonlets. If so, then it was the first instance of rings around a planetary satellite. Impacts on Rhea probably lobbed the boulders and dust into orbit. Simulations showed that under the combined influence of Rhea and Saturn, rings at the observed altitudes could persist for a very long time.[212,213]

Departing Rhea, Cassini turned to make observations of Saturn and Enceladus. Images of the backlit crescent of Enceladus, a little over 100,000 km away, clearly showed what scientists had been seeking. The bright patch previously seen over the south pole was not a camera artefact, it was real. This illumination revealed there to be a dozen fountains of icy particles that blended as they rose to create a larger but fainter plume. By backtracking the fountains, it was found that they were erupting from the hottest parts of the tiger stripes.[214]

On reaching apoapsis on 11 December, Cassini obtained excellent images of the half-illuminated Saturn transected by the thin line of the rings seen edge-on. It then returned to periapsis on the 24th and imaged Epimetheus and Janus close together against a background of Saturn and the rings. This was part of a plan to monitor

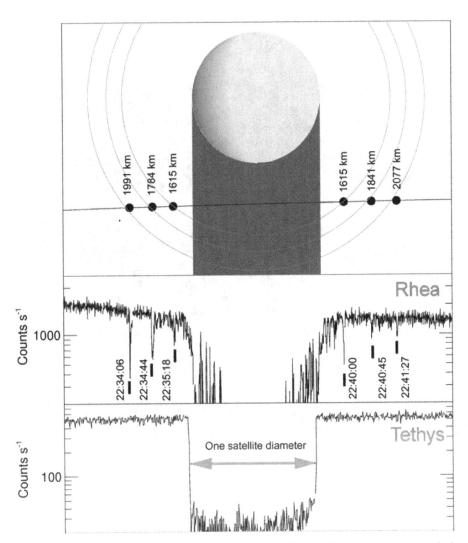

This graph show the tantalizing detection of particle depletion events symmetrical relative to the flyby of Rhea, which hinted at the existence of a ring around the moon. None would be found during the rest of the mission, however.

these co-orbital satellites as they swapped orbits; the swap itself would occur on 21 January 2006.

Outbound from Saturn on 26 December Cassini crossed the orbit of Titan for the T9 encounter. This was an unusually distant equatorial pass with a closest point of approach at 10,411 km. This high altitude enabled the remote-sensing package, and in particular the mapping spectrometer, to make lengthy observations of large areas of the surface with relatively good resolution. The camera imaged Aztlan, Quivira, Bazaruto, Elba Facula and Omacatl Macula. Some of these views were designed to

The dark side of Enceladus showing backlit plumes over the south pole. (NASA/JPL/ Space Science Institute)

overlap earlier radar swaths. For the first and last time on the primary mission, the ultraviolet spectrograph took spectra using a cell designed to measure the isotopic ratio of hydrogen to deuterium. Cassini made a diametric crossing of the plasma tail downstream of Titan – the only time this would happen. The plasma instruments collected data for direct comparison with that supplied by Voyager 1 when it made a similar crossing in 1980.

Cassini had its T10 encounter on 15 January 2006, this time inbound to Saturn on the 20th revolution. The closest approach was at an altitude of 2,043 km over the equatorial zone of the night-side. The flyby geometry allowed the magnetometer to sample the ionosphere of the moon for comparison with data obtained the previous August and October. The camera took pictures of Xanadu and the Huygens site for stereoscopic analysis in combination with the pictures from December 2004. The mapping spectrometer took high-resolution cubes of Xanadu, Shangri-La, and the Tortola Facula 'snail'. About 40 minutes prior to closest approach Cassini entered the moon's shadow. The ultraviolet spectrograph used this first solar occultation of

This gigantic storm was photographed in 'storm alley' on 27 January 2006. The image was taken over the night-side of the planet, illuminated by light reflecting off the rings. (NASA/JPL/Space Science Institute)

Titan to identify constituents of the atmosphere, particularly nitrogen, methane and other hydrocarbons. Finally, the composite spectrometer was aimed at the limb over a region where the atmosphere was expected to transition to the winter polar vortex.

Receding from Saturn in a 39-day equatorial orbit, Cassini observed the planet and documented a bright spiral-shaped thunderstorm in 'storm alley'. Although the storm was on the night-side, it was dimly illuminated by sunlight reflecting off the rings. It had first been spotted by amateur astronomers and was about the size of Australia. Cassini detected bursts of radio waves which were attributed to powerful lightning bolts.

On 25 February Cassini reached periapsis at 5.58 Rs, and two days later the T11 flyby had a closest point of approach at 1,812 km just beyond the dawn terminator. This was to be the first of four encounters of the primary mission mainly dedicated to precise radio tracking to investigate the internal structure of the moon, seeking in particular evidence for a subsurface ocean. To conduct a comprehensive survey of Titan's gravity, two of the four passes would occur over the equator and two over the poles, with one pair occurring when Titan was at periapsis in its orbit and the other pair at apoapsis. This pass was equatorial with Titan at apoapsis. The reaction wheels were used to control the attitude of the spacecraft, eliminating any onboard source of trajectory perturbation.

Cassini returned to Titan on 19 March inbound to Saturn on the next revolution. The main item for this equatorial T12 flyby was a radio-occultation of the southern hemisphere. This was the first such experiment since the Voyager 1 flyby of 1980. Remote-sensing was permitted only at long range. Clouds over the south pole were imaged and the radar collected only radiometry and scatterometry. About 1 hour 25 minutes prior to closest approach Cassini turned to bounce its radio waves off the surface of the moon at an almost grazing angle while terrestrial antennas 'listened' for the reflection. In principle the strength and polarization of the reflected signal would provide data on the physical nature and roughness of the surface. However, no specular reflection attributable to standing liquid was recorded. In fact, no echo from this first bistatic radar experiment of the mission was captured on Earth at all. Starting 20 minutes before closest approach Cassini transmitted a radio carrier over all three of its frequency bands. Terrestrial antennas recorded the manner in which the signal was attenuated as it passed through the atmosphere of the moon at ingress and at egress from occultation. The three wavelengths behaved quite differently, the Ka-band was absorbed and the S- and X-band only marginally so. Perhaps Ka-band radio waves were absorbed by some particular kind of molecule, or by aerosols in the clouds. Furthermore, the attenuation profiles at ingress and egress were slightly different. While out of communication, the spacecraft flew over the night-side with the closest point of approach at 1,949 km.

Two days later, just after periapsis, Cassini made a non-targeted encounter with Rhea at a range of 82,000-km during which it managed to image the day-side and also the night-side illuminated by Saturn-shine. Another quiet orbit followed during which Cassini passed periapsis on 28 April. Two days later, heading out, it had the equatorial T13 flyby. Approaching Titan's day-side the camera monitored clouds.

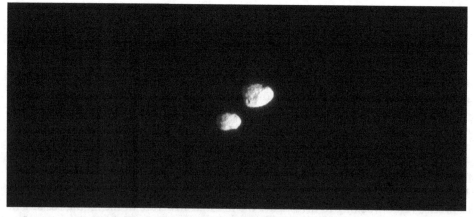

Janus and Epimetheus on 20 March 2006, two months after having swapped their orbits. Janus (appearing smaller) was actually farther from the camera than its co-orbital partner, Epimetheus. (NASA/JPL/Space Science Institute)

One hour 15 minutes before closest approach the ultraviolet spectrograph observed an occultation of beta Orionis. Receding from the moon the camera was to search for lightning and auroras over the night-side. The priority for this encounter was the radar, which had not imaged Titan since 28 October. The swath crossed the central portion of Xanadu and areas for which good camera coverage existed to assist in interpreting the results. Scientists were particularly eager to find out what made the 'continent' so bright. Meanwhile, Cassini reached its minimum altitude of 1,856 km near the terminator and then departed over the night-side. Unfortunately, when the spacecraft was to turn to Earth and send the data that it had stored, it failed to make contact. Communications were re-established on command from the ground. It was decided that a stray cosmic ray must have switched off the transmitter's oscillator. Luckily, only a small portion of the data from the flyby had been overwritten by particles and fields data collected afterward, and this data-taking was halted as soon as contact was regained. The first 8 minutes of the radar swath at the eastern end of Xanadu had been lost.

The recovered radar data provided a 2,500-km swath showing Xanadu to have a very rugged surface with a hummocky texture. It was transected by long, branching channels which drained into the adjacent dark terrains. Some of the channels were the longest yet seen on the moon. Several small craters were present, one of which possessed a central peak. However, with there being no unambiguously volcanic or tectonic features the nature of the continent remained a mystery. Some interesting results were provided by the radar in altimetry and radiometry modes. The altimeter showed Xanadu to stand above the adjacent dark areas, making it a true continental landmass. On the other hand the radiometry suggested that the surface of the rough terrain in the interior was relatively clean ice. It is possible that these elevations had been swept clean by methane rain, producing an eroded surface of almost bare ice. There were large dune fields in Shangri-La, west of Xanadu, and a feature named Guabonito. This had first been seen by the camera as an 80-km ring of bright terrain and was suspected of being either a buried impact crater or a volcanic feature. The radar swath continued across Kerguelen Facula and Shikoku. Most of the mapping spectrometer data was lost owing to the downlink anomaly, but what was returned included the best cubes to-date of the crater Sinlap.[215,216]

Twenty days later, on 20 May, Cassini had the T14 encounter on the inbound leg of its orbit. In order to give priority to another radio-occultation experiment the spacecraft made remote-sensing observations only at long range. Fourteen minutes after Cassini slipped behind the limb of the southern hemisphere, seen from Earth, it reached the minimum altitude of 1,879 km over the night-side, then, on the way out, reappeared from behind its bulk. The bistatic radar experiment was performed at both ingress and egress and this time a reflection was received, albeit a faint one. In addition to profiling the ionosphere and atmosphere of the moon, the two radio-occultations observed to-date by Cassini provided a better estimate of the radius of Titan than was possible using the single radio-occultation by Voyager 1.

Over the next few days Cassini observed occultations of beta, epsilon, and zeta Orionis by Saturn and imaged Polydeuces from a range of 64,000 km, resolving it as an elongated object some 3 km long.

A portion of the T13 radar swath over the radar-bright Xanadu 'continent' (top). Dark spots in this portion of the T16 radar swath (bottom) are probably lakes of hydrocarbons whose flat surface reflected radio waves away from Cassini. (NASA/JPL/Cassini RADAR team)

More than a month later, on 2 July, following periapsis on the 25th revolution, Cassini had the T15 encounter. Like the previous flybys this, too, would be almost precisely equatorial. Distant global observations were made over the day-side as the spacecraft approached, using the camera and the infrared spectrometer. Five hours prior to closest approach the high-gain antenna was pointed at Titan and the radar began to take radiometry data. It then took 15 minutes of scatterometry. Finally, an hour from closest approach, the particles and fields instruments were given priority and the vehicle flew through the plasma wake downstream of the moon. The flyby was at an altitude of 1,906 km. Remote sensing resumed half an hour later with the camera, infrared spectrometer and ultraviolet spectrograph all observing as Cassini receded over the dusk terminator.

This was the final encounter of the equatorial part of Cassini's primary mission. Starting with the next encounter, later in the month, the orbital inclination would be increased in steps to about 60 degrees. This would also mark the onset of the so-called pi-transfer involving shorter and more inclined orbits until Cassini encountered Titan on the opposite side of the planet. Equatorial orbits had provided insight into the kilometric-radio emissions, whose modulation had originally been taken to be the rotation of Saturn's core. Since Cassini's arrival in the Saturnian system the period of this emission had varied by as much as 1 per cent over several months. It seemed likely to be caused by the rotation of plasma and magnetic fields deep in the planet's magnetosphere, with the source of the emission located in the morning-to-noon sector of the magnetosphere and the fluctuations being caused by instabilities in a disk of plasma near the orbit of Enceladus. Two mechanisms were advanced to explain the variations in periodicity. Short-term variations appeared to correlate well with parameters such as the speed, density and pressure of the solar wind that impinged on the planetary magnetosphere. Larger variations might be due to the geysers of Enceladus injecting large quantities of matter into a plasma disk and the interaction between this disk and the ionosphere of the planet.[217,218,219,220,221] An independent estimate of the rotation of Saturn was obtained from atmospheric models by assuming the observed eastward distribution of the winds and jet streams with latitude was a stable configuration. The rotation period thereby obtained of 10 hours and 34 minutes was compatible with the results of other analyses but several minutes shorter than the period of the kilometric-radio emissions. If this period were true, then the winds at the observable cloud tops would not always blow eastward; as on Jupiter they would alternate between eastward and westward.[222,223] Although a difference of a few minutes in the rotation period of the planet might seem minor, it would actually imply an internal structure and a core mass fraction more like that of Jupiter. This would in turn have important implications for theories of how the solar system formed.[224,225] Another outcome from the equatorial part of the primary mission was a refinement of the equations of motion of Saturn, revealing some subtle effects that have yet to be adequately explained.[226]

Descending to periapsis on the 26th revolution Cassini flew by Titan on 22 July and had its T16 encounter. This pass over the north polar regions had the effect of increasing the inclination of the orbit to almost 15 degrees. From the scientific point of view the radar had priority because its imaging ran from the polar regions to the tropics. The north pole was still in winter darkness and hydrocarbon lakes, if they existed, were most likely to be present there. The issue of the minimum permissible altitude for flybys was still unresolved. There were indications that the atmosphere bulged over the equator. This implied the density at high altitudes would be greatest over the equator and decline at higher latitudes. Accordingly, for this encounter the flyby altitude was lowered to 950 km in order to more efficiently adjust the orbit of the spacecraft.

During the approach the radar obtained scatterometry data for Dilmun, northern Shangri-La, and the northern border of the Xanadu continent, and then switched to imaging mode for a swath which ran for 6,130 km and reached 83°N. Northward of 70° were at least 75 radar-dark patches ranging in size from the resolution limit of a few kilometers to over 70 km. The fact that dark sinuous channels discharged into these patches strongly implied that they were indeed lakes of liquid hydrocarbons, their mirror-smooth surfaces giving them the appearance of dark spots. Some filled steep-sided depressions and resembled terrestrial lakes occupying karst depressions or the calderas of volcanoes. Some depressions were not completely filled. Others had rims resembling those of the lakes but were dry. It seemed that only the polar regions provided a relative humidity of methane and ethane sufficient for bodies of liquid on the surface. Possibly they filled in winter and evaporated in the summer. Methane evaporating from lakes could produce the fog found globally, including at the Huygens site.[227,228] The ion and neutral mass spectrometer collected data during this low pass in order to determine the structure and composition of the atmosphere and ionosphere. It identified benzene molecules and ions which could polymerize to form the more complex and heavier molecules of the permanent high-altitude haze layers. The instrument accurately profiled the density and concentration of benzene. At ionospheric altitudes where the instrument was directly sampling, this molecule appeared 1,000 times more concentrated than in the stratosphere where it had been detected earlier by terrestrial telescopes and by the composite infrared spectrometer on the spacecraft.[229,230] Cassini survived the low pass, but the fact that the attitude control system was active for 60 per cent of the time meant that the density over the poles was still substantial. The simple latitude-dependent model of the atmosphere was inadequate. Thirty minutes after closest approach the ultraviolet spectrograph observed an occultation of the bright star Spica by Titan. During the encounter the mapping spectrometer had sought mid-latitude clouds, and for the first time it had utilized the 'noodle' technique of taking long, zigzagging spectra instead of single multispectral snapshot 'cubes'.

Two days after encountering Titan Cassini had a distant, non-targeted encounter with Dione and discovered that some of the features glimpsed at low resolution by the Voyagers did not exist. Two quiet orbits then followed, with solar conjunction between 5 and 9 August occurring just after the beginning of the 27th revolution. On returning to periapsis the spacecraft made long-range observations of the backlit plumes of Enceladus.

The next surge of activity was the T17 encounter on 7 September. This occurred two orbits after the previous one and at almost the same point in the moon's orbit. As the closest point of approach would be at only 23°N it had been decided to raise the flyby to 1,000 km. The mass spectrometer was to collect atmospheric data that would help in designing future passes; in particular, whether to raise the one in late October that was scheduled to occur at 1,030 km. The camera took pictures both prior to and after closest approach and the mapping spectrometer observed a stellar occultation 15 minutes before the flyby, but it was the mass spectrometer and radar which had priority. Operating only for a few minutes the imaging radar traced a short swath over the fully illuminated day-side that mainly contained a sea of dunes in the Fensal region but also happened to include, near one edge, the 29-km crater Ksa whose central peak and well-defined ejecta blanket indicated its relative youth. Dunes had obliterated the western part of the blanket. The rarity of craters on Titan was evidence of ongoing rapid resurfacing and erosion. It was estimated that craters would last for only a few hundred million to a billion years. But given their paucity and the uncertain rate of cratering in the Saturnian system it was impossible to draw meaningful statistical conclusions. The fact that circular crater-like features were more common in Xanadu suggested that this continent was one of the more ancient exposed surfaces. As the attitude control system had less difficulty than expected in holding the craft stable during its penetration of the atmosphere, project managers confirmed 1,030 km as the altitude of the next encounter.

Two days later Cassini had non-targeted encounters with Methone at 15,000 km and Enceladus at 40,000 km. That same day magnetospheric instruments recorded two drops in energetic electrons at positions some 3,000 km apart in the vicinity of Methone's orbit. It seemed unlikely that such a small moonlet could have produced

Ksa, imaged during the T17 flyby, is a rare case of a Titanian impact crater. Note how dunes are diverted around the crater and its ejecta field. (NASA/JPL/Cassini RADAR team)

such a broad depletion, so it was taken to represent the signature of a ring co-orbital with the moon and this was designated R/2006 S5.[231]

As Cassini climbed to apoapsis on the night-side of Saturn it spent 12 hours on 15 September transiting the shadow cone of the planet. It was 2.2 million km from Saturn and 15 degrees out of the ring plane, viewing the unilluminated face of the rings. With backlighting opaque regions looked dark, particularly the dense B ring, and tenuous rings stood out clearly, as did material within the Cassini division. The situation was ideal for studying faint details of the known rings and for discovering new ones. The diffuse G ring was seen to have a distinct inner edge, and structures were seen that traced from Enceladus into the E ring, possibly streams of particles 'feeding' the ring. This eclipse also netted four new ringlets designated R/2006 S1 to S4. The first one, a diffuse ring made of small dust grains, was related to the co-orbital moons Epimetheus and Janus. The second one shared its orbit with Pallene. Both rings were thousands of kilometers wide. The third, located in the outer gap of the Cassini division, was dubbed the Charming Ringlet. It was so bright under this illumination that it saturated some of the images. It had to be densely packed with small dust motes. The infrared mapping spectrometer later showed that this material resembled the F ring and Encke division, and was different from the adjacent rings. The fourth feature was a narrow, tenuous and discontinuous ringlet located between two broad bands in the Cassini division.[232,233] Also visible in images taken from the shadow cone was the night hemisphere of Saturn illuminated by sunlight reflected from the rings. Because the eclipse occurred only a month after solar conjunction, Earth and the Sun were close together in Saturn's sky and the pictures included a view of Earth 10 AU away seen through the outer part of the main ring system. And when the spacecraft was occulted by Saturn from the viewpoint of Earth, the radio-

A mosaic of images taken by Cassini as it transited through the shadow of Saturn on 15 September 2006. The faint rings exterior to the 'classical ones', as well as the interior of the Cassini division are much more evident when backlit, as in this case, than in reflected light. The bright 'star' visible at 9.30 o'clock between the classical rings and the G ring is the Earth. (NASA/JPL/Space Science Institute)

occultation was used to probe the atmosphere at near-equatorial latitudes at ingress and at mid-southern latitudes at egress.

Cassini reached apoapsis two days later. The first part of the 29th revolution was spent on a 70-picture survey of the G ring. This faint ring was unique because, being located tens of thousands of kilometers from the nearest known satellite, its existence could not readily be explained. Images taken on 19 September revealed a bright arc within the ring apparently trapped by a resonance with the moon Mimas, which completed six orbits in the time required for the ring to complete seven. This held material in an arc that spanned 60 degrees in longitude and was only 250 km wide. It looked very similar to the arcs in the ring system of Neptune discovered by Voyager 2 in 1989. The magnetospheric imaging instrument found a localized drop in the electron flux near the G ring. This indicated that the ring must also contain centimeter-to-meter sized bodies. These would not be easily captured by the camera in back-illumination. Proof of the existence of largish particles within the ring had been obtained in September 2005 when a grain at least 0.1 mm in size hit Cassini's dust detector as the spacecraft passed near the outermost edge of the ring. From its magnetospheric signature the total mass of the material in the arc was equivalent to a moonlet 100 meters across. Collisions and erosion of the largest bodies of the arc were invoked as the source for the entire ring.[234]

The closest point of approach of the T18 encounter on 23 September occurred at 960 km above latitude 71°N. As previously, Cassini's night-side approach to Titan provided an opportunity for the camera to seek lightning and auroras. But, as on the 7 September flyby, the priority instruments at closest approach were the radar and mass spectrometer. The radar expanded the imaging coverage of the northern polar regions, revealing many more small lakes. In addition, the mapping spectrometer detected a large ethane cloud at the rim of the polar vortex. The camera wrapped up

The irregular moon Janus transiting over the Saturnian cloud tops on 25 September 2006. (NASA/JPL/Space Science Institute)

the encounter 10 hours after closest approach with distant observations of the dusk side of the atmosphere.

The frequent flybys of Titan were rapidly increasing the inclination of Cassini's orbit from 15 to 25 to 38 to 47 degrees, reducing its eccentricity and increasing its periapsis to 5.5 Rs. But the orbital period was held at around 16 days, the same as Titan, in order that each orbit would produce a flyby. At the same time encounters with the other satellites, targeted or non-targeted, were becoming rarer. As a result, most observations were devoted to Saturn, its atmosphere and its rings, and to Titan itself. The T19 encounter on 9 October was another high-latitude pass and, due to the resonance in the orbits of the spacecraft and the moon, it occurred in a similar orbital position to the previous three. On this occasion Cassini passed 980 km over the night-side. The radar operated around closest approach and the swath included some bright featureless patches and mysterious circular patterns. More lakes were found, as well as extensive channel networks. The extent of a particularly large lake could not be determined since only part of it was within the swath. In addition, the radar collected radiometry, scatterometry and altimetry.

At periapsis on 11 October Cassini observed Spica being occulted by the rings. A narrow ringlet was observed in the Cassini division that had been absent on other occasions. The camera was aimed for 3 hours at the south polar region of Saturn to image a dark eye-like vortex centered on the pole, in which the overall airflow was downward. This provided a view of the clouds at deeper levels that were normally masked by the top cloud deck. Because the vortex exposed the interior, the infrared radiometer saw it as an anomalously warm spot. The 'eye' was surrounded by high clouds and hazes which formed two concentric 'eye-walls'. Because the walls cast shadows it was possible to measure their heights: the circular outer one was 40 km high and the more oval inner one was 70 km high.[235]

The T20 encounter on 25 October was to be almost equatorial. After the test in September, the altitude planned for the closest point of approach at 7.5°N had been at 1,030 km. The primary instrument for this flyby had been vigorously debated by the scientists, resulting in the decision to allow the infrared mapping spectrometer to collect high-resolution data. The approach started with a radar imaging swath of Tortola Facula from a range of over 20,000 km, which was less than ideal. Then the spectrometer started taking 'noodles' over Fensal, targeting in particular dune fields in order to determine their composition. The noodle swath included erosional and depositional features and what appeared to be cryovolcanic flows. The instrument wrapped up by taking high-quality moderate-resolution cubes of the south pole. Its data revealed the existence of a 150-km mountain range in the southern hemisphere whose peaks were 1,500 meters tall. Meanwhile, the camera took images of Senkyo and Tseighi. This flyby increased the inclination of Cassini's orbit to 55 degrees but trimmed its period from 16 to 12 days, with the result that the spacecraft would not encounter Titan again for some time. The next two and a half orbits provided only a non-targeted encounter with Enceladus on 9 November at a range of 90,000 km.

On 23 and 24 November Cassini took a long sequence of images covering one orbit of Prometheus to observe in detail how it affected the F ring. Both the ring and the moon had slightly eccentric orbits so their interaction was maximum when

Prometheus wandered near to the ring at apoapsis, drawing particles toward it. This created bridges of material that linked the two. However, the ring and the moon had different orbital periods and the streamers lagged behind Prometheus. On the next orbit, Prometheus interacted with a different part of the ring, drawing out another streamer a few degrees ahead of the previous one, and so on. Over time this created a series of dents which gave the inner edge of the ring a 'saw-tooth' appearance.[236] Other observations were devoted to a search for 'propellers' in the main rings, and a complete longitudinal scan at modest resolution located nineteen of them.

When far away from the ring plane Cassini cooperated with the Hubble Space Telescope to study auroras on Saturn. While the Earth-orbiting telescope monitored the ultraviolet auroral oval, Cassini took infrared cubes and measured the magnetic field parameters and energetic particles fluxes. The oval dynamics had already been studied from Earth for limited viewing angles and geometries. When the solar wind upstream was calm the oval was known to form a circle at a latitude of 75 degrees, but when the wind pressure increased, it contracted poleward and could also adopt a spiral-like morphology. Cassini, from its vantage point, had a view over the entire auroral oval. For example, on 11 October the spiral shape was recorded and on 10 November the main aurora consisted of a weak oval with several brighter spots, the brightest of which was in the morning sector. These observations proved that unlike the auroras of other planets that of Saturn had unique characteristics and structures, including regions of emission located both equatorward and poleward of the main oval. Some of these structures seemed to involve magnetospheric processes which were not fully understood.[237] The high-inclination orbits also revealed the existence of a second component of the kilometric-radio emissions, at times as strong as the first component but with a slightly shorter period. They appeared to originate from the auroral regions, with the northern ones having a slightly different period than the southern ones. The Voyager instruments had mainly observed the northern radio emissions. No such phenomenon had been found on any other 'magnetic' planets. It could be a manifestation of the mechanism which transferred the internal rotation of the core of the planet to the plasma in the magnetosphere.[238,239]

Cassini passed periapsis on its 34th revolution on 2 December and observed an occultation of Spica by the F ring. The starlight was twice momentarily interrupted by opaque objects that spanned 1 km. They were dubbed 'Butterball' and 'Fluffy'. It is possible that the latter was the ring core itself.[240]

On 12 December, on the inbound leg of its 35th revolution, Cassini had the T21 encounter. Since the previous flyby Titan had completed three orbits of Saturn and Cassini four orbits. The closest point of approach was at 44°N and at an altitude of 1,000 km. With 1 hour 10 minutes to go, the ultraviolet spectrograph observed an occultation of alpha Persei to measure a vertical profile of hydrocarbons and haze. Once again, however, the main instruments were the radar and mass spectrometer. The radar imaging at mid-northern latitudes featured Belet, a region on the trailing hemisphere that had been poorly observed. It showed dark featureless plains, dunes, and scattered bright patches. Half of the terrain consisted of linear hills interspersed with dunes. This flyby restored Cassini's period to 16 days and produced the T22 encounter one orbit later, on 28 December. On this occasion the camera took high-resolution images of Tsegihi and the composite infrared spectrometer measured the spatial distribution of carbon monoxide, water and hydrogen cyanide and made two temperature maps of the surface. But the primary task was to continue the survey of the gravity field by radio tracking. As on 27 February Titan was near apoapsis, but this time Cassini reached a higher latitude of 40°N. The closest point of approach at 1,297 km was the closest thus far to be controlled using the reaction wheels. Cassini made the T23 flyby on 13 January 2007, on the inbound leg of the next revolution. The geometry of the 1,000-km flyby was very similar to the previous one but this time the primary task was radar imaging. The swath overlapped some of the terrain surveyed by the imaging radar and the radar altimeter during the very first flyby in 2004, as well as in February and October 2005. In particular, it provided a second view of Ganesa Macula. This was suspected of being a cryovolcanic dome until a 3-dimensional model based on data from this pass and from October 2004 revealed it not to be domical in shape and so probably not a volcano. The swath included the northern half of a large circular structure that had a bright rim 180 km across which looked like an impact crater. The dark, flat floor of this basin seemed to have been flooded by a smooth deposit of ice. Fields of longitudinal dunes were also imaged, in this case with small circular features superimposed that could be craters. A bright feature on the northern edge of the swath appeared to be a cryovolcanic flow. The swath ended near some mountains previously observed by the infrared mapping spectrometer. On the way out, the ultraviolet spectrograph observed an occultation of eta Ursae Majoris by the moon, which was then more than 20,000 km from the spacecraft.

On 29 January 2007 Cassini was to perform the first part of the 'pi-transfer' which would rotate the periapsis of its orbit 180 degrees around Saturn. This encounter would be on the inbound leg, but the next one, on 22 February, would be on the outbound leg. The ensuing encounters, all on the outbound leg, would gradually reduce the inclination of the vehicle's orbit from its current 59 degrees until it was back in the equatorial plane. Later encounters would increase the inclination again, until by the end of the primary mission it was at an angle of about 75 degrees.[241]

For the final inbound encounter, T24, on 29 January, Cassini made its approach to Titan over the night-side and receded over the day-side with a minimum altitude of 2,631 km. The visible, infrared and ultraviolet remote-sensing instruments were assigned priority. In particular, the camera extended its high-resolution coverage to the west and the infrared spectrometer observed an occultation of gamma Crucis. The encounter held the orbital inclination at 59 degrees and increased its period to 18 days.

On 13 February Cassini observed a rare eclipse of Iapetus, 2 million km distant. Traveling high above the sunlit side of the rings provided a number of opportunities for observing Saturn's most impressive feature. A number of azimuthal scans of the rings were made. These were divided into shadow observations measuring the fall in temperature as the ring particles passed into the shadow cone of the planet, and observations to measure the influence of the particle size, rotation, and dynamics on their temperature. By February 2007 a total of 48 scans had been made, of which 29 targeted the shadow zone. From this data, bodies in the rings were determined to be slowly spinning and to have large temperature differences between the day-side and night-side. On entering the planet's shadow the particles in the thin C ring showed a rapid fall of up to 20 degrees. However, particles in the dense B ring did not show any such temperature change.[242] It was also proved that the innermost D ring had vertical structures and was actually warped "like a corrugated tin roof", with the corrugations winding themselves tighter over time. This hinted at a recent impact of a meter-sized object in the mid-1980s, just after the Voyager flybys. Moreover, the D ring itself had been displaced inward some 200 km since it was imaged by those missions.

Cassini, outbound, returned to Titan on 22 February for the T25 encounter with a minimum altitude of 1,000 km above 30°N. Taking advantage of the illumination during the approach the remote-sensing instruments observed both the south polar regions, including Ontario Lacus, as well as the north polar regions. The mapping spectrometer observed a northern polar lake but the spectral data was too noisy to draw conclusions about the composition of the liquid filling it. However, the radar was the main instrument. Its 10 hours of activity included 30 minutes of imaging centered on the moment of closest approach. This produced a long swath across the Saturn-facing hemisphere from near the north pole to 30°S in the process crossing six previous swaths, which was a unique feat. It increased to almost 15 per cent the portion of the surface that had been mapped by radar. The southernmost part of the swath contained linear hills and dune fields. A feature proved to be surrounded by deposits that could be of volcanic origin. Other circular features could be craters. Also imaged was Sotra Facula, a 40-km dome which had released a 180-km flow to the south that was partially covered by dunes. Departing from Titan the radar also tried to observe for 5 minutes at reduced resolution in order to image Hotei Arcus. These images showed flow-like features covering surrounding terrains which were believed to be cryovolcanic flows, but other origins, such as sedimentary deposits, could not be ruled out. The most amazing portion of the swath, however, was in the high-latitude lake regions, above Senkyo and Belet, where the radar found a body of liquid hydrocarbons 400 km across, later dubbed Ligieia Mare. A second 'sea' of hydrocarbons with a large island that resembled Sardinia in shape was marginally included in the swath. Initially dubbed the 'Caspian Sea' for its size, it was later officially named Kraken Mare. This flyby was also of interest to the particles and fields scientists because if the magnetosphere were sufficiently compressed by the solar wind they might gain their first observations of Titan in interplanetary space, or more particularly in the planetary magnetosheath just downstream of the bow shock. However, this did not happen. Five hours after closest approach the camera started to take medium-resolution images in order to monitor northern mid-latitude clouds that had developed earlier in the month. The T25 flyby put Cassini back into a 16-day orbit.

When Cassini was near apoapsis, well above the equatorial plane, the composite infrared spectrometer scanned the north polar regions of Saturn to monitor changes in its atmosphere as the end of winter neared and that pole began to catch sunlight.

After passing periapsis on 7 March Cassini had its T26 encounter on the 10th. It flew by at an altitude of 981 km with the closest point at the temperate latitude of

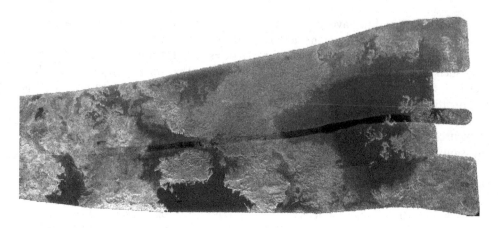

Part of the T25 radar swath showing Ligeia Mare at right and a portion of Kraken Mare with a large island at the center. (NASA/JPL/Cassini RADAR team)

32°N. This time the most important observations were by the mass spectrometer. Its inlet was pointed forward at closest approach to scoop up material and determine its composition. About 45 minutes earlier, spectrometers had monitored the limb near the south pole of Titan occult the Sun. The geometry of the flyby facilitated more imaging of regions in the northern trailing hemisphere, north of Belet and Adiri, as the spacecraft withdrew. This imagery included kilometer-resolution frames of the terrain that was observed by the radar the previous December. The high-inclination orbit did not provide many opportunities to observe the other satellites, so the next major event was the T27 encounter on 26 March. This had a minimum altitude of 1,010 km. It provided the first radio-occultation over the northern hemisphere. This started 23 minutes prior to closest approach and cut from high southern latitudes to temperate northern ones. Bistatic radar experiments were attempted on both ingress and egress, with the region surveyed at ingress being situated just east of Ontario Lacus. Although this encounter left the period of the spacecraft's orbit at 16 days it reduced the inclination to 52 degrees.

Cassini reached apoapsis on 31 March. Heading back in on the 42nd revolution the infrared spectrometer made a series of scans of the north pole of Saturn at high spectral resolution, and detected the hexagonal atmospheric pattern which had been discovered by Voyager 2 in 1981 and then also detected in infrared images taken from Earth shortly before Cassini arrived in the system. The hexagon consisted of a warm cyclonic vortex rotating rigidly with the planet. A similar eye-shaped feature was over the south pole, but what was unexpected was its persistence over the north pole as this had not received heat from the Sun since the 1995 equinox. 'Hot' polar spots were probably formed by gases plunging into the deep atmosphere and being warmed as they were compressed. Unfortunately, details of the precise atmospheric mechanism causing this were not understood, nor was why this effect produced the hexagonal shape.[243,244] Spectroscopy showed small amounts of carbon monoxide, phosphine and other compounds which convection currents had transported to the cloud tops.

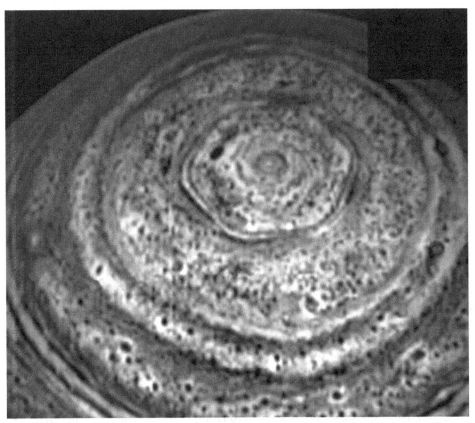

This mosaic of images of the northern polar hexagon of Saturn was obtained by Cassini's visible and infrared mapping spectrometer in October and November 2006. (NASA/JPL/University of Arizona)

The T28 encounter on 10 April saw Cassini pass Titan at a minimum altitude of 991 km on a trajectory that once again allowed the radar to obtain an imaging swath that covered similar terrain to the February pass, including the dark seas discovered on that occasion at high northern latitudes. The radar was therefore able to view this region from a different perspective to enable 3-dimensional views to be created of the areas around Kraken Mare and Ligeia Mare, respectively the largest and second largest hydrocarbons seas. Putative volcanic features were seen at the southernmost end of the swath, as well as bright terrain. As in February the swath crossed Sotra Facula, an isolated mountainous area that had two peaks rising to over 1,000 meters and a profusion of non-circular craters and finger-like flows that bore similarities to terrestrial mountains like Etna in Sicily and Icelandic volcanoes. This was the best candidate for a cryovolcano and the mapping spectrometer established the flows to have a different composition to the adjacent terrains. It was not clear whether Sotra Facula was still active but it was decided to monitor it routinely.

The anticipated hydrocarbon ocean had proved noteworthy for its absence. In its place, a complex hydro-geological system had been revealed by the radar. The most striking features were riverbeds and drainage channels. There were rivers at every latitude ranging from the equator (where Huygens had touched down) to the poles. They showed considerable variety, some being more than 1 km wide, others being deep canyons, some having radar-dark (and hence smooth) floors and others having radar-bright floors littered with debris. In contrast to the channels of Mars which appeared to have been caused by either catastrophic floods or the sudden melting of underground reservoirs of ice, the rivers on Titan had all of the right characteristics for having been formed by methane and ethane rainfall, although at the time Cassini inspected them they were dry. The spacecraft had yet to catch a thunderstorm in the act of dumping rain onto the surface. There were many short-lived clouds which drifted in longitude but only a few persisted in one place, the best example being one that was often present over the south pole. It was possible that conditions in the atmosphere could cause 'ghost rain' without clouds when turbulence in the altitude range 20–30 km caused hydrocarbons to condense. In fact, terrestrial telescopes had observed liquid methane spectroscopically in this atmospheric layer.

When Cassini arrived in the system so little was know about Titan that it had to be presumed that its axis of rotation was perpendicular to the plane of its orbit and its rotation synchronized with its orbital period. These parameters would usually be measured from imagery but the moon's extensive haze and atmospheric refraction made it ill-suited to such precise observations. However, it was soon realized that overlapping radar swaths could be used. The first 14 radar passes had produced two or more overlapping images for 19 areas of the surface but radar scientists were not able to exactly match features from one swath to the next; there were discrepancies of up to 40 km. Most of this could be explained if the spin axis of Titan were tilted at 0.5 degree from the plane of its orbit. However, features could still be offset by several kilometers from one pass to the next. It was concluded that the spin period was marginally faster than the orbital period. A possible explanation for this, was that the winds were altering the rotation period. In fact, the winds did not have to be affecting the entire moon: if the icy crust, tens of kilometers thick, were uncoupled from the core by an ammonia-rich liquid water ocean then that would be influenced while the core continued to rotate synchronously. Subsurface oceans were already strongly suspected for Europa, Callisto, and possibly also Ganymede in the Jovian system. But there remained discrepancies between the model and observations that could signify other causes. In particular, non-synchronous rotation could also be the result of a recent large impact, but no trace of that site had been seen in any image or radar swath.[245,246,247]

As Cassini climbed toward apoapsis it took sequences of images of Hyperion to study its chaotic rotation and distant observations of the bright trailing hemisphere of Iapetus. At periapsis it made some observations of the dark side of Enceladus to observe its plumes and jets forward-scattering sunlight. Triangulating images of the moon's limb taken over time from a variety of viewing directions allowed scientists to trace the jets back to their sources. At least eight discrete sources were identified in this manner and it turned out that all four tiger stripes were issuing jets, with the strongest ones from Baghdad and Damascus. Moreover, most of the geyser sources matched the sites of temperature anomalies and hot spots discovered by the infrared radiometer.[248]

Cassini returned to Titan on 26 April for the T29 encounter with a geometry that was similar to the previous flybys over the northern hemisphere. It also included a 22-minute solar and a 21-minute Earth occultation just before closest approach. No particular observations were planned for them, however. The minimum altitude was 981 km. The radar was again the primary instrument, taking an imaging swath from the eastern part of Fensal through Elpis Macula to the northern polar region, filling in some gaps in the map of the great northern lakes. The camera started to operate 6 hours after closest approach by taking low-resolution global maps and then on the way out it resumed monitoring clouds. The encounter reduced the inclination of the spacecraft's orbit from 46 to 39 degrees.

Around periapsis on 10 May Cassini had a second radio-occultation of Saturn's atmosphere, ingressing at 75°N and egressing near the equator, then cutting through one ring ansa.

The T30 encounter occurred two days later. The closest point of approach was at 69°N and an altitude of 959 km. Since the ground track passed over regions already imaged by the radar it was decided to obtain a long altimetry swath in the run up to closest approach in order to determine the topography of the landscape. In addition, scientists experimented with a technique that performed altimetry using the central radar beam while the adjacent beams obtained imagery. Altimetry would be able to be taken at the same time as imaging in future encounters to rapidly increase the coverage by the altimeter. At the end of this sequence, 4 seconds were devoted to scatterometry over the clouds at a high northern latitude where it was still winter night. The idea was to measure the methane droplets in the clouds and thus find out whether it was raining. The results showed the droplets to be either extremely small or very sparse, indicating that it was not raining. The radar then made high-resolution observations of the northern hemisphere to locate more lakes and channels, collecting the data to expand the 3-dimensional map of this region. The high-resolution swath included a portion of Kraken Mare in order to investigate whether it was radar-dark because it was naturally flat or because it was a depression holding liquid hydrocarbons. The flyby was also of interest to the magnetometer scientists because Titan was on the relatively unexplored dawn-side of the magnetosphere. Moreover, there was again the possibility that at the time of the encounter the moon might be in the solar wind upstream of Saturn; it wasn't.[249,250]

On the next revolution Cassini took an imaging sequence designed to track the moonlet Pan for an entire 9.5-hour revolution in order to refine the parameters of its orbit. On 26 May it made a non-targeted flyby of Tethys at a range of 103,000 km, observing the Saturn-facing hemisphere that featured Ithaca Chasma. The infrared spectrometer was trained on a streak of dark material west of the canyon in order to determine its composition. Later that day Cassini observed Mimas in the shadow of Saturn. During such eclipses the infrared spectrometer measured the thermal inertia of the moon by recording how its surface cooled. The next day, at periapsis, Cassini made distant observations of Enceladus and Dione. In particular, by passing nearly between the Sun and Enceladus it was able to view the moon in an illumination that was ideal for measuring how it reflected sunlight. Then, while the camera targeted the Saturn-facing hemisphere of Dione, the ultraviolet spectrograph observed this moon occult a star in an observation to seek traces of an atmosphere or outgassing. After that, distant observations were made of Rhea.

Continuing away from Saturn, Cassini had the T31 flyby on 28 May. Occurring at the fairly high altitude of 2,299 km the encounter was not used for radar studies. Instead it used a radio-occultation to make atmospheric and ionospheric profiles. A 32-minute, essentially diametrical occultation profiled the atmosphere at 75°S and 75°N. After closest approach a bistatic radar experiment was performed in an effort to investigate the boundary between a dark and a bright area just north of Belet. The camera took images of Dilmun, Shangri-La and Adiri at varying resolutions, as well as pictures of the night-side in order to search for lightning and to monitor auroras.

Approaching apoapsis Cassini discovered a new small satellite. It first appeared in a 6-hour imaging sequence taken on 30 May. The 2.2-km body orbited between Methone and Pallene, occupying an orbital resonance with Mimas. Initially named S/2007 S4 it was later named Anthe. Along with Methone and Pallene it formed the Alkyonides family of the smallest satellites known in the Saturn system.[251] Starting in May 2007 many of the distant, small irregular moons were observed by Cassini using long imaging sequences to improve knowledge of their orbits and their sizes. Given their large distances from the planet and their eccentric orbits the rotations of these satellites were not synchronized with their orbits, so it was hoped to be able to determine their rotation periods. The observation distances ranged between 6 and 22 million km depending on the body, so of course none was resolved as anything more than a faint speck of light. However, Cassini could observe them at angles and geometries that were impossible from Earth. The objects targeted in this way in the ensuing years included Albiorix, Bebhionn, Bergelmir, Bestla, Erriapus, Fornjot, Greip, Hyrrokkin, Ijiraq, Jarnsaxa, Kari, Kiviuq (suspect of being binary), Loge, Mundilfari, Narvi, Paaliaq, Siarnaq, Skathi, Skoll, Suttungr, Tarqeq, Tarvos, Thrymr, Ymir and the unnamed S/2004 S12 and S13.

The 46th revolution was fairly quiet. In addition to another non-targeted flyby of Mimas it provided a view of Atlas at a range of 34,000 km which allowed imaging of its southern hemisphere at a resolution of 320 meters per pixel. Given the small size of the moon, it spanned a mere 120 pixels in the field of view of the wide-angle camera.

The T32 encounter on 13 June was the eighth outbound flyby in succession. The closest point of approach was at 84.5°N and 965 km. The primary instrument in the 15 minutes before and after closest approach was the mass spectrometer. However, the flyby included Sun and Earth occultations. The former was observed at ingress in the extreme-ultraviolet to study the haze layers and produce vertical temperature and composition profiles. At long last, the flyby occurred while Titan was traveling outside the magnetosphere. This provided a rare opportunity to study how the moon interacted with the interplanetary magnetic field and solar wind. Cassini flew from the night-side of Titan into the ionosphere and then over the day-side, anti-Saturn hemisphere. In the process it left the magnetosphere at 15.4 Rs, as this contracted with changing solar wind conditions upstream, and again 20 minutes before closest approach as it flapped back and forth. The particles and fields instruments observed a disturbance in the interplanetary magnetic field created by Titan, suggesting that the moon retained a 'memory' of the Saturnian field to which it had been exposed a few hours earlier.[252]

This flyby reduced the inclination of Cassini's orbit from 18 to just 2 degrees, in essence returning it to the equatorial plane of the Saturnian system. This orbit was not particularly well-suited to observing the rings, which would be nearly edge-on, but it provided for frequent targeted and non-targeted encounters with the satellites as well as opportunities to observe their mutual occultations and conjunctions. Near periapsis Cassini had a non-targeted encounter of Tethys at a range of 18,400 km that provided imagery of the Saturn-facing hemisphere and Ithaca Chasma. It then sped by Mimas at a range of 103,000 km and Enceladus at a range of 88,600 km, with the camera taking low-resolution pictures of the anti-Saturn hemisphere of the latter.

Two days later, on 29 June, after so many high-latitude encounters, Cassini had the last in a series of nine outbound encounters. This T33 flyby was at an altitude of 1,933 km with the closest point at the almost equatorial latitude of 8.1°N. The main scientific objective was a 3-hour period of radio tracking. Unlike previous 'gravity flybys', which occurred when Titan was at apoapsis, this time it was at periapsis. The manner in which the moon responded to tidal effects at different distances from Saturn would hopefully reveal whether it had a subsurface ocean. The particles and fields instruments took advantage of the vehicle's position upstream of Titan in the magnetosphere. The camera started imaging an hour prior to closest approach and covered dune fields in central and western Adiri that had been imaged by the radar in October 2005. The camera team would then attempt to match the radar features with the visible ones and correlate by analogy surrounding terrain not seen directly by the radar. The infrared spectrometer took high-resolution images and spectra of the cloud streaks in the northern polar region. These proved not only to be at much lower altitudes than the ethane polar clouds but also of a different composition. By terrestrial analogy they were interpreted as methane that had evaporated from lakes and was condensing as fog.[253]

This Titan flyby put Cassini into a 22-day orbit, starting the process of stretching its apoapsis to reach the distant Iapetus in September. After apoapsis on 9 July the T34 encounter on the 19th, inbound to Saturn, occurred at an altitude of 1,332 km. This flyby provided a unique opportunity to image and investigate some rarely seen terrain of the Saturn-facing hemisphere centered on Senkyo. As seen by the radar in July 2006 the region included dry channels, riverbeds and dune fields. The camera and mapping spectrometer observed the moon for over 24 consecutive hours, taking advantage of the geometry to obtain high-resolution multispectral measurements of northwestern Shangri-La and Belet at resolutions of the order of several hundred meters. Individual dunes were targeted in an attempt to resolve their spectrum and identify their composition. Bistatic radar observations were carried out over central and western Adiri, an area that looked quite rough and mountainous, in the hope of observing specular mirror-like reflections from bodies of liquid. The flyby further increased the orbital period to almost 40 days and the apoapsis to 69 Rs. This orbit had the longest period since that on which the orbiter released the Huygens probe.

The next day Cassini had a non-targeted 38,000-km flyby of Helene, the leading Trojan companion of Dione. Images at a resolution of several hundred meters, plus ultraviolet and infrared spectra and thermal measurements revealed Helene to be an angular 32-km object extensively pitted with craters; just as low-resolution Voyager pictures had suggested. The geometry of the ring-plane crossing enabled Cassini to view the rings back-lit by the Sun and this was exploited to try to re-observe faint ringlets. A faint streak near Pallene was taken to be its co-orbital ring. The distant day-side apoapsis allowed Cassini to make movies lasting an entire rotation of the planet in order to track clouds. Science activity was halted on 16 August as Saturn entered solar conjunction, then resumed on the 29th just in time for the periapsis at 5.4 Rs. On that occasion Cassini imaged the F ring and then had a non-targeted encounter with Tethys at a range of 55,400 km that facilitated moderate-resolution imaging of the leading hemisphere featuring the large crater Odysseus. A relatively close non-targeted flyby of Rhea at a range of 5,727 km followed in which Cassini passed downstream of the moon. Once again it sensed a broad electron depletion region with short-period dips hinting at the possibility that this large satellite had its own rings. Observations of the Saturn-facing hemisphere of the moon, only partly illuminated by the Sun, were made during the approach. Longer exposures easily revealed features lit by Saturn-shine. At closest approach Cassini's remote-sensing instruments examined the youthful 75-km crater Inktomi, which had been narrowly missed in the single previous targeted encounter.

On 31 August, the day after flying by Rhea, Cassini had the T35 encounter. This flyby started another series of outbound encounters that would run to the end of the primary mission. As a result of the 3,324-km flyby altitude, this time most of the observations were by the remote-sensing instruments. The composite spectrometer observed Titan during the approach to measure the composition of its atmosphere. Then an hour prior to closest approach the ultraviolet spectrograph observed sigma Sagittarii occulted by the moon. A second stellar occultation half an hour later was monitored by the infrared spectrometer. The mapping spectrometer then observed the night-side of the north pole, and two hours after closest approach it took high resolution cubes of the northern polar regions and of Dilmun and Shangri-La on the anti-Saturn hemisphere. In particular, it obtained good multispectral views of Adiri, western Shangri-La and the crater Selk. It also targeted the bright spots of Nicobar and Mindanao Faculae in central Shangri-La. Meanwhile, the camera examined a number of circular features and other candidate impact craters and took images of Adiri at high-resolution, including the northeastern part on which Huygens landed. The plasma-wave instrument continued to seek lightning. Remarkably, Cassini had yet to detect radio bursts unambiguously associated with lightning. This contrasted somewhat with observations by Huygens of possible distant storms. It was possible that Titan actually had no lightning; or that it was rare and had simply not occurred during the total of 300 hours that Cassini had spent in the vicinity of the moon thus far; or that the radio waves produced by lightning could not escape the ionosphere. Another hypothesis requiring investigation was that unlike clouds of water vapor in the Earth's atmosphere, clouds of methane drops in Titan's atmosphere would not become electrostatically charged. In fact, the northern temperate belt of clouds did not show the convective motions that would create lightning.[254] The radio package would continue observations but more than 600 hours of data during 70 flybys only confirmed the negative result. Bursts were indeed detected on a single flyby, but the fact that they disappeared when Saturn was occulted by Titan suggested that they originated from the planet.

Two days after T35, Cassini made the first of two maneuvers to target the only Iapetus encounter of the primary mission. Meanwhile, the camera had a view of the fully illuminated planet and so monitored its atmosphere, clouds and rings. Cassini started to observe Iapetus early in September. Because the flyby was 4 days before apoapsis, it occurred at the fairly modest relative rate of 2.4 km/s, allowing a longer period of less frantic observations than was the case for other satellites. The primary scientific objective was to determine whether the two-sided appearance of the moon was endogenous or exogenous. The encounter sequence started early on 9 September, some 33 hours prior to closest approach. In approaching the night-side and the dark hemisphere, the inbound images showed the crescent phase. But long exposures captured details of the night-side lit by Saturn-shine, including the poorly observed south polar regions. Downlink periods interspersed with the observations were exploited to collect tracking data to measure the gravity and internal structure of the moon. Inbound the ultraviolet instrument obtained spectra of the surface and observed an occultation of sigma Sagittarii to search for a tenuous envelope; none was expected and none was found. As the range reduced, the camera and imaging spectrometer targeted the equatorial ridge. At around closest approach the camera took an 8-frame mosaic along the length of the ridge with a best resolution of about 12 meters. Stereoscopic analysis revealed that in some places the slopes leading up to the ridge had angles exceeding 30 degrees.

The flyby occurred at 14:16 UTC on 10 September. The minimum altitude was 1,622 km with the closest point of approach at a latitude of 4°S. Shortly thereafter Cassini crossed from the dark to the bright hemisphere and used its instruments to examine the transition. Even at the highest resolution the heavily cratered surface appeared to be either black or white with no nuances in between. No evident differences in cratering frequencies existed between the two hemispheres, meaning that they were of more or less the same age. Within the dark Cassini Regio was Turgis, which at 580 km in diameter was the largest impact crater on the entire moon. Other notable features included a 60-meter crater, later dubbed Escremiz, at the center of a very bright ray network. Intriguingly, a number of dark patches were found on the bright trailing hemisphere corresponding to depressions, dark-floored craters and troughs, and craters at high latitudes mostly sported dark floors and equator-facing rims as well as bright pole-facing rims. The existence of craters with bright rims in Cassini Regio suggested that the dark blanket was at most a few meters thick. The thinness of the dark veneer had already been suspected from terrestrial radar observations which could not distinguish between the two hemispheres, and it was confirmed by radiometry and scatterometry from Cassini's radar. The temperature and texture of the surface on both the day-side and night-side was measured at varying resolutions by the infrared radiometer and spectro-meter. The day-side temperature seemed to be correlated with surface brightness. At

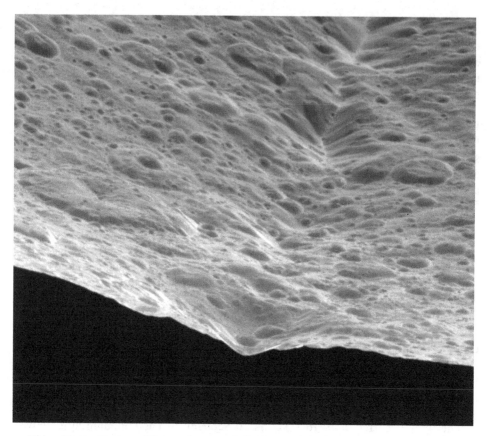

This wide angle view of Iapetus' equatorial ridge was obtained during the September 2007 flyby. (NASA/JPL/Space Science Institute)

−144°C the dark terrain was 'warmer' than the bright terrain at −160°C. The mapping spectrometer sought polycyclic aromatic hydrocarbons to explain the nearly-black color of the dark terrain. In fact, Cassini Regio proved to be essentially ice-free. In addition to scatterometry and radiometry of the surface at large distances the radar made a short low-resolution imaging scan of the leading hemisphere, the first ever of an icy moon, and made an altimetry run over the equatorial ridge.

The new data indicated that the leading hemisphere had been 'painted' with dark material that originated from elsewhere. The icy crust was mantled by a thin layer of dust. Sunlight shining on the chocolate-dark material would warm it, resulting in increased sublimation of the underlying ice. Water molecules would leak from the dark hemisphere and, if too slow to escape the moon, would migrate and eventually refreeze on the colder surface of the bright hemisphere. This 'thermal segregation' phenomenon would be particularly effective on Iapetus because of the fortuitous combination of its size, distance from the Sun, long orbital period and synchronous rotation. Simulations showed that during the 79-day period the effect would build up heat on the dark hemisphere sufficiently to create the ying-yang appearance in at

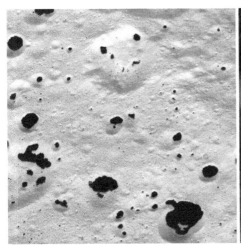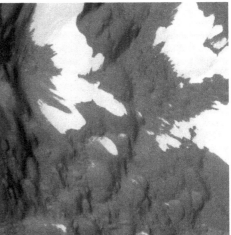

Two close-up views of the transition terrain between dark and bright terrain on Iapetus. (NASA/JPL/Space Science Institute)

most several tens of millions of years. A millimeter of ice on the dark hemisphere would completely sublimate in less than 10,000 years.[255],[256] It remained a mystery for some time, however, how the leading hemisphere became dark in the first place. The 'usual suspects' were the outer satellites and in particular Phoebe, or Ymir, the second largest irregular moon at some 18 km across. This mystery was solved a few years later when the Spitzer Space Telescope scanned the region of space in which Phoebe orbits. These infrared observations discovered Saturn's largest ring, a broad sheet of dust that extended in toward the planet from the orbit of retrograde Phoebe. The smallest particles in the ring would be free to slowly migrate inward to Iapetus, Hyperion or Titan.[257] Of course, the chaotic rotation of Hyperion tends to even out the coating, and the high winds of Titan's atmosphere capture in-falling dust. Only Iapetus had developed the dichotomy.

At apoapsis on 14 September Cassini began its 50th revolution. Unfortunately, the vehicle entered safe mode while transmitting the last Iapetus data. Only a few minor observations were lost before the new sequences started on 16 September. At that time many millions of kilometers from Saturn, Cassini took several movies of the planet to monitor its atmosphere and clouds over an entire 10-hour rotation. The mapping spectrometer observed the planet for 20 hours, recording the south polar auroras. The spacecraft also monitored the rings and the now far off Iapetus to view parts of the bright hemisphere that had not been visible during the flyby, including several large craters with dark floors.

On 30 September, passing periapsis, Cassini had non-targeted encounters with Dione, Enceladus and Tethys. The Enceladus encounter provided the best views yet of the leading hemisphere. This area was known to be crisscrossed with the lightly cratered remains of what appeared to be 'extinct' tiger stripes.

As Cassini drew away from Saturn it had the T36 encounter on 2 October. This flyby was at an altitude of 973 km on a trajectory designed to reduce the period of

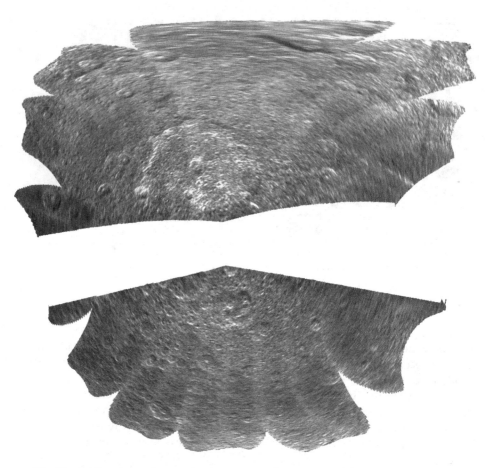

The Cassini imaging radar swath of Iapetus' leading dark hemisphere. (NASA/JPL)

the spacecraft's orbit to 24 days, draw the apoapsis closer to Saturn and away from Iapetus, and restart the process of cranking up its orbital inclination to again be able to view the ring system from vantage points far out of plane. During the approach to Titan the infrared spectrometer scanned the atmosphere, and the radar collected radiometry and scatterometry over Fensal and eastern Tseighi. At closest approach the mass spectrometer had priority in order to directly sample the upper atmosphere for hydrocarbons. The radar took an imaging swath of the leading hemisphere down to 70°S. This showed the existence of three liquid-filled lakes at southern latitudes. Their surfaces appeared so flat and the liquid they held so absorptive that the radio reflections were mostly of 'noise'. The swath also confirmed the crater Selk, whose existence had been suspected from earlier imaging spectrometer data. At the same time as it was imaging, the radar got altimetry of the Huygens site. The camera and other remote-sensing instruments resumed monitoring Titan about 6 hours after the flyby, viewing, as on so many occasions, the bright Adiri region, eastern Belet and western Shangri-La on the anti-Saturn hemisphere.

Strands in the F ring caused by the gravitational action of Prometheus imaged on 25 October 2007. (NASA/JPL/Space Science Institute)

Around apoapsis Cassini received a new upload of flight software. It made few, if any, observations during this time. Scientific activity resumed on 18 October with a movie of the Encke gap. Three days later saw a rare non-targeted encounter with Hyperion at a range of 122,000 km. Then there was Titan at 440,000 km. In spite of the long range some good observations were made of poorly characterized areas on the trailing hemisphere, including the northern part of Adiri. Also covered were a bright area between Belet and Senkyo, a candidate impact crater with a dark floor, and a large sea-like dark spot in the north. No targeted encounters occurred around periapsis but long-range observations were made of the icy satellites. In addition the fully illuminated rings were observed from slightly south of their plane, with a radial scan being made from the D ring through the B and A rings, out to the F ring. On 24 October Cassini observed Enceladus, some 636,000 km distant, occult zeta Orionis.

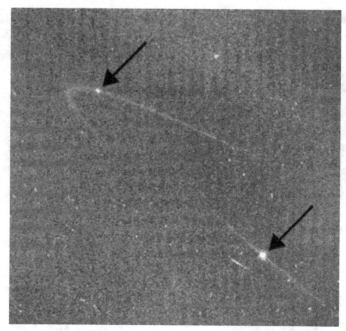

A heavily processed image of the ring arcs accompanying Anthe and Methone (the bright 'star' at lower right). (NASA/JPL/Space Science Institute)

The star ran almost parallel to the limb of the moon and through the south polar plumes. The results revealed the presence of four distinct gas jets some 10 km in width, superimposed on the plume. Ultraviolet spectra indicated a water density double that present at the time of the 2005 occultation of gamma Orionis. However, based on modeling of when the tiger stripe fissures should be closing, the opposite effect had been expected! The occultation implied that the source of the plume was a reservoir of liquid water rather than gases trapped in crystalline ice. The spectra did not show the carbon monoxide that was suspected from in-situ sampling by the mass spectrometer; it was concluded that the mass spectrometer must actually have detected molecular nitrogen which, since it had the same mass as carbon monoxide, would have given the same reading.[258] Later in this revolution the spacecraft made another radio-occultation of Saturn, this time crossing the mid-latitudes from 41°S to 39°S. On 29 October long-exposure images to recover the small satellite Anthe serendipitously found extremely faint narrow arcs surrounding both it and Methone, which was also in the field of view. These images showed that the two moons were accompanied by ring arcs several tens of degrees long and confirmed the existence of the Methone ring, R/2006 S5. The Anthe ring was designated R/2007 S1. These rings were probably constrained to short arcs by resonances with Mimas.[259,260]

The periapsis of 17 November was also encounter-free. Saturn's atmosphere and the rings were the focus of attention. But the small satellites were imaged to study their orbits and motions, in particular how their orbits were perturbed by the larger moons. There were non-targeted encounters of Enceladus and Rhea. This view of a crescent-phase Enceladus from a range of 170,000 km provided an opportunity to monitor the south polar geysers. Cassini again passed through the shadow cone of the planet. Being closer in and traveling more rapidly, this pass did not last as long as that near apoapsis in September 2006. This time the main interest was the upper atmosphere. Just after the eclipse the cameras again targeted Anthe to get a better fix of its orbit, shape and size.

Two days later, outbound from Saturn, Cassini had the T37 encounter. This time the flyby was at a minimum altitude of 999 km with the closest point of approach at 22°S. The two infrared instruments had priority during the approach, observing the night-side and collecting data on the composition and chemistry of the atmosphere. The mapping spectrometer collected a long 'noodle' covering the western edge of Shikoku Facula, eastern Senkyo and Mindanao Facula. The primary instrument for closest approach was the mass spectrometer, which gathered atmospheric samples. Ultraviolet scans of the disk of Titan were made to map auroras and hydrocarbons. Because the vehicle penetrated the ionosphere the particles and fields instruments were particularly busy. On the way out the camera obtained a view of Adiri on the trailing hemisphere.

On 27 November, as Cassini was heading back in from apoapsis, a new storm on Saturn was detected by the plasma-wave experiment. It was 21 months since the previous storm. When both Voyagers detected storms at intervals of a few months it had been presumed that they occurred continuously, but that was evidently not the case. Optical observations would have to wait until Cassini was properly positioned in relation to the planet. Fortunately the storm lasted until mid-July 2008.[261]

Just an hour prior to periapsis on 3 December Cassini had a 9,200-km flyby of Epimetheus; the closest yet of any of the small moons. Epimetheus was showing its night-side at closest approach, so the best images were taken at longer range as the spacecraft withdrew.

Less than two days later, outbound from Saturn, Cassini had the T38 encounter. The minimum altitude was 1,298 km and the closest point of approach was at 79°S. The vehicle approached the night-side with radio tracking measuring the moon's gravity and mass. The remote-sensing instruments had priority at closest approach. The mapping spectrometer and camera started operating over Adiri, examined the Huygens site, and ran south to the lakes. On departing they imaged the crater Selk and dark linear markings in Dilmun. The most interesting measurements were made over the south pole where the mapping spectrometer surveyed Ontario Lacus at a spatial resolution of about 500 meters. It found a number of absorption bands in the spectrum of the lake that were not present on the surrounding terrain, evidently due to liquid ethane mixed with compounds such as propane and butane. Methane was also expected to exist in liquid form in the lake but it was not possible to isolate its spectrum from methane in the atmosphere. These observations proved Ontario to be a lake and made Titan only the second celestial body on which standing bodies of liquid were known to exist. Multispectral images of Ontario revealed its contour to comprise two separate concentric rings. The intermediate-brightness inner ring was interpreted as a 'beach' of wet sediments carried by tides of seasonal hydrocarbon cycles. The brighter outer ring appeared drier and seemed to represent the ethane-impregnated mud flats of an ancient shoreline.[262,263,264]

The T39 encounter on 20 December flew a similar ground track to the previous one, with a minimum altitude of 970 km. Remote sensing on the way in and out at relatively long ranges monitored the chemistry and meteorology of the atmosphere. The radar had priority for the 5 hours centered on the closest point of approach and was to produce the first radar imaging swath of the south pole. It made a 5,600 km swath from 30°S over the pole to almost the same latitude on the other hemisphere. It also took altimetry of Tseighi inbound and of northern Adiri outbound. Eroded mountainous terrain and mottled plains were in evidence throughout the swath and, just as in the north, there were no dune fields in the southern hemisphere outside of the narrow equatorial band. Closer to the south pole there were channels associated with several dark spots that were evidently lakes. Overall, however, the swath had far fewer features that could be interpreted as lakes than were present near the north pole. Furthermore, the steep-sided depressions that held lakes were rare, or totally absent. Instead there were flat-floored valleys filled with smooth material that could be deposits of sediments or dry lakes. This asymmetry may be due to longer-period phenomena involving the orbits of Saturn and Titan, perhaps including inclination and eccentricity cycles with periods of the order of 45,000 years. In particular, the northern winters now fall around Saturn's perihelion, making them shorter than at the south pole and allowing ethane lakes to be better preserved at the north than at the south. Interestingly, the swath located fewer lakes at the south pole than were seen in long-range optical observations in 2004 and 2005. A simple explanation could be that those lakes had evaporated or drained as the seasons evolved in the intervening years. (In this case it would be interesting to monitor the remaining lakes during an extended mission to see whether their shorelines changed.) Supporting the seasonal explanation for the disappearance of the southern lakes was the fact that since 2004 the changing weather patterns had prevented cloud formation over that pole.[265]

Although all of the lakes discovered so far at the two poles of Titan represented about 600,000 square kilometers, or 1 per cent of the surface, this was insufficient to provide the source of methane that replenished the atmosphere over geological time scales.[266,267]

Two other observations were made on this revolution: another radio-occultation enabled Cassini to profile the atmosphere of Saturn at latitudes 18°S and 68°S, and just after the T39 flyby it made long-range observations of the poorly characterized northern regions of Rhea. After apoapsis on 27 December the vehicle returned to the inner system. On 3 January 2008, the day before periapsis, it used many of its remote-sensing instruments to observe the bright star alpha Scorpii (Antares) being occulted by the rings to probe the opacity, fine structure and density of the rings. In particular, viewed by the infrared spectrometer and the ultraviolet spectrograph, the occultation suggested the existence of additional objects near the core of the F ring that ranged in size from tens to thousands of meters. Just before periapsis at 3.97 Rs there was a non-targeted encounter of Dione at a range of 127,000 km showing its illuminated leading hemisphere. Tethys was observed from a point that was almost directly between it and the Sun, ideal for measuring surface reflectance.

Outbound from Saturn, Cassini had the T40 encounter on 5 January. In this case

Antares shining through the rings on 3 January 2008. (NASA/JPL/Space Science Institute)

the minimum altitude was 1,014 km and the closest point of approach was near the Huygens site. Approaching the night-side, remote-sensing continued the search for lightning. With 44 minutes remaining to closest approach the infrared spectrometer observed alpha Bootis being occulted by Titan and 21 minutes later the ultraviolet spectrograph observed alpha Lyrae (Vega) being occulted. Priority was then given to the mapping spectrometer to take high-resolution spectral cubes in the vicinity of the Huygens site. It also obtained the best spectral scans to-date of the crater Selk. Receding from Titan the camera imaged features in western Xanadu and Shangri-La, including the candidate cryovolcano Tortola Facula. This encounter put Cassini into a 12-day orbit inclined at 47 degrees to the equator of the Saturnian system. As a result, the next three revolutions had no targeted encounters with any of the major satellites.

Around periapsis in mid-January Cassini observed another radio-occultation by Saturn, with the point of egress near 70°S. On the 16th it made distant observations of Rhea in opposition to the Sun in order to collect information on the texture and nature of the icy surface. Near apoapsis Cassini had a non-targeted flyby of Titan at a range of over 850,000 km, and observed the northern polar regions in search of small lakes in an area surveyed by the radar the previous February. Despite being so distant, this encounter provided the best views to-date of the crater Menrva. And the

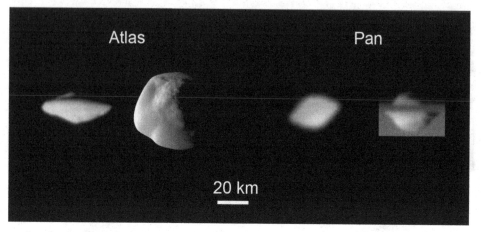

Seen by Cassini Atlas and Pan, the two moonlets embedded in the rings were revealed to have a solid core and a 'flying saucer'-like equatorial disk of debris. (NASA/JPL/Space Science Institute)

periapsis passage of the 58th revolution provided a radio-occultation by Saturn and non-targeted encounters with Atlas and Pan which were revealed to be small ellipsoidal bodies of 39 × 18 km and 33 × 21 km, respectively. Both had prominent symmetrical smooth equatorial bulges which gave them the appearance of 'flying saucers'. The most likely explanation for this shape was that the 'core' of the body had swept up ring particles until they completely filled its minuscule gravitational 'sphere of influence'. Whereas the core seemed to be solid ice and fairly rough, the smooth bulge was probably no denser than freshly laid snow. The low densities of these two moonlets had been suspected from the masses inferred by the manner in which they imposed scalloped waves on the rings.[268]

From the point of view of their densities the Saturnian system was thus realized to possess two distinct satellite populations: dense moons from Mimas outward and fluffy, powdery ones from Janus inward. All of the fluffy moons likely represented the accumulation of ring matter on solid cores only half their present sizes. On the other hand the Trojan satellites of Dione and Tethys (Helene, Polydeuces, Calypso, and Telesto) may well be small chips off their parent moons. Numerical simulations implied that as moonlets accreted from ring particles, they would be driven outward by the perturbations that they exerted on the rings themselves. The other side of this process of momentum transfer meant that the rings would simultaneously contract toward Saturn. Further analysis explained the presence of propellers in only those regions of the rings that were fairly distant from the planet. Gravitational accretion and disaggregation processes recycling matter in the rings could shed light on the paradox of their age. Although the tidal destruction of a large satellite to produce a ring would be likely only in the early years of the solar system, the recycling would sustain bright and ice-rich rings which gave the impression of being young.[269,270,271] On the basis of Cassini observations and simulations a theory for the origin of the rings was proposed that explained most of their characteristics. It began with a large satellite the size of Titan possessing a silicate core and a crust of ice. This satellite would have migrated inward during the final phases of the formation of the solar system. At a certain small distance from the proto-Saturn tidal stresses would have stripped off the ice but left the core intact. The icy particles would have formed the rings. The rocky core would have continued to spiral in and merged with the parent body. In fact, simulations raised the possibility that the icy material formed not only the rings but also all the inner satellites out to Tethys. This theory nicely explained why, if the rings were the result of the destruction of an ancient satellite, they were almost completely 'rock-free'.[272,273]

Cassini started its 59th revolution at apoapsis on Valentine's Day by making long-range observations of the thin crescent of Iapetus in order to refine models of the moon's shape. At periapsis it had a non-targeted encounter with Janus at a range of 144,000 km and made observations to better identify not only the moon's shape and geology but also its surface properties and texture. A solar occultation was used to observe the backlit rings and characterize their dust population.

Cassini returned to Titan for the T41 encounter on 22 February. The track of the 1,000-km flyby was similar to the previous one, passing close to the Huygens site. A thin crescent was visible during the approach. The composite spectrometer took atmospheric and surface temperature measurements over Hotei Arcus. Five hours to closest approach the radar initiated 7 hours of observations. It obtained altimetry of Tsegihi on the way in and of northern Adiri on the way out. In between it took an imaging swath which ran almost parallel to the southern boundary of Xanadu. The high-gain antenna was pointed to the right of the ground track for the first portion of the swath and then to the left, in order to image Hotei Arcus and a region south of Xanadu in the first portion and far southeastern Adiri and the area surrounding the Huygens site in the second portion. When combined with data obtained during the October 2005 pass the second portion of this swath would provide stereoscopic views of the landing site and its surrounding terrains. On the first portion, the one dedicated to Hotei Arcus, a lobate feature just to the north matched the position of a possible cryovolcanic flow identified by the mapping spectrometer. The ultraviolet spectrograph observed several stellar occultations in the hours following the flyby. The first was eta Canis Majoris crossing the haze layers over the dawn terminator. This was followed by epsilon Canis Majoris, one of the brightest ultraviolet stars, in a peculiar observing geometry that allowed Cassini to see the star pass in and out of the atmosphere twice, the first time 22 hours after the flyby and the second 6 hours later, in both cases getting high signal-to-noise data on the haze layers of the upper atmosphere.

There were no targeted encounters on the next 11-day revolution and the most notable observation was an imaging sequence to discern clouds at varying depths in Saturn's northern hemisphere. A non-targeted flyby of Titan on 10 March occurred at a range of almost 1 million km. Two days later, several hours before periapsis on its 61st revolution, Cassini had the E3 encounter. After an almost 3-year hiatus this targeted flyby of Enceladus had initially been designed to pass 1,000 km from the moon. Spurred by scientists, engineers examined the possibility of skimming by as low as 30 km to inspect the tiger stripes in detail either optically or by radar. One adventurous proposal was to directly sample the geysers, but there was concern that this material might contaminate the camera optics, ruining future research.[274] In the end it was decided to compromise.

Cassini started to observe Enceladus 15 hours out at a range of 600,000 km. The mapping spectrometer and other remote-sensing instruments stared at the northern anti-Saturn and trailing hemisphere, collecting data for many hours to ensure a high signal-to-noise ratio in the spectra for the non-ice components on the surface. Two scans by the infrared radiometer were made to find any 'hot spots' on the northern hemisphere. The camera got its first good view of the north polar regions and took pictures with a best resolution of 200 meters. In particular, scientists were eager to know whether the tiger stripes at the south pole were unique, or whether there were 'fossil stripes' elsewhere on the moon. The plasma spectrometer had priority for 26 minutes, including the moment of flyby. The mass spectrometer, dust detector, and particles and fields instruments were also operating. The flyby was 48 km above the intermediate latitude of 20°S. With a relative speed of over 14 km/s, imaging was impractical. As Cassini pulled away it continued on a southerly track to sample the plumes by traveling more or less parallel to the issued material. Thirty seconds after closest approach the vehicle was over the polar regions, and 30 seconds later, at an altitude of 650 km and climbing rapidly, it passed through the heart of the plumes. The ion and neutral mass spectrometer recorded a gas density in the jets 20 times greater than predicted. The gas was carrying with it tiny ice particles which caused frequent, albeit minute, attitude disturbances when they hit the vehicle. The mass spectrometer took advantage of the deep penetration of the plumes to undertake an analysis with a high signal-to-noise ratio. In addition to water ice, nitrogen and carbon dioxide, the plumes were found to contain acetylene, hydrogen cyanide and complex organics such as ethane, propane and formaldehyde. Other molecules in trace amounts included deuterium and complex hydrocarbons such as benzene. There may also have been some argon. The data often showed sudden spikes in the signal, probably when ice crystals entered the spectrometer's inlet and melted. The magnetometer used this close flyby to investigate how the moon interacted with the magnetosphere and to seek evidence a subsurface ocean.

A mere 3 minutes after the time of Cassini's closest approach Enceladus began a 2-hour passage through the shadow cone of Saturn. Exploiting the close range to achieve a high spatial resolution the composite infrared spectrometer measured the thermal inertia of the surface as it cooled in eclipse and made temperature scans of the southern hemisphere and tiger stripes in darkness to identify the locations of all the major hot spots. It found that the highest temperatures of the entire moon were along the tiger stripes. The camera was also trained at the dark moon to detect any 'non-thermal' emissions such as auroras or radiation-induced glows. The infrared spectrometer then observed Enceladus as it emerged from Saturn's shadow.

Outbound from Saturn on the next revolution Cassini had the T42 encounter on 25 March. The minimum altitude was at 999 km. As before, the spacecraft viewed a thin crescent on the way in and its departure was over the trailing hemisphere. The mass spectrometer had priority at closest approach. The imaging spectrometer made a long zigzagging 'noodle' which began south of Hotei Arcus, crossed Texel and

Mindanao Faculae and ended up west of Dilmun. This encounter was the second of a series of four which occurred with Titan sunward of Saturn, and scientists once again hoped to find it outside the magnetosphere but this was not the case. The June 2007 encounter was destined to remain the only one of the primary mission to occur with Titan immersed in the solar wind.

There followed five relatively quiet 9-day orbits that provided only non-targeted flybys of the inner moons. But the high inclination enabled Cassini to observe the rings and the atmosphere of Saturn and to conduct long-range imaging to refine the orbits of the smaller moons. For example, during the 63rd revolution in early April it observed an occultation of alpha Arae by the rings. The next time around another ring occultation featured the giant southern star beta Centauri (Agena). In addition, pictures of the Anthe ring clearly resolved its double-stranded structure some 200 km wide.[275] Between 6 and 9 April, around apoapsis, Cassini collected a 22-frame 5-color wide-angle mosaic covering the night-side of the classical system of rings, finding 16 propellers in a narrow band in the mid-A ring. Later, propellers were also observed by the shadow their central body projected onto the ring plane under progressively more edge-on illumination and were found to be several hundred meters in size. Another 'propeller belt' was discovered between the Encke gap and the outer part of the A ring. This belt was populated by fewer but larger objects. Eleven propellers in this second belt were tracked for several years and were nicknamed in honor of aviation pioneers. One, named Blériot after the French aviator Louis Blériot, showed up in pictures from 2005 to 2009. On one occasion, by sheer chance, it was even detected during a stellar occultation. Thanks to the number of observations obtained, it was possible to study its motion. Although it was essentially a normal Keplerian orbit, it displayed drifts of several thousand kilometers evidently due either to gravitational interactions with the larger moons or perhaps to being jostled by impacts. Over the ages perturbations from the rings should drive the moonlet embedded in a propeller outward. This may well have happened to Atlas, orbiting just beyond the outer edge of the A ring. As a unique small-scale case of a celestial body embedded in a debris ring, the dynamics of propellers constituted a potential source of insight into how planets form around stars.[276] Other observations were made of the 'morning shadow boundary' where ring particles emerged from the shadow of Saturn, in an attempt to catch spokes in the act of forming.

Around periapsis on 10 May Cassini had a non-targeted flyby of Tethys with a view of the southern part of its Saturn-facing hemisphere, featuring Ithaca Chasma in particular. The T43 encounter was two days later, climbing away from Saturn. In mid-April terrestrial telescopes monitoring Titan in the infrared had recorded the eruption of a vast cloud system at 'tropical' latitudes and it was hoped this marked the onset of a seasonal change in the weather.[277] On the approach Cassini observed the night-side using its infrared spectrometer. It then switched to radar. After some radiometry of the Saturn-facing hemisphere the radar took altimetry of Hotei Arcus and north of Belet. The northwest to southeast imaging swath started southwest of Hotei Arcus, over the western central part of Xanadu, northern Shangri-La, central Dilmun, and on into northwestern Adiri. Included for the first time was the western

The bright propeller 'Earhart' (top left), named after the American aviatrix Amelia Earhart near the Encke gap of the A ring seen on 11 April 2008. (NASA/JPL/Space Science Institute)

part of Tui Regio, located southwest of Xanadu, which was an anomalously bright and compositionally different patch suspected of being either a cryovolcanic plain or a dry sea bed. The track over Xanadu partially overlapped the coverage from the April 2006 pass, enabling stereoscopic analysis of the terrain. Remarkable features identified in Xanadu at this time included a circular structure 20 km wide and three parallel lines which resembled faulted ridges. In the Dilmun region, about 1,000 km north of the Huygens site, was a crater 115 km in diameter. Named Afakan, it was only the fifth *bona fide* impact identified on Titan. It had a flat floor with a hint of a central peak and one section of its rim appeared to have collapsed. About halfway through the swath, Cassini reached the minimum altitude of 1,001 km. The closest point of approach was at 17°N, with the Sun directly overhead. The spacecraft then receded over the illuminated disk with the camera monitoring clouds.

This flyby had further shortened the period of Cassini's orbit to just over a week. The next periapsis on 17 May provided non-targeted flybys of Dione, Janus and Tethys. In particular, Dione was monitored as it entered the shadow cone of Saturn. There was also a radio-occultation of Saturn with the points of ingress and egress at 22°N and 58°S respectively, and an occultation by the planet of alpha Centauri, the closest stellar system to the Sun. After returning to periapsis on 25 May Cassini again flew by Titan on the 28th. The 1,400-km T44 encounter was the last northern hemisphere pass of the primary mission. The camera got some of its best images so far of Hotei Arcus as Cassini approached the terminator. The mapping spectrometer covered an area from eastern Tui Regio to Hotei. However, observations unrelated to Titan were piggybacked onto the schedule. In particular, early in the encounter the composite spectrometer had a unique opportunity to make radial scans of the lit and unlit faces of the A, B and C rings. Once again the main instrument for Titan was the radar. Thirty minutes before closest approach it started to take altimetry of Hotei Arcus and other regions which had been surveyed on previous passes. Using radar data collected thus far, scientists had measured the true shape of the moon and established it to be slightly oblate, flattened at the poles, but basically spherical to within a few hundred meters. The highest areas were less than 1,000 meters above the reference sphere, and in fact most of the radar and visible surface features were not really topographic structures. What was interesting was that the shape of Titan was more compatible with a body that formed closer to the planet than its present location, implying that it had 'migrated'. Another interesting speculation was that if an essentially continuous 'methane table' existed several hundred meters below the surface in the low-lying poles, this could explain the presence of hydrocarbon lakes only in those regions.[278] The radar imaging swath started in southern Xanadu, crossed its southwestern part, continued across eastern Tui, Eir Macula, central Shangri-La and its dune fields, Nicobar and Oahu Faculae, the southern part of Dilmun, and ended in northwestern Adiri, again partially overlapping the April 2006 data. The images showed bright channels up to 5 km wide on the dark plains at the edge of Xanadu. The orientation of dune fields was used to chart how the prevailing winds dealt with the elevated landmass of the continent. Cassini departed over the fully illuminated day-side. Data collected in this part of the encounter was deemed expendable if the antennas of the Deep Space Network were required by the Phoenix mission, which was arriving at Mars at that time.

The T44 encounter concluded the primary mission in terms of targeted flybys. A few more non-targeted encounters of Titan and other moons would occur during the next five revolutions but the perspective provided by the high inclination was used to observe Saturn's atmosphere, auroras and rings. By June 2008 the approaching equinox illuminated both poles of Saturn, enabling observations of the entire planet. Changes in cloud patterns were seen on both Saturn and Titan. Imagery and spectra of Titan to-date indicated that its clouds clustered around three latitudes: the north pole where it was winter, the south pole where it was summer, and a narrow band at southern temperate latitudes. The latter appeared to be due to a global atmospheric circulation system, this latitude matching the one where the southern polar haze cap ended. This distribution was expected to change with the seasons and indeed as the mission progressed small bright clouds at the south pole had become rarer and long streaks had begun to appear in the northern hemisphere.[279]

At periapsis on 1 June 2008 the high-inclination orbit gave an almost diametric radio-occultation in which Cassini passed behind Saturn at 37°N and reappeared at 68°S. On that revolution there was a non-targeted encounter with Titan at a range of 365,000 km during which the camera imaged the northern lakes to measure their perimeters. At over 1,000 km around, Kraken Mare, at its fullest extent, would have an area in excess of 400,000 square kilometers. Other imaging sessions observed eclipses of Dione and Rhea and attempted to detect a ring around the latter.

On 26 June Cassini started its 75th revolution around the planet, the final one of its primary mission. On approaching periapsis the composite infrared spectrometer measured the surface temperature and thermal inertia of Mimas as it passed into the shadow cone of Saturn. Just before crossing the ring plane there was a non-targeted flyby of Enceladus at a range of 84,000 km. This provided a view over the northern portion of the Saturn-facing hemisphere during the approach, and the southern part afterwards. On the way out the camera was trained on a number of poorly mapped fractures, including Isbanir Fossa and the belt of folded terrain which surrounds the region that contains the tiger stripes and the south pole. Shortly thereafter the moon went into eclipse. At about the same time as the Enceladus encounter Cassini was 31,000 km from Janus. After inspecting the southern hemisphere of Mimas, the spacecraft wrapped up the periapsis activities by turning its attention to the rings.

At midnight UTC on 1 July 2008, precisely four years after orbit insertion, the primary mission officially ended. Since its arrival the spacecraft had returned more than 140,000 images.

## THE EQUINOX MISSION

As happened so often, at the end of its primary mission Cassini was still in good shape, had an abundant reserve of propellant, and all its instruments were functional apart from some components of the magnetometer which had failed in 2006. It was therefore decided to extend the mission in order to collect more scientific data. The plan for a 2-year extended mission to be undertaken at the same operational pace as the primary one was approved in February 2007. It would consist of 60 orbits at the time of Saturn's northern spring equinox. The equinox itself would occur in August 2009. In the run up eclipses of satellites by Saturn, mutual satellite eclipses and the transits of the shadows of satellites across the planetary disk and on the plane of the rings would all become more frequent, facilitating unprecedented observations. As a result the extension was named the 'equinox mission'.

During these 2 years Enceladus would get seven additional flybys at 2,000 km or nearer, including two extremely close ones. It was deemed particularly important to measure the small gravitational field of the moon in order to investigate its internal structure. Titan would get 26 flybys that would extend the radar imaging coverage by an additional 8 per cent to a total of about 30 per cent of its whole surface. In particular, these flybys would yield seven dusk encounters, three high northern latitude ground tracks and a mid-tail crossing of the moon's wake. Dione, Rhea and Helene would each receive a flyby at less than 2,000 km. Mimas, which had been neglected during the primary mission, would also get a fairly close flyby. However, no Iapetus encounters were factored in because it was so difficult to reach. The mission would provide a large number of mid-latitude atmospheric occultations over the northern hemisphere of Saturn and three 'ansa-to-ansa' occultations by the rings.

Cassini would begin the extended mission at a very high orbital inclination and then descend to an equatorial orbit, remaining sufficiently out of the ring plane near equinox (when the rings would be illuminated edge-on by the Sun) to provide many opportunities to monitor not only the rings but also the poles of the planet. To begin with the spacecraft was in a 2.7 × 20.8 Rs orbit inclined at 75 degrees with a period of 7 days. Many observations during the early orbits would be devoted to the rings, where spokes were becoming ever more common, to the atmosphere of the planet, and to the icy moons. As Cassini climbed to apoapsis on its 74th revolution, which it reached on 3 July, it made observations of ring spokes and the minor satellite Anthe and its associated ring. The first month was relatively quiet, with no targeted flybys. At periapsis on 14 July Cassini crossed the shadow cone of the planet and used this geometry to image some of the icy moons, including Enceladus, Janus, Mimas and Rhea, as they entered eclipse. Moreover, the ultraviolet spectrograph and infrared spectrometer observed a solar occultation by the planet to monitor the highest haze layers, their density, altitude and temperature. Receding from Saturn over its south pole Cassini targeted the polar vortex to gather data in readiness for the equinox, when the Sun would set there. While approaching periapsis the vehicle was able to exploit its position far above the north pole to observe several southern hemisphere stars being occulted by the rings, followed by other stars after periapsis.

The first burst of activity came on 31 July, outbound from Saturn, with the T45 encounter. The minimum altitude was 1,614 km with the closest point of approach at 43°S. During the approach the camera imaged the leading hemisphere, observing southwestern Xanadu over the crescent, continued its coverage of Hotei Arcus, and monitored clouds. The primary activity was radio tracking to further investigate the shape and internal structure of the moon. This began 12 hours out and lasted for 16 hours. The four gravity survey flybys of the primary mission had provided a precise determination of the moon's mass and density, which averaged 1.8 times that of ice. Estimates of its moment of inertia implied that the interior was best fitted by a two-layer model consisting of an outer shell of almost pure water ice and a core of either hydrated rock or a mixture of rock and ice. In either case the interior would be only partially differentiated. This made it intermediate between Callisto, which was not differentiated, and Ganymede, which had an iron-rich core capable of sustaining a weak dipolar magnetic field. This meant that the interior of Titan had never been hot enough for the ice and rock to fully separate. However, some of the parameters determined on this and previous gravity flybys seemed inconsistent with an entirely solid object. An alternative model assumed the presence of an ocean of liquid water that isolated an icy crust from an icy mantle and a core made of a mixture of ice and rock.[280,281,282] The existence of subsurface oceans in the Jovian satellites had been established by the signature of a magnetic field induced in the moon by the planet's field. The magnetic field of Jupiter had a significant tilt relative to the planet's spin axis, creating a periodic 'forcing' on the moon and its electrically conductive ocean. But Saturn's magnetic field was almost coaxial with its spin axis and any forcing would be very weak and would in turn generate a weak response in an electrically conductive ocean. Furthermore, for a flyby of Titan that was safely above the outer haze layers of the atmosphere the presence of the ionosphere could easily mask the signal of an induced magnetic field.

On the 79th revolution, in early August, Cassini searched for additional satellites in the region between Mimas and Enceladus where the Alkyonides family orbited, and also observed gamma Crucis being occulted by the ring system. At periapsis on 4 August there was a non-targeted encounter with Mimas, which at that time was in the shadow of Saturn. Images were taken after the moon emerged into daylight to fill in gaps in the coverage of its Saturn-facing hemisphere. In addition, there was a radio-occultation of the rings lasting 30 minutes. Cassini returned to periapsis one week later and on 11 August had the E4 encounter, the first Enceladus flyby of the equinox mission.

The high-inclination orbit meant that Cassini approached Enceladus from above the north pole and withdrew over the south pole. The remote-sensing instruments had priority to investigate the tiger stripes and surrounding terrain just after closest approach but the pass through the plumes was an opportunity for the dust sensor to analyze small particles. Observations 17 hours prior to closest approach monitored the interaction between the plumes and the E ring, but encounter observations proper began 9 hours out with radio tracking to study the moon's gravity. Four hours later, and with 280,000 km to go, the infrared mapping spectrometer began to observe the northern trailing hemisphere. Its high signal-to-noise, near-infrared spectra verified the presence of non-ice on the surface. Previous flybys had detected carbon dioxide, hydrogen peroxide and lightweight organics at southern latitudes. Routine imaging began immediately thereafter with the moon filling the field of view of the narrow-angle camera. Meanwhile, the ultraviolet spectrograph searched for emissions from the dissociation of water molecules in the plumes. Cassini then flew by at a relative speed of 17.7 km/s at an altitude of 49 km above a point on the night-side at 28°S. It emerged into sunlight almost exactly over the south pole. At closest approach the dust analyzer and plasma spectrometer were sampling particles from the E ring and plumes. Flying over the south pole, the camera undertook a so-called 'skeet-shoot' sequence in which, in order to compensate for the high relative speed at such close range, Cassini slewed at its maximum possible angular rate in the same direction as the moon's apparent motion and took a sequence of seven short-exposure images in rapid succession as the moon crossed the field of view of the camera. The resulting high-resolution pictures had pixel scales ranging from 7 meters to 28 meters and covered the areas where the hottest temperatures had previously been recorded by the infrared spectrometer, including three hot spots and known emission centers within the tiger stripes. The first image was taken exactly one minute after closest approach. It was still on the terminator between the day and night. The first fully-lit image came 33 seconds later over Cairo Sulcus. The footprint then moved over hot spots near Baghdad, over a known eruption site, and finally over Damascus Sulcus. Seen close up the tiger stripes had a distinctive 'V' profile about 300 meters deep, with adjacent ridges about 150 meters tall. They were littered with icy blocks. Spots of brighter material could be fresh snow fields of coarse-grained ice. The fact that the active sources did not appear distinctive implied that geysers could erupt at any point along a sulcus.

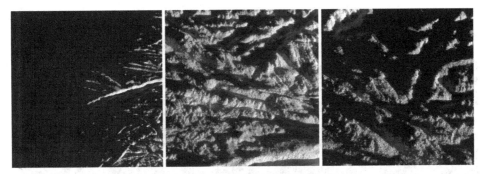

The first three images of the 11 August 2008 'skeet-shoot' sequence of very high resolution images of the south pole of Enceladus. The first picture (left) was taken by the wide angle camera, and the later ones by the narrow angle one. (NASA/JPL/Space Science Institute)

On departure the remote-sensing instruments again sought trace compounds on the surface, only this time over the southern hemisphere and, owing to the small but rapidly increasing distance, at high resolution. The spectrometers were also used to locate amorphous ice, whose crystalline lattice had been destroyed by exposure to solar and cosmic radiation, and so would indicate areas of ancient ice. Less than 30 minutes after closest approach the composite infrared spectrometer took priority to observe Enceladus slipping into eclipse and measure its temperature and thermal inertia. A few minutes later Cassini reached periapsis at 3.94 Rs. Observations by the remote-sensing instruments continued well after Enceladus left eclipse and then 3-frequency radio tracking resumed. Tracking revealed Enceladus to be the densest of the major moons in the system. Some 60 per cent of it was rock, probably in the form of a core. Then there was a crust of ice tens of kilometers thick. The relatively high density could indicate that the moon had lost a large fraction of its low-density material such as water ice. It was calculated that if the geysers had been active since the moon's formation, then it may well have lost 20 per cent of its original mass. The encounter was wrapped up 16 hours after closest approach by ultraviolet spectra and multispectral imaging. The particles and fields instruments had collected data to better determine how the moon interacted with the E ring. In particular, scientists wished to distinguish between particles coming from the plumes and those ejected from the surface by impacting dust.

A few minutes prior to closest approach Cassini had maneuvered to position its ion and neutral camera and the electron spectrometer optimally to sample along the lines of the planet's magnetic field. It had recorded powerful beams of ions aligned with the magnetic field and electrons over a wide range of distances from the moon but these fluxes abruptly disappeared with 1 minute remaining. The beams could be expected to form a 'footprint' in the auroral regions at the poles of the planet where the field lines linked the moon to the planet, and there excite hydrogen molecules to glow faintly in the ultraviolet.[283]

On reaching apoapsis after the Enceladus flyby Cassini took infrared scans and images of the faint E and G rings. Two images taken on 15 August to track arcs in

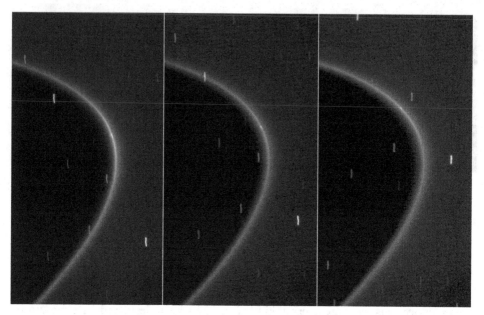

The tiny moon Aegeon first revealed itself as a brighter, denser portion of the G ring. (NASA/JPL/Space Science Institute)

the G ring revealed a new satellite; increasing to 61 the number of satellites known in the system. Less than 1 km in size, this moonlet was dubbed S/2008 S1 and later named Aegeon. Because it was within a bright G ring arc some scientists suspected that it was a transitory clump of dust but observations over the next 2 years showed it to have all the characteristics of a solid body.[284,285] On the same day, 15 August, Cassini had a non-targeted encounter with Titan. Although observing at a range of 300,000 km it obtained some of the best views yet of northwestern Xanadu and the crater Menrva. And, at long last, clouds were forming in the northern polar regions.

At periapsis on 19 August Cassini flew by Pallene at a range of 44,000 km and, being a mere 4 km in size, the moon spanned only a dozen pixels in narrow-angle images. Descending to periapsis on its 82nd revolution a week later, the ultraviolet spectrograph had a direct view of auroral activity over the north pole of Saturn that, at times, filled the circle from 82 degrees of latitude to the pole. Owing to the weak magnetic field of the planet, auroras on Saturn changed with variations in the solar wind upstream much more rapidly than was the case for Jupiter. On this occasion the spectrograph recorded three views of the northern polar regions showing (just as observations two revolutions earlier had indicated) a slightly brighter spot precisely where the magnetic field linked the planet to Enceladus, the main source of ions in the system. The variability of this glowing spot was probably due to changes in the plume activity. A similar but much more powerful 'flux tube' linked Io and Jupiter, with the volcanoes of Io providing the plasma.[286]

September was another quiet month for Cassini, still orbiting Saturn once every week, because the planet was in solar conjunction for the first week. In addition to routine particles and fields observations the spacecraft performed communications tests through the solar corona. During the remainder of the month it completed four orbits, none involving targeted encounters. Eclipses of Tethys and Enceladus were observed, Aegeon was recovered, the formation of clouds on Titan and changes in the shoreline of Kraken Mare were monitored, propellers and spokes were imaged, etc. On 1 October the camera and ultraviolet spectrograph observed an occultation of alpha Crucis by the rings and by Saturn. On the same revolution, the 87th, distant observations targeted Rhea while the vehicle was crossing the orbital plane of that moon in order to search for possible rings.

Returning to periapsis, on 9 October Cassini had the E5 encounter, the first of two Enceladus flybys a mere three orbits apart. As in August it approached on an inclined fast trajectory which came in over the northern hemisphere and departed over the southern hemisphere. This time the optical instruments would not operate at closest approach, yielding priority instead to the mass spectrometer and particles and fields instruments. The vehicle was to use its thrusters rather than its reaction wheels for attitude control because it had suffered significant attitude disturbances whilst in the plumes. The encounter sequence began 8 hours prior to the flyby with the radar collecting scatterometry data to assess the roughness of the surface at high northern latitudes and radiometry to measure the energy balance. The other remote-sensing instruments were not activated until the range had reduced to 265,000 km. One hour before closest approach attitude control was switched to thrusters and the vehicle was turned to face the mass spectrometer inlet forward, an attitude that was also ideal for the dust sensor and plasma spectrometer. It zoomed by the moon at a relative speed of 17.7 km/s with the minimum altitude of 25 km at 28°S. Less than 30 seconds after closest approach it flew directly over the south pole at an altitude of 200 km and penetrated the plumes. The mass spectrometer produced the highest signal-to-noise ratio spectra to-date and identified complex trace molecules such as benzene, methanol and formaldehyde. A particularly significant result was the first certain detection of ammonia, which made up about 0.8 per cent of the plume. This made a subsurface ocean more plausible, since ammonia would act as anti-freeze. The instrument also detected argon-40. The ratio of hydrogen to deuterium proved to be almost twice the terrestrial value and comparable to cometary values. For this isotopic signature to have survived, the primordial objects from which the satellites of Saturn formed could not have been subjected to substantial heating.[287]

The dust analyzer was configured for a high sampling rate in order to produce a spatial profile of the density and composition of the plumes. Prior to the detector being saturated by impacts its observations provided the best evidence to-date for a subsurface ocean. During the frequent passages through the E ring this instrument had revealed there to be three populations of particles: those of almost pure water, those of water and silicate material, and those of water and salts of potassium and sodium. However, more than 99 per cent of the particles in the plumes were ice rich in sodium and potassium salts. Sodium chloride (table salt) and sodium bicarbonate were both present. This composition was consistent with an ocean that had been in contact with the rocky core of the moon. Salt would remain dissolved in the water as a crust of pure water ice formed above. Therefore, the detection of salt made the presence of a subsurface ocean almost a certainty. The abundance of carbon dioxide in the plumes also hinted that 'soda' would be found just below the surface of the moon.[288],[289] The plasma spectrometer found particles intermediate in size between vapor and dust grains. During this and other 2008 flybys and passages through the E ring, the ion and neutral mass spectrometer sought evidence of the existence of a torus of water molecules sharing the orbit of Enceladus but in-situ measurements were inconclusive. In 2010 the Herschel infrared space telescope operated by the European Space Agency confirmed the existence of this torus. A small percentage of the water would escape from the torus to coat neighboring moons and create the anomalous water concentrations in the upper atmosphere of the planet that had been discovered in the 1990s. This makes Enceladus the only moon in the solar system known to influence in a substantial way the chemistry of its parent planet.[290]

The optical instruments resumed observing 15 minutes after closest approach, by which time the resolution of the narrow-angle camera was about 140 meters. Once again, 46 minutes after the flyby, Enceladus entered the shadow cone of the planet. The camera imaged the moon in eclipse and the infrared instruments measured the thermal inertia of Baghdad and Damascus Sulci. These observations continued for hours after the moon emerged from eclipse.

During its lengthy series of excursions over both poles of Saturn, Cassini could study the kilometric-radio emissions and their sources. The northern and southern emissions were traced back to sites of auroral radio emission, known also to exist in the Earth's magnetosphere. Moreover, the two slightly different periods were seen to vary over time and were expected to 'match' in April 2010, some 8 months after the equinox. The fact that long-range studies by Ulysses had shown the two periods to match 9 months after the 1995 equinox implied a definite link with equinoxes. It could be related to seasonal variations or solar illumination. One proposal was that it varied with seasonal changes in the electrical conductivity of the high atmosphere of Saturn. On 17 October, some 5 Rs from Saturn, the spacecraft flew through the southern kilometric-radio source. The plasma-wave experiment, magnetometer and plasma spectrometer all took data. In fact, this area would be studied in more detail at the end of the second extension of the mission, when Cassini was in a 'proximal orbit' which skimmed very close to the planet.[291],[292]

On 24 October, during its 90th revolution, Cassini had its closest encounter with Mimas to-date, passing by at a range of 57,000 km, but no imaging was carried out because it flew over the night-side of the moon and only a thin crescent was visible. While over the night-side of the planet the camera started a campaign to search for lightning flashes, which ought to be easier to see without the clouds being brightly illuminated by sunlight reflecting off the rings which, this being an equinox, were edge-on to the Sun. At about the same time, during a routine orbit-trim maneuver, the 169th since orbit insertion, one of the 1-N thrusters had a significant underburn. Subsequent maneuvers confirmed the problem, characterized as degradation such as would be expected for a thruster nearing the end of its life. It was therefore decided to transfer small thrust propulsion to the backup system. This was accomplished in March 2009, at a time when no maneuvers and high-priority scientific observations were underway.

Meanwhile, the E6 encounter was on 31 October 2008. Cassini flew more or less the same trajectory as in August and early October, this time reaching a minimum altitude of 169 km with the closest point of approach again at 28°S. The encounter sequence started more than 8 hours prior to closest approach with remote sensing across the spectrum from infrared to ultraviolet. As previously, the objective was to investigate water in the vicinity of the moon and non-ice compounds on its surface. The spacecraft executed another skeet-shoot maneuver over the south pole just after closest approach with the field of view of the camera observing three hot spots and active sites within the tiger stripes. On this occasion a total of nine pictures were taken, achieving a resolution of 10 meters for Baghdad and Damascus Sulci. At the same time the temperatures of the floors of several sulci were measured using the composite infrared spectrometer. After the vehicle had flown through the plumes as it receded it took image mosaics of decreasing resolution of the south pole, as well as ultraviolet spectra and thermal maps. Fifty minutes after the flyby the moon was eclipsed by Saturn for two and half hours. The gravity survey was supplemented by radio tracking both before and after the encounter.

On 3 November Cassini returned to Titan for the first time in 3 months. Setting up this encounter a mere 3 days after that with Enceladus posed quite a challenge, and in the end the Enceladus flyby was some 15 km lower than planned. By now the Titan encounter sequences included a real-time update to specify the instrument orientations. The T46 encounter at an altitude of 1,105 km was almost equatorial, with the closest point of approach at just 3.5°S. During the approach the remote-sensing suite had a view of a partially illuminated Titan with southwestern Xanadu visible. The camera took one global and one regional map, the latter centered upon Hotei Arcus. Frequent imaging of this 'smile' had been requested by scientists who

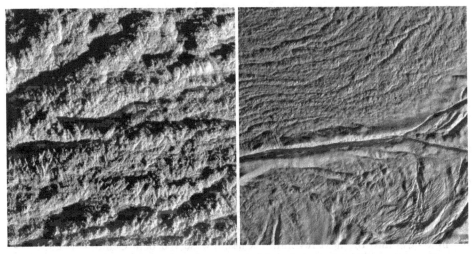

One of the highest resolution images of Enceladus, showing boulders in the southern polar region (left). A more distant image of Damascus Sulcus on this moon (right), with a resolution of about 24 meters per pixel. (NASA/JPL/Space Science Institute)

hoped to see changes which would confirm its cryovolcanic origin. The objective of the flyby, however, was radio science and an hour before closest approach priority was given to tracking. In addition, there were radio-occultation and bistatic radar experiments on both the way in and the way out. As seen from Earth the spacecraft disappeared 14 minutes before closest approach over the mid-southern latitudes and after an almost diametrical crossing reappeared 33 minutes later over mid-northern latitudes.

The T46 flyby slightly increased the period of Cassini's orbit to 8 days, so the T47 encounter occurred precisely two revolutions later on 19 November with the same geometry. On the way in multispectral images were obtained of Hotei Arcus close to the terminator with a best resolution of 12.5 km. When studied along with radar imagery from T41 the data suggested a geological interpretation. The central part of Hotei Arcus appeared to be a basin filled with anomalously bright flows. To the radar these areas had a variable roughness although their edges were smooth. At the southern edge of the basin was a sort of mountain range almost 1 km tall which appeared to be coated with dark hydrocarbons. The radar images showed channels discharging into dark blue depressions that were visible only in the infrared. Two dark spots on the northern margin were calderas. The structures seemed to be very young, perhaps only a few thousand years old.[293] About one hour prior to the 1,023-km flyby the ultraviolet spectrograph had priority to observe an occultation of eta Ursae Majoris to measure the opacity of the upper atmosphere and detached haze layers. Around the time of closest approach the mapping spectrometer collected a 'noodle' across the anti-Saturn hemisphere that included, for the first time, data for the Huygens site at a resolution of 1 km to assist in determining the composition of the areas the probe imaged during its descent. There was an ultraviolet occultation of beta Canis Majoris on the way out.

Shortly prior to the next periapsis, on 24 November Cassini had a non-targeted encounter with Tethys at a range of 25,000 km and took a pair of imaging mosaics of the southern part of the Saturn-facing hemisphere for stereoscopic analysis of its topography. It also snapped pictures of Helene at a range of 65,000 km.

Cassini had the T48 encounter on 5 December. The closest point of approach was at 10.3°S and the altitude of 961 km was near the lowest survivable value for a near-equatorial flyby. It was the 12th post-periapsis encounter in succession and, as previously, the remote-sensing instruments observed the leading hemisphere on the way in. One hour before closest approach priority was given to the radar and mass spectrometer. Starting 20 minutes before closest approach the radar made the first of two high-resolution imaging swaths. This cut through Eir Macula and Tui Regio. Then while the mass spectrometer sampled the day-side equatorial ionosphere the radar took its second imaging swath, this time covering southeastern Xanadu, dune fields in central Shangri-La, and southeastern Dilmun. Forty-three minutes after the flyby the optical instruments observed the moon occult epsilon Canis Majoris. To wrap up, Cassini reached apoapsis at the end of its 95th revolution.

Three days into the next revolution Cassini observed beta Centauri occulted by both the rings and Saturn. This provided an opportunity to measure the ultraviolet opacity of the upper atmospheric hazes of the planet. Occultations of alpha Crucis and beta Centauri were observed later in the month.

Cassini returned to Titan on 21 December for the T49 encounter and this time the 971-km flyby took it over high southern latitudes. On the way in the camera imaged the southern regions of the leading hemisphere while the infrared mapping spectrometer attempted long-range observations of Ontario Lacus. After almost 12 hours of optical remote-sensing the radar was assigned priority 45 minutes prior to closest approach. It took a west-northwest to east-southeast imaging swath that ran to high southern latitudes. There was a dark feature that had not been present when Cassini arrived at Saturn. It appeared to have been formed by hydrocarbon rains in late 2004. It was not evident whether there was any liquid in what seemed to be a transient lake. At this point the radar switched to altimetry for a track southeast to northwest through Ontario Lacus to measure the slopes that drained into the lake. The results showed the lake to lie in a topographic depression a few hundred meters deep. Cassini resumed radar imaging as it flew over Belet. The pass was wrapped up with altimetry, scatterometry and radiometry, and by visual monitoring of clouds over the northern hemisphere. Like previous flybys the geometry of this encounter was such that if the solar wind were sufficiently active the sunward perimeter of the magnetosphere would be inside Titan's orbit and the moon would be encountered in true interplanetary space but yet again this was not the case. Just 4 minutes after its closest approach to Titan Cassini reached apoapsis 1.18 million km from Saturn. The T49 encounter had increased the period of the spacecraft's orbit to slightly less than 10 days.

Traveling high above the ring plane, on 24 December Cassini observed Saturn and its rings occult delta Centauri and beta Crucis, and the following day the A and B rings occulted alpha Arae. On the last day of 2008 Cassini completed what was officially designated its 99th revolution, but was actually its 100th due to the early modification to the tour in order to reschedule the Huygens mission. On 8 January 2009, for the first time, the shadow of one of the moons (Epimetheus) was cast onto the ring plane. This was the first of many such 'shadow play' events which Cassini monitored.

In late 2008 and early 2009 Cassini took deep backlit images of space around Rhea to try to detect some of the putative ring material. A number of images were taken on either side of the moon and over a variety of illumination angles but with negative results. The electron depletion events were real, but were not due to rings. One possibility was weak geysers but long-exposure images failed to show any. All the evidence pointed to Rhea being inert.[294] In late January and early February there were ring and planet occultations of beta Crucis, gamma Crucis, beta Centauri and alpha Trianguli.

Cassini returned to Titan on 7 February for the T50 encounter. In the intervening time the spacecraft had made five revolutions while the moon had made only three. The moon had not been ignored though, because all along its orbit the camera had monitored it for clouds. During the very low pass which dipped to 967 km with the closest point of approach at 33.7°S, almost antipodal to the Huygens site, the mass spectrometer had priority but the radar was also able to collect a short swath on the trailing hemisphere cutting through the northern part of Mezzoramia, which was a semicircular dark area in the southern hemisphere suspected of being a dry sea, and

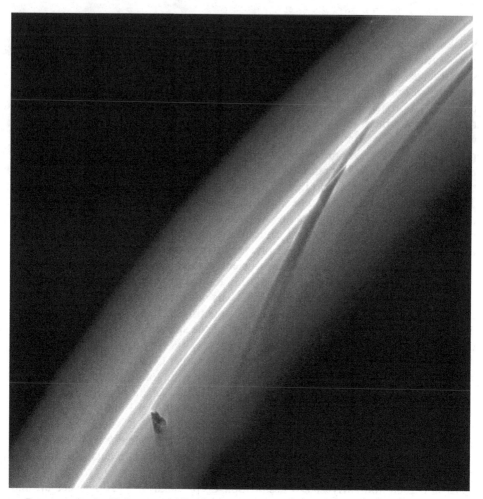

Streamers in the F ring caused by the passage of Prometheus (bottom, left of center). (NASA/JPL/Space Science Institute)

through an intermediate-latitude dark spot that looked like it could be either a lake or a cryovolcanic caldera. Mezzoramia would be the target of optical observations during a non-targeted encounter on the next orbit, although the range would exceed 1 million km. The radar swath continued through a number of bright streaks located south of Senkyo which were believed to be mountain ranges and ended in a field of dunes. Altimetry was collected prior to and after imaging, as was lower resolution spot-imaging data. Soon thereafter Cassini reached apoapsis. The flyby had further increased the period of its orbit by over 2 days and raised the periapsis to between the orbits of Rhea and Titan, well away from all of the inner moons.

Another rare phenomenon in 2009 was Prometheus reaching the apoapsis of its slightly elliptical orbit very near the periapsis of the F ring, and causing the moon to pass near the inner edge of the F ring. The alignment proper would not occur until

December but on 9 February the first pictures were taken to monitor the event and caught the shadow of the moonlet projected on the ring. The shadows of the main moons also started to transit across the ring plane. On 20 March Cassini observed a transit of the shadow of Tethys on the A ring.

The T51 encounter was on 27 March. This was the 15th and last of a long series of outbound flybys and, together with T52 half an orbit later, was to complete a second pi-transfer that shifted the periapsis from the night-side to the day-side. On the approach the optical instruments had priority. The camera imaged the southern leading hemisphere, in particular southwestern Xanadu, western Tseighi and Hotei Arcus. The imaging spectrometer operated during closest approach, observing the south polar region. When the vehicle flew directly above Ontario Lacus 12 minutes prior to closest approach this instrument took multispectral cubes to be compared with data taken in December 2007. At the same time the camera attempted some high-resolution imaging of the lake and surrounding regions. The flyby occurred at an altitude of 963 km. The mass spectrometer had priority on the way out over the terminator and the mid-to-high latitudes of the day-side. A bistatic radar experiment was performed 15 minutes after the flyby. Meanwhile, a mere 8 minutes after the time of closest approach, Cassini reached the apoapsis of its 107th revolution. The spacecraft had been in a low-eccentricity high-inclination orbit which took it from 13.6 to 19.7 Rs every 16.5 days, but T51 exchanged the periapsis and apoapsis by further circularizing the orbit. The post-T51 orbit ranged between 19.7 and 20.9 Rs with a period of 16.8 days and an inclination of 65 degrees.

Two days later the camera captured an amazing sequence of the rings occulting Omega Centauri, a globular cluster of several million stars. Some of the stars were observed transiting the Keeler gap. Later the shadow of Mimas was seen crossing the A ring.

On reaching the opposite side of its near-circular orbit Cassini approached Titan for the second time in 8 days. The 4,147-km flyby of the T52 encounter on 4 April was one of the spacecraft's most distant targeted encounters yet. During the approach the optical remote-sensing instruments were active. In addition, the radar took long-range radiometry over the north pole. About 3 hours prior to closest approach the ultraviolet spectrograph monitored a very slow occultation of alpha Eridani to study the structure and chemistry of the polar vortex. Radio science had priority at closest approach. Cassini slipped behind the disk of Titan at mid-northern latitude and 10 minutes later reappeared near the equator. This probed the equatorial atmosphere at a high vertical resolution for the first time since the Voyager 1 flyby in November 1980. If the early orbits of the primary mission tour had not been redesigned there would have been a near-equatorial occultation during the February 2005 encounter. There was a bistatic radar experiment over the northern Saturn-facing hemisphere at ingress and over the southern trailing hemisphere at egress. During the time that the vehicle was out of contact with Earth it was transiting the moon's shadow cone and reached the closest point of approach in darkness. The departure trajectory over the day-side allowed the camera to view previously unseen regions of the southern trailing hemisphere at high resolution. It imaged a dark region east of Mezzoramia, which was itself near the limb. In the ensuing days the camera resumed long-term monitoring of clouds.

By now, transits of the moons' shadows on the rings had become frequent. The grazing illumination of the rings even revealed the existence of unseen clumps or moonlets several hundred meters across. These objects projected long shadows on the thinner surrounding rings. The edge-on illumination enabled Cassini to observe the shadows cast by particles deviating from the ring plane. For example, it saw the waves along the edge of the Keeler gap oscillating as much as 1,500 meters above and below the plane in response to the small orbital inclination of Daphnis. And, in the reverse process, the saw-tooth shadows at the edges of the Encke and Keeler gaps were monitored in the hope of finding undiscovered embedded moonlets that caused them.

Traveling on its near-circular 16-day orbit Cassini had a non-targeted encounter with Titan at a range of 300,000 km and imaged Hotei Arcus, Ganesa Macula and the crater Menrva. The resonant orbit returned the spacecraft to Titan on 20 April. T53 had similar geometry and illumination to the previous encounter and, owing to the relatively distant flyby range of 3,599 km, priority was given to optical remote-sensing with the radar merely performing radiometry on the way out. Three hours to closest approach the ultraviolet spectrograph observed another slow occultation of alpha Eridani to investigate the upper atmosphere and 11 minutes before closest approach it observed a 9-minute solar occultation that produced a density profile of atmospheric nitrogen. Such profiles could be compared with data from the attitude control system which was enduring greater than predicted torques during the very close flybys. Another non-targeted encounter of Titan occurred on 27 April, on this occasion at a range of 700,000 km. On 5 May, inbound on the next revolution, the T54 encounter was a 3,242-km flyby with the closest point of approach at 14.1°S. The composite spectrometer had priority at closest approach and observed the limb to produce low-latitude vertical profiles of composition, temperature and aerosols. Starting at about 9,000 km on the way out the camera took high-resolution images of a region of the trailing hemisphere at high southern latitudes where dark spots of unknown nature had been seen previously. The mapping spectrometer obtained data to enable known lakes to be monitored and new ones to be sought. Two days later, long-exposure color images were taken of Titan in eclipse to study its atmosphere dimly illuminated by Saturn and by the other satellites.

On 13 May Cassini observed the first eclipse of one moon by another, when the shadow of Enceladus crossed the disk of Mimas.

At 966 km the T55 encounter on 21 May was very close. The radar was assigned priority. On the way in it took distant scatterometry and radiometry data prior to a low-resolution scan of regions north of Xanadu with simultaneous altimetry. It then switched to imaging, starting in northern Xanadu, crossing Shangri-La and ending at 70°S. It included Crete Macula, dune fields south of Shikoku Facula and Hobal Virga, which was a dark lineament running east-to-west at mid-southern latitude. It imaged canyons and tributary channels carved by liquid methane draining from the highlands to lowlands at high-southern latitudes. While the radar was operating, the mass spectrometer sampled the night-side atmosphere at low latitude. There was an 18-minute Earth occultation and a slightly shorter solar occultation but neither was exploited for scientific observations. On the way out the radar collected altimetry at

A long-exposure view of Titan in eclipse, taken on 7 May 2009, when Cassini was approximately 667,000 km from the moon. (NASA/JPL/Space Science Institute)

high-southern latitudes as well as low-resolution images near to Ontario Lacus. The optical instruments then resumed monitoring clouds.

T56 on 6 June was the fifth of ten inbound flybys in which Cassini encountered Titan at the same point in the moon's orbit. Whilst the period of the vehicle's orbit was held at 16 days, it was becoming progressively more eccentric. Five hours out the radar started to collect radiometry of the visible disk, including the northern lakes. It then switched to scatterometry and, in turn, to low-resolution imaging with simultaneous altimetry of northern Xanadu and some areas already mapped in July 2006. For 19 minutes around closest approach it took a swath running northeast to southwest from just north of Xanadu, across Shangri-La to about 66°S. Apart from a partial overlap the line of this swath was to the southeast of the previous pass and almost parallel to it. The radar again imaged the bright Crete Facula in northeastern Shangri-La, Kerguelen Facula, the lineaments Bacab and Perkunas Virgae, and the southern mid-latitude Hobal Virga. The 968-km flyby occurred at 32.1°S, half way through the swath. As on the previous encounter the mass spectrometer sampled the ionosphere as the spacecraft flew over the dusk side of the moon. An hour after the flyby the ultraviolet spectrograph observed an occultation of eta Ursae Majoris.

The T57 encounter on 22 June was a 955-km flyby, this having been judged the lowest safe altitude for latitude 42.2°S. The main instruments at closest approach were the radio science package, mass spectrometer and radar. A radio-occultation started 24 minutes prior to the flyby and ended at the moment of closest approach. However, because other instruments took priority only the ingress was exploited to investigate the atmosphere at 79°N, near the edge of the polar vortex. This was the highest-latitude radio-occultation of the mission. A slightly shorter solar occultation occurred at about the same time but it was not observed because, while behind the moon, the mass spectrometer sampled the ionosphere above the southern night-side.

Meanwhile the radar was collecting an imaging swath that ran parallel to the tracks of the two previous encounters, starting in southwestern Xanadu, running south of Shangri-La to the northwestern portion of Ontario Lacus near 75°S. Afterwards it acquired altimetry as well as lower resolution scans. Scatterometry and radiometry of the southern hemisphere departing from Titan wrapped up radar observations for this encounter. Optical remote-sensing had lower priority but after closest approach the areas observed included mid-southern latitudes of the trailing hemisphere. This flyby had the effect of further increasing the eccentricity of Cassini's orbit, so much so that it put the periapsis between the orbits of Dione and Rhea. At the same time the inclination was slowly being reduced.

On 7 July Cassini took a wide-angle mosaic of the entire inner system, covering not just the crescent Saturn but also the night-side of the rings, Mimas, Enceladus and Tethys.

The T58 encounter the following day was a 966-km flyby with the closest point of approach at 52.2°S. During its approach Cassini viewed Titan as a thin crescent and the mapping spectrometer took data for 2 hours at a range of about 200,000 km in order to search for clouds, observe Tui Regio in darkness, and search for thermal

Imaging spectrometer observations during the T58 encounter provided the most visually striking proof of the presence of liquid hydrocarbons at the poles of Titan, as it took this picture of sunlight reflecting off the surface of Jingpo Lacus. (NASA/JPL/University of Arizona/DLR)

signatures of possible cryovolcanoes. Meanwhile, as Titan progressed into northern spring, an increasing number of the polar lakes started to catch direct sunlight. As a result the mapping spectrometer captured views of the glint of sunlight reflecting off liquid. Although the radar coverage of this region was incomplete the source of this glint appeared to be the southwestern shore of Kraken Mare. The area was later appropriately named Jingpo Lacus after 'Mirror Lake' in China. To reflect sunlight in this manner the waves on the lake must have been a hundred times flatter than is typical for a terrestrial ocean. Reflection of sunlight was still faintly detectable over land, indicating that the shores of Jingpo were either wet mud flats or covered with puddles.[295,296] A 27-minute solar occultation prior to closest approach was observed by the ultraviolet spectrograph for data on the density of the north polar vortex, to be compared with measurements by the attitude control system during high-latitude passes. These occultations characterized the structure of the layers of haze and the way in which they changed from one pass to the next. While the mass spectrometer sampled the ionosphere the radar imaged a swath that started at the western edge of Xanadu and ran south of Shangri-La to end near Ontario Lacus. Comparing infrared images taken in July 2005 with July 2009 radar data the shores of Ontario seemed to have retreated several kilometers, reducing its depth by a few meters. This could be a seasonal effect but scientists cautioned that it could be merely a side-effect of comparing data from different instruments. Forty-seven minutes after the flyby the ultraviolet spectrograph observed an occultation of eta Ursae Majoris. On the way out the camera imaged Senkyo and Tsegihi at high to medium resolution.

For the T59 encounter on 24 July Titan would be downstream of Saturn in the magnetosphere and the particles and fields instruments would be able to sample a number of interaction regions. Because the flyby geometry would not provide much opportunity to observe illuminated regions of the moon the camera and mapping spectrometer had few observations scheduled. Nevertheless, during the approach it was possible to observe sunlight glinting off the northwestern part of Kraken Mare, on the limb of the unilluminated hemisphere. The plasma spectrometer had priority at closest approach and it monitored the interaction between the moon's ionosphere and the magnetosphere, looking for molecules originating from the moon that had been ionized and 'picked up' by the magnetosphere as it swept by the moon. For 40 minutes around closest approach the mass spectrometer exploited the high southern latitude to collect data to fill gaps in the latitudinal coverage of the atmosphere and night-side ionosphere. Meanwhile, the radar collected another in a series of parallel swaths across the southern regions of the trailing hemisphere. The 955-km flyby at 62°S occurred while Cassini was out of contact with Earth. The encounter reduced the periapsis distance to a point inside the orbit of Enceladus.

As Cassini reached periapsis on 26 July it examined the southern portion of the trailing hemisphere of Janus at a range of 100,000 km using its camera and infrared spectrometer and then studied the plumes of Enceladus while that moon was being eclipsed by Rhea. It was during these observations that it discovered a new satellite.

The moonlet, dubbed S/2009 S1, was only 300 meters in size and orbited 700 km interior to the outer edge of the B ring. It was spotted by the 40-km-long shadow it

Out-of-plane structures within the rings of Saturn clearly stand out with edge-on illumination in this 26 July 2009 image (left). Daphnis and the waves at the edges of the Keeler gap casting shadows on the adjacent rings (right). (NASA/JPL/Space Science Institute)

The moonlet S/2009 S1 embedded in the B ring. It was discovered by the long shadow it projected with the ring illuminated edge-on. The bright spot to the left is a cosmic ray hit. (NASA/JPL/Space Science Institute)

cast in the near-edge-on illumination. It was probably similar in size to the objects that form the propellers in the A ring.[297],[298]

Cassini also produced results pertaining to the physics of the solar system. Using data collected by the ion and neutral camera up to July 2009 scientists created 'all-sky' maps of the neutral atoms whose high energies were presumed to be the result of interactions with energetic protons at the boundary of the heliosphere. If this was so, then the data could be used to map the shape, pressure, magnetic field, and other characteristics of the region in which the solar wind interacted with the interstellar medium. This heliopause was at the same time being explored in-situ by the two Voyagers, which crossed the inner termination shock at 94 and 84 AU in 2004 and 2007 respectively, and from Earth orbit by energetic neutral atom 'cameras' on the IBEX (Interstellar Boundary Explorer).[299] The Cassini data was from neutral atoms with energies greater than those that were 'imaged' by IBEX and revealed a broad belt of energetic protons that was not predicted by models.[300]

The T60 encounter on 9 August produced a 971-km flyby with the closest point of approach at a high southern latitude. However, the observation sequences were relatively short because scientists assigned a higher priority to observing the rings a few hours from the precise moment of equinox. As Cassini approached the night-side of the northern anti-Saturn hemisphere the infrared spectrometer was seeking changes in the circulation of the northern polar vortex. Starting 5 hours prior to the flyby the radar made a series of observations that culminated in imaging swaths of central Shangri-La at low resolution and a long altimetry scan that ran from Xanadu to Ontario Lacus. The mass spectrometer got its first sample of the atmosphere over the south pole. On this occasion Cassini spent 54 minutes occulted by the moon and out of sight of Earth. After making a few observations on the way out the spacecraft turned its attention to the ring system, in particular seeking spokes and the shadows projected by moonlets and propellers.

At 00:15 UTC on 11 August the Sun passed through the equatorial plane of the Saturnian system, marking the start of spring in the northern hemisphere. Pictures taken several days later showed the northern face of the rings faintly illuminated by Saturn-shine. This ghostly portrait was possible because the Sun was not yet able to fully illuminate the rings. The edge-on illumination revealed the 'shadow play' of a subtle corrugation covering the entire C ring just like ripples on a pond. The waves were several tens of meters tall and were tens of kilometers apart. Backtracking the corrugation revealed that it developed from a tilted ring which had strayed from the plane in 1983. Perhaps Saturn had encountered a cloud of debris from a comet or an asteroid and this had produced the initial tilt that had spread out and been glimpsed in this very shallow illumination.[301]

In August Cassini began to study a light-toned 3,000-km storm in the temperate southern 'storm alley' on Saturn which lasted for most of the remainder of the year. Images of the storm on the night-side of the planet captured for the first time a long series of lightning flashes. It was the equinox that made these observations possible because with the edge-on rings unable to reflect sunlight onto the planet the night-time hemisphere was completely dark. The localization of lightning 125 to 250 km below the cloud tops meant that the storm was in an ammonium hydrosulfide layer

Lightning on the night-side of Saturn photographed by Cassini's narrow-angle camera on 30 November 2009. (NASA/JPL/Space Science Institute)

or a deeper water vapor layer of the atmosphere. Another reason why lightning was so difficult to photograph on Saturn was that it apparently occurred deeper than on Jupiter and so was masked by the haze layers.

The T61 encounter on 25 August was the last of a series of ten inbound passes spaced 16 days apart. It produced a 970-km flyby with the closest point of approach at 19°S. As was the case for most of the recent close flybys the radar was assigned priority. It first acquired low-resolution images of southwestern Shangri-La, east of Shikoku Facula and fields of dunes imaged on two previous passes. It then started a high-resolution swath running east to west across Dilmun, Adiri and Belet, through the dune fields of the trailing hemisphere. This swath was parallel to, and partially overlapped, that of October 2005. Repeated observations of the 'dune seas' of Belet at a wide variety of radar illumination angles would permit scientists to produce a 3-dimensional reconstruction of the topography. The closest approach was over the terminator heading into daylight. Starting 1 hour after closest approach the imaging spectrometer had priority. The camera 'rode along' snapping pictures of western Senkyo, dark streaks south of it, and the southern part of the trailing hemisphere. This flyby was used to increase the period of the orbit to 24 days, which meant that it would not encounter Titan again until the moon had completed three revolutions around Saturn and the spacecraft had completed two.

In the second week of September Cassini had a non-targeted flyby of Iapetus at a range of 1 million km, which was a welcome opportunity to collect further data. At the same time distant observations of Titan imaged an area of western Belet where it was thought that storms seen in April may have rained onto the surface. From 15 to 20 September was solar conjunction. Shortly thereafter Cassini, now approaching apoapsis, had a non-targeted flyby of Titan at a range of 300,000 km and inspected the space near Iapetus in search of faint rings or plumes. As it returned to periapsis of its 119th revolution Cassini used its ultraviolet spectrograph, infrared mapping spectrometer, camera and plasma spectrometer to investigate the auroras on Saturn. From 5 to 8 October almost 500 pictures were taken and the equinox illumination provided the very first long-exposure visible-light images of the northern aurora. At latitudes beyond 70°N, an auroral curtain was seen to rise to 1,200 km above the limb of the planet.

The T62 encounter on 12 October was at an altitude of 1,300 km with the closest

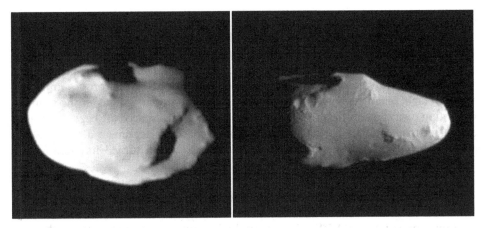

The small moon Telesto, seen on 27 August 2009 (left). Cassini took its best images of Tethys' Trojan moon Calypso on 13 February 2010, at a distance of approximately 21,000 km (right). (NASA/JPL/Space Science Institute)

point at 64°S. This inbound flyby completed the transition to a low inclination and produced the conditions required for encounters of Enceladus and some of the other inner moons. Cassini approached the night-side and the anti-Saturn hemisphere and then, just before reaching the closest point, penetrated the shadow of the moon. On exiting solar occultation the ultraviolet spectrograph studied the upper atmosphere and detached haze layers at high southern latitudes. The mapping spectrometer also observed the occultation. Optical remote-sensing was performed in all the available wavelengths as the spacecraft receded over the day-side.

Speeding through periapsis Cassini had non-targeted flybys of Rhea, Tethys and Mimas. It also monitored the plumes of Enceladus in preparation for flybys on two consecutive revolutions. The first was on 2 November, just 3 hours after periapsis. Designated E7, it was the slowest Enceladus flyby to-date at a relative speed of 7.7 km/s; in fact, it was some 10 km/s slower than in 2008. As a result in-situ analysis instruments such as the mass spectrometer could collect data at an unprecedented high signal-to-noise ratio. Approaching the night-side the moon was so close to the Sun from Cassini's perspective that it was too risky to use the optical instruments. Instead, the radar acquired scatterometry of the night-side, to measure the roughness of the surface and radiometry to measure the energy balance. At periapsis similar observations were made of the mid-southern latitudes of Saturn. The ultraviolet spectrograph took spectra of the plumes of Enceladus as they transited the disk of the planet. With an hour to go to the closest approach to the moon Cassini slewed to align the inlet of the mass spectrograph to collect particles from the plumes and the E ring. At the same time, the dust instrument was analyzing the impacting particles. As Enceladus dashed through their fields of view, the wide-angle and narrow-angle cameras both took two pairs of images targeting an area at intermediate latitude and a south polar region in the vicinity of the terminator. At closest approach Cassini passed almost directly over the south pole at an altitude of just 99 km. This was the deepest passage yet through the plumes, with the path crossing those of Alexandria and Baghdad. The rationale for this low polar pass through the plumes was to allow the mass spectrometer to sample ammonia and complex hydrocarbons before they were dissociated by solar ultraviolet. Moreover, by recording the torques imparted to the vehicle scientists would be able to estimate the density of gas in the plumes. Departing over the anti-Saturn hemisphere in daylight the spacecraft began to use its optical instruments. In particular, the infrared spectrometer made a temperature map of the south pole. Leaving Enceladus behind Cassini turned its attention to Tethys some 300,000 km away.

On its next revolution Cassini had its E8 encounter on 21 November. As before it approached over the night-side, zoomed almost directly over the south pole and departed over the day-side but this time the range had been increased to 1,603 km with the closest point at 82°S. The camera had a good perspective of the night-side and plumes on the way in and observed structures within the jets at high resolution, mapping their sources and searching for new ones. Seven minutes prior to closest approach Cassini turned to enable the infrared spectrometer to measure the thermal emission from several discrete points on the surface and as the spectrometer looked at each target the camera took a matching mosaic. The first stop was over Baghdad Sulcus, which was imaged to facilitate a search for changes since it was viewed 11 months previously. Images taken during this pass had a best resolution of 15 meters per pixel. A number of other multispectral, color, and high-resolution observations were made on the way out. In particular, this was the first opportunity to image at high resolution the poorly observed young-looking parts of the leading hemisphere.

Shortly after leaving Enceladus, Cassini had a non-targeted encounter with Rhea at a range of 24,000 km. It examined the equator to determine whether some faint streaks were the remains of a transient ring which had 'rained' on the moon; the streaks proved to be lines of very small young-looking craters. Cassini also imaged intriguing uplifted blocks which sliced through heavily cratered plains.

Cassini returned to Titan on 11 December. T63 was the first in a series of four outbound encounters at 16-day intervals designed to briefly raise the inclination of the vehicle's orbit to 20 degrees before returning it to the ring plane. It was a fairly distant flyby at 4,650 km with the closest point of approach at 33°N. The geometry was complementary to the Voyager 1 and December 2005 Cassini flybys, which occurred respectively in the noon and midnight regions of the magnetosphere. The plasma spectrometer had priority for 5 hours while flying through the wake of the moon in the dusk region of the magnetospheric tail in order to study the interactions between the moon, its ionosphere, and the magnetosphere. On the way out Cassini acquired mosaics of northern Adiri, Shangri-La and Belet on the sunlit hemisphere, and then resumed monitoring clouds at low resolution.

On 25 December, a day before periapsis on the 123rd revolution, Cassini passed through the shadow cone of Saturn. Taking advantage of the absence of the Sun it imaged the almost fully darkened hemisphere of Enceladus from a range of 600,000 km in order to monitor the polar plumes. In this case it also managed to capture the shadow of the moon projected on the diffuse E ring. A non-targeted encounter with Prometheus at a range of just 59,000 km the following day produced the first high-resolution (350 meters per pixel) views of its leading hemisphere. Prometheus was elongated, 119 km on its major axis. It had a very smooth surface and by virtue of being an F ring shepherd it was 'painted' with a thick layer of dust which made its craters and other features rather subdued.

During the T64 encounter on 28 December Cassini approached the night-side of Titan more or less from the direction of Saturn. The closest point of approach of the 955-km flyby was at 82°N. It departed over the anti-Saturn hemisphere in daylight. On the way in the radar took scatterometry data, a short swath of altimetry near the mid-northern Elpis Macula, and then imagery running from the north pole to a point in the vicinity of Adiri. This was the only swath over the northern polar lake region of the extended mission and targeted Punga and Ligeia Mare in search of seasonal changes in response to the changing weather patterns. As with previous low-altitude flybys the mass spectrometer was assigned priority at closest approach, this time to collect data for direct comparison with that from when Cassini repeatedly passed over the northern polar regions earlier in the mission. On the way out the imaging spectrometer filled in some gaps in the spectral mapping of Adiri.

T65 on 12 January 2010 was the third of the series of four resonant encounters. For 4 hours around the 1,073-km flyby the mass spectrometer had priority in order to sample the atmosphere over the southern pole. The closest point at 82°S allowed instruments to make observations to complement those of the previous revolution. At closest approach the radar collected two short swaths, the first of Mezzoramia and the second of Ontario Lacus with simultaneous altimetry to monitor changes in its liquid level. Unfortunately some data was lost due to heavy rain at the Madrid Deep Space Network station, to which Cassini was downloading. The south polar flybys to-date had provided significant overlapping radar imagery. Changes were detected in three regions. Two were short-lived partially filled lakes, or perhaps just mud flats. The third, Ontario Lacus, was confirmed to have a retreating shoreline. Between 2005 and 2010 it had retreated as much as 20 km by either evaporation or ground infiltration. The discovery of fog nearby indicated that the first explanation was viable, since fog would only exist in the presence of evaporation. On the other hand, infiltration had been directly observed in the damp soil at the Huygens site. In contrast to these observations, seven overlapping northern polar passes had shown only minor signs of surface changes.[302] The plasma spectrometer and magnetometer observed the magnetospheric wake of the moon for hours prior to and following the flyby. On the way out the mapping spectrometer took multispectral mosaics of the Huygens landing site, western Shangri-La and Adiri with a resolution of 25 km.

On 16 January Cassini observed a close conjunction of Janus and Epimetheus in which the co-orbital satellites approached within 12,000 km of each other. A series of similar conjunctions culminated on the 21st with them swapping orbits. For the next 4 years Janus would be closer to Saturn than Epimetheus. On the 26th, during the 125th revolution, Cassini had a unique 7-hour ansa-to-ansa radio-occultation by the rings and the planet. The next day it imaged Prometheus from about 30,500 km and Aegeon from over 13,000 km. In spite of the 100-meter resolution, Aegeon still

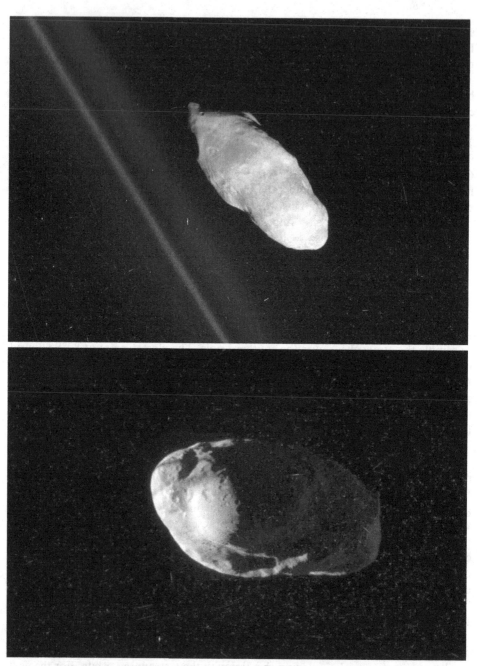

Two of the best images of Prometheus, one of the F ring shepherd moon, seen sidewise on 26 December 2009 (top) and face-on on 27 January 2010 (bottom). Note in the latter the shadow of the ring at upper left. (NASA/JPL/Space Science Institute)

spanned only a handful of pixels. Although no surface features were visible it was possible to estimate its shape. The images of Prometheus, on the other hand, were the best to-date and included the shadow of the F ring projecting onto its surface. Cassini viewed Prometheus along the longest axis, giving it a shape reminiscent of a heart. Although the resolution was better than in December, less of the irregular object's surface was illuminated. A non-targeted encounter of Dione at a range of 45,000 km provided the best images yet of its north pole.

T66 on 28 January was a fairly high 7,490-km flyby. During the approach the mapping spectrometer observed an occultation of alpha Tauri (Aldebaran). Priority was given to the remote-sensing instruments. The imagery taken at closest approach covered the trailing hemisphere at high southern latitudes and the near-equatorial region of Adiri, with extensive mosaics of a feature near 40°S that was interpreted as a dry lake. As the last of the four outbound encounters, this flyby deflected the spacecraft into an 18-day orbit that would not cross the path of Titan for some time.

On 13 February, shortly before periapsis, Cassini spent 4 hours in the shadow of Saturn. Images of the sunlit limb of the planet were taken to monitor its haze layers. It then had a non-targeted flyby of Calypso at a range of 21,000 km. A sequence of 25 images revealed thin parallel landslides. These were particularly evident near the terminator where the low-angle illumination emphasised surface relief. On the same day Cassini had its closest-ever encounter with Mimas, passing by at an altitude of 9,520 km. One high-resolution mosaic targeted the large crater Herschel which had prompted the 'Death Star moon' moniker. There was dark material clinging to the walls, and the floor was pitted by only a few small craters. The composite infrared spectrometer made a temperature scan of the Herschel-centered hemisphere. There was an unexpected temperature distribution with a warm region forming a peculiar 'C' shape and the remainder of the hemisphere with the exception of a warmer spot in Herschel being 15°C colder. This could indicate a textural or density variation in the ice on the surface, possibly corresponding to ejecta from Herschel or to material acquired from the E ring. There were no isolated 'hot spots' such as on Enceladus and it remained a mystery why Mimas, similar to Enceladus in terms of dimensions and surface composition, and having a distinctly more eccentric orbit, did not show comparable internal activity. The encounter was wrapped up with some images of the fully illuminated disk of Mimas passing in front of Saturn.

On 2 March Cassini flew over the northern polar region of Rhea at an altitude of just 101 km. The particles and fields instruments sought traces of the putative ring but detected nothing. The mass spectrometer confirmed the presence of the gaseous exosphere previously detected by the plasma spectrometer. It was a hundred times denser than the envelopes of Mercury or the Moon but still a thousand-billionth the density of the terrestrial atmosphere. It consisted of oxygen and carbon dioxide, and was the only such envelope in the solar system apart from the atmosphere of Earth. The oxygen probably derived from surface water ice decomposed by the impacts of the energetic particles that circulate in Saturn's magnetic field. The origin of carbon dioxide, which was only detected over the day-side as the spacecraft withdrew, was more difficult to explain unless there were carbon-bearing minerals or organics on the surface of the moon.[303] The camera acquired a color mosaic of the Saturn-facing

A close-up view of the giant crater Herschel on Mimas. (NASA/JPL/Space Science Institute)

hemisphere featuring wispy terrain faults and an impact basin. Some slightly bluer areas could imply either compositional variations or differences in the texture of the ice. About 1.5 hours after the flyby Rhea entered Saturn's shadow and the infrared spectrometer searched for any 'hot spots' and measured its thermal inertia. The next day Cassini sped through periapsis and made its closest approach yet to Helene, the Lagrangian companion of Dione. On the way in it first imaged the night-side of the moonlet, dimly illuminated by Saturn-shine, and then, during the 1,820-km flyby, it performed a 7-frame skeet-shoot imaging sequence over the southern portion of the anti-Saturn hemisphere. Unfortunately, the pointing at closest approach was slightly off and Helene only appeared in a corner of the highest resolution images, most of which had Saturn in the background. Evidently this problem derived from uncertain knowledge of the mass and gravity of Rhea, encountered only 30 hours earlier, as

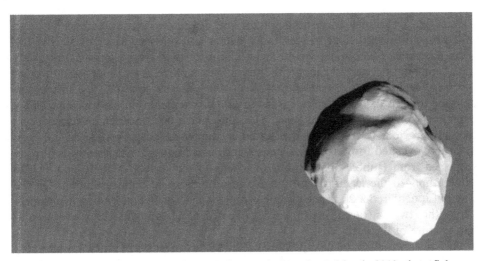

The small moon Helene gliding over Saturn during the 3 March 2010 close flyby. Unfortunately, higher resolution pictures did not manage to capture the whole moonlet because of a pointing problem. (NASA/JPL/Space Science Institute)

well as by the complicated and not fully understood orbital motion of a Trojan satellite like Helene. On the way out Cassini continued imaging the partly lit moon using both cameras and filters in an effort to determine the composition of its surface. At high resolution Helene displayed extensive striations, gullies and landslide networks on the walls of its craters.

After these close flybys, revolution 128, which lasted from 12 to 29 March, was relatively relaxed. When amateur astronomers discovered a bright feature in 'storm alley' on Saturn, Cassini examined it using its composite infrared spectrometer and found traces of phosphine drawn from deep inside the atmosphere. The bright-toned cloud was due to ammonia, which formed ice crystals on reaching the low pressures of the upper atmosphere. Near apoapsis on 27 March a single narrow-angle picture caught a number of moons, including Iapetus, Helene, Epimetheus, Calypso, Titan and Telesto.

On 5 April Cassini returned to Titan. This T67 encounter was a single inbound equatorial pass scheduled between two sets of outbound encounters. It was another relatively distant flyby at an altitude of 7,462 km and hence was primarily devoted to remote sensing. The vehicle approached the night-side with only a thin crescent visible and left over the day-side. The composite infrared spectrometer scanned the temperature of a slice of atmosphere at 70°N looking for possible dissipation of the polar vortex with the onset of the spring season. The study of the polar atmosphere of Titan was interesting to Earth scientists because, being isolated by a vortex from the rest of the atmosphere and having a unique chemistry and dynamics, it bore a similarity to the Earth's ozone hole. As the orbiter passed the terminator and moved to the day-side the camera was the prime instrument, taking high-resolution images on a track from eastern Belet across the trailing hemisphere into western Senkyo.

A day and half after passing Titan, and a few hours before periapsis on 7 April,

Cassini flew by Dione at an altitude of 503 km. Four hours out the camera began to take pictures, in particular to seek faint plumes and other evidence of activity but there was none. The plasma spectrometer had priority at closest approach to study the interactions between the moon and the magnetosphere as Cassini passed through the wake region downstream of the moon and near its equator. The instrument detected ionized oxygen molecules, a sign of an extremely thin neutral atmosphere containing oxygen, like that of Rhea. The atmosphere had not been detected during the only previous (October 2005) targeted encounter because Cassini had not been favorably oriented for detection. The presence of an oxygen exosphere was already suspected from Hubble observations of ozone on the surface of the moon.[304] The camera imaged the leading hemisphere and terminator at high resolution. On the way out it acquired a 20-frame mosaic of the Saturn-facing and leading hemisphere while the two infrared instruments collected data on the composition and temperature of the surface.

As Cassini reached the next periapsis of its 21-day orbit on 28 April a solar and a radio-occultation by Saturn provided a rare chance to investigate the low northern latitudes unhindered by the rings. The E9 encounter followed several hours later, the objective of which was to collect radio-science data to investigate the southern hemisphere of Enceladus and ascertain whether there were density anomalies in the polar region. In particular, scientists wished to determine whether the plumes were associated with a subsurface ocean or with diapirs of warm ice. A total of 26 hours were devoted to this experiment, during which the high-gain antenna was pointed at Earth. Although no remote-sensing was done the particles and fields instruments collected data. Halfway during the experiment the spacecraft skimmed the surface of the moon at an altitude of 100 km, within 500 meters of the planned distance, at a relative velocity of 6.5 km/s.

As Cassini receded to apoapsis it took some long-range pictures of Iapetus. As the spacecraft's orbit was entirely within that of Iapetus, it could observe only the Saturn-facing hemisphere.

At periapsis on 18 May Cassini had the E10 encounter. This time the altitude was 435 km and the priority was optical remote-sensing. The approach was over the night-side and the departure over the day-side. On the way in the spacecraft passed through the shadow of Enceladus and the ultraviolet spectrograph used the solar occultation to analyze the plumes for the molecular nitrogen that would result from the decomposition of ammonia. A precise measurement of the amount of ammonia in the plumes would impose constraints on the subsurface temperature of the moon. In the run up to closest approach Cassini snapped six high-resolution images which revealed very fine structures within the polar jets. The final image of this series was a stunning view of distant Titan seen through the rings with the dark limb and faint jets of Enceladus in foreground. Over the day-side the camera was trained on the sources of the jets. On the way out mosaics of decreasing resolution were obtained of the Saturn-facing hemisphere of the moon.

Two days later, on 20 May, Cassini flew by Titan at a relative speed of 5.9 km/s during the T68 encounter. There was a solar occultation at about the same time as the closest point of approach, which was at an altitude of 1,400 km over the night-

Cassini took this image during the 18 May 2010 flyby of Enceladus. It shows the dark south pole of the moon in the foreground and distant Titan cut by the edge-on rings. (NASA/JPL/Space Science Institute)

side at the intermediate latitude of 49°S. The priority was radio tracking for gravity data but images and spectra were also obtained. On the way in only a crescent was visible and the remote-sensing observations were mostly of clouds and the detached haze layers. At closest approach the composite infrared spectrometer profiled the temperature and composition of the atmosphere in the vicinity of the south pole. On the way out the camera monitored the anti-Saturn hemisphere for several days but no clouds were visible.

This flyby returned Cassini to an inclined orbit with a period of 16 days and so, for the first time in 5 months, it was able to view the rings from out of plane. Many observations inbound from apoapsis in late May were of the rings in progressively improving illumination. As the spacecraft passed through the ring plane just before periapsis on 3 June its path provided an occultation of omicron Ceti (Mira) and the camera imaged the star passing behind the F ring. Other observations were made of propellers and spokes. At periapsis there was a non-targeted flyby of Pandora at a range of 100,000 km which provided a view of the hemisphere opposite to that seen in September 2005.

Outbound from Saturn on 5 June Cassini had the T69 encounter, which was a 2,044-km flyby at high northern latitude. On approaching the night-side the radar collected radiometry and scatterometry of the Saturn-facing hemisphere. It followed up with a short low-resolution imaging swath of equatorial and north polar regions

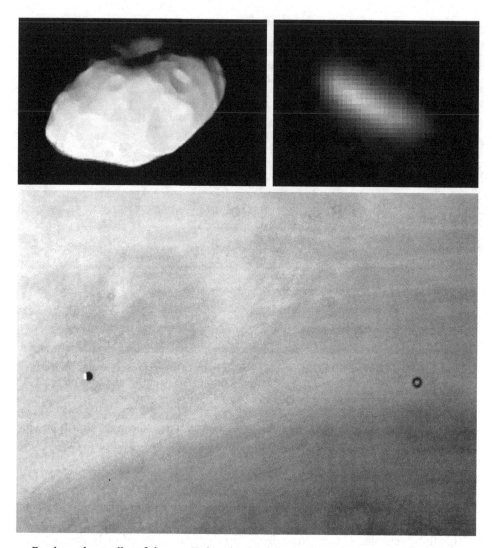

Pandora, the smaller of the two F ring shepherd moons, seen on 3 June 2010 (top left). A highly enlarged picture is one of the best views of the minuscule moon Aegeon (top right). An overexposed view of Pallene (below, at left) hovering over Saturn's clouds; the dark spot to its right is a dust speck on the camera optics. (NASA/JPL/Space Science Institute)

of the leading hemisphere. Fifteen minutes prior to closest approach the mapping spectrometer took spectral cubes of the polar region. Scientists hoped that the polar hood of haze and clouds would have faded sufficiently to permit this instrument to detect the spectroscopic signature of hydrocarbons in the Kraken and Ligeia seas and the runoff channels that fed them. Remote sensing continued on the way out. In particular, the camera got a 9-frame mosaic spanning Belet, western Shangri-La and northwestern Adiri. This high-latitude encounter left the period of the spacecraft's orbit at 16 days but cut its inclination from 12 to just 2 degrees.

T70 on 20 June was to be the final encounter with Titan of the equinox mission. In geometrical terms it was similar to its immediate predecessors, with the vehicle approaching from the direction of Saturn viewing the night-side, and receding over the illuminated anti-Saturn hemisphere. But with a minimum altitude of 880 km at a latitude of 82°N it was the closest Titan encounter ever. The focus of attention was radio tracking to investigate whether the moon possessed a subsurface ocean. The spacecraft was to travel through the upper atmosphere in an aerodynamically stable 'weathervane' attitude which, by sheer good fortune, was within several degrees of that ideal for pointing the high-gain antenna at Earth. Moreover, the very low flyby was a good opportunity to seek the possible signature of a magnetic field induced in a subsurface ocean. The thrusters fired about twice as much propellant as expected but otherwise the flyby was extremely smooth and provided good data. Soon after the radio-science experiment, control was given to the ultraviolet spectrograph for a scan of the atmosphere along a meridian located just east of the Huygens site. Next, this instrument and the infrared mapping spectrometer used an occultation of Spica to investigate the composition and chemistry of the upper atmosphere. Armed with occultation profiles for a range of latitudes and longitudes scientists would be able to reconstruct the dynamics of the processes of polymerization and dissociation of the various molecules in great detail. On the way out the imaging spectrometer and camera made observations of Ching-Tu, Adiri and Dilmun. As a result of this flyby the inclination of the spacecraft's orbit was increased to 19 degrees.

On 26 June Cassini started its 134th revolution, and four days later the equinox mission was officially concluded.

## THE SOLSTICE MISSION AND FINALE

Cassini scientists, engineers and managers started to discuss a second extension to the mission even before the first one got underway. A number of end-of-life options were studied, governed by the constraints that the vehicle must not cross the rings lest it suffer damage, and it must not impact Enceladus or Titan in order to avoid contaminating them with terrestrial biological matter. At the end of its mission the Galileo spacecraft had dived into Jupiter to preclude the possibility of it impacting on Europa. One option for Cassini was therefore to dive into Saturn, either from a short-period orbit of 6 to 10 days, or from a longer-period orbit due to gravitational perturbations by Titan and by the Sun. Another option was to enter a stable orbit, possibly just beyond that of Titan or beyond that of Phoebe. Alternatively, Cassini could escape Saturn to conduct further investigations in the outer solar system. One adventurous possibility was a 'Grand Tour' of Jupiter, Uranus and Neptune but it was open to debate whether the spacecraft would still be usefully functional when it reached Neptune in 2061. A flyby of one of the Centaur objects was of potential scientific interest but would still involve a decade or more of travel.[305,306,307] In the end, scientists and engineers came up with plans to extend Cassini's useful life by no fewer than seven years to the northern hemisphere summer solstice in 2017. An orbital mission from 2004 to 2017 would essentially span half of a Saturnian year and ought to reveal seasonal effects on Titan's lakes and the atmospheres of Saturn and Titan. As in the case of the Galileo mission, however, support would have to be reduced. One way to enable a smaller team to run the spacecraft was to restrict the encounter sequences to just a few days.

The second mission extension, dubbed the 'solstice mission', was to conduct an additional 160 orbits with 56 targeted encounters of Titan, 38 of which would be at altitudes below 2,000 km; a dozen Enceladus flybys at less than 5,000 km, some as low as 50 km; three flybys of Dione at 500 km or less; and two Rhea flybys, one at 75 km. The solstice mission would also provide observations of some the smaller moons, including some poorly observed ones. For example, there would be a flyby in May 2012 of Methone, which Cassini had discovered, at a range of 1,900 km. Unfortunately the fuel budget would not permit a return to Iapetus. There were to be two equatorial orbit phases from 2010 to 2012 and then again in 2015 and early 2016 mostly to provide flybys of the icy satellites, and two high-inclination phases which would focus on the magnetosphere and the rings, in particular the propellers. Four equatorial periapsis passes undisturbed by satellite flybys were to be used for gravity measurements. And, of course, there would be many radio, solar and stellar occultations by Saturn, by the rings and by Titan. At the conclusion of the equinox mission about 80 per cent of the fuel reserve had been expended. Consequently, the solstice mission would be able to make fewer maneuvers in order to set up targeted encounters. Operational planning would be simplified by devoting each periapsis to a single scientific objective: either the planet, its magnetosphere, its rings, or its icy satellites. Of course, Titan flybys would continue to be handled differently.

The solstice mission was to pursue questions which arose during the first half of the orbital mission. On Enceladus, Cassini would continue to study the geysers seeking variability in their activity over long or short timescales for insight into the controlling mechanism. Encounters could search for the signature of a subsurface ocean, better constrain the composition of gases in the plume, and further study the gravitational field. The mission would extend the seasonal coverage of Titan to the solstice, hopefully observing the northern lakes directly illuminated by the Sun and obtaining spectra to study the dissipation of the polar winter vortex in the north and its development in the south. At the same time, atmospheric models predicted that mid-latitude clouds would "migrate" from the south to the north. The hydrogen-to-helium ratio of the atmosphere of Saturn, vital for understanding the evolution of the planet, would be measured using stellar occultations and infrared spectra. It would also be interesting to see whether 'storm alley' remained fixed at its southern latitude or showed any seasonal effects. The mission would conclude with a series of 'proximal' orbits in order to obtain an unprecedented perspective on the planet and its phenomena. By flying low over the planet's cloud tops, these would take the spacecraft through the innermost radiation belt, auroral zones and magnetosphere for particles and fields studies in search of possible departures of the magnetic field from its apparent symmetry. The proximal orbits would also map the gravity field of the planet in detail, yielding information on its internal structure and on the mass and rate of rotation of its core. Moreover, they would refine the estimated mass of the rings as well as provide information on seasonal phenomena like spokes and on moonlets in apparently empty gaps.[308]

In February 2010 NASA and ESA approved a budget of $60 million to extend Cassini's mission to 2017, although the Science and Technology Facilities Council in the United Kingdom withdrew support.

At its first solstice mission periapsis on 5 July Cassini snapped several narrow-angle images of Daphnis at a range of 73,000 km that showed the small moon as an object spanning 20 pixels.

Two days later, heading out from Saturn, the spacecraft encountered Titan. The geometry and illumination for T71 were similar to recent encounters, providing a 1,005-km flyby with the closest point of approach at 56°S. Cassini flew through the bow of the region in which the moon's atmosphere and ionosphere interacted with the pre-dusk side of the magnetosphere. As a result the mass spectrometer was the prime instrument at closest approach, both to study this interaction and to monitor how the atmosphere was responding to the change of season and the variability of the solar wind during the 11-year cycle solar cycle. The magnetometer and plasma spectrometer were also operating during the flyby. The radar was riding along at closest approach and acquired a short swath of Mezzoramia and the poorly mapped western part of the southern hemisphere. On the way out over the day-side optical imagers monitored clouds and the other spectrometers measured the chemistry and composition of the atmosphere. The flyby had the effect of increasing the period of the spacecraft's orbit to 18 days, ending the series of resonant Titan encounters.

On its next revolution Cassini made some observations of Saturn's atmosphere that took advantage of the fact that the Sun was still barely illuminating the rings. It undertook another intensive imaging campaign of the night-side in order to search for lightning in 'storm alley'. On 24 July it provided a radio-occultation at a near-equatorial latitude that would normally be ruined by the rings being present on the line of sight. On 13 August, reaching periapsis, the vehicle flew within 100,000 km of Dione, within 37,000 km of Tethys, and observed beta Orionis (Rigel) occulted by the rings. But Enceladus was the focus of attention.

The E11 encounter was a 2,550-km flyby at a relative speed of 6.8 km/s and it provided a view of the south pole for the remote-sensing instruments that would not be repeated until much later in the mission, and was all the more welcome because as autumn set in darkness was increasingly hiding this region. The encounter started with almost ideal long-range imaging of the plumes over the night-side and, owing to the relatively distant encounter, the composite infrared spectrometer could stare at the moon for long periods of time to obtain particularly good data, including one temperature scan running along the Damascus tiger stripe which was in darkness at the time. The flyby yielded the highest-resolution temperature maps yet, showing a number of spots at temperatures several tens of degrees warmer than any previously measured. The spatial resolution of 800 meters allowed scientists to identify warm areas on the flanks of the Damascus trench itself. High-resolution scans of the ends of Alexandria and Cairo were also obtained showing branching warm features and isolated warm spots. Images taken at the same time showed the complex tectonics of the tiger stripes. On the way out the camera took mosaics of progressively lower resolution of the Saturn-facing hemisphere. During the encounter, engineers were carrying out tests to determine whether it was possible to use the low-gain antenna rather than the high-gain antenna for Doppler radio tracking, so that during flybys devoted to gravity studies the vehicle could adopt attitudes more favorable for other instruments.

The next periapsis on 3 September was devoted to studying the rings as Cassini flew through the shadow cone of the planet with a view of their unilluminated face similar to that of September 2006. A non-targeted encounter with Dione at a range of 39,000 km provided some of the highest-resolution coverage to-date of the north polar regions of this moon.

During its 138th revolution Cassini returned to Titan. T72 on 24 September as the first in a series of high-altitude encounters outbound from Saturn, in this case an 8,175-km flyby with its closest point of approach at 15°S. It was primarily devoted to cloud formation studies and seasonal effects including a possible break-up of the northern winter polar vortex using the camera, the two infrared instruments and the ultraviolet spectrograph. At closest approach the mapping spectrometer was aimed at the boundary between Belet and Senkyo and took cubes at a resolution of 5 km. On the way out Cassini switched to cloud monitoring, and three days after the flyby it photographed the largest example yet seen on Titan: an arrow-shaped white cloud several hundreds of kilometers across. Even more interestingly the cloud was in the equatorial region where they had been extremely rare; possibly indicating seasonal changes. Unfortunately, Saturn was very close to the Sun as viewed from Earth and

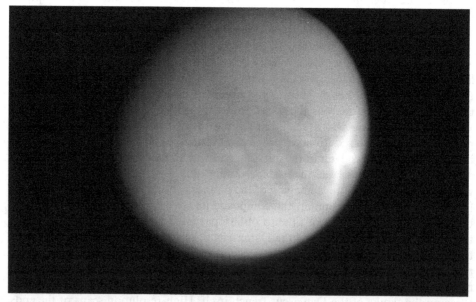

Titan's equatorial 'arrow storm' marked the beginning of spring rains. (NASA/JPL/ Space Science Institute)

it was not possible to follow the evolution of this cloud by telescope. Several days later Saturn entered solar conjunction.

Inbound to Saturn on 14 October Cassini passed Titan at a range of 172,000 km as the prelude to a series of non-targeted encounters over the next three days. First was Mimas at 70,000 km, then Pallene at 36,000 km, Dione at 31,000 km, and finally Rhea at 39,000 km. Along the way there were opportunities to inspect some of the smaller moonlets. In addition to imagery of all these satellites, the composite infrared spectrometer made thermal scans of Mimas, Dione and Rhea. The bright 'arrow storm' was still visible on Titan. Mimas was imaged just as it was about to slip into eclipse behind Saturn. Pallene was viewed in front of Saturn. It spanned fewer than 30 pixels, the pictures were overexposed, Cassini was looking down the longest axis and almost no surface detail was visible. However, having the bright planet in the background allowed the camera to see the night-side of the moonlet as a dark silhouette, revealing its complete shape. The best pictures yet of the southern part of Dione's leading hemisphere were obtained. Other tasks at this time included an infrared scan of the faint rings; two stellar occultations by Saturn, one of alpha Ceti at a southern latitude and one of alpha Hydrae at a northern latitude; imaging the plumes of Enceladus; and a search for rings around Rhea. Cassini also made observations of some moons requested by schoolchildren under JPL's 'Scientist for a Day' educational program.

Starting in October 2010, Cassini undertook astronomical research, in particular using its remote-sensing instruments to make observations of extrasolar planets by monitoring the characteristic drop in light as a planet transited across the disk of its parent star. During this and the following months, the 'arrow storm' on Titan was monitored by long-range observations. Early images revealed that the brightness of the southern boundary of Belet, to the east of the cloud, had changed. By the end of the month, however, most of it had resumed its previous appearance and only a few darker patches persisted. The most likely explanation was that this terrain had been wetted, possibly even flooded by methane rain stimulated at equatorial latitudes by conditions at the equinox. Such events were expected to be rare but intense, producing enough rain to sculpt the creeks and other river-like features observed by Huygens and Cassini.[309],[310]

On 2 November Cassini entered safe mode for the sixth time since launch. This time it was due to a 'bit flip' in the command and data management computer that was almost certainly induced by a stray cosmic ray. Contact with Earth was lost at the onset of the problem. Even if control were regained in time to make a planned orbital trim maneuver five days later controllers were not confident that they would be able to restore the spacecraft to a full operational status in time for the encounter with Titan on the 11th. This became the first time an encounter had been skipped as a result of the vehicle having entered safe mode. No science data was collected on the 7,921-km flyby of the T73 encounter. The lost sequence was to have acquired temperature maps of the stratosphere and made multispectral measurements around the Huygens site, Adiri and Shangri-La. Amongst other observations that were lost were attempts to detect the shadow of Saturn being cast on the 'Phoebe ring', non-targeted encounters with Dione and Enceladus, a stellar occultation by Saturn, and further long-range monitoring of the 'arrow storm' on Titan. The vehicle resumed regular observations on 26 November as it headed to periapsis on revolution 141. Two days later it had a non-targeted flyby of Hyperion at a range of 72,000 km.

A little over 2 hours after periapsis on 30 November the E12 encounter saw Cassini fly past Enceladus at an altitude of 48 km with the closest point of approach at 61°N. The composite spectrometer made a thermal map of the night hemisphere at low latitudes on the way in and another of the day-side on the way out in search of hot spots away from the south pole. On the way in the camera imaged the backlit polar plumes. The main objective of the flyby, however, was to collect gravity data over the northern hemisphere for comparison with similar data from the southern hemisphere to further investigate the internal structure of the moon and the putative subsurface ocean. The experiment consisted of a 3-hour tracking session at closest approach and 3-hour 'wings' on both the inbound and outbound legs. Accordingly, the spacecraft turned to point its high-gain antenna to Earth for closest approach. In this orientation Enceladus would briefly cross the field of view of the camera and ten narrow-angle and wide-angle images were taken of small portions of the leading hemisphere. Afterwards remote-sensing observations of the illuminated anti-Saturn hemisphere targeted the boundary of the smooth young-looking plains and the older cratered terrain.

Climbing to apoapsis Cassini was able to recover one of the observations lost on the previous orbit by inspecting the leading Lagrangian point of Rhea to determine whether the second largest satellite in the Saturnian system had Trojan companions to match those of smaller Tethys and Dione. Over the next few revolutions similar searches were made for Iapetus, Titan, Enceladus, Mimas and Rhea, as well as for additional Trojans of Dione.

On 5 December the plasma-wave experiment detected radio emissions from a new storm brewing in Saturn's atmosphere. On the same day the camera observed an inconspicuous white spot 1,300 × 2,500 km located at 32°N. Less than two hours after the storm was detected by Cassini, amateur astronomers also reported it as a barely visible feature. Over the next three days a rising plume of bright material created a string of white clouds. Within a week the cloud had stretched eastward to a length of 8,000 km. Within a month it occupied a longitude range in excess of 90 degrees at that latitude. This was only the sixth such 'great white spot' to have been observed during the last 135 years. The previous one was in 1990 and was studied by the newly launched Hubble Space Telescope. The last 'great white spot' to occur in the northern hemisphere was in 1903; all the others were at southern latitudes. It was interesting that this storm occurred early in the Saturnian year, since previous storms had appeared near the summer solstice. In any case this was the first chance to study one from Saturnian orbit. Such storms were caused by insolation allowing water and ammonia to condense, making moisture cells deep inside the atmosphere. In time, the cells became giant upwelling plumes of warm 'air' that injected bright ammonia ice crystals into the upper atmosphere to be caught by the jet stream and progressively drawn around the planet at that latitude. Jupiter, the other gas giant of the solar system, had nothing comparable, possibly because its spin axis was more perpendicular to the plane of its orbit around the Sun and had less extreme seasonal cycles.

While controllers and scientists were busy creating sequences that would allow the remote-sensing instruments to monitor this storm into early 2011 the plasma-wave experiment had no difficulty continuing to follow its evolution. Lightning was occurring at a frequency many times greater than for any storm observed since the spacecraft arrived in 2004. On 12 December the lightning rates peaked at in excess of 10 bursts per second – it was impossible to count them because the receiver was saturated.[311,312,313,314]

Having made another revolution Cassini returned to periapsis on 21 December. The previous day, in a non-targeted encounter with Dione at a range of 100,000 km, it had imaged the thin crescent in search of plumes and geysers but saw nothing. At periapsis it flew past Enceladus.

The E13 encounter was a virtual repeat of its predecessor. The approach allowed the infrared spectrometer to make a temperature map of the night-time hemisphere and the camera to image the backlit plumes, capturing an amazing view of Mimas beyond the jets. Midway through the approach phase Cassini was briefly turned in order to witness a rare occultation by Saturn of alpha Orionis (Betelgeuse). It then skimmed over Enceladus at an altitude of 48 km with the closest point of approach at 62°N. This occurred while crossing the dawn terminator. At closest approach, the mass spectrometer and dust analyzer studied the composition of the icy particles on the spacecraft's path. On the way out the camera obtained a 10-frame mosaic of the illuminated hemisphere of the moon.

Receding from Saturn, Cassini finally achieved a good view of the storm in the planet's northern hemisphere. Apoapsis on 31 December drew revolution 143 to an end. On 9 January 2011 the imaging spectrometer obtained cubes of the storm over the night-side showing cold clouds as dark shadows silhouetted against the warmth of the interior. At periapsis the next day this instrument observed the dark limb of the planet occult first Sirius, the brightest star in the sky, and then a fainter star in Leo. Two days later the spacecraft made its fourth close flyby of Rhea, this time at an altitude of 76 km with the closest point of approach at 76°S. Each camera took a pair of pictures at the moment of flyby. Mosaics of the anti-Saturn hemisphere were also obtained, targeting a line of small craters suspected of being impacts from the demise of an equatorial ring. The plasma spectrometer again searched for evidence of a faint ring and the camera attempted to image this over the dark limb (in several cases capturing Dione and the foreshortened rings of Saturn in the background) but again finding no trace of a ring around Rhea.

The day before apoapsis the composite infrared spectrometer made a scan of the 'great white spot' on Saturn in cooperation with the Very Large Telescope in Chile operated by Europe. By then the storm was an elongated meandering tail of white clouds with weaker cells and encircled the planet. The features at this latitude had last been observed by Cassini in late October, before the storm appeared, so it was fairly easy to conclude that it had completely disrupted the slow seasonal evolution of the weather system, as well as the atmosphere's normal structure and chemistry. The northern hemisphere had been seen to steadily warm with the progression from winter to spring, but the storm had reversed that trend. The storm itself consisted of a cold vortex 5,000 km across with a warmer periphery and a cooler tail. As the 'air'

The great storm in Saturn's northern hemisphere imaged by Cassini's wide-angle camera on 26 December 2010. (NASA/JPL/Space Science Institute)

in the disturbance was displaced upward through several tens of kilometers it was cooled by 10°C. However, streaks dubbed 'stratospheric beacons' rose above the cold vortex and were as much as 20°C warmer than their surroundings. Despite being the warmest features on the entire planet the 'beacons' had no visible-light counterparts. Finally, the storm had changed the chemistry of the upper atmosphere, both drawing phosphine and ammonia up from the warmer atmospheric layers and causing a depletion of the normally abundant acetylene.[315]

There were no targeted encounters on the 144th revolution but Cassini made long-range observations of Titan, Dione, Enceladus, Mimas and Helene. In the case of Enceladus it monitored the plumes and took mosaics covering plains, ridges and sulci. Moreover, distant images of the night-side of the moon captured some of the most dramatic views of geysers and plumes. On 27 January Cassini passed 15,000 km from Aegeon and took 18 narrow-angle images with a resolution of better than 100 meters. The moonlet was revealed to be an extremely elongated reddish object between 1.2 and 1.6 km in length but just 300 to 600 meters wide.[316]

Returning to periapsis one revolution later on 18 February Cassini had the T74 encounter. The plasma spectrometer had priority near closest approach during this 3,650-km flyby to study the interactions between Titan and the magnetosphere in the vicinity of the dusk terminator. On the way out the camera imaged the Saturn-facing hemisphere. Fourteen hours after the flyby, and 240,000 km out, Cassini was finally able to view the area directly under the massive September 2010 storm. The narrow-angle camera achieved the reasonable resolution of about 1.4 km and saw a number of possible changes in the appearance of the surface.

Revolution 146, from 6 March to 3 April, was largely devoted to observing the effects of the 'arrow storm' on Titan and the 'great white spot' on Saturn. The spacecraft reached periapsis on 17 April and two days later had the T75 encounter. At 10,053 km, this flyby of Titan resembled that of Voyager 1 in that it passed through the moon's tail on the dusk side of the magnetosphere. Moreover, almost exactly one Saturnian year had passed between the Voyager flybys and T75, and the atmospheric seasonal features were also taken to be similar. The prime instruments at closest approach were the plasma spectrometer and the magnetometer. In particular, the vehicle was orientated to align the field of view of the spectrometer with the predicted direction of plasma flow. Cassini returned to Titan less than a revolution later on 8 May, two days before periapsis. The T76 encounter involved an approach over the night-side, a minimum altitude of 1,873 km, and exit in daylight. On the way in the mapping spectrometer and the camera studied high-altitude detached hazes and the formation of the southern polar hood. High-resolution cubes of northern Adiri and dune fields in northern Senkyo were also obtained. At periapsis Cassini made a movie of the auroras over the south pole and the night-side of Saturn, monitored the plumes of Enceladus at long range, and an Enceladus eclipse sequence with Helene present in the background. On 7 June Cassini made one of the closest approaches to Iapetus of the extended mission but still came no closer than 860,000 km. One week later the plasma spectrometer had to be switched off when it caused a series of voltage shifts and short circuits. The fault was traced to capacitors in the instrument; otherwise it was operating normally. About 4 hours before periapsis on 18 June Cassini passed Helene at a range of 7,000 km. It approached from the night-side, flew directly over the pole and departed over the poorly imaged Saturn-facing hemisphere. This time the image frames were perfectly centered on the moon and showed gully-like slides and flows. The anti-Saturn hemisphere proved to be heavily cratered but, apart from old eroded craters, there were only a few dozen small craters on the Saturn-facing side.

On 20 June Cassini had the T77 encounter. It was a 1,358-km equatorial flyby. Approaching over the night-side the ultraviolet spectrograph made measurements of the atmosphere and the camera monitored high-altitude hazes at the limb and the polar hoods. The radar had priority for 2 hours at closest approach. It started with a long altimetry run across Xanadu and the boundary to Shangri-La. There was then a high-altitude imaging swath over northwestern Fensal between the craters Menrva and Ksa, followed up by a short high-resolution swath of southwestern Xanadu and western Tui Regio. Combining these measurements with data from September 2006 provided stereoscopic images of Ksa and northern Xanadu. The encounter was also interesting for a noon equatorial pass through Titan's wake in the magnetosphere, where the particles and fields instruments could map particles escaping downstream of the moon.

In July the 150th revolution was mostly devoted to studying Saturn, particularly the storm in the northern hemisphere, but it also imaged some of the moons at long range. By this time, the tail of the storm had developed a dark central area that approached and collided with the head. The head then split and became increasingly difficult to see and track.

On 1 August, at periapsis on the next revolution, Cassini had a non-targeted encounter of Rhea at a range of just under 6,000 km. The ultraviolet spectrograph observed an occultation of epsilon Orionis to monitor the oxygen exosphere over both the day-side and night-side while the camera took pictures as the occultation progressed. On 25 August the orbiter had a non-targeted encounter with Hyperion at a range of 25,000 km, the closest in years. By chance the chaotic rotation of the moon placed the Bond-Lassell crater and its central mound in plain view. Returning to periapsis on its 153rd revolution, on 12 September Cassini had its T78 encounter. This 5,821-km flyby was devoted to optical remote-sensing. In particular, there was an occultation of chi Aquarii and a solar occultation. The ultraviolet spectrograph obtained the second-ever measurement of the ratio of hydrogen to deuterium in the atmosphere of Titan. On passing periapsis Cassini made long-range observations of Enceladus, Pallene and Tethys. In the case of Tethys, Cassini confirmed the existence of a long-suspected thermal anomaly, which resembled that of Mimas, with a cooler band at equatorial latitudes. It then had another flyby of Hyperion, this time at a range of 58,000 km.

On 1 October, 3 hours before periapsis, Cassini returned to Enceladus for the E14 flyby, the first of a closely-spaced series of three encounters that would take it just 99 km over the surface of the active moon. Particles and fields instruments were in control. In particular, the mass spectrometer and plasma-wave instruments measured the composition of the material sprayed by the vents as the spacecraft passed directly over the south pole. After the encounter the composite infrared spectrometer and camera imaged the day-side of Enceladus as it entered and then exited the shadow of Saturn, with Titan in the background. The encounter was wrapped up by distant, full-disk color imaging of the leading hemisphere.

Cassini then underwent a 5-day solar conjunction in mid-October, during which its radio carried out a three-wavelength corona survey. Three hours before periapsis on 19 October it had the E15 encounter at a range of 1,231 km from the moon. As on the previous occasion, the spacecraft approached from over the night-side. This time, remote-sensing instruments were assigned priority during the closest approach, with the ultraviolet spectrometer observing an occultation of two stars in the Orion's belt by the plumes in order to monitor the density and the localization of the single jets. On the other hand, before Enceladus again went into eclipse, the camera had a view of Isbanir Fossa, a long groove discovered by

Voyager 2, with suitable lightning to show it to be a conspicuous crack. Also carried out during this revolution, the 155th, was distant imagery of Aegeon to document the dynamics of the arcs of the G ring.

Finally, on 6 November Cassini had the E16 flyby of Enceladus, the last of three encounters one revolution apart, this time passing by at a range of 496 km. The radar took not only radiometry and scatterometry data, but also low-resolution and high-resolution imagery; one of the very few times this was done for a satellite other than for Titan and the first occurrence for Enceladus. Besides yielding scientific data about the moon, the experiment allowed calibration of Titan radar images on a target that was far easier to image using cameras. The 25-km-wide swath was taken over the southern trailing hemisphere, cutting through the folded ridges bounding the southern polar region. The best resolution was obtained at about 62°S, but the swath did not go sufficiently far south to include any tiger stripes. Again, around closest approach, Enceladus slipped into eclipse in the shadow cone of the planet.

On 12 December, passing periapsis of the 158th revolution, Cassini made its third targeted flyby of Dione. It approached the night-side, scanning the trailing hemisphere with instruments to measure the thermal inertia of the grooved and fractured terrain. But since the range was only 98.8 km, the closest the spacecraft would approach this moon during the entire mission, most of the encounter was devoted to radio tracking in order to determine whether it was differentiated like Enceladus, with a rocky core, or a more or less homogeneous mix of rock and ice like Tethys. The average density of Dione was higher than for Tethys, so some difference was expected. The mass spectrometer meanwhile attempted to sample Dione's oxygen exosphere. Departing over the day-side hemisphere, the camera obtained mosaics of the hemisphere opposite to Saturn. A little less than 36 hours later, the vehicle had its T79 encounter with Titan, this time passing by at a relatively distant 3,585 km. On the inbound leg, the infrared spectrometer made temperature scans over the night-side, as well as limb scans of the atmosphere over the north pole and southern hemisphere. At closest approach, the mapping spectrometer acquired high-resolution data of dune fields in Belet and sand-poor dunes in Ching-Tu.

One of the ultra-stable oscillators of the radio system malfunctioned in December. While engineers worked to determine whether this could be brought back into service, the spacecraft used a backup unit for communications. Use of the backup during the remainder of the mission would result in scientific data of lower quality during occultation experiments, but otherwise the mission, and in particular the gravitational surveys, would not be affected.

As 2011 closed, Cassini was approaching Titan for the 2 January 2012 distant T80 flyby, which was a very high altitude inbound encounter at 29,416 km. Due to the large distance, imaging and remote-sensing instruments had priority during the encounter, in particular observing occultations of the two infrared-bright stars CW Leonis and R Leonis, and on the outbound leg making radiometric scans to map the hitherto poorly covered southern trailing hemisphere. Speeding through periapsis 2 days later, Cassini observed an occultation of theta Orionis, the Trapezium cluster of stars in the Orion nebula, by the Saturnian atmosphere, and monitored the jets on Enceladus to determine whether their activity changed with the cyclic tidal effects once per orbit. The spacecraft returned to Titan on 30 January for the T81 distant flyby at a range of 31,131 km. Again, given the high-altitude encounter, the remote-sensing instruments were prime, targeting in particular Ontario Lacus to monitor its liquid level. Cassini returned to Titan less than 3 weeks later, on 19 February, for the T82 encounter at a range of 3,803 km, which set up the conditions for three very close flybys of Enceladus.

Cassini had non-targeted flybys of Enceladus at over 9,000 km on 9 March, and of Rhea, almost 42,000 km off, the next day, as it prepared for the 74 km E17 flyby of Enceladus on the 27th, the first of the series of three close flybys. The second followed on 14 April and the third on 2 May. The latter had a geometry similar to that of the E9 flyby. Another flyby of Titan on 22 May then put the spacecraft into a 16-degree orbit in order to observe the well-illuminated rings.

This inclined orbit will prevent further icy satellite encounters until March 2013, when there will be a 1,000-km flyby of Rhea. Cassini will not return to Enceladus until mid-October 2015, when the south pole will be completely in the shadow of the winter night. As for Titan, south polar radar passes are scheduled in October 2013 (T95), February 2014 (T98) and July 2016 (T120), and north polar passes and radar imaging in May and September 2012 (T83 and T86), May, July, and October 2013 (T91, T92 and T95), August 2014 (T104) and February 2015 (T109).[317]

By November 2016 Cassini will be in a high-inclination orbit with its periapsis near the orbit of Enceladus. A Titan flyby on the 29th will drop it to a position just 10,000 km from the F ring and not very far from the edge of the main rings. Twenty such orbits will be flown to study the F ring and the A ring in great detail. Crossing the ring plane close to the orbits of Janus and Epimetheus will require the vehicle to face its high-gain antenna forward to serve as a dust shield, just as it did at the time of Saturn orbit insertion.

On 22 April 2017 Cassini will perform its final Titan flyby, T126. The gravity-assist will dramatically shrink the periapsis distance, leapfrogging the entire span of the main rings to the gap between the D ring and the planet, a mere 3,500 km above the clouds. There was a controversial proposal to fly Pioneer 11 through this gap in 1979 before it was decided to use that spacecraft to test crossing the ring plane just beyond the 'classical' rings, which was what Voyager 2 would have to do in order to pursue the Grand Tour beyond Saturn.[318]

A total of 23 'proximal' orbits will be flown, during which Saturn will reach the northern summer solstice on 24 May 2017. These orbits will have an inclination of about 63.4 degrees with their plane essentially fixed in space, and all periapses will occur near local noon to allow imaging of the illuminated face of the rings in great detail and for continuous tracking by the Deep Space Network. Such orbits will also yield frequent Earth and solar occultations by the rings. The main scientific interest will be to measure the gravitational field of the planet to determine its moment of inertia to a high level of accuracy, its internal structure, and the size and mass of its core. By virtue of passing between Saturn and the rings the gravitational pull of the rings will be separated from that of the planet to enable the rings to be accurately 'weighed' for the first time. The magnetic field will be mapped in detail, as will be the dust environment. In fact, particles from the ring system should extend down to the cloud tops. It is even possible that molecules from the upper atmosphere may be intercepted and analyzed in-situ by the mass spectrometer, providing the first direct measurement of their composition. If the magnetometer can measure small periodic asymmetric components of the magnetic field then it might be possible to impose constraints on the period of rotation. Interestingly, at about the same time, the Juno mission (to be described in Part 4 of this series) will be making similar observations at Jupiter.[319] And, like Juno, Cassini will investigate how the composition of the atmosphere varies with depth by using its imaging spectrometer and the radar operating as a microwave radiometer.

A final flyby of Titan on 11 September 2017 at a range of 87,000 km will nudge Cassini's orbit so that four days later, at the periapsis of its 293rd revolution, it will dip below the level of the cloud tops. Like Galileo at Jupiter, therefore, Cassini will destroy itself by plunging into the atmosphere of the planet that it came to study. So will end one of the most productive and successful planetary exploration missions ever flown.

## REFERENCES

1 Cruikshank-1972
2 For Eos see Part 1, pages 289–291
3 Martin Marietta-1976
4 For Purple Pigeons see Part 1, page 256
5 Fink-1976
6 Lorenz-2009
7 For the Solar System Exploration Committee see Part 2, page 96–102
8 JWG-1986
9 For Mariner Mark II see Part 2, pages 98–102
10 For Vesta and CAESAR see Part 2, pages 104 and 120–123
11 Beckman-1986
12 Owen-1986
13 Withcomb-1988
14 Lebreton-1988
15 AWST-1989
16 For CRAF see Part 2, pages 106–114
17 ESF-1998
18 Sanford-1992
19 For Saturn's radio bursts see Part 1, page 377
20 Jaffe-1997
21 Murray-1992
22 Southwood-1992
23 Dornheim-1996
24 Smith-1997a
25 Smith-1997b
26 Ratcliff-1992
27 Calcutt-1992
28 Coates-1992
29 Flamini-1998
30 Somma-2008
31 For Mars 6 and Martian argon see Part 1, pages 167 and 226
32 Lorenz-1997
33 McCarthy-1996
34 For Pioneer Venus' descent see Part 1, pages 274–276
35 Jäkel-1996
36 Lorenz-1994
37 Owen-1999
38 Tomasko-1997
39 Zarnecki-1992
40 Lebreton-1997
41 Hassan-1997
42 Jones-1997
43 Sparaco-1996
44 Schipper-2006
45 Smith-1998
46 Wolf-1998

47  Kohlhase-1997
48  Covault-1997
49  Dornheim-1998
50  Gurnett-2001
51  Zarka-2008
52  Clark-2009
53  Dornheim-2001a
54  Mitchell-2000
55  Hsu-2009
56  Dornheim-2001b
57  Porco-2003
58  Hill-2002
59  Gurnett-2002
60  Bolton-2002
61  Kurth-2002
62  Krimigis-2002
63  Gladstone-2002
64  Mauk-2003
65  Nixon-2010
66  Mitchell-2002
67  Tortora-2004
68  Bertotti-2003
69  Deutsch-2002
70  Strange-2002
71  Schipper-2006
72  Fischer-2006
73  Pryor-2008
74  Griffith-2003
75  West-2005
76  Clarke-2005
77  Crary-2005
78  Kurth-2005
79  Bagenal-2005
80  Kempf-2005a
81  Kempf-2005b
82  IAUC-8389
83  Jacobson-2006
84  Porco-2005a
85  Flasar-2005a
86  Jewitt-2007
87  Kelly Beatty-2004
88  Esposito-2005
89  Clark-2005a
90  Johnson-2005
91  Brad Dalton-2005
92  Sanchez-Lavega-2005
93  IAUC-8401
94  Dougherty-2005
95  Gombosi-2005
96  Gurnett-2005
97  For Voyager 2 ring plane crossing see
     Part 1, page 389–390
98  Spahn-2006a
99  Tiscareno-2006
100  Tytell-2004
101  Tiscareno-2007
102  Porco-2005b
103  Waite-2005a
104  Krimigis-2005
105  Mitchell-2003
106  Dornheim-2004a
107  Baines-2005
108  Dornheim-2004b
109  Young-2005
110  Gombosi-2005
111  Dougherty-2005
112  Baines-2005
113  Baines-2005
114  IAUC-8432
115  Spitale-2006
116  Waite-2005b
117  Schaller-2006
118  Porco-2005c
119  Sotin-2005
120  Barnes-2009a
121  Elachi-2005
122  Covault-2004a
123  Beckes-2005
124  Garnier-2007
125  IAUC-8432
126  Showalter-2005
127  Spitale-2006
128  Mitchell-2004
129  Griffith-2005
130  Porco-2005c
131  Barnes-2005
132  Rodriguez-2009a
133  Flasar-2005b
134  Shemansky-2005
135  Dandouras-2009
136  Levison-2011
137  Porco-2005b
138  Porco-2005c
139  Porco-2005d
140  Flasar-2005a
141  Esposito-2005
142  Ostro-2006
143  Morring-2005a

144 Hueso-2006
145 Tokano-2006
146 Griffith-2006a
147 Lebreton-2005
148 Tomasko-2005
149 Niemann-2005
150 Fulchignoni-2005
151 Zarnecki-2005
152 Bird-2005
153 Karkoschka-2007
154 Soderblom-2007a
155 Kazeminejad-2007
156 Owen-2005
157 Morring-2005b
158 Schipper-2006
159 Morring-2005a
160 Elachi-2006
161 Murray-2005a
162 Jacobson-2006
163 IAUC-8524
164 Cuzzi-2010
165 Charnoz-2005
166 Murray-2008
167 Déau-2009
168 Esposito-2008
169 Postberg-2008
170 Kargel-2006
171 Kivelson-2006
172 Porco-2006
173 Spencer-2006
174 Dougherty-2006
175 Tokar-2006
176 Jones-2006
177 Spahn-2006b
178 Waite-2006
179 Hansen-2006
180 Brown-2006
181 Nimmo-2006
182 Kieffer-2006a
183 Nimmo-2007
184 Hurford-2007
185 Dombard-2007
186 Porco-2008
187 Kerr-2006
188 Kieffer-2008
189 Lakdawalla-2009
190 Sremcevic-2007
191 Esposito-2008
192 For Saturn spokes see Part 1, pages
379–380
193 Mitchell-2006
194 Mitchell-2005
195 Lunine-2008
196 Paganelli-2007
197 Griffith-2006b
198 For Ithaca Chasma and Odysseus see Part 1, pages 387–388
199 For Hyperion's rotation see Part 1, pages 385–386
200 Jacobson-2006
201 Thomas-2007
202 Cruikshank-2007
203 Ostro-2006
204 Soderblom-2007b
205 Lorenz-2006
206 Lancaster-2006
207 Lunine-2008
208 Paganelli-2007
209 Lorenz-2010
210 Wald-2009
211 Burch-2007
212 Jones-2008
213 Kerr-2008
214 Porco-2008
215 Lorenz-2008a
216 Paganelli-2007
217 Gurnett-2007
218 Bagenal-2007
219 Giampieri-2006
220 Zarka-2007
221 Kivelson-2007
222 Read-2009
223 Showman-2009
224 Anderson-2007
225 Podolak-2007
226 Iorio-2008
227 Stofan-2007
228 Sotin-2007
229 Waite-2007
230 Atreya-2007
231 IAUC-8773
232 IAUC-8759
233 Tiscareno-2007
234 Hedman-2007
235 Dyudina-2008
236 Charnoz-2009
237 Stallard-2008
238 Kurth-2008

239  Gurnett-2009
240  Esposito-2008
241  Mitchell-2007
242  Leyrat-2008
243  Fletcher-2008
244  Morring-2008
245  Stiles-2008
246  Lorenz-2008b
247  Sotin-2008
248  Spitale-2007
249  Lopes-2010
250  Lorenz-2008c
251  IAUC-8857
252  Bertucci-2008
253  Brown-2008a
254  Fischer-2007
255  Denk-2010
256  Spencer-2010
257  Verbiscer-2009
258  Hansen-2008
259  Hedman-2009a
260  IAUC-8970
261  Fischer-2008
262  Brown-2008b
263  Barnes-2008
264  Barnes-2009b
265  Schneider-2012
266  Turtle-2009
267  Stofan-2008
268  Charnoz-2007
269  Porco-2007
270  Charnoz-2010
271  Burns-2010
272  Canup-2010
273  Crida-2010
274  Morring-2007
275  Hedman-2009a
276  Tiscareno-2010
277  Schaller-2009
278  Zebker-2009
279  Rodriguez-2009b
280  Iess-2010

281  Sohl-2010
282  Baland-2011
283  Pryor-2011
284  IAUC-9023
285  Hedman-2009b
286  Pryor-2011
287  Waite-2009
288  Postberg-2009
289  Postberg-2011
290  Hartogh-2011
291  Lamy-2011a
292  Lamy-2011b
293  Soderblom-2009
294  Buratti-2009
295  Stephan-2010
296  Barnes-2011
297  IAUC-9091
298  Spitale-2009
299  For the heliosphere, heliopause,
     Voyager and IBEX see Part 1, pages
     443–451
300  Krimigis-2009
301  Hedman-2011a
302  Hayes-2011
303  Teolis-2010
304  Tokar-2012
305  Yam-2007
306  Davis-2007
307  Kloster-2009
308  Spilker-2010
309  Turtle-2011
310  Schneider-2012
311  Sanchez-Lavega-2011
312  Fischer-2011
313  Read-2011
314  Sanchez-Lavega-1989
315  Fletcher-2011
316  Hedman-2011b
317  Hayes-2011
318  See Part 1, page 146
319  Helled-2011

# 8

# Faster, cheaper, better continues

## A LANDMARK YEAR

The year 1996 had been a landmark year for planetary exploration and in particular for the United States of America. As the development of the Cassini spacecraft was nearing completion, NASA launched the first three missions which had been built in accordance with its new 'faster, cheaper, better' strategy. This concept, intended to provide frequent, more focused and less expensive deep-space missions had given rise to two different programs. Mars Surveyor, a JPL program, was to use small and inexpensive spacecraft to recover from the loss in 1993 of Mars Observer, an orbiter which was to have investigated the geology and climatology of that planet. Starting with Mars Global Surveyor, these new missions were to seek evidence to determine whether there were times when the planet's climate was suitable for life. They were also to pave the way, technologically and scientifically, for human missions at some indeterminate future date. Initiated in 1993, the Discovery program sought to exploit the concept of NASA's decade-old Small Explorer series of Earth-orbiting scientific satellites to make planetary missions cheaper and more frequent, and it was open to participation from NASA centers, universities and industries. The first two missions were the Near Earth Asteroid Rendezvous (NEAR) and Mars Pathfinder. NEAR, designed by the Applied Physics Laboratory (APL) of Johns Hopkins University, was to enter orbit around near-Earth asteroid (433) Eros for a reconnaissance lasting a year. JPL's Mars Pathfinder was to demonstrate a rough landing system and test a prototype planetary rover. These two programs promised a new 'golden age' in solar system exploration and, as will be related in this and the next chapter, were largely successful.

But 1996 also saw the disappearance of one of the oldest players in deep-space exploration. Russia, which had inherited most of the expertise and infrastructure of the former Soviet Union, saw the launch of its first probe end in dismal failure. This, and financial constraints, meant that Russia remained more or less in the shadow for the next decade.

On the other hand, as other space agencies developed 'flagship class' missions of their own (e.g. the European Rosetta comet orbiter), they were also giving thought to smaller and more flexible missions.

## STAR TREK PROPULSION

As NASA switched its deep-space missions from the 'flagship' approach which had produced such giants as Galileo and Cassini in favour of the Discovery program, it became clear to JPL that other programmatic changes would be required. In the past, it had used the flagships to fund the development of new technologies – for example the development of solid-state data recorders for Cassini to supersede troublesome tape recorders. A downside of this approach had been that in order not to jeopardize missions, only new technologies which had a more traditional and less risky backup available would be utilized. This meant that some 'mission enabling' but untested technologies for which no alternative existed stood little chance of flying in order to prove itself. The most important example was ion propulsion, also known as Solar Electric Propulsion (SEP), which was proposed in the 1970s to perform slow-speed encounters of comets and asteroids – e.g. the Halley/Tempel 2 International Comet Mission. Although the US had flown two Space Electric Rocket Tests (SERT), the second of which operated its thrusters for a total of 3,781 hours before the engine failed owing to an electrical short, and it had used this technology on experimental satellites, ion propulsion was 'immature', in particular because nothing was known about its interactions and possible idiosyncrasies with scientific instruments and it was not clear how to design and navigate an ion-propelled mission. Future applications of this technology not only involved solar electric propulsion, of course, whose use was mainly in the inner solar system, but also the combination of RTGs with ion engines (Radioisotope Electric Propulsion, REP) and nuclear reactors with ion engines (Nuclear Electric Propulsion, NEP) in order to reduce the durations of missions to the outer solar system. Building on these needs, Charles Elachi, the head of the Space and Earth Science Programs Directorate at JPL, proposed to NASA's Administrator Daniel Goldin a new program of technology oriented missions in the spirit of 'faster, cheaper, better' that would flight test and qualify 'mission enabling' technologies for use in future missions, whilst also making scientific observations as a bonus. As a result, in July 1995 Congress gave NASA the go-ahead to initiate the New Millennium program of short-development missions. Deep Space 1 (DS1) was to evaluate using ion propulsion for deep-space missions, and Deep Space 2 was to evaluate miniaturized planetary penetrators with Mars as the target. Deep Space 3, authorized later, was meant to fly three spacecraft in separate orbits around the Sun to demonstrate the ability to form an optical interferometer capable of resolving planets of other stars. Unfortunately, this ambitious idea, which would have required extremely precise control and knowledge of the relative positions of the spacecraft, never got off the ground.

Among the options considered for Deep Space 1 was a relatively conventional, 'small bodies' mission using chemical propulsion, with the innovation being the use of miniaturization to achieve a launch mass of only 100 kg. It soon became clear, however, that an enabling technology for a small-bodies mission would be the use of solar electric propulsion and Deep Space 1 became an ion-propelled comet and asteroid demonstration mission. Another technology to be tested would be an autonomous navigation system which would photograph the sky, identify the targets against the background of stars, and use their relative positions and motions to compute its own location in the solar system, plan its path and even decide the necessary trajectory corrections. A dramatically shortened development phase of just 36 months was decided early on in order to demonstrate JPL's ability to manage a fast-paced mission, with launch scheduled for July 1998. In 1995 a small company in Arizona, Spectrum Astro, was chosen as an industrial partner. Other experimental technologies were also chosen for demonstration – including a miniature integrated camera and spectrometer based on a study of miniaturized instruments for the Pluto Fast Flyby mission proposal, a 'remote agent' that would effectively operate the spacecraft when not in contact with Earth, and an advanced solar panel fitted with small lenses to concentrate sunlight and thereby boost the power output of each cell. These technologies were chosen on the basis of their becoming available within 2 years and either enabling new kinds of mission or reducing mission costs. The technologies accepted for flight consisted of those that were *essential* to the mission in the sense of there being no conventional alternative; those that were *fundamental* to the mission but could be replaced during development with a conventional alternative; and those that were merely *enhancing* and could be discarded without impact on the mission. An example of a fundamental technology was an advanced 3D-processor that was replaced by a standard processor when it became evident that the advanced system would not be ready in time.[1] The total cost of the mission was $138.5 million, including the launcher but excluding flight operations and technologies which had been developed by separate programs or other agencies – most notably the ion propulsion system and solar concentrators. Including operations and grants for the analysis of scientific data, the total cost exceeded $150 million.

Deep Space 1 had a box shaped body 1.1 × 1.1 × 1.5 meters in size, to which were affixed a pair of gimbaled solar panels, each comprising four 113 × 160 cm sections, that increased the total span to 11.8 meters. On the base was the ion engine, with external booms providing access to batteries and fuel lines after the spacecraft was mated to the launcher. Most of the instruments, antennas etc were on the other end of the box. The high-gain antenna was a spare 'hockey-puck' from Mars Pathfinder, and was the most visible heritage from the Mars program. Attitude determination was by a star tracker, a solar sensor and an inertial platform with laser gyroscopes. As part of the rationale of the low-cost technology demonstration, it was decided *not* to provide redundancy for most of the conventional technologies. Hence, attitude control was to be achieved entirely by hydrazine thrusters, although control around two axes would still be possible by directing the thrust of the ion engine. The launch mass was just 486.3 kg, including 81.5 kg of xenon and 31.1 kg of hydrazine for attitude control and course corrections. It was estimated that to accomplish the same mission using just standard technologies and components would have required a spacecraft mass of 1,300 kg.

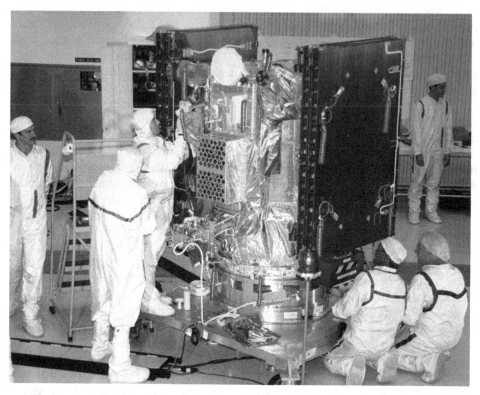

The ion-propelled Deep Space 1 probe undergoing final preparations before launch.

The ion thruster utilized by Deep Space 1 had been developed jointly by JPL and the propulsion-oriented Lewis Research Center (now the Glenn Research Center) for the NASA Solar Electric Propulsion Technology Application Readiness (NSTAR) program that was initiated in 1993 specifically to make this technology available for space exploration. In contrast to earlier experimental ion thrusters which had used mercury or cesium as fuel, NSTAR used xenon because this was gaseous at room temperature, rather than liquid or solid, and its exhaust would not contaminate the spacecraft. The 30-cm engine accelerated xenon, ionized by electron bombardment, through negatively charged molybdenum grids with an exit velocity of 40 km/s to provide a low but sustainable thrust. At 'full throttle', the 92 mN of thrust was equivalent only to the weight of a typical sheet of paper on Earth. The weakest of 112 thrust levels was just 19 mN. After being ejected, the ion beam would be neutralized by injecting into it a stream of electrons in order to prevent electrically charging the rest of the spacecraft. Although its thrust was weak, the speed of the ion 'plume' was very high and the total amount of thrust that could be obtained from a given mass of fuel (i.e. its specific impulse) was ten times greater than that of a chemical engine. It was this that made ion propulsion more efficient and more suitable to applications such as a tour of the main asteroid belt that involved large total velocity changes. Despite the minuscule thrust, Doppler tracking by antennas

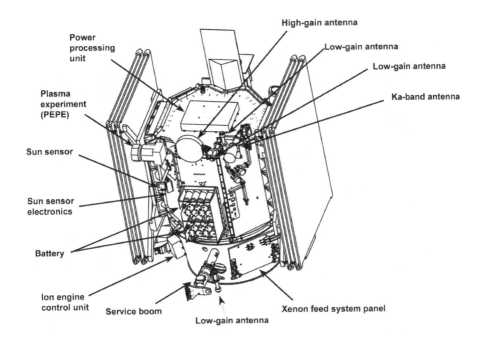

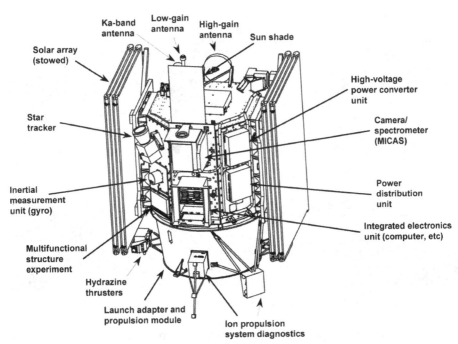

Deep Space 1 (solar arrays are not deployed).

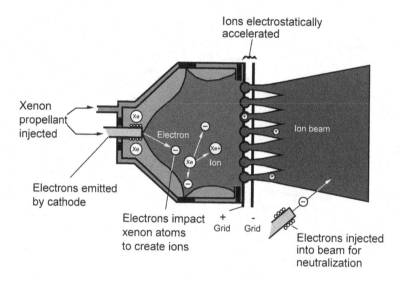

Ions electrostatically
accelerated

Xenon
propellant
injected

Electron

Ion beam

Electrons emitted
by cathode

Electrons impact
xenon atoms
to create ions

+
Grid

-
Grid

Xe+
Ion

Electrons injected
into beam for
neutralization

The working principle of an ion engine.

The NSTAR ion engine mounted on Deep Space 1.

on Earth would be sufficiently sensitive to measure the spacecraft's acceleration. The engine could be gimbaled in a cone of 5 degrees radius to obtain thrust vector control. In fact, when the engine was on, the hydrazine thrusters were never used for attitude control. Extensive tests were carried out in which a prototype demonstrated, indeed exceeded, its design life of 8,000 hours at full power. As Deep Space 1 was being developed, in 1997 a Russian commercial Proton launched the PanAmSat 5 telecommunications satellite built by Hughes, which had an experimental xenon ion engine for station-keeping in geostationary orbit.

The second important technology to be tested by this mission used 720 Fresnel lenses to concentrate sunlight on 3,600 solar cells for about 20 per cent more power than a conventional array. Deep Space 1 required 2,400 W, mainly to operate the ion engine. Since solar concentrators would enable a given power to be provided by a sparser coverage of cells, less radiation protection would be required and the panels would be less massive. Solar concentrators were a technology under study by the BMDO (Ballistic Missile Defense Organization). They were to have been tested by the METEOR satellite, but its launch on a privately developed Conestoga launch vehicle dramatically failed in October 1995. The military therefore provided them to JPL free of charge in exchange for in-flight engineering data.

A number of experimental scientific instruments were also part of the payload, the goal being to test them realistically in the space environment. A multi-wavelength imager/spectrometer that would feed pictures to the automatic navigation software integrated two cameras and shared optics with an infrared and an ultraviolet imaging spectrometer. The optical system had an aperture of 100 mm, and the visible channel had a focal length of 677 mm. One of the cameras had an unfiltered 1,024 × 1,024-pixel CCD sensor with a field of view of 0.78 × 0.69 degrees, and the other incorporated an experimental 256 × 256 CMOS (Complementary Metal–Oxide Semiconductor) active pixel sensor. The whole structure of the instrument was made from silicon carbide and, as was the case for the optics, had no moving parts. The low thermal expansion coefficient of silicon carbide enabled the instrument to be assembled and focused at room temperature and then remain focused even in the coldness of space.[2] The Plasma Experiment for Planetary Exploration (PEPE) electrostatic integrated ion and electron spectrometer was to measure the solar wind and assess the plasma environment created around the spacecraft by the ion thrusters. It had many of the same functions of an analogous instrument on Cassini, but at 5.6 kg was a fraction of the mass. Its delivery was so late that 6 weeks before launch it was still uncertain whether it would fly. Whilst the cameras and plasma instrument were mounted on the opposite end of the spacecraft from the ion engine, diagnostic packages were set less than a meter from the engine in order to monitor its operation. These included a retarding potential analyzer and a pair of Langmuir probes for measuring the plasma environment. Contamination was monitored by two quartz microbalances and two calorimeters. Electromagnetic noise was measured by a 2-meter-long dipole antenna and a search-coil magnetometer. The magnetic field itself was measured by a pair of 'extra small' German fluxgate magnetometers that were prototypes of an instrument intended for ESA's Rosetta comet mission. They were installed on a 50-cm boom to position them close to the ion plume they were to diagnose.

The autonomous navigation system could steer Deep Space 1 by changing the thrust and/or direction of the ion engine, or by making specific firings of it or the hydrazine thrusters. Orbital data for 250 asteroids were stored on board, together with a baseline trajectory for the spacecraft and ephemerides of all the major planets of the solar system in order to be able to determine the position of the spacecraft at any time. The 'remote agent' would plan and manage many activities with minimal intervention from the ground, including deciding when or whether to perform some actions and monitoring the response of the other systems. The navigation system and remote agent software were to be implemented by a non-redundant main computer. A novel small deep-space transponder was to transmit to Earth a simple 'tone' that would inform controllers whether the spacecraft considered itself to be healthy or in need of assistance. The use of such a 'semaphore' promised enormous savings on operating costs of future missions. Moreover, receiving tones would require smaller dish antennas on Earth, only several meters across. Other technologies included a small solid-state amplifier to evaluate transmissions in the microwave Ka-Band, low power electronics, a smart power switch and a structure that combined load-bearing elements with thermal control, heat transmission and electric circuitry.[3,4]

But there were substantial hardware and software development problems which pushed the launch back by several months. One of these concerned the remote agent software, which the engineering team finally decided to do without at launch and to upload in-flight. The development time of other software was greatly reduced by adapting it from Mars Pathfinder. However, the biggest issue was the hardware, and in particular the high-voltage converter that was to feed low voltage systems off the high-voltage ion engine's power line. Its development was marked by the death of its principal designer who, as was customary for 'faster, cheaper, better' missions, had made very little written documentation of his work, with the result that its delivery was a year behind schedule. In March 1998 the team acknowledged that they would not finish their system tests in time for the planned July launch, and it was slipped to October. While this gave engineers some breathing room, it also forced a switch in the flyby objectives.[5]

As the purpose of the Deep Space 1 mission was an engineering demonstration of the ability to use ion propulsion to perform encounters, the specific targets were a secondary concern. The original plan was for the spacecraft to depart Earth during a month-long launch period running through the whole of July 1998, start thrusting in August and continue until late October. It would then encounter asteroid (3352) McAuliffe in mid-January 1999 at a relative speed of 6.7 km/s and go on to periodic comet 76P/West–Kohoutek–Ikemura in June 2000 at a relative speed of 15 km/s.[6] A flyby of Mars was to be performed in April 2000, with a possible close encounter of Phobos. McAuliffe was one of the asteroids named after the seven astronauts who perished in the Challenger accident. It was a near-Earth object believed to be about 2.5 km across. West–Kohoutek–Ikemura was a Jupiter-family comet with a period of about 6 years that was discovered in 1974. Its nucleus was estimated to be about 3.8 km across. Calculations indicated that it passed just 0.01 AU from Jupiter in March 1972, changing its orbit from one with a period of 30 years and a perihelion distance of 5 AU to the present one.[7] The 3-month delay in the launch date required the selection of new targets. As the primary target the unnamed near-Earth asteroid 1992KD was selected from a list of over 100 candidates. This had been discovered in May 1992 by JPL astronomers and asteroid hunters Eleanor Helin and Kenneth

Lawrence. Orbiting the Sun in a plane inclined at 28 degrees to the ecliptic, it would yield a flyby at a relative speed of 15.5 km/s. Although classified as an Amor object orbiting between Mars and Earth with its aphelion in the asteroid belt, simulations showed its orbit to be rapidly evolving, and that within a few thousand years it will become an Earth-crosser. Observations by some of the largest telescopes on Earth indicated it to be no larger than 3 km in its longest dimension, elongated by a ratio of at least two in its axes, and to have a relatively long rotation period of 9.4 days. Early spectral data suggested it was one of the rare V-class of asteroids similar to Vesta, and as such probably chipped off Vesta by an impact. Such asteroids are also believed to be the sources of the rare eucrite and diogenite basaltic meteorites. But follow-up spectra showed a similarity with the rare Q-class of asteroids, and to be almost indistinguishable from certain types of chondritic meteorites. Remarkably, whilst Q-class asteroids are well represented among the near-Earth objects, only a few have been identified in the main belt. The fact that 1992KD was probably a chip broken off a larger object and that its unstable orbit would soon cause it to cross that of our planet made the encounter of great scientific interest because it offered data and details of the mechanism for the transport of asteroids and debris from the main belt to the inner solar system.[8,9] To raise public interest in the mission, the Planetary Society held a competition to name 1992KD, with the discoverers making the final choice. Hence 1992KD became (9969) Braille, after Louis Braille, the inventor of the raised dot language designed to enable the blind to read by touch.

Although the small all-solid Athena 2 had been one of the possible launchers for a lightweight Deep Space 1, it was switched to the Delta 7326 when the focus of the mission became the ion propulsion system. With three strap-on solid boosters, this was the least powerful form of the Delta II. It would be the first launch of NASA's Med-Lite program. The launch period ran from 15 October to 10 November 1998. The closing of this window was not dictated by celestial mechanics, but simply by the need to clear the pad at Cape Canaveral for the first of two Mars Surveyor missions due for launch in December. It lifted off on 24 October. After releasing the third stage into parking orbit, the second stage of the launch vehicle maneuvered to release a small communication and remote-sensing satellite, SEDSat 1 (Students for the Exploration and Development of Space Satellite). Meanwhile, at the appropriate point the third stage fired to inject Deep Space 1 into a solar orbit ranging between 1.0 and 1.3 AU.[10] Some of the key technologies were to be tested during the first 2 months of the flight, and some already on the first day.

Hardware issues developed immediately after injection, with the spacecraft being heard of many minutes later than expected as a result of a Van Allen radiation belt-induced problem with the 'off-the-shelf' star tracker used to determine the attitude of the spacecraft. Finally, the Deep Space Network antenna at Canberra made contact, in the process confirming the correct working of two of the technologies being evaluated: the solar array concentrators and the deep-space transponder. When controllers switched on the ion engine on 10 November for what was to be a 17-hour test, it shut down after just 4.5 minutes and remained in standby. All efforts to restart it were unsuccessful. Next, the star tracker misbehaved again and put the spacecraft into safe mode. After control was regained, it was decided to subject the engine to thermal expansion and contraction cycles by exposing it to the Sun and to space. It was thought that the firing had been curtailed by an electrical short caused by debris in the molybdenum grids, as had been observed on other ion engines. It was hoped that thermal cycling would dislodge the debris. While the team was busy working to restart the engine, it had to omit the planned calibration and characterization of the camera responses by observing the Earth and Moon as the spacecraft departed, and this lost opportunity would reduce the readiness of the instrument to observe Braille.[11] Meanwhile, on 16 November the Palomar Observatory succeeded in imaging the spacecraft at a range of 3.7 million km. After 2 weeks of recovery, new software was uploaded and the engine was restarted. The fact that it worked faultlessly thereafter enabled the team to declare on 2 December that the 'minimum success criteria' for the mission had been achieved. The first 10 days were spent thrusting along the Earth-spacecraft line in order to maximize the Doppler shift of its transmissions and enable the engine's thrust to be precisely measured. Then, the spacecraft was turned to the optimal attitude for thrusting to intercept Braille.

Calibration images were taken in the visible and infrared range, with the camera optics still protected by a cover that was opaque at ultraviolet wavelengths and was to be released later in the mission. Unfortunately, when the cover was opened it was discovered that the ultraviolet channel had some malfunction in the signal chain that rendered its data meaningless. The greatest issue, however, was that the visible and infrared channels were particularly prone to stray light scattered off the illuminated portions of the vehicle. Because this included the sensors required by the navigation system, this delayed comprehensive tests. Meanwhile, in January 1999 Cassini and Deep Space 1 were within 0.5 AU of each other, and 36 hours of solar wind data was obtained in a cooperation between the two missions funded at opposite ends of JPL's cost spectrum.

In a 40-minute exercise on 22 January the ion engine was stepped through several thrust levels, ending at the highest level yet achieved, and was then turned off. This exercise provided some of the best data on plasma interactions with the spacecraft, the plasma instrument on the far end of the spacecraft conclusively recording xenon ions only at the highest thrust. In any case, solar wind protons could still clearly be distinguished in the spectra, and their distribution appeared to change little when the ion engine was operating.[12] This (and later) tests showed that an ion engine could be used without significantly impairing particles and fields observations. Furthermore, the flux of xenon ions returning to the plasma instrument allowed the charging of the spacecraft to be measured. Another test was whether transmissions were affected by the radio signal passing through the 'plume' of the engine; they were not.

In February, Deep Space 1 received a 4.1-Mbyte upload of completely new flight software which included navigation software revised to deal with the scattered-light problem. The entire software was replaced in May to include the remote agent and again in June to restore the full functionality of the navigation system and enhance the capability of its image-processing. Tests of the autonomous navigation software after the upgrade were extremely successful, with the system obtaining about one asteroid image per week, determining its position and then implementing corrections to the engine's throttle and thrust direction. Comparison of the position derived onboard with that calculated from ground tracking showed a difference no greater than 1,000 km, which at this phase of the mission was excellent. The time remaining until the encounter was devoted to further tests and to resolving some issues. High-data-rate contacts were held on a weekly basis, with thrusting paused and new navigation pictures taken. Briefer contacts were also often held between these longer sessions. When thrusting concluded on 27 April, Deep Space 1 was on a ballistic trajectory to encounter Braille. A total of 1,800 hours of thrusting over 34 firings had consumed 11.4 kg of xenon and delivered a total velocity change of 699.6 m/s. The engine's performance was established to be within 2 per cent of that predicted.[13] On three occasions in May, the infrared spectrometer took spectra of Mars at ranges varying between 105 and 110 million km. Spectral features were found that could have been due to previously undetected minerals exposed at the surface.[14]

In addition to using the multispectral camera, Deep Space 1 was to study Braille using the plasma instrument to seek any disturbance in the solar wind that might show the asteroid to possess an intrinsic magnetic field. Moreover, as the solar wind impinged on the surface of the asteroid it could free charged particles which would provide hints as to its composition. The plan was to subject the spacecraft to a very stringent navigational test by having it close to within 10 km of the asteroid – and possibly even as near as 5 km – to make the closest interplanetary encounter to-date. The spacecraft was traveling in the plane of the ecliptic, and the flyby would occur at the ascending node of the asteroid's inclined orbit, 1.3 AU from the Sun. In addition to taking navigation images to refine the position of the asteroid, during the approach the camera was to take spectra and pictures for scientific purposes. At the time of closest approach, however, the angular rate would exceed the maximum slewing rate of the attitude control system, and at that time only the plasma instrument would be able to collect data. The best images were therefore expected to have a resolution of 30 to 50 meters. About an hour after closest approach, the spacecraft was to turn to face Earth and start to transmit its data.

Starting 30 days before the encounter, Deep Space 1 collected navigation images and the autonomous system performed trajectory corrections using the ion thruster. A fortnight before the event, a full rehearsal was conducted in which the spacecraft's software successfully tracked a 'virtual' target. One issue that surfaced early on was that Braille was much fainter than expected, and so failed to show in the navigation images. Nevertheless, it was thought the camera would spot the target about a day before the flyby, which would provide time to make the final adjustments. However, a number of problems marred the encounter, and it proved to be a disappointment for the scientists.

When Braille was finally spotted, some 40 hours prior to the encounter, it was so faint that it was discernible only after the images were enhanced by processing on Earth. It was found to be about 430 km from its predicted position, and a navigation software simulator at JPL was used to compute the trajectory correction because the additional image processing required to identify it was not onboard. This was performed several hours later. In the last 2 days of the approach, corrections were made using the hydrazine thrusters rather than the ion engine in order to save time. With 17 hours to go, the onboard software located the asteroid using its own image-processing routines. But an hour later the navigation system put the spacecraft into safe mode, causing the computer to lose the data it was processing. A correction that had been scheduled for 5 hours later was also lost. As the team was coaxing Deep Space 1 out of safe mode, three navigation images that had been taken prior to the glitch were downlinked and used by JPL to design a correction that was executed a mere 6 hours prior to the flyby. As 16 images had originally been intended for this final correction, the orbit determination was inferior and the encounter distance was greater than desired. The camera tracked the target as the asteroid loomed, and 27 minutes before the encounter it correctly switched over from the CCD to the CMOS sensors. When it had been discovered months prior to launch that the CCD would have trouble in tracking an illuminated extended target, it was decided to use the CMOS for the last minutes of the flyby requiring a rapid change in the autonomous navigation software. Unfortunately, since the asteroid was much fainter than expected, the system was unable to locate the target in any of these 23 images. Moreover, something bright (probably a cosmic ray that hit the detector) distracted the navigation processing and caused it to aim the camera away from the target. As a result, Deep Space 1 was not pointing toward Braille when it started the scientific imaging sequence. The spacecraft reached the minimum distance of 26 km over the night-side of Braille at 04:46 UTC on 29 July 1999. Shortly afterwards, it resumed tracking Braille with the CCD using the last position determination available.[15] Circumstances had therefore conspired to make Deep Space 1 fail to obtain detailed pictures of the surface of the smallest asteroid yet encountered. When the final pictures of the approach phase were taken some 70 minutes before the encounter, Braille was 40,000 km away and covered a mere four pixels. Then, owing to the limited onboard memory, distant images were not stored in order to leave room for the expected high-resolution ones. As a result, almost all of the observations of the asteroid were lost. When the camera imaged it again, 914 seconds after the flyby, Braille was 14,000 km away and only hints of its shape were discernible. In the two best images, obtained 18 seconds apart, the half-illuminated asteroid appeared to be fairly elongated, 2.2 km on the longest axis and 1 km on the shortest. This raised the possibility of it consisting of several bodies in contact. The CMOS images taken at the same time showed a smudge barely distinguishable from the sky background.[16,17,18] A dozen good infrared spectra were obtained 17 minutes after closest approach and showed a general similarity to Q-class asteroids, with the presence of both pyroxene and olivine. Since the spectral features of these minerals seem to fade as the surface of an asteroid is exposed to the space environment, this suggested that perhaps the surface of Braille was relatively young and unweathered. Photometry showed Braille to be a relatively bright body since it reflected more than one-third of the light that it received from the Sun, although its complex shape made it almost disappear during the approach phase.[19]

Although the data from the magnetometers installed to monitor the ion plume of the engine was severely affected by the frequent activation of magnetic actuators in the attitude control thrusters, by the switching on and off of onboard apparatus, and by a large residual field caused by permanent magnets in the engine, an extensive analysis revealed a suspicious 'bump' in the magnetic field occurring at exactly the

This is one of the best images of asteroid (9969) Braille taken by Deep Space 1's CCD camera about 15 minutes after the close flyby.

time of closest approach. This may correspond to a small but measurable magnetic dipole moment intrinsic to the tiny asteroid.[20]

A review board at JPL investigated the less-than-nominal encounter and came to the conclusion that the problems were probably unavoidable, being simply a case of theory not conforming to the 'real world'. However, team members admitted that the encounter had not been as thoroughly planned as it would have been had this been a required part of the mission, their intention having been to meet the criteria that defined a technical demonstration of success. The flyby had been prepared only several weeks before the event, the rehearsal had been quite late, and obtaining scientific data had been a secondary consideration. In fact, the scientific data from the Braille encounter had not even been documented or archived in the same manner as for a 'scientific' mission because funding for this was never made available.[21]

The Deep Space 1 mission was nominally to end in mid-September, but engineers and scientists hoped that NASA would grant an extension. One possible target was 107P/Wilson–Harrington, with an encounter in January 2001 at a relative speed of 15.8 km/s. It was first seen as a faint comet in 1949, but when recovered in 1979 it had ceased to show cometary characteristics and was catalogued as (4015) Wilson–Harrington. The nucleus was estimated to be about 4 km across. It could be either a normal comet which has somehow switched off its activity, or a dormant comet that suffers occasional outbursts.[22,23] As Deep Space 1 would be near solar conjunction at the time of the encounter, making communications from Earth difficult, extensive use of automatic navigation would be needed.

Another possibility was an encounter with 19P/Borrelly within a few days of the comet's perihelion passage in September 2001. It was a fairly typical comet in the

Jupiter family, with an orbital period oscillating around 6.9 years. It was discovered by Alphonse Louis Nicolas Borrelly from Marseille in December 1904, and with the exception of 1939 and 1945 it had been seen at every return. Its orbit was quite well defined, and experiencing only distant encounters with Jupiter it "evolves smoothly and experiences insignificant changes" over the centuries. It was a moderately active and relatively carbon-poor comet. Observations by terrestrial telescopes and by the Hubble Space Telescope in 1994 indicated a well-defined coma and short tail, with an elongated nucleus about 8.8 km long and 3.6 km wide which rotated every 25 hours or so. Borrelly was also well known for its asymmetric coma, which was more elongated toward the Sun. Its orbit had a relatively high inclination of 30 degrees and the perihelion was between the orbits of Mars and Earth.[24,25]

Within 36 hours of leaving Braille behind, Deep Space 1 had resumed thrusting to keep these options open. On 18 September the primary mission was completed with all of the technology demonstration objectives successfully achieved. NASA funded a 2-year extended mission that eventually cost about $9.6 million. This shifted the focus from technology demonstration to science, and called for encounters with both Wilson–Harrington and Borrelly. On 20 October the spacecraft ceased thrusting and spent several months coasting. At this point, the ion engine had been in operation for a total of 3,571 hours, delivered a total velocity change of 1.32 km/s and had used 21.6 kg of xenon.

In late October and early November pictures were taken of Mars from a range of 55 million km and the infrared spectrometer obtained a total of 48 spectra covering two full axial rotations of the planet. When the plasma instrument was powered up on 1 November it suffered an internal discharge that limited its performance. Before any data could be returned, disaster struck on 11 November when the ailing star tracker failed. As there was no backup system, the spacecraft, denied 3-axis attitude control, entered safe mode, turned to face its solar panels to the Sun, and initiated a once-per-hour roll on this axis whilst it awaited instructions from Earth, which was now 1.6 AU away. The plasma instrument was automatically powered off by the diagnostic software, and it could not be recovered by simply powering it back on. It is not known if there was any connection between the short circuit that crippled this instrument and the failure of the tracker, but there is no evidence to indicate such a connection. With Deep Space 1 fixed in solar-inertial attitude, the high-gain antenna was not pointing at Earth. In the absence of a high-data-rate link, there was no way to receive full telemetry and ascertain whether the tracker could be resuscitated. Nor was it clear what could be done to recover the mission without using the tracker. As the primary mission had been completed, serious consideration was given to shutting down the spacecraft. But in mid-January 2000 a nutation was added to its rotation so that the coning action would intermittently cause the beam of the high-gain antenna to sweep across Earth, whence the strength of the signal received was used to infer the spacecraft's orientation and the rotation was stopped. Then a difficult process of controlling the attitude from Earth was begun in an effort to maximize the high-gain antenna data rate. At this point, the full telemetry stored onboard since the tracker failure was downloaded, along with the images of Mars and other data which had been taken in November – these spectra were some of the best yet obtained of

Mars in this particular infrared range, and appeared to confirm the probable presence of previously undetected minerals.[26]

Engineers were then able to exploit the spacecraft's unique architecture to regain control. Whereas previously the scientific instruments and engineering devices such as attitude control sensors were separate, Deep Space 1 was able to feed its science camera imagery to the main computer for processing by the autonomous navigation system. The reduced engineering team therefore decided to try to implement a novel attitude control regime in which the orientation would be measured not by the wide-field-of-view star tracker but by the science camera, whose field of view was much narrower. It took more than 4 months to write the necessary software, during which the spacecraft had to be controlled from Earth through weekly sessions and frequent attitude maneuvers. However, the time spent in recovering the spacecraft from safe mode and implementing the software modification left insufficient time for thrusting to reach both Wilson–Harrington and Borrelly, and scientists found the latter to be a more interesting target. The software that was uploaded during 10 days in late May and early June was designed to use a suitably bright reference star for the engine thrusting attitude (referred to as the 'thrustar'). It would command the spacecraft to turn to take an image of this star in the science camera, process the image by removing background and scattered light, and then pass the positional data to the attitude control system to determine the orientation. This routine still required extensive ground intervention to confirm that the correct star was being tracked. Moreover, it was not possible to integrate the new software with the autonomous navigation system, with the result that the duration of thrusting, throttle settings, orientation etc had to be specified by sequences uploaded from the ground. The uploading of the 267 files required was concluded on 8 June, and the main computer was then rebooted to complete the process. The testing had to be carried out promptly, since less than a month remained before thrusting would have to resume to intercept Borrelly. Deep Space 1 acquired its first star on 12 June, enabling the mission to continue. It was one of the most complex and brilliant in-space recoveries of a robotic mission. Nine days later, the engine was turned on to ensure that it was still operational after months of inactivity, and to check the thrust control routines. The tests went so well that thrusting for Borrelly was resumed on 28 June, one week earlier than planned. The development of the new software had caused the original budget to overrun by about $800,000. The low-thrust trajectory to the comet had been redesigned to take into account the constraint that the thrusting attitude had to allow the thrustar selected at any particular time to fall within the narrow field of view of the science camera.

Deep Space 1 had plenty of xenon remaining to reach Borrelly, but a great deal of hydrazine had been used in maneuvering to re-establish the high-gain link after the star tracker failure. In fact, less than one-third of the original supply remained, and it was expected to be used before the Borrelly encounter. When the spacecraft was weighed prior to launch and a weight margin was found to exist, 4 kg of hydrazine had been added in topping up the tank as an extra margin, and this would turn out to be vital. Fortunately, when the ion thruster was operating the engine gimbal could direct a proportion of its thrust to 2-axis attitude control, to reduce hydrazine consumption. It was decided that when Deep Space 1 was thrusting it would use the maximum throttle level, and when it was not needed it would run at a low setting ('impulse power') alternately in a direction normal to the orbit and then against it, for a zero sum, whilst continuing to minimize hydrazine usage. The engine would therefore be operating almost all of the time, including during high-gain antenna communication sessions. As a result, in the summer of 2000 Deep Space 1 beat SERT 2's record for the longest-operational ion propulsion spacecraft.[27]

From late October to late November 2000, the spacecraft passed through solar conjunction. During this time, it continued to thrust at low power to derive attitude control and maintained its high-gain antenna more or less pointing at Earth in order to check whether communications were possible and to sound the solar corona with radio signals. Two communication sessions were successfully held, one of which, on 14 November, was when the spacecraft was very close to the Sun as viewed from Earth and verified that the ion engine had been operating at impulse power for most of the time. On 2 January 2001 the engine throttled up to maximum power. By early May the spacecraft had achieved the ballistic Borrelly intercept orbit, and throttled back to impulse power and continued to thrust in alternating directions to maintain attitude control.

Unlike the Braille encounter, the flyby of Borrelly was thoroughly prepared, with the specifically written software being uploaded in early March. In particular, after the camera had recognized the nucleus of the comet it was to center it in each image and both cope with the rapidly changing appearance of the nucleus during the flyby and ignore the distractions of jet activity, the diffuse glow of the coma, cosmic-ray strikes etc. The nucleus-tracking routines were exercised using Jupiter as a target, and two encounter rehearsals were conducted in May and June – although both were compromised by errors. Since Deep Space 1 had not been designed for a cometary flyby, it was not protected from dust and therefore its large solar panels were at risk of being damaged by particles in the coma. A flyby range of 2,000 km on the sunlit side of the nucleus was chosen as a compromise between a relatively large but safe distance and a closer pass for improved scientific results. Although 100 impacts with particles exceeding 40 micrometers in size could reasonably be expected, the chance of a disastrous impact was minimized by having the spacecraft travel most of the time with its solar panels edge-on so that any impact on the arrays would occur at a grazing angle.[28] The plasma wave antenna part of the engine diagnostic package was to be used to detect the bursts of plasma produced as dust specks were vaporized on striking the vehicle's body. A similar technique had been used by the International Cometary Explorer (ICE) and serendipitously on Voyager 2 as it crossed the rings of Saturn and later deliberately at Uranus and Neptune. The plasma instrument, which had been given software patches to maximize its data return, would monitor ions, electrons and magnetic fields. The integrated camera was to obtain relatively high-resolution images and infrared spectra of the nucleus and coma. Because the CMOS camera had proved too insensitive for faint targets, this time only the CCD would be used.

Several ground-based telescopes were to monitor Borrelly before and during Deep Space 1's encounter, as would the Hubble Space Telescope and the Swedish Odin astronomy satellite. Setting up a comet flyby was dependent on accurate knowledge of the position of the nucleus. The initial estimates were made by telescopic studies. On 25 August Deep Space 1 began its own imaging search, although spotting it from a range of 40.3 million km required extensive image processing on Earth. A total of 11 imaging sessions were held until 10 hours before the encounter on 22 September, by which time the nucleus was 600,000 km away. A lot of hydrazine was consumed in switching the spacecraft's attitude from that for thrusting to that for imaging and then to point the high-gain antenna at Earth, hence only a limited number of images were obtained. Corrections were performed by the ion engine. In one particular case, on 11 September, this was successfully achieved despite the interruption of work at JPL owing to the terrorist attacks on New York and Washington. Two days later the fifth imaging session was held, and after the sixth session two days after that the ion engine was switched off. The flight team then faced the problem of exactly where to target Deep Space 1, since images of the closing comet showed two bright centers in the coma, about 1,000 km apart. In particular, to pass 2,000 km from the brighter one would cause it to pass dangerously close to the fainter one if that were to prove to be the nucleus; whilst to target the fainter one would cause it to pass too far from the brighter one to achieve the planned scientific observations if that were to prove to be the nucleus. Worryingly, if the two centers were due to the nucleus having just split, then the spacecraft would very likely fly into the cloud of debris and dust produced by the fragmentation. In the end, it was decided to stick to the plan. Another issue was the unknown brightness of the nucleus. Pictures were taken about 11 hours out and promptly relayed to Earth to enable a rough determination, so that the exposure times for the flyby imaging could be better specified. Three different encounter sequences were uploaded to the spacecraft to cope with the unknown brightness of the nucleus, but the default settings were found to be probably the best. Despite its complexity and the many unknowns, this time the spacecraft's activity during the encounter was faultless.

Scientific observations started 12 hours before closest approach, using the plasma instrument and the engine diagnostic sensors. Cometary ions that had been 'picked up' by the solar wind were first detected at 588,000 km inbound, as Deep Space 1 adopted the optimal attitude for such observations. The bow shock in the solar wind was encountered at a range of 152,000 km, some 2.5 hours before closest approach. Scientific imaging began 83 minutes before the flyby. This also yielded a sighting of the nucleus to refine the image tracking. Next were infrared spectra of the coma, but stray light contamination rendered these useless. Observations were interrupted from time to time to use the camera to measure the attitude of the spacecraft, the last such check being only 35 minutes before closest approach. The first images showed a narrow jet of dust several kilometers wide and 100 km long, within 30 degrees of the direction of the Sun. Starting 32 minutes before closest approach, two images of the nucleus were taken every minute for tracking and science, but owing to the limited data storage only the pixels around the nucleus (which was located automatically) of a few of these images were retained. In fact, a total of 52 images were collected over a period of 90 minutes, detailing the coma, nuclear jets and finally the nucleus itself. To ensure that the nucleus would be captured in the event of the automatic tracking losing the target, two wide mosaics were taken of the sky around the spacecraft.

The moment of closest approach was 22:30 UTC on 22 September 2001, with the spacecraft passing over the sunlit side of the nucleus at a relative speed of 16.6 km/s. The comet was just 8 days after perihelion and near the ascending node of its orbit.[29] The range of 2,171 km was slightly greater than planned owing to uncertainty in the nucleus ephemeris, but still perfectly suitable for science. No outbound images were taken, primarily because to view back along its velocity vector Deep Space 1 would have had to perform a rapid rotation which would have consumed a lot of hydrazine. Some 30 minutes after closest approach, it made a more efficient turn to point its high-gain antenna at Earth and restarted the ion engine for stabilization whilst it sent its encounter data. An outbound shock was crossed at 96,000 km from the nucleus, but it was less distinct than that inbound.

Although coming 15 years after the Halley flybys, this was only the second time that a cometary nucleus was imaged. Extensive processing was needed to eliminate motion blurring. The best image was taken 170 seconds before closest approach, at a distance of 3,556 km, and yielded a resolution of just 47 meters. This facilitated the first geologic assessment of the surface of a comet. The dark nucleus was seen to be the shape of a bowling pin, some 8 km in length and 3.2 km in width, with a number of different terrains and features. Its 3-dimensional shape was inferred from pairs of high-resolution pictures obtained at ranges between 4,400 and 3,500 km. It could be broadly divided into a smooth and a mottled terrain. There were many circular dark cavities 200 to 300 meters across, but their characteristics did not seem compatible with impacts; they were probably sublimation or collapse features. The images of Halley's nucleus provided by Giotto had been fuzzier owing to the denser coma, but in that case too there was only one tentative impact crater. It seemed that impacts on cometary nuclei were rapidly eroded. Slightly brighter smooth terrain characterized the central region of Borrelly's nucleus, with darker flat mesas possibly marking the sources of the jets. Certainly the source of the narrow collimated jet of dust appeared to be in this region. Scattered pits, hills, ridges, streaks and bumps were evident that might have been created by the sublimation of ice. Such features might be the sites of previous activity. Consistent with telescopic estimates, only about 10 per cent of the surface appeared to be currently active. Overall, the nucleus was extremely dark, reflecting on average just 3 per cent of the light that it received from the Sun, with small patches reflecting as little as 1 per cent – about as dark as the powdered toner used in printers and copiers, making this some of the darkest material yet observed in the solar system. Such a dark body would absorb heat easily and release volatile molecules, leaving behind a 'lag' that would make it even darker.

Images of Borrelly indicated that whilst the process that dominates the surface of asteroids is bombardment, the nuclei of comets (or at least members of the Jupiter family) are geologically active bodies shaped by sublimation and by activity during their frequent perihelion passages. The smaller end of Borrelly was canted at a small angle toward the spacecraft, with parallel fractures marking the neck. This could indicate that the nucleus was actually a pair of objects that collided at a low relative speed and stuck together. This may have occurred whilst in the Kuiper Belt, before the orbit of the comet was perturbed by Jupiter's strong gravitational field.

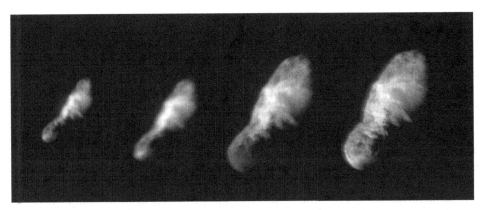

The Borrelly approach sequence. Images of the nucleus were taken respectively (left to right) at: 8,858 km, 6,616 km, 4,387 km, 3,556 km. The last is also Borrelly's highest resolution image.

A processed image of Borrelly showing jets of dust emerging from the nucleus. The horseshoe-shaped 'loop' is also visible just beside the 'neck' between the two components of the nucleus.

Long-duration exposures revealed two sets of jets. The primary one was that seen in the days leading up to the encounter, and the other, which was offset at an angle to the first, comprised several smaller jets each of which was a few kilometers long and several hundred meters wide. Their sources seemed to be dark features in the smooth terrain. A sunward-aimed fan seemed to originate from the small end of the bowling pin. The direction of the main jet did not seem to change. The fact that this remained fixed for at least 34 hours, which was much longer than the rotation period of the nucleus, indicated that its source was near the pole. In fact, the nucleus was rotating along its minor axis, which was pointing more or less sunward at perihelion. What would have happened to Deep Space 1 if its trajectory on the illuminated side of the nucleus had taken it through the jet is open to speculation. (Interestingly, telescopic observations from Earth showed that when the primary jet shut down as the Sun set over its source around the end of 2001, a weaker jet became active on the opposite hemisphere.) A mysterious feature seen during the encounter was named 'the loop'. It was not evident whether this was a feature of the nucleus or of the coma. It first appeared some 9 minutes before closest approach as a bright spot near the night-side of the neck of the bowling pin and then rapidly changed in shape, becoming a fuzzy saber shape before reverting to a spot. It is possible that it was an active jet that was viewed from a head-on perspective.

After the imaging run, by now 157 seconds before closest approach and at a range of 2,910 km, two long exposures were started with the field of view of the infrared spectrometer sweeping around the predicted position of the nucleus in the hope of obtaining spectra of it. This time luck was forthcoming and the very first exposure provided no fewer than 46 spectra of 165-meter-wide swaths along the length of the nucleus. No evidence of water ice was detected, but an absorption band which was present in all the spectra could have been due to a hydrocarbon mix – although the precise composition was never identified and this particular band was not evident in the spectra of any other solar system body. The spectral results were consistent with the temperature measurements which showed the surface to be between about 30°C near the terminator and 70°C near the subsolar point.

The peak magnetic field was detected by the engine diagnostic sensor when 5,000 km off from the Sun-nucleus axis. This probably marked the cometopause boundary where the solar wind plasma was slowed on encountering the comet and piled up on its sunward side. One interesting plasma observation was that every coma plasma feature and boundary encountered, both inbound and outbound, was asymmetric with respect to sunward and with respect to the closest approach distance, corresponding to a displacement 1,500 km northwest of the nucleus. This asymmetry had not been observed for any other comet, and perhaps the large, constant, northward tilted and collimated jet seen erupting from the nucleus had something to do with it. Abundant water-related ions were found, with the concentration reaching its peak about 1,500 km before closest approach, by which time they constituted more than 90 per cent of the total detected ions. In fact, the gas production rate was estimated to be about four times greater than for comet Grigg–Skjellerup, which was visited by Giotto, or the same order of magnitude as that measured by the ICE spacecraft upon encountering comet Giacobini–Zinner. A total of 17 impacts were recorded by the plasma-wave antenna during four separate 0.5-second sampling periods. A number of other slower 'events' were recorded that could either have been larger particles or strikes on the other end of the vehicle.[30,31,32,33,34,35,36,37]

On 8 October Deep Space 1 initiated a 2-month 'hyper-extended mission' during which some of the tests made during the primary mission were repeated in order to assess degradation of the systems after 3 years in space, and the ion engine was put through tests at xenon flows and electrical power levels that could have jeopardized earlier phases of the mission. Plasma data, obtained only intermittently in previous years owing to the short circuit in November 1999, was now taken continuously for a full 2 months. Meanwhile, options for ending the mission were being considered, including the possibility of an additional asteroid flyby about a year later, but it was determined that the remaining hydrazine was sufficient for only about 2 months of operations, and xenon for only 3 months even running at impulse power. Another possibility was to hand the spacecraft over to the US Air Force to assist in training students in deep-space operations, but, true to the 'faster, cheaper, better' style, the team had not generated the documentation that would have been needed to make this feasible, if at all. Finally, on 18 December 2001 Deep Space 1 was commanded to turn off its transmitter, thereby effectively ending the mission and leaving the spacecraft in a 1.22 × 1.46-AU orbit extensively modified from its original insertion orbit as a result of 16,265 hours (or 678 days) of ion thruster operation over an interval of 3 years. A total of 73.4 kg of xenon had been consumed, providing a total velocity change of 4.3 km/s. The engine had ignited correctly 199 of 200 times, with the only anomaly being its initial test firing. But this was not the end of the story, as engineers at JPL again tried to contact the spacecraft at opposition in March 2002 to put the Ka-Band transmitter through some tests. The schedule reflected the fact that in the attitude in which the spacecraft had been left, at opposition Earth would be near the high-gain antenna beam. Attempts were made on 2 and 6 March, but there was no response. It is likely that while it was in hibernation the attitude control system had expended the remaining hydrazine, after which it would have drifted until the solar panels were no longer pointing at the Sun.[38]

Despite all of the technical problems and the risks the engineers took to overcome them, Deep Space 1 was a complete success as a technology demonstration mission. It pioneered many of the techniques to be used by subsequent missions, in particular by the coma sample-return Stardust. Dust density profiles in a cometary coma were better calibrated, and nucleus tracking software was tested and certified in use. The autonomous navigation and the nucleus recognition and targeting routines would be exploited by Deep Impact, a mission to study the effects of smashing a projectile into a cometary nucleus. Other deep-space missions, beginning with the 2001 Mars Odyssey orbiter, would use the small transponder tested by Deep Space 1. And then, of course, the Dawn main belt asteroid orbiter would use solar electric propulsion on a mission to rendezvous with first Vesta and then Ceres in the main belt by flying a profile that would have been prohibitive if reliant on chemical propulsion.[39]

Of course, NASA was not alone in devising technology demonstration missions to facilitate future planetary and scientific missions. At about the same time as JPL was readying Deep Space 1 to fly, the European Space Agency started its Small Missions for Advanced Research in Technology (SMART) program, the aim of which was to bring balance and flexibility into the agency's science plans. As one of the European cornerstone missions envisaged an ion propulsion Mercury orbiter, the first SMART mission (as in the case of NASA's New Millennium program) would be dedicated to flight testing electric propulsion. It would use xenon as its fuel, but a different type of engine to the NSTAR gridded ion thruster engine. The Hall thruster used a radial magnetic field and a circular current to accelerate the whole plasma, not just positive ions, to high speed. In principle, this technology could yield larger and more powerful engines.

From the beginning, SMART 1 was designed as a very small spacecraft of about 350 kg in mass which could either hitch a ride on a commercial Ariane 5 launcher and be deposited in a geostationary transfer orbit with an apogee of 36,000 km, from which it would boost, or ride a former-Soviet ballistic missile converted for use as a space launcher and fly a direct ascent to escape velocity. Mission profiles included placing SMART 1 into a lunar polar orbit, or flying by or rendezvousing with near-Earth asteroids and/or comets. But then it was realized that no low-speed asteroid rendezvous would be feasible by leaving from a geostationary transfer orbit. In fact, flybys would be possible only using very high performance engines and would occur at relatively high speed and after flights of at least 3 years. Nevertheless, possible targets included three small near-Earth asteroids and comets Tempel 2 or Haneda–Campos. A direct entry into heliocentric orbit offered more interesting options that included a rendezvous with Orpheus and the 'killer asteroid' (35396) 1997XF11. This latter object caused a sensation in 1998 when it was computed that in October 2028 it would pass within 45,000 km of Earth and the orbital uncertainties allowed the possibility of an impact which, because the asteroid was 1 or 2 km in size, would cause a global catastrophe. But precise computations of the possible positions with respect to Earth in 2028 showed that there was a vanishingly small probability of its passage being less than 30,000 km. When the orbit was refined using pre-discovery observations dating back to 1990, this not only ruled out an impact but also shifted the point of closest approach to almost 1 million km.[40,41] In the end, it was decided to operate SMART 1 in lunar orbit, with the possibility of later redirecting it to an asteroid. The spacecraft hitched a ride with a commercial payload, was released into a highly elliptical orbit in September 2003 and used its ion engine to spiral out to the Moon for its scientific mission.

## GOING WILD

In 1995, even before NEAR and Mars Pathfinder were launched, NASA selected the third mission of the Discovery program. Lunar Prospector, as it was called, would be the agency's first lunar probe since Explorer 49 in 1973, and was to conduct detailed geological and mineralogical mapping of the Moon. With a budget of less than $60 million, it would be a real bargain. At the same time, three options were selected for further study as the fourth mission. In fact, two of the three candidates were already fairly well-defined when the program was initiated in 1993. These were the Venus Multi-Probe Mission (VMPM) and the Suess–Urey solar wind sample-return.[42] The third mission was a new proposal by JPL, the University of Washington in Seattle, and Martin Marietta (soon to become Lockheed Martin Astronautics). Building upon a decade of studies which included the Halley Earth Return (HER), the Planetary Observer-style Comet Intercept and Sample Return (CISR), Giotto 2, the European CAESAR and the US–Japanese Sample Of Comet Coma Earth Return (SOCCER), JPL proposed the Stardust mission to return cometary coma material to Earth. The Principal Investigator would be Donald Brownlee of the University of Washington, credited with the discovery of 'Brownlee particles' – the small, fluffy specks of dust which were collected by aircraft researching the stratosphere and believed to be the remnants of cometary dust swept by our planet.[43]

Like its predecessors, Stardust would fly through the coma of a Jupiter-family comet to expose a sample collector made of 'underdense' material, namely aerogel, which is a silicon dioxide-based solid foam made mostly of vacuum and as such the lowest-density solid known. Invented in the 1930s, it found space applications in the 1980s and 1990s, in particular, as a result of its extremely low thermal conductivity as a thermal insulator. Upon striking such a collector, dust particles, gases and other volatiles that had just been released into the coma by the nucleus would be almost instantaneously brought to a halt from a speed of several kilometers per second, yet be preserved intact owing to minimal heating of the aerogel. On being stored in a sample-return capsule the collector would later be returned to Earth, where it would be possible to analyze the captured material in laboratories. This mission was to provide a look at cometary material that was believed to have remained more or less unchanged since the formation of the solar system, and might even contain some unprocessed material from the pre-solar cloud out of which the Sun and planets condensed. After the Ulysses mission discovered in 1992 the presence of streams of very fine interstellar dust entering the inner solar system, Stardust acquired a new objective, namely the collection of some of these grains. To tell the interstellar dust from the cometary dust a double-faced collector was to be used, with one face for each sample. This was a good example of a 'rapid response' in which the Discovery program could study a phenomenon within years of it becoming known. Recovering interstellar dust would advance this topic from the domain of astronomy – where it was observed by its extinction, scattering, polarization and infrared emission – to a laboratory analysis of its chemistry and mineralogy. In particular, scientists wished to test the hypothesis that an anomalous extinction 'bump' implied the presence of graphite in interstellar dust.

Unlike earlier studies, which had focused on developing sample-return techniques and then cast around for a suitable cometary target, the Stardust researchers knew from the start that they wished to visit 81P/Wild 2, a Jupiter-class comet which had a period of 6.4 years and a nucleus estimated to be about 4 km across. The history of Wild 2 was particularly interesting, and explains why it simply *had* to be the target. On 10 September 1974 the comet passed within 900,000 km of Jupiter, which was inside the orbit of the largest moon Ganymede. Prior to this, its orbit had a period of 42 years, was inclined at 19 degrees to the ecliptic and, in ranging between about 5 and 19 AU, traveled between the orbits of Jupiter and Uranus.* According to a recent statistical study, its orbit had remained more or less stable for at least 8,000 years, although simulations showed a possibility that until a few thousand years ago it was a long-period comet. The Jupiter encounter reduced the inclination of the comet's orbit to about 3 degrees and gave it a perihelion just beyond the orbit of Mars. It was discovered in this orbit in January 1978 by Paul Wild (pronounced 'Vilt') of the Astronomical Institute of Berne, Switzerland. With some uncertainty due to the unpredictable cometary rocket effects, forward projection shows that it should remain in the inner solar system for several millennia, with a relatively high possibility of it becoming an Earth-crosser for a few hundred years before being perturbed back beyond Jupiter.[44,45] The 2003 return, which was the one for the Stardust encounter, would be only Wild 2's fifth perihelion as a short-period comet.

---

* A different value of the aphelion distance is usually cited, but that was based on the backward evolution of an orbit derived from only two apparitions of the comet and hence had large inherent uncertainties. The value given here is based on an orbit derived from five apparitions.

(It would be a relatively poor apparition for a terrestrial observer, as it would remain close to the Sun in the sky.) The hope of the Stardust team was that, having undergone little or no sublimation until recent times, Wild 2 would still have an essentially pristine surface. Remote sensing of the surface of the nucleus and an analysis of the samples from the coma could provide unique data on the processes it underwent in forming and evolving in the Kuiper Belt.

In astronautical terms the parameters of Wild 2's current orbit, and in particular its inclination, were relatively benign for arranging an encounter. In fact, it was this aspect of the comet which had led to its being considered as a possible target for previous cometary proposals, including the 1980s CRAF mission. For Stardust, the astrodynamicists had designed a remarkable orbit that made the opportunity simply too good to miss. It was to be launched in February 1999, make an Earth flyby in January 2001 to boost its aphelion, and fly past Wild 2 in January 2004 at a range of about 150 km, some 99 days after the comet's perihelion in September 2003. The spacecraft's 2.5-year orbit would allow it to deliver its sample to Earth in January 2006. The trajectory was a compromise between an energy requirement at departure small enough for the spacecraft to be launched by a Delta II, and a slow encounter speed in order not to compromise the sample as it was captured. In fact, the use of orbits resonant with that of Earth (such as planned for the CONTOUR mission, to be described later) would have allowed an even smaller launch energy at the cost of a greater encounter speed. For example, SOCCER could have ridden a smaller rocket and returned a sample from Wild 2 in 5 years instead of 7, but at an encounter speed over 30 per cent greater. The plan selected for Stardust would actually produce the slowest cometary flyby yet. Another constraint in designing the trajectory was the requirement to encounter the comet when it was near perihelion to stimulate activity, yet sufficiently far from perihelion that the coma would not be so dense as to put the spacecraft at risk or to obscure the nucleus from view. Additionally, the collection of interstellar dust had to be timed for when the spacecraft was sufficiently far from the Sun, and at a point in its orbit when it would be moving in the same direction as the interstellar dust passing through the solar system, so that the particles traveling at a speed of about 26 km/s would strike the collector at a relative speed of 10–20 km/s. Finally, in comparison to earlier cometary sample-return proposals, which relied on the Space Shuttle for retrieval, making them unnecessarily complex, Stardust was designed from the start to deliver its samples by releasing a capsule that would make a direct entry into the Earth's atmosphere.

Given its scientific significance, NASA selected Stardust as the fourth Discovery mission in November 1995 to demonstrate comet sample-return technologies. The spacecraft was based on Lockheed's 'Space Probe' bus designed to support a variety of deep-space missions. Its main body was shaped roughly like a telephone booth, 1.7 meters in length with a 66 × 66-cm cross-section. The side walls were all carbon fiber with aluminum honeycomb. Two 4.8-meter-long solar panels with a total area of 6.6 m$^2$ supplied between 170 and 800 W of power depending on the distance from the Sun. At the encounter, which was to occur 1.86 AU from the Sun, they would provide at least 300 W. The original intention was to use gimbals with one degree of freedom, but it was later decided to fix the panels parallel to the main axis to give the spacecraft an 'H' configuration. A 15-W transmitter and a 0.6-meter-diameter body-fixed parabolic antenna allowed a maximum data rate of 22 kbps, although at encounter the rate would be only about 7.9 kbps. In addition, there was a medium-gain antenna and three low-gain antennas for near-Earth phases of the flight. It had a 128-Mbyte memory for housekeeping and management/storage of scientific data. In fact, the data acquired by the instruments during the encounter would not be able to be returned in real-time, as at that time the high-gain antenna would not be pointing at Earth. Redundant star trackers, sun sensors, inertial platforms and accelerometers would provide attitude determination. Two strings of four main ultra-pure hydrazine monopropellant thrusters rated at 4.4 N were to be used to adjust the trajectory, and four smaller 0.9-N thrusters were to provide stabilization and orientation control. To avoid contamination, the thrusters were mounted in four pods on the side opposite to both the solar arrays and the door of the sample capsule. Although this configuration balanced the torques on the spacecraft, when the thrusters were fired the unbalanced forces slightly perturbed its orbit and these effects had to be taken into account for precise navigation.

During the flyby, Stardust was to maintain its main axis within 2 degrees of the direction from which the dust would arrive in order to minimise the area in need of protection by a Whipple shield. This consisted of a front composite 'bumper' shield, three panels of ceramic cloth spaced 5 cm apart, and a composite 'catcher shield' which doubled as the end panel of the body's primary structure and was to dissipate

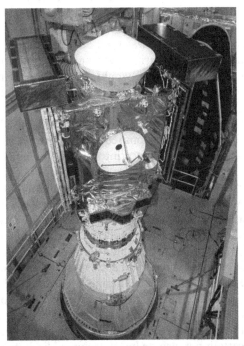

Stardust about to be closed inside the protective fairing of the Delta 2 launcher. The white object at top is the heat shield of the sample return capsule.

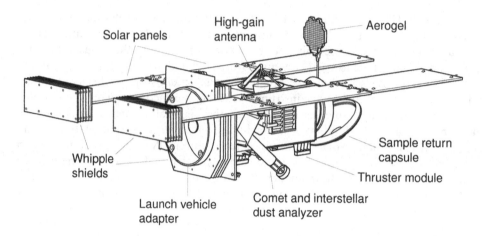

The Stardust spacecraft as it would appear during sample collection.

and spread debris. The circular interface to the launch vehicle was at the center of the bumper shield. Overall, the shielding was designed to halt a 1-gram pebble impacting at 6 km/s. The solar panels were also protected, but with only two cloth layers.[46] The 0.5-meter-tall sample-return capsule was at the rear at encounter in order to minimize dust hits that could damage its heat shield. It comprised a front 120-degree cone-sphere and a rear truncated cone with a maximum diameter of 0.81 meter. The main heat shield was a novel phenolic impregnated carbon ablator (PICA) and the rear was a lightweight material which was used on Viking, Mars Pathfinder and the Space Shuttle External Tank. Extensive tests and computational fluid dynamics simulations ensured that the capsule would be stable during the hypersonic, supersonic and transonic to subsonic flight regimes that would occur during re-entry.[47]

The racquet-shaped sample collector was on a boom that used wrist and shoulder joints to provide the 2 degrees of freedom required to extract it from the aluminum canister in the capsule and raise it to face into the impinging stream. (This scheme had been adopted instead of the original idea for a square collector to slide out in the manner of a drawer from a cabinet.) The collector's aluminum frame comprised two sides, referred to as 'A' and 'B'. Each side contained 130 blocks of aerogel, each 2 × 4 cm, and two smaller trapezoidal blocks for a total exposure area of 1,039 cm$^2$. The blocks were 3 cm thick on the 'A' side to collect cometary material, and 1 cm thick on the 'B' side for interstellar dust. Aluminum foil was wrapped around the aerogel cells both to ease their removal and also to act as a secondary collector material for small particles. Three chips of different materials were mounted on the deployment arm to safeguard against contamination of the collectors and to monitor the space environment. Initially, the capture of cometary particles in space at speeds of several kilometers per second was a poorly understood process, in part because the nature of such material was uncertain but also owing to technical limitations on testing. But it was thought that a particle should be halted within a few microseconds of striking an aerogel block, penetrating it several millimeters, with the energy melting the aerogel to trap the particle. The high-velocity capture of particles in space using aerogel was successfully demonstrated by an experiment on Shuttle mission STS-47 in 1992.[48] The sample collected by Stardust was to be only 1 milligram, comprising particles ranging from 1 to 100 micrometers in size. The minimum criteria for success was to obtain at least 1,000 cometary particles of 15 micrometers or larger, and at least 100 interstellar particles. After a preliminary analysis by mission scientists, the samples would be made available to the scientific community around the world. A portion of the samples were to be stored for analysis one decade or so later using techniques which were not available at the time of the mission. The Stardust spacecraft carried two chips etched with the names of people who had expressed their support for the mission: one, bearing 136,000 names, was in the sample capsule and the other, with more than a million names, was mounted on the racquet arm.

On returning to Earth, the capsule would enter the atmosphere at very high speed and would finally land within a 10 × 33 km ellipse in the desert US Air Force Utah Test and Training Range (UTTR). Compared to the last time the US had recovered spacecraft from deep space – namely when the US Navy deployed large numbers of ships for the manned lunar Apollo missions – this was a smaller target, with a much smaller recovery team. Whilst on the one hand this meant that the cost of recovery could be minimized by making use of helicopters and off-road vehicles, on the other hand it imposed tighter constraints on return navigation and the re-entry trajectory. The parachute system included a deployment mortar that would be fired by a gravity switch and a timer, a supersonic drogue chute that would be opened at an altitude of about 30 km by which time aerodynamic drag would have slowed the capsule to Mach 1.4, and the main 8.2-meter-diameter canopy to be deployed at about 3 km to slow the final descent. Also in the capsule was a radio beacon to aid in locating it, and batteries with sufficient power to ensure many hours of operation after landing. The return of cometary material was deemed to pose no risk to life on Earth – after all, cometary matter and meteorites rain down continuously – so no particular sterilization or contamination protection requirement was imposed.

Compared to the sample collectors, all of the other scientific instruments carried by Stardust had a secondary role. A camera would provide both optical navigation and scientific images of the comet's nucleus. The camera combined f/3.5 wide-angle Voyager-heritage optics with a sensor head from Galileo, a space-qualified 1,024 × 1,024 CCD from Cassini, electronics from Deep Space 1, and an 8-position Voyager filter wheel for color imaging. A periscope would be used in taking pictures of the comet in order to protect the main optics. A flat mirror was mounted in front of the camera to deflect the light 90 degrees into the optics. In addition, the mirror could be rotated 220 degrees along the axis of the camera in order to track the cometary target at closest approach, when its relative motion would be greatest. In the final minutes of the approach phase, the nucleus would also be able to be tracked by rotating the spacecraft.[49] A German mass spectrometer dust analyzer based on the design flown in 1986 on Giotto and the Vegas was to provide measurements of the composition of dust that entered the instrument. The analyzer had been modified to be able to detect and analyze interstellar dust, and to accommodate the much slower flyby than in the case of Giotto. At the Halley flyby speed of in excess of 70 km/s, the instrument had been able to analyze only atoms, since even complex molecules were 'atomized' on coming into contact with the detector, but the fact that the Wild 2 flyby would be 11 times slower meant that it would be able to detect and analyze molecules. Moreover, by using very similar instruments for Halley and Wild 2, the data obtained could be more readily compared. A dust flux monitor was based on the only US instrument carried by the Vegas, and also from Cassini's high-rate dust detector. This used two polarized plastic film sensors that issued an electrical pulse whenever a particle as small as $10^{-11}$ grams penetrated them, plus a pair of quartz acoustic sensors mounted on the Whipple shielding to detect the largest specks of dust – in order to trigger the second sensor a particle would first have to pass through the bumper shield. This monitor would collect data on the number, mass and frequency of dust particles in the coma. It would also be capable of detecting interstellar dust hits and the crossing of meteor streams, including the stream dispersed along the orbit of Halley, which the spacecraft would cross 2 months after launch. In fact, the dust flux monitor was an addition to the payload, mainly to provide a means of assessing the health of the spacecraft. Although the mass spectrometer dust analyzer and the dust flux monitor were to operate at every available opportunity to obtain data on interplanetary and interstellar dust, they would have to be switched off around aphelion when the solar panels supplied the least power. Finally, radio tracking would yield an estimate of

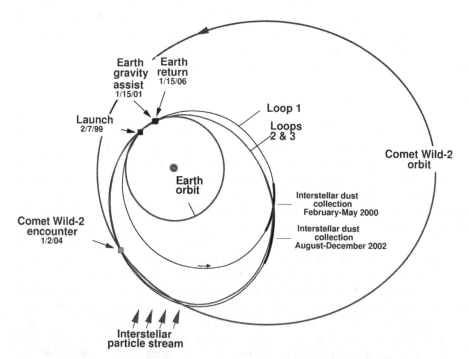

An overview of Stardust's three loops around the Sun. Midway during the second interstellar sample collection, in November 2002, it would fly by asteroid Annefrank.

the mass of the cometary nucleus by how it perturbed the spacecraft's orbit, and the manner in which the vehicle was slowed by impacts would indicate the density of the coma.

The total mass of the Stardust spacecraft at launch was just 385 kg, including the 45.7-kg sample-return capsule and 85 kg of propellant. The total cost of the mission was to be $168.4 million, in addition to $45 million for the launch vehicle.[50,51,52,53]

Stardust was delivered to Cape Canaveral in November 1998 for integration with its launch vehicle, which was to be a Delta II with three solid boosters and an escape stage. The 20-day launch window for Wild 2 opened on 6 February 1999, but if this could not be achieved a second opportunity existed between April and June with an encounter prior to perihelion instead of after it. In fact, Stardust was launched on 7 February, having been delayed the previous day by a beacon telemetry problem that occurred with only 1 minute remaining on the clock.

Remarkably, after more than 30 years of developing mission scenarios, this was NASA's first true cometary mission to be launched! It was also the first deep-space sample-return since 1976 (the previous one being the Soviet Union's Luna 24), the first ever to go beyond the Moon, and the first US robotic sample-return mission. It was a great moment for sample-return missions, because as Stardust left Earth the cometary nucleus sample-return Deep Space 4 mission was in the design phase and NASA was studying missions to return samples of Mars's moons, of Mars itself and of the solar wind; both NASA and ESA were studying returning samples of Mercury and Venus; and Japan was in the process of preparing a near-Earth asteroid sample-return mission for launch.[54]

Signals from Stardust to indicate that it had separated from the escape stage as scheduled, halted its inherited spin, deployed its solar panels and faced them towards the Sun were received a mere 51 minutes after liftoff. It was in a 0.99 × 2.20-AU orbit with period of 2 years. One day after launch, it flew by the Moon at a range of 53,100 km. The launcher's accuracy was so good that the option of a burn to 'clean up' the inherited trajectory was canceled, as was the midcourse correction option in October. The dust analyzer and dust flux monitor were powered on soon after launch to collect some early data. Initial plans called for the camera to image the Moon for calibration in late February, but it was not turned on until the next month, at which time it suffered a series of software problems that delayed its full activation. In early May the sample-return capsule was unlatched and slightly opened for components to outgas freely in vacuum. It would generally be maintained in this position, but the lid was to be closed during engine firings in order to preclude the attitude control logic having to accommodate a flexible appendage. Calibration pictures of the bright star Vega were taken through every filter and at two different exposure times in October and transmitted to Earth the next month. These showed that a contaminant, probably gas issued just after launch, had coated the cold optics of the camera and was scattering light and blurring the image. It was decided to try to use the CCD and mirror motor heaters to clear the contaminant off the lens. If it remained, the blurring would not impair optical navigation until close to the target, but image processing would be required to obtain clear images of the comet's nucleus – as had been necessary when NEAR's camera was heavily contaminated following the missed Eros rendezvous.[55] Another issue was that a few months into the mission the dust flux monitor began to suffer thermal problems. After confirming that it could operate normally for at least the 30 minutes or so that would be required for the closest encounter phase, it was switched off. This left only the mass spectrometer dust analyzer operating during the cruise.

In January 2000, near aphelion in the main asteroid belt, Stardust made a 159-m/s deep-space maneuver in three parts between the 18th and 20th. This, together with a 'throat-clearing' burn in late December, set up the Earth flyby for 2001. Soon after making the burn, the spacecraft went through solar conjunction. On 22 February the collector was deployed for the first time, operating the wrist actuator to position its 'B' side almost perpendicular to the interstellar dust stream. Every few weeks as the flight progressed, the motor would change the angle by several degrees. It remained exposed until 1 May, for a total of 69 days, and was then stowed back in the capsule. Meanwhile, some early data on the chemical analysis of five specks of interstellar dust reported by the mass spectrometer dust analyzer during the first year in space was published. There were long periods when the attitude of the spacecraft could be selected so that only interstellar dust would intercept the narrow field of view of the analyzer. Being almost entirely a mix of complex organic molecules, possibly highly polymerized, these particles resembled "tar-like substances". No minerals, graphite, or any other form of carbon (diamond or amorphous) were detected. Nor were there the expected very small particles that ought to have been able to penetrate the solar system unhindered.[56] Due to the limitation on power around aphelion, the cruise was passive, with few if any observations being made. In November 2000, the Sun, near the maximum of its activity cycle, produced showers of high-energy particles in one of the largest storms of recent times. The energetic protons striking the detectors of the star trackers produced a number of false star detection alarms. This put Stardust into safe mode, requiring ground intervention to restore it. Meanwhile, progress was made in clearing the camera lens by switching on the CCD and mirror motor heaters and orienting the spacecraft so that the Sun illuminated the camera radiator for half an hour or so, substantially improving the situation.

Stardust's first orbit of the Sun concluded with the Earth flyby that would reshape its trajectory. A rehearsal of the navigation procedure for its eventual return to Earth was performed as it approached the night-side at about 6.5 km/s. The closest point of approach on 15 January 2001 was 6,008 km above a spot just southeast of Africa's southern coast. It was imaged by professional and amateur astronomers alike from California, Hawaii, Australia, Hungary and Mexico. In theory the flyby could have been used to gather new data on the 'flyby anomaly' that affected both Galileo and NEAR, but frequent use of the thrusters to control the spacecraft's orientation made the trajectory data too 'noisy' for such a precise analysis.[57] The flyby had the effect of increasing Stardust's orbit to 0.99 × 2.72 AU and the period from 2 to 2.5 years to ensure a return to Earth exactly 5 years later in order to deliver its precious load. The inclination of its orbit was also increased to 3.6 degrees, to almost match that of its target.

Some 17 hours after the flyby, the camera took 23 calibration images of the north

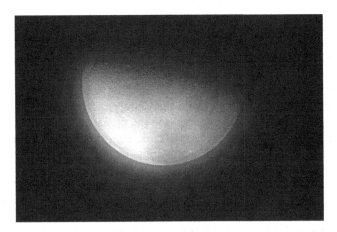

A processed Moon image taken by Stardust during its Earth flyby. Note that despite the processing, a halo of scattered light is still present.

pole of the Moon from a range of 108,000 km. Whilst these low-resolution pictures had an extensive halo of scattered light around the bright target, they represented a considerable improvement in quality and proof of the strategy of heating the optics to clear the contamination off the lens. Weeks later, just when the problem seemed to be solved, the camera was hit by another problem. The filter wheel appeared to be stuck at its last position. Fortunately, it was at the clear filter. If the wheel remained stuck, it would rule out color imaging of the comet but as the nucleus was expected to be essentially black this was not a serious loss. An optical navigation rehearsal for the encounter was carried out in June when the spacecraft was at the same part of its orbit as it would be when it intercepted Wild 2 one revolution later.

Cruise activities were wound down as Stardust headed towards aphelion, with the only active instrument, the mass spectrometer dust analyzer, being turned off. The major events during this period were solar conjunction in late December 2001 and a 2.65-m/s deep-space maneuver on 18 January 2002. The 2.72-AU aphelion on 18 April set a new record for the heliocentric distance of a solar-powered spacecraft. At that distance, the energy in sunlight was only 14 per cent of that at 1 AU. In early August, Stardust was again in a suitable position to sample interstellar dust and on the 5th the 'B's side of the racquet was again exposed.

Early in the mission, an analysis of the orbits of in excess of 50,000 asteroids had shown that 1,300 would be within 0.1 AU of the spacecraft at one point or another. One month before the opportunity arose, NASA decided to exploit a relatively close flyby of asteroid (5535) Annefrank to make engineering tests on the way to Wild 2. This small asteroid, discovered on 23 March 1942 by Karl Reinmuth of Heidelberg, was dedicated in 1995 to Anne Frank, who wrote a poignant diary of her experience hiding in Amsterdam for 2 years during the Second World War and died in a Nazi concentration camp in 1945, aged only 15. Annefrank is a rather mundane main belt object with an orbital period of 3.3 years. The best fit to its 'light curve' suggested a rotation period of 16 hours. Spectra obtained in support of the encounter showed it to belong to the most common taxonomic type of S-class objects, resembling stony meteorites. The 7.4-km/s flyby would test sequences to be used when encountering Wild 2. In fact, the simultaneous working of the attitude control system and camera in conditions similar to the comet encounter had not yet been tested. Because orbital uncertainties were rather large and successful imaging at a closer range could not be guaranteed, and also to avoid any dust in the vicinity, a flyby range of 3,100 km was chosen; more than 10 times that intended for the comet.

The fact that Stardust approached Annefrank from the night-side complicated the task of the recognition and tracking software – in fact, even after the navigation images that were taken on the inbound leg 38, 32, 26 and 18 hours before closest approach were processed on the ground they failed to show the target! There was no data on how bright the night side of an asteroid would appear and so scientists had to make guesses. Both dust instruments were active, although, as expected owing to the range, neither detected anything. The window for imaging was restricted to no more than 30 minutes, since longer would require the spacecraft to turn away from the Sun and rely on its batteries. Although this flyby was strictly an engineering test, certain scientific data was expected.

Stardust came within 3,078 km of Annefrank on 2 November 2002. Over a period of 15 minutes it took 72 images at various phase angles. An additional 34 pictures were taken for camera-pointing purposes, but not saved for transmission to Earth. In the absence of optical navigation images, pointing was undertaken based on the best ephemerides for both Stardust and Annefrank. The first four pictures had the asteroid close to the edge of the frame, and then the pointing software adjusted the aim to place it nearer the center of the field. Owing to the need to give the tracking software a bright target, many of the pictures were saturated. Although the lens contamination caused substantial blurring, about 40 per cent of the total surface of the asteroid was imaged at a resolution of 185 meters at best. At 6.6 × 5.0 × 3.4 km, it was larger than expected. Its irregular, angular profile bore a resemblance to a triangular prism. The flat surfaces may have formed when it broke off a parent body, but another theory suggests they were caused by "abrasive" impacts by smaller objects.[58] The pointed end marked the longest axis. Several rounded excrescences could indicate smaller bodies that had been accreted, with a dark line tracing the contact. With the data available, shape reconstruction was difficult and it was not possible to identify the spin axis and rotational period. There were a number of craters 0.5 km in size. The surface was somewhat darker than predicted, which explained why Stardust could not locate it before the encounter.[59,60]

The sampling of interstellar particles resumed after this brief interruption, and continued to 9 December. At the end of its two sessions the 'B' side of the racquet had been exposed for a total of 195 days.

A 71-m/s maneuver was executed in two steps on 17 and 18 June 2003 to set up the Wild 2 encounter based on the latest ephemeris. It would be followed by a series of refinements in the days leading up to the encounter. However, an issue that was noted early in the mission was that the performance of the thrusters was difficult to predict, with the result that each correction had a significant uncertainty. Whilst this was acceptable for the cometary encounter, it would make it difficult to establish the

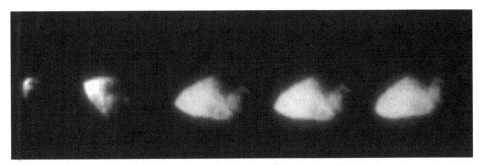

A sequence of images of main belt asteroid Annefrank taken at distances of 6,168, 3,848, 3,087, 3,079 and 3,143 km.

correct trajectory for returning the sample capsule to Earth. To better characterize the thrusters and the perturbations imparted by the attitude control system, Stardust performed three rehearsals of the entry maneuver in late June and early July, when it was about 1 AU from the Sun on a course matching the geometry of the actual event and using a 'phantom' Earth.[61] Following a course correction on 16 July, the dust analyzer was powered back on. It was to remain on through to the cometary flyby 6 months hence. Meanwhile, after telescopic studies of the dust density in the coma of Wild 2 made during the comet's 1997 perihelion were digested and validated by an international team of 'comet experts', the mission's navigation and science teams decided to increase the flyby distance from 150 to about 300 km due of concern that the inner coma would prove too dense to fly through safely.

The encounter period began 88 days before the flyby, and was to run through to 31 days afterwards. In the first phase, navigation images were to be taken twice per week to try to detect the comet as soon as possible and refine its ephemeris. The first such images were taken on 13 November 2003, just after the heaters were applied to dissipate the contaminant that had been building up again on the lens. Remarkably, despite still being 25 million km away, the comet was detected as a faint blob on the very first attempt. Its identity was confirmed by the second batch of pictures, several days later. But the optical navigation data was quite 'noisy' owing to the recurring issue of the lens contamination. Starting 2 weeks prior to the encounter, navigation imaging became a daily exercise. On 24 December the 'A' side of the racquet was deployed to start to sample cometary particles. Another heating cycle was applied to the camera to maximize the clarity of the observations. The comet was observed by some of the world's largest telescopes as it emerged from solar conjunction over the 3 weeks leading to the encounter. In addition to refining the comet's position, these observations provided thermal and photometric data and facilitated estimates of the dust and water production rates. The Stardust team then decided to reduce the flyby distance to 250 km. This was set up on 31 December, by the 12th course correction of the mission. The flight plan included two more inbound corrections, but excellent navigation allowed these to be canceled. On the same day, as Stardust was about to penetrate the coma, it turned to face the Whipple shields forward. "Just like in Star Trek", joked the program manager, "we have our shields up."

The close-encounter phase began 5 hours before the flyby, at which time Stardust was about 100,000 km from Wild 2. With about 30 minutes to go, the camera started to take pictures at 30-second intervals and a total of 72 were stored onboard over the next 38 minutes for later replay to Earth. These included long and short exposures to investigate the nucleus, coma and jets of dust and gas. The dust monitor was turned on 15 minutes before closest approach. About 6 minutes out, the spacecraft rolled to keep the nucleus in the camera's field of view, maintaining its Whipple shields and racquet exposed to cometary dust. During this phase, which lasted about 12 minutes, the spacecraft switched to its medium-gain antenna and issued only a carrier signal. Five minutes out, the navigation camera started to take a picture every 10 seconds to precisely track the nucleus. Although this activity continued for 10 minutes, most of these images were not stored in memory. Stardust passed the nucleus at 19:22 UTC on 2 January 2004 at a relative speed of 6.12 km/s. Its trajectory came from slightly above the comet's orbit plane and across the sunward side of its nucleus. On the way out its view was of the dark side of the nucleus and imaging was no longer feasible. About 5 hours after the flyby Stardust started the 30-minute process of stowing the racquet, ending with the lid of the capsule being sealed. The plan to carry out a final interstellar dust sampling run during the third orbit was canceled in order not to put the cometary sample at risk.

The data from the Wild 2 flyby was downloaded over the next few days using the high-gain antenna. The camera provided some very good images of the nucleus at a much finer resolution than for either Halley by Giotto or Borrelly by Deep Space 1. The nucleus proved to be 3.3 × 4.0 × 5.5 km in size, and probably rotating on the shortest axis. The fact that it seemed to be more or less round suggested that, unlike Borrelly, it was not an aggregation of smaller fragments. However, like Borrelly it appeared to have an albedo of only about 3 per cent. The most striking characteristic was the presence of circular depressions up to 2 km across, some of which had soft profiles while others were bounded by almost vertical cliffs hundreds of meters tall. There were also two oblong depressions that were nicknamed the left and right feet. The soft-profile cavities could be impact craters that formed in a porous and poorly cohesive material, with the energy of the strike releasing part of the volatile content. The flat-floored structures suggested impacts in a porous silicate which had greater cohesion. The evident absence of craters smaller than 500 meters could be because they would be readily erased by outgassing, sublimation and other cometary activity. In fact, porous, volatile-rich ejecta from a major impact would probably disintegrate into powder before it could produce secondary craters. There were small depressions with some of the characteristics of the flat-floored ones, but they were not circular in shape and might be the product of sublimation. A large basin informally named after Gene Shoemaker, the renowned planetary geologist who died in an accident in 1997, was remarkable for its pinnacles, spires, mesas and other fine structures. In fact, the most unexpected features were pinnacles and spires up to 100 meters high, some of which sported pointed ends. They were best seen along the terminator and limb. One possibility was that they marked ancient vents at which water vapor rich in volatiles emerged from the surface and turned to hard ice. Alternatively, they could simply be the remnants of small mesas that were eroded away on all sides. In any case, the fact that they had not slumped indicated that the surface had a significant strength. The rigidity of the nucleus was also implied by several zigzagging scarps of up to 2 km long. An intriguing possibility hinted at by the pictures was that the nucleus had a layered structure. The sharp features and absence of debris on the floors of craters indicated that there was no covering of fine regolith such as commonly observed on asteroids.

Although the nucleus was usually saturated in long exposures taken for tracking purposes, such images were well suited to the mapping of jets of dust and gas. There were no fewer than 20 active jets, two of which, as in the case of Halley, appeared to originate on the night-side. In fact, Wild 2 appeared to have the highest percentage of active surface of all the comets visited by spacecraft thus far. Remarkably, where jets could be traced back, most appeared to emerge from slopes that faced the Sun. The 1.2-km crater named Mayo contained a region from which dark lines originated that looked as if they were the source of one of the five active jets which could be traced to it. The main sources of activity on Borrelly had been mesas, but these structures were much smaller on Wild 2. Intriguingly there were a number of bright spots several hundred meters across that appeared to show no surface relief in stereoscopic imagery.

The mass spectrometer dust analyzer recorded only 29 spectra over 18 minutes. In fact, although the instrument had been designed to handle hundreds of impacts per

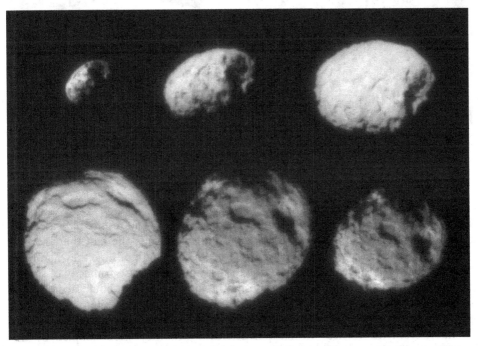

The Wild 2 encounter sequence. Notice how rugged the nucleus appeared, especially when compared to that of Borrelly. The last two pictures of the sequence clearly show the two large sinks nicknamed the 'left foot' and the 'right foot'.

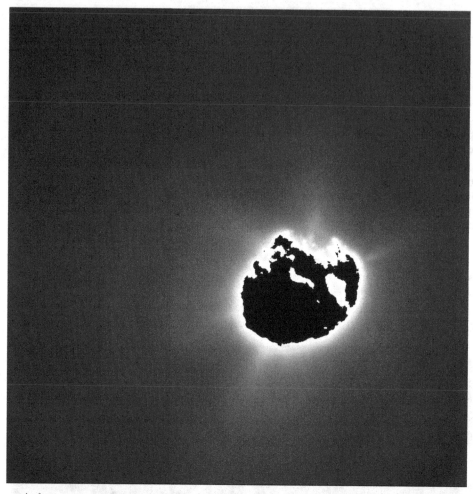

A long exposure around closest approach shows jets of gas emerging from the overexposed nucleus of Wild 2. Some of these jets would be crossed by the probe and recorded by its instruments.

second, it never recorded more than one hit per second. The results indicated the composition of Wild 2 to be very similar to Halley, despite the fact that the two objects had different ages and activities in terms of the length of time they had been making close approaches to the Sun. The dust seemed to consist primarily of simple organic compounds. Only one spectrum showed any feature that could arise from a mineral. No volatiles such as water and carbon monoxide were detected, but this was probably because these had been lost in the tens of minutes required for the particles to travel from the nucleus to the position at which they were sampled by the spacecraft. Neither did the spectra show any evidence of the complex organic molecules such as amino acids that some people believe to exist in comets. There were, however, hints of solar radiation-driven sulfur chemistry and a nitrogen-rich chemistry. The dust monitor sampled for more than 33 minutes centered on closest approach, and its results were surprising different to those of the dust analyzer. It reported a non-uniform distribution of impacts. A fast rise in the number of hits was recorded inbound, and then a decrease as the range from the comet increased again. Interestingly, this profile comprised a number of bursts and spikes, each of which lasted for at most several seconds, suggesting that the particles arrived in collimated bursts with up to 1,200 impacts per second and few if any impacts in between. This testified to the 'clumpiness' of dust in the coma. In fact, some of these bursts could be due to concentrated jets only a few hundred meters in width which had just been ejected by the nucleus. Remarkably, the times of the dust bursts matched the times calculated for intercepting some of the jets seen in the pictures. Seven dust particles even managed to pierce the front part of the Whipple shield. The largest of these was probably about 4 mm across – making it the size of a small pebble. From these rates, it was calculated that the 'A' side of the racquet should have captured about 3,000 particles 15 micrometers or larger in size.

Surprisingly, the impact rates reported by the dust monitor surged again about 700 seconds after closest approach. This extremely narrow and focused burst could not have been caused by jets coming all the way from the nucleus, which was then 4,000 km away, as in the hours since it was issued the fine dust would have dispersed into broad fans. Instead, scientists posited that the comet was releasing clumps of matter up to a meter or so in size which upon sublimating created their own jets, and that the spacecraft must have flown close to one of these clumps. Observations of splitting nuclei, large objects detected by radar in cometary comas, and other phenomena, all lent credence to this proposal. This would receive a dramatic demonstration in the 2010 Deep Impact flyby of comet Hartley 2. It was difficult to reconcile the data from the two dust instruments in terms of the coma itself. Perhaps the discrepancy was simply due to their different sensitivities, since the ratio of the impact count of one and that of the other remained essentially constant.

Later, the spacecraft team mused that if they had known how dense the coma was, they would have targeted a more distant pass. Although no damage was inflicted to the spacecraft's structure and solar arrays because these were shielded, there was a suspicion that the periscope of the camera had been significantly degraded by the extensive 'sandblasting'.

Doppler tracking of Stardust during the flyby not only provided an upper limit to the mass of the nucleus, but also estimates of the masses of particles whose impacts caused the attitude control system to take action to restore the vehicle's stability. For example, a particle that struck 15.5 seconds prior to closest approach was estimated to have a mass of between 20 and 80 milligrams.[62,63,64,65,66,67]

A small deep-space maneuver in early February directed Stardust back to Earth. The 2-year return flight was mostly eventless, with activity picking up only when it was nearing home. Course corrections were performed 60 and 10 days out to refine the aim point for the Earth encounter. The final refinement, made 29 hours and more than 700,000 km out, targeted a landing on the Utah Test and Training Range. This was accurate within a few kilometers and the velocity turned out to be correct within a few millimeters per second. Stardust crossed the orbit of the Moon some 16 hours prior to entry. Then, 12 hours later, the umbilical to the capsule was severed and a mechanism on the main spacecraft spun it up to 13.5 rpm for stability in free flight. Three spring-loaded bolts released it at 05:57 UTC on 15 January 2006. The images intended to be obtained to show the capsule breaking away were not taken. Minutes later, Stardust made a deflection burn to move off the trajectory that had been set up to cause the capsule to enter the atmosphere. It made a 258-km flyby to resume solar orbit.

At 09:57 the capsule hit the atmosphere at a point some 20 km east of the coast of northern California, traveling at 12.9 km/s. This was the highest speed ever flown by a spacecraft trying to re-enter the Earth's atmosphere. The very shallow trajectory caused the deceleration to peak at 33 g. Teams of scientists flying over Nevada took measurements of the heating, heat fluxes and luminosity of this artificial meteor in order to calibrate their models of the re-entry of natural meteoroids and other space debris. A winter blizzard was occurring during entry, with strong winds that pushed the capsule northward. As soon as the deceleration declined to 3 g, which occurred at the planned altitude of 32 km, a gravity switch triggered the deployment of the drogue chute that was to stabilize the capsule as it passed through a phase of aerodynamic instability. Video cameras recorded the descent from 35 km to the ground. The descent went completely as predicted, although the drogue parachute appeared to have opened a little later than planned, and the main one opened at about 3.4 km commanded by a backup pressure sensor. It was night over the Utah range, but metal foil strips on the parachute enabled radars to track the capsule to pinpoint its landing site. On touching down on damp soil some 8.1 km north-northwest of the aim point at 10:10, the capsule bounced several times, rolled and finally came to rest on its side.[68] Fifteen minutes later, helicopters homing on the capsule's radio beacon landed nearby and the recovery team found it to be in almost perfect condition. The entry should have been very nearly nominal, as the front heat shield showed the predicted symmetrical heating pattern and the aftbody was barely charred.[69] After the disaster of Genesis in September 2004 (of which more later) when the parachute failed and the capsule broke open upon striking the ground, the safe recovery of the Stardust capsule was a great encouragement for future planetary automatic sample-return missions. After assessing its overall status on-site, the capsule was placed into an environmentally controlled container and flown to the Johnson Space Center in Houston, where NASA maintains most of the Apollo lunar rocks and its collection of meteorites. When the sample canister was opened in a clean room on 17 January it was seen that many of the impacts on the aerogel, which were expected to be of microscopic size, where actually readily visible to the naked eye as small foggy spots.[70]

It turned out that the mission had collected more than 10,000 particles larger than 1 micrometer which were believed to be representative of the nonvolatile material of Wild 2. A picture was released several days after the recovery which showed the first sample to be removed, dubbed the 'quickstone', with a bifurcated track cored into the aerogel. This was a dramatic visual confirmation that samples had indeed been successfully collected. However, when removed from the aerogel block using an ultrasonic blade, the quickstone was shown to have created a 'pentafurcated' or five-spiked track, as if the particle itself had exploded on penetrating the collector. It

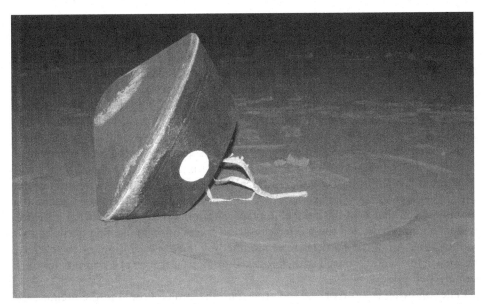

It came from outer space! Stardust's sample return capsule resting safely on the ground after its successful re-entry over Utah.

Bingo! Three captured particles are plainly visible in this single image as foggy spots in the aerogel bricks.

transpired that the impacts had produced deep root-shaped cavities in the aerogel. Whenever a particle had not fragmented, its track was in the shape of a carrot, many times longer than it was wide. In cases where a particle did explode, it produced a bulbous track with thinner, multiple spray-like spikes. Bulbous tracks were probably carved by 'traveling sand piles' – particles made of a mix of fine compact minerals embedded in a poorly cohesive low-density organic matrix that tended to evaporate explosively on striking the aerogel. Only the mineral grains survived more or less intact to form the spikes. Moreover, often the inner walls of the tracks were lined with molten aerogel mixed with cometary elements and beads of refractory minerals. A third, rarer, type of cavity consisted of a bulb without roots. The deepest track had penetrated 2.19 cm into the aerogel, which was a little more than two-thirds of the thickness of a cometary sample block. A number of craters had also been dug in the aluminum foil that held the aerogel in place within the racquet, the largest of which was 0.68 mm across, and these often preserved traces of the impacting particles as a molten slug. All of the particles were modified by their deceleration, often severely, and partly melted and mixed with aerogel. But grains larger than 1 micrometer were usually well preserved.

The following is more or less the procedure that was followed in the preliminary elemental assessment of the Stardust particles. After the tracks were cut out of the aerogel blocks as wedges called 'keystones', these were probed by focused beams of X-rays that caused fluorescence of some elements. The single tracks were then split open and the walls analyzed by ion mass spectrometry. The residues of the particles

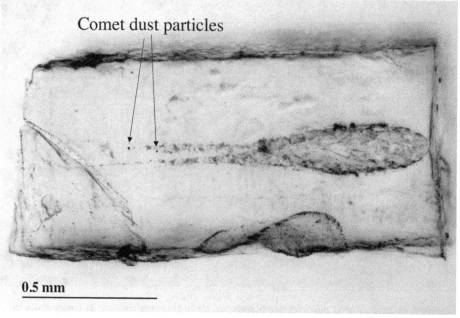

A typical bifurcated track lined with cometary particles after being cut from an aerogel block.

that had struck the aluminum foil were identified by a scanning electron microscope and then analyzed by X-rays and mass spectrometry. The elemental composition of the material recovered from Wild 2 mostly confirmed the results of the impact ion mass spectrometers operated by Giotto and the Vegas. However, whereas the Halley results were derived from less than a nanogram of dust, Stardust provided a larger sample and this was able to be analyzed using more advanced and less constrained techniques in state-of-the-art terrestrial laboratories. Different spots along the walls of a single track often differed in elemental composition, implying that the particle was an aggregate of grains of various compositions. Indeed, this was confirmed by mineralogical analyses. None of the particles analyzed in the preliminary scientific assessment consisted of a single mineral. They were aggregates of sub-micrometer-sized grains. The most abundant minerals in these grains were crystalline silicates such as olivine and pyroxene, as well as compounds of iron and sulfur like troilite. This was surprising, as silicates in interstellar particles were expected to be mostly in the non-crystalline, amorphous phase. However, it is possible that the amorphous material was destroyed, or at least disguised, by mixing into the silica glass that was created in the aerogel by the heat of the high-speed impacts. Remarkably, analysis of the microscopic craters made in the aluminum foil indicated that even the smallest impactors, some only tens of nanometers across, shared the same mineralogy. There were no traces of carbonates or hydrated silicates. It seemed unlikely that these were destroyed by the act of capture. Most likely, the process involving water that created such minerals was inoperative on Wild 2. This was in contrast with the results from another Discovery mission, Deep Impact, that concluded several months prior to the recovery of the Stardust capsule and detected silicates and carbonates in the interior of the nucleus of comet Tempel 1.

Isotopic analyses indicated that although the ratios of hydrogen, carbon, nitrogen and oxygen varied widely among the particles in the Stardust sample, there were few examples of anomalies sufficient to imply an origin outside the solar system. In this sense, Wild 2 turned out not to have a raw 'pre-solar' composition and therefore not to comprise unprocessed 'stardust'. In terms of oxygen isotope anomalies, only three grains were found which could confidently be said to predate the formation of the solar system, and these were all in the residue of impacts on the aluminum foil. The particles from Wild 2 were expected by researchers to resemble Brownlee particles. These interplanetary particles had anomalous organic and inorganic materials which were evidently inherited from the interstellar medium by the pre-solar cloud of dust and gas. Unexpectedly, there were large differences, particularly in the abundance of carbon and isotopic anomalies – so much so, that a relationship between Wild 2 and the 'parents' of the Brownlee particles appeared to be a tentative hypothesis at best. In fact, the Stardust particles more closely resembled the composition of main belt asteroids. It is possible that at some time bodies which formed in the warmer, inner solar system found their way into the Kuiper Belt, and that Wild 2 was one of these. Overall, the mission found that, as a scientific paper put it, "the distinction between comets and asteroids is, in many cases, simply a matter of aging (loss of volatiles) and orbital parameters". A similar conclusion was reached by the identification of numerous particles, including some containing refractory calcium-, aluminum- and titanium-rich minerals that could only have formed by melting at temperatures in excess of 1,000°C (namely in the innermost solar system) whilst their compositions were similar to chondrules found in meteorites believed to come from the middle-to-outer asteroid belt. The fact that Wild 2 proved to be a mixture of refractories from the inner solar system and volatiles from the coldest outer fringes of the solar system implied that magnetic fields and other violent phenomena for which young stars are known must have transported and intermixed material in the pre-solar cloud. Isotopic analyses of a small refractory particle provided a dating of similar material from Wild 2. The particle appeared to have formed close to the Sun at least 1.7 million years after the oldest known solar system solids found in meteorites. It must be cautioned, however, that the data for refractory vanadium-bearing titanium nitride is considered controversial, as these particles could have been made by the reaction of the titanium of the spacecraft's fuel tanks, which included vanadium for strength and protection against corrosion by the hydrazine; indeed, hydrazine is commonly used to produce titanium nitride in laboratories.[71]

Scientists found the first evidence that Wild 2 had once possessed liquid water, in the form of cubanite, an iron and copper sulfide which can form only in its presence. Liquid water on a comet could be short-lived if it was due to the heat released by an impact, or longer lived if maintained by heat from the radioactive decay of nuclides. Interestingly, because cubanite would decompose if heated above 210°C this also set an upper limit to the temperature encountered by the comet during its existence.

As expected, no volatile-rich material was found in the samples, as sunlight would have evaporated it from the particles during the flight from the nucleus and any that remained would have been released by the heat of striking the aerogel of the racquet. A search for organics was performed by a variety of analytical techniques, but their identification was difficult due to the presence of oxygen, nitrogen and hydrocarbon contaminants in the aerogel. In any case, a puzzling result was the relative paucity of organic matter in Wild 2 in comparison to interplanetary Brownlee particles. This could indicate either that the comet was poor in organics (as in-situ observations had suggested) or just that they did not survive collection. However, polycyclic aromatic hydrocarbons were present. Some of these were undoubtedly formed by the heating of particles at impact, but others, which resembled those found in interplanetary dust and meteorites, were probably indigenous to Wild 2. Simple amino acids were also found embedded in the aerogel, possibly having been deposited by cometary gas or delivered by microscopic particles. In fact, molten aerogel even preserved traces of cometary noble gases. From this point of view, the most important result (after more than 2 years spent in developing the analytical techniques) was the identification on aerogel and aluminum foil by scientists at NASA's Goddard Space Flight Center of glycine, a simple amino acid; the first such molecule to be found on a comet. This followed decades of speculation that these objects carried the basic building blocks of biology to Earth. At first, researchers were skeptical of its cometary origin, suspecting it to be the result of handling contamination, but isotopic analyses ruled against this possibility.

In summary, the results of the extremely successful Stardust mission showed just how little we knew about comets (each probably has its own different 'personality') and about the phenomena that accompanied the formation of the solar system and of our star.[72,73,74,75,76,77,78,79,80,81,82,83]

At the time of writing this book, interstellar particles are being recovered from the racquet, the initial effort having been spent on the cometary sample. Having struck the aerogel at speeds typically three times greater than was the case for the cometary particles, these are expected to have been even more extensively modified. They left much smaller tracks. To locate them, JPL created a 'citizen science project' in which microscope images of the collector blocks were distributed to volunteers, and to date some fifty candidates have been proposed.

The project located 28 tracks on the interstellar particle collection side. However, in most cases they had angles consistent with small debris chipped off the probe's solar panels. Only seven, dubbed 'midnight' tracks, had orientations consistent with an interstellar dust stream origin. The most promising particle was found at the end of track I1043,1,30. Known as "particle 30", this had split in two on being braked by the aerogel, and the distinct chemistries of its pieces testified to their heterogeneity. They consisted of iron- and nickel-rich grains embedded in a matrix rich in calcium, silicon and magnesium, along with other heavy atoms such as aluminum, chromium, manganese, copper and gallium.[84]

While the capsule delivered the precious samples to scientists, the spacecraft itself remained alive and healthy. On 30 January 2006 it was told to power down most of its systems. Then, as had the cometary spacecraft that preceded it, Stardust went into hibernation. Since its $0.92 \times 1.70$-AU orbit had a period of 1.5 years, it would make a 1-million-km flyby of Earth in mid-January 2009, at which time, if the necessary funding were forthcoming, it could be reawakened for redirection to one of several possible targets. The only extended mission to receive any publicity involved a flyby of comet 9P/Tempel 1. The main objective of this encounter would be to see if the nucleus of that comet had changed since it was struck by a projectile released by the Deep Impact spacecraft on 4 July 2005 (of which more in Part 4 of this series). Scientists were particularly interested to obtain images of the crater dug by that projectile. It was to have been imaged by the Deep Impact mothership, but the dust issued by the impact had obscured the view. The rotational state of the nucleus would have to be refined so as to arrange the timing of the Stardust flyby to view that particular location. The flyby would also be able to measure the composition, size distribution and flux of dust particles in the coma – measurements which Deep Impact had not been equipped to undertake. In preparation for a possible mission extension, Stardust was awakened from hibernation on 5 February 2007 to assess its health and that of its instruments. In 2006 it had survived its closest perihelion to date and suffered the effects of two solar flares, but it proved to be in remarkably good condition. In view of the scientific interest, in July 2007 NASA selected the NExT (New Exploration of Tempel 1) proposal as a $25 million Discovery 'mission of opportunity'.

In the months after resurrecting the spacecraft, a few software issues were solved, the renewed contamination on the camera's lens was characterized, and (in the first major maneuver since the deflection that followed the release of the sample capsule) a 5-m/s burn refined the Earth flyby in 2009. Meanwhile, the attitude control system was reprogrammed to ensure that the remaining propellant would facilitate the entire mission. While the instruments were re-calibrated, the trajectory was further refined by three corrections, the last one occurring on 5 January 2009. The lens was 'baked' to evaporate some of the contamination. Since it was suspected that the camera's periscope had been damaged by the Wild 2 encounter, this was exercised 55 hours prior to the flyby by taking pictures of the Moon from a range of 1.1 million km in order to determine whether it could be used at Tempel 1. Other pictures were taken 'off periscope'. The results showed that the periscope would be usable. The closest point of approach of the Earth flyby on 14 January was at an altitude of 9,157 km above the Pacific Ocean just west of Baja California. The trajectory was so accurate that no exit correction was required. A 24-m/s correction performed more than one year later, on 17 February 2010, lowered the perihelion by 75,000 km and delayed the encounter by 8 hours 21 minutes, to nearer the time when Deep Impact's crater ought to be visible. Stardust further refined its aim by 0.33 m/s on 20 November.

The Tempel 1 flyby would be the first time that a comet was revisited to examine it on successive perihelia. Images would hopefully show how cometary features had changed in between. Comets are known to lose the equivalent in mass of a surficial layer centimeters to meters thick. How this affected and modified the appearance of the nucleus, however, was not yet known.

In order to be sure of flying over the same side as Deep Impact, astronomers had worked hard to make sure the rotation period of the nucleus was pinned down and its changes due to jets and outgassing were understood. The nucleus was spinning every 41.9 hours, but there remained a possibility that the calculations had gotten the phase wrong and that the opposite side would be seen. The objective was to image at least 25 per cent of the area already seen by Deep Impact. Apparently layered terrain seen by Deep Impact would be targeted, as this could yield information on how the nucleus had formed in first place and accreted. Scientists also wished to scrutinize an apparent smooth flow crossing the plateau near where the impactor had hit. Of course, images would be used to determine the size and shape of the crater formed when Deep Impact hit, and its ejecta pattern and possible layering. The flyby would occur on 15 February 2011, with the comet 39 days past its perihelion point. In the time since the sample capsule had been returned to Earth, Stardust had made 4 revolutions around the Sun. Moreover, it would occur only 3 months after Deep Impact itself flew by a second comet, 103P/Hartley 2.[85]

The first optical navigation images of the comet were taken on 16 December 2010 and then twice weekly until 4 January 2011, at which time the imaging rates picked up. However, these sets of images failed to detect the comet, still too faint to show. This made designing course correction more difficult, as the precise position of the target was not yet known and the small amount of fuel remaining imposed strict constraints. Meanwhile, the spacecraft was suffering a series of electronics latch-ups and safe-mode episodes that required a reboot early in January. Tempel 1 was finally detected at a range of only 26.3 million km on 18 January. Three maneuvers were planned for the approach phase. The first, on 31 January, was a 130-second, 2.6-m/s burn that consumed about 300 grams of fuel and moved the predicted position at the time of encounter by almost 3,000 km. The next correction on 7 February tweaked the speed by a mere 56 cm/s and targeted a point 200 km from Tempel 1. After this, Stardust started to take eight images of the comet every 2 hours. Imaging for science began 7 days out. Three days later, the camera mirror was 'baked' for the final time to evaporate any recent contamination. The third correction, made 2 days out, lasted 50 seconds and targeted a point approximately 170 km from the nucleus. Finally, 42 hours before the encounter, Stardust took its final navigation images. If it had looked like the flyby would be too close, a final maneuver would have been made but this was not necessary.

The encounter sequence proper began 24 hours before closest approach. The dust analyzer was activated 3 hours before the flyby. One hour later Stardust turned to face its dust shields forward. Autonomous tracking of the nucleus began 30 minutes out. With 20 minutes to go, the dust flux monitor was activated. Five minutes out, the spacecraft started a long rolling maneuver designed to keep the comet in the field of view of the camera. Imaging began 1 minute later. There was a break in contact with Earth while these observations were made. The flyby of Tempel 1 occurred at 4.40 UTC on 15 February at a distance of 178 km. During a 10-minute period the camera obtained 72 images, the number being limited by the memory available on board. The first 12 images were taken at 8 seconds interval, followed by 48 every 6 seconds and finally again 12 every 8 seconds. An hour after the flyby, by which time data collection by the dust monitor had been terminated, the spacecraft was turned to face its high-gain antenna to Earth in order to download the data.

Stardust had once again survived unscathed its dive into a cometary coma, storing a total of 78 Mbytes of data. The only glitch of the encounter was that the command to start the transmission with the pictures taken at closest approach failed, and they were downloaded in the order in which they had been taken, with the most distant first. Each image took about 15 minutes to transmit. The preparatory work paid off nicely when the highest resolution pictures were received, with the same hemisphere as viewed by Deep Impact being visible in the inbound leg, together with part of the previously unseen hemisphere. Actually, models had guessed the rotation phase of the nucleus within a mere few degrees. The best resolution was 12 meters per pixel. By comparison with Deep Impact imagery, there were clear signs of erosion along the scarp at the edge of a smooth 'tongue-like' flow, several kilometers long. The cliff seemed to have retreated tens of meters in places and its outline was also seen to have changed. Remarkably, although the impact site was in plain view, there was no obvious crater. However, this was later identified as a subdued depression 150 meters across almost straddling the edge of one of two large craters. The new feature also had a small mound at its center from ejecta which had rained back down. This gave some information on the strength of the cometary nucleus, as it implied a surface resembling dry, loose snow. Stardust then flew over entirely unmapped terrain. On the unseen side, three terraces of different elevation were seen, separated by banded scarps that were at most 2 km across. The lowermost terrace sported two 150-meter circular features. Layered terrain was seen, with layers a few meters thick, as well as heavily pitted areas and other regions that seemed to have undergone extensive sublimation. There were pits that could have been formed by explosive outbursts.

This is one of the highest resolution images of the nucleus of Tempel 1 taken by Stardust. The hemisphere at right had already been seen by the Deep Impact spacecraft in 2004.

Jets of matter were faintly visible on the limb of the nucleus, but overall the comet seemed to be less active than in 2005.[86]

Other instruments had measured the composition, size and flux of dust particles within the coma. The flux monitor recorded over 5,000 impacts, with the particles seemingly coming in bursts instead of forming a continuous, uniform stream. This had already been observed in previous cometary flybys. Most impacts were recorded after closest approach, as Stardust headed into the tail. The spacecraft suffered a dozen impacts that managed to breach the front layer of its Whipple shield. Several dozen particles had hit the dust analyzer, enabling it to detect organics, including simple carbon as well as cyanogen.

Almost 24 hours after the encounter, with all the pictures safely returned to Earth, departure imaging began. Stardust continued to take pictures of the receding comet, initially every 5 minutes, and then every 12 minutes until 10 days post-encounter. Imaging by the navigation camera was terminated on 24 February. It was estimated that less than 3 per cent of the original fuel reserve remained, effectively negating a second extended mission. Engineers at JPL proposed ending the mission with an engine burn that would completely empty the tank, to gain a valuable opportunity to validate the models used to estimate the quantity of fuel remaining on board. There

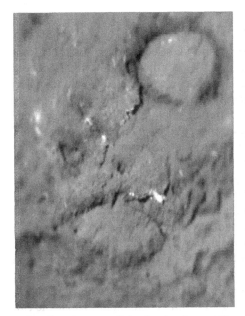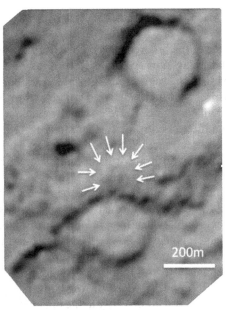

Stardust images allowed scientists to identify the depression created when the Deep Impact impactor hit the nucleus of Tempel 1.

being no fuel gauges for use in weightlessness, intrinsically inaccurate mathematical methods that modeled fuel consumption were used. The depletion burn was initially planned for April, but there were indications that there was so little fuel remaining that pressurizing gas was finding its way to the propellant tanks. For this reason, the maneuver was brought forward two weeks to 24 March. Stardust was commanded to point its medium-gain antenna to Earth so that telemetry could be monitored in real time and then initiate a burn that was programmed to last for 45 minutes. Of course, given the fuel situation, it was estimated that the burn would last no longer than 10 minutes and change the velocity by 35 m/s at most. In fact, the burn lasted only 146 seconds, which was near the lower end of the expected duration range. With the fuel completely drained, the thrusters then vented the helium pressurant. Twenty minutes after the scheduled conclusion of the burn, the spacecraft, now unable to control its orientation to maintain its solar panels facing the Sun, executed a pre-programmed sequence that turned off its transmitter at 23.33 UTC, effectively ending the mission for good.

Some proposals have been made to build other spacecraft using the Stardust bus. One Discovery candidate named Ice Clipper envisaged slamming a 10-kg spherical projectile into the Jovian satellite Europa while the main spacecraft passed overhead to collect material from the resulting cloud of debris. Although some instruments would have provided real-time analysis, the real point of the mission was to return a sample to Earth. Remote sensing by the Galileo mission showed the surface of this moon to be laced with a variety of chemicals, and a sample would assist in assessing the possibility of there being life in the ocean which is believed to exist beneath the thin icy shell.[87]

## FASTER, CHEAPER, WORSE

One of the original eleven Discovery mission proposals selected for study in 1993 was the Solar Wind Sample Return submitted by JPL in cooperation with Lockheed Martin. It was to spend several years beyond the Earth's magnetosphere and collect samples of the solar wind for return to Earth. If possible, the spacecraft was to be small enough to share a launch vehicle with another satellite.[88] The mission was one the three chosen for further study in 1995 out of a total of 28 candidates. By then it had been renamed Suess–Urey, after Hans E. Suess, a professor of chemistry at the University of California at San Diego, and Harold C. Urey, who had won the Nobel prize for chemistry in 1934. In 1956 they jointly published a seminal paper on the abundances of the elements based on geochemical and astronomical data.[89] But it lost out to the Stardust mission.[90] It was renamed Genesis and resubmitted to the next round, when five out of 34 proposals made it to the final selection: Genesis; the Comet Nucleus Tour (CONTOUR); the Mercury Surface, Space Environment, Geochemistry and Ranging mission (MESSENGER); the Venus Environmental Satellite (VESAT); and the Aladdin mission.

Proposed by APL, Brown University, the Johnson Space Center and Lockheed, Aladdin was to investigate the evolution of the Martian system, in particular the two small satellites – which were possibly captured 'small bodies'. After entering orbit around Mars, the spacecraft would make slow flybys of both Phobos and Deimos for preliminary remote-sensing observations, and then further passes to collect samples from specific geological units. The sampling would involve firing a small projectile to strike a selected area at a speed of at least 1 km/s. As the spacecraft flew through the resulting cloud of debris it would collect at least 3 mg (and possibly much more) of dust and pebbles from the regolith using a flexible fiber trap that would be reeled out from, around and back into the return capsule like a carpet – hence the name of the proposal. Repeated encounters with both moons offered redundant opportunities for sampling. There were to be five projectiles: two for each moon, plus a spare. A different section of the carpet would be exposed for each sampling run. Hence each sample would be able to be analyzed separately, and with knowledge of its context. In addition, the spacecraft would carry a dust detector to confirm the catch, a suite of three multispectral cameras for navigation and science, and a visible/near-infrared spectrometer for geological studies of the moons' surfaces and selected areas of the planet to obtain data that would dovetail with that from the existing orbiters.[91,92,93] The samples would shed light on whether these moons have a common origin with the planet, or originated independently; and if the latter, whether they are vestiges of primitive objects from the outer solar system that delivered volatiles and organics to the inner solar system.

The MESSENGER Mercury orbiter will be described in detail in Part 4 of this series, as it was subsequently selected and successfully flown as a Discovery mission.

VESAT was proposed by JPL, Ball Aerospace, and the Universities of Wisconsin and Oxford for a simple and inexpensive mission which would enter a high circular orbit of Venus and use three instruments that spanned a range of wavelengths from ultraviolet to the mid-infrared and address several long-standing mysteries. It was to map on a global basis over a period of at least one planetary rotation the cloud-top pressures, cloud particle sizes, wind fields and temperature fields at different levels of the atmosphere down to the surface, and the abundance of various chemically important species. From this, researchers hoped to explain the global circulation of the atmosphere and its meteorology and chemistry, including the interactions with the rocky surface.[94]

In October 1997 NASA approved two Discovery missions: Genesis would be the fifth of the series and CONTOUR the sixth.

The original plan for Genesis was to place a simple spinning spacecraft into a halo orbit around the L1 Lagrangian point about 1.5 million km sunward of Earth, where over a period of 2 years it would expose a series of collectors to the flux of the solar wind.[95] When it finished this sampling, it would head back to Earth and release the sample-return capsule.[96] The bus was a rectangular 2 × 2.3-meter platform on which were mounted the launch vehicle interface, the main batteries, star sensors, thrusters and a pair of 55-cm-diameter spherical tanks for 142 kg of hydrazine for attitude control. Also mounted on the Earth-facing side was a single medium-gain antenna for telemetry transmission. Two radial solar panels took the total span to 6.5 meters and generated at least 281 W. Four low-gain antennas and solar sensors for attitude determination were mounted on the solar panels. At the center of the Sun-facing side was a truss made of three bipods and a central six-legged and spring-loaded circular support for the 1.31-meter-tall, 1.62-meter-wide sample-return capsule. The capsule had an aerodynamic configuration based on that of Stardust, but with certain major differences to the rear and was almost twice as large. It was to be opened and closed in space like a clamshell using a massive hinge that would remain attached to the main deck. Inside the capsule was the collector canister, miscellaneous electronics and batteries, a GPS receiver and a radio beacon to enable the recovery team to localize the capsule after re-entry. The rear section held a 1.6-meter-diameter supersonic drogue chute and a 4 × 10-meter parafoil for the main descent.[97]

The main platform carried two scientific instruments: a solar wind ion monitor and an electron monitor to provide data on the solar wind environment around the vehicle. (These were essentially copies of instruments flown on the Ulysses out-of-ecliptic mission.) Their data would be fed to the flight software, where an algorithm would recognize and characterize the impinging solar wind.[98],[99] The remainder of the scientific payload was entirely enclosed in a cylindrical aluminum canister that was 78 cm in diameter and 35 cm high and installed in the return capsule. Inside this canister was a rotary actuator on which was mounted a stack of four circular metallic trays. Each tray had an area of approximately 0.3 m$^2$, and was covered with an array of 54 or 55 full-hexagonal tiles 10 cm across and six half-hexagonal tiles that were made of highly pure materials (including various forms of diamond, silicon carbide, sapphire and gold) chosen to suit the particles that they were intended to collect. In addition to cleanliness, 'analyzability' and thermal requirements, the materials from each array had to be uniquely identifiable in case tiles were dislodged or damaged during recovery. The array on top of the stack was to be continuously exposed to the solar wind, and so was referred to as the 'bulk sample' array. The three lower arrays would be exposed depending on the solar wind characteristics, as

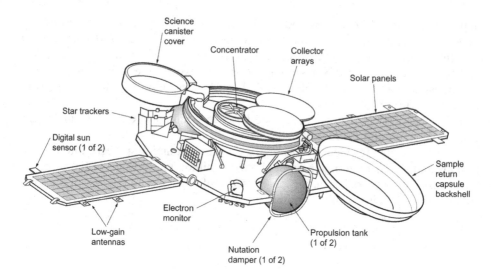

The Genesis spacecraft.

measured by the real-time monitors: one for solar wind from coronal holes, another for coronal mass ejections and the last for normal 'interstream' wind. In addition, there was a second bulk sample installed on the interior of the canister's cover. Finally, other exposed portions of the canister and sample-return capsule (such as the inside of the lid of the capsule) exposed collectors such as metallic glass and gold foil.[100,101] Since the small size of the collector arrays precluded measuring the isotopic ratios of oxygen and nitrogen in this way, a 46-cm cylindrical electrostatic concentrator on the floor of the canister used an electric field to deflect solar wind particles. The light elements (such as hydrogen) would be rejected, but more massive ones would be concentrated and implanted on a 6.2-cm target which consisted of sectors of diamond and silicon carbide.[102,103,104,105]

The orbital design for Genesis was particularly interesting. Once in the halo orbit around the L1 point, it was to spend at least 22 months collecting solar wind samples and characterizing their sources. It would rotate at 1.6 rpm for stability, with its spin axis pointing 4.5 degrees ahead of the Sun to face the apparent direction of arrival of solar wind. It would have to make daily attitude corrections in order to maintain this orientation. Attitude determination used different sets of solar sensors and two star trackers. Attitude corrections would be performed using two clusters of four 0.88-N thrusters, but larger trajectory corrections would use four 22-N thrusters. All of the thrusters used highly pure hydrazine and were mounted on the anti-sunward side of the spacecraft. Nevertheless, the orbit would be corrected only when the lid of the capsule was closed, in order to preclude contaminating the sample-collection arrays. After this phase of the mission, Genesis would close the capsule and inject itself into a so-called 'heteroclinic connecting trajectory' connecting the L1 point upstream of Earth to the L2 point downstream of Earth. On the way, the spacecraft would cross the orbit of the Moon in such a way that on its return from the L2 point it would be

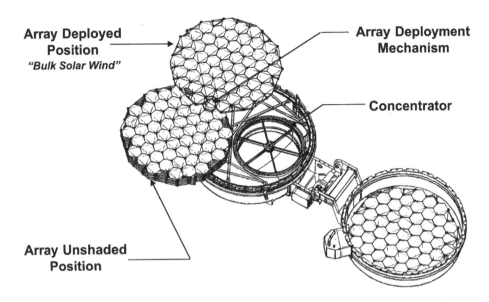

**Array Deployed Position**
*"Bulk Solar Wind"*

**Array Deployment Mechanism**

**Concentrator**

**Array Unshaded Position**

The sample canister of Genesis consisted of "bulk" collectors, including the one inside the lid of the canister, movable trays and the solar wind concentrator.

on a collision course with Earth. This complex trajectory was dictated by the need to have the samples returned in daylight – a direct return from the L1 point would have caused re-entry to be in darkness.[106] If the weather in the recovery area was deemed unsuitable, a last-minute correction would put the spacecraft into a 24-day orbit that would offer an opportunity for sample-return at each perigee.

After separation, the sample-return capsule was to hit the Earth's atmosphere over the Pacific just off the coast of the United States at a speed of 11.04 km/s, traveling east towards the Utah Test and Training Range, where it was to be recovered. It was originally intended that the bus would burn up over the Pacific. At a height of some 33 km the capsule would deploy its supersonic drogue chute, and then the parafoil at 6.7 km in order to slow the rate of descent to 5 m/s. A hard landing was considered impractical owing to the extreme fragility of the collector arrays, so the plan was for a pair of helicopters, one prime and the other in backup, to maneuver into position to snatch the capsule in mid-air by snagging the parafoil using a hook. The Air Force had used fixed-wing aircraft to make hundreds of such recoveries between the 1960s and 1980s to retrieve film capsules from several types of US spy satellites. In those cases, the load imposed by the recovery had reached 4 g. This was excessive for the fragile payload of the Genesis capsule. It was decided to use the method introduced during the Vietnam war to enable a helicopter to recover a reconnaissance drone, for which the loads did not exceed 2 g. In rehearsals, the test capsule was successfully snatched on the first pass.[107] After recovery, the helicopter would land immediately so that the high-drag parafoil could be cut off the capsule, and then it would lift off again and fly the capsule to a temporary clean room in which it was to be inspected, opened and purged with highly pure nitrogen. If all

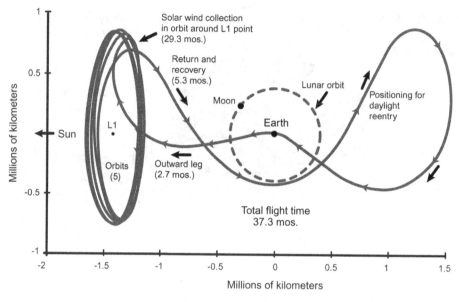

Genesis' orbit design.

went to plan, the sample canister would arrive at the Johnson Space Center in Houston about 11 hours after landing. There, the samples would be documented, archived and studied. The total mass of solar wind recovered was expected to be between 10 and 20 micrograms. The cost of the mission was quoted as $264 million.[108,109,110,111]

Genesis was to have been launched in December 2000 or January 2001, but it was delayed to July–August to ensure that it would not interfere with preparations for the Mars Odyssey mission. A fifth orbit around the L1 point was added to arrange a late summer re-entry, when weather at the Utah Range would be most favorable. On the positive side, this would increase the total solar wind collection time to 29 months. The new launch window opened on 30 July and lasted to 14 August. Thunderstorms and another launch meant the start had to be scrubbed three times, but it finally lifted off on 8 August 2001. The Delta II had three solid boosters and a solid-fuel upper stage. After 48 minutes in parking orbit, the second and third stages fired in succession to insert Genesis into an orbit which had its apogee at 1.2 million km in the direction of the L1 point. The spacecraft made a correction 2 days later to refine its trajectory.

On 17 August the sample capsule was slowly opened over a period of 30 minutes, then the doors of the ion and electron monitors were also opened. With the capsule lid open, an anomalous temperature rise was seen on several components, including the avionics, the concentrator, some mechanical components of the canister, and the lithium batteries for re-entry. There was a strong possibility that the white thermal protection that coated these components had somehow become contaminated and the thermal characteristics had altered as solar ultraviolet polymerized the contaminant. Particular concern was caused by the battery, which rose in temperature by 10°C to reach 23°C after a fortnight, since a failing battery could prevent the deployment of the parachutes. On 15 September, therefore, the capsule's lid was closed almost all of the way pending a resolution of the thermal issue. Hundreds of battery cells like those on Genesis were put through tests at temperatures up to 60°C, and the results showed that they should be able to meet all mission requirements. Meanwhile, the lid was completely closed on 13 November in preparation for the 268-second burn on the 16th that placed Genesis into a halo orbit with a radius of 800,000 km and a period of 6 months around the L1 point. The lids of both the capsule and the canister were opened on 30 November, exposing the bulk sample collectors for the first time. The deployment mechanism of the arrays was successfully tested on 3 December, and then put under the control of the software that monitored the solar wind in real-time. The concentrator was tested the next day.[112,113] As Genesis flew its five halo orbits, it performed a total of 15 station-keeping maneuvers.

Although the main scientific portion of the mission would not start until after the recovery of the samples, some results were obtained from the solar wind monitors. Early in the mission, a relatively high percentage of solar wind from coronal mass ejections was evident whilst that from coronal holes was almost absent, equating to only 5 per cent of the collection time. This was not unexpected, as Genesis had been launched near the maximum of the solar activity cycle of June 2000. Starting from October 2002, coronal hole flow became the most frequent state, even surpassing interstream flow. The particles from coronal holes were the most desired, since they ought to be representative of the composition of the photosphere.[114] The bulk arrays were exposed for 887 days, the interstream collector for 334 days, the coronal hole collector for 313 days and the coronal mass ejection collector for 193 days. To put this into context, the longest exposure of a solar wind collector on the Moon was by the Apollo 16 mission in 1972 and was a mere 45 hours.[115] The trays were stowed on 1 April 2004, and the canister and capsule were sealed the next day. Although the thruster firing on 22 April imparted a change in velocity of less than 1.5 m/s, it was sufficient to initiate the return leg of the mission. On 29 April Genesis flew by the Moon at a range of just over 250,000 km. On 1 May it passed perigee at 392,300 km and began to climb towards the L2 point, which it reached in July. Thruster firings on 9 and 29 August, and again on 6 September, set up the trajectory for atmospheric entry on 8 September.

The capsule was commanded to jettison some 5.5 hours prior to re-entry, and then in a change to the original plan the bus executed a deflection burn to enter a parking orbit with a 1.28-million km apogee as a precaution in case the capsule had failed to separate. In fact, the capsule did release. As the bus had just shed almost one-third of its mass, retained a significant amount of propellant and carried the electron and ion monitors, alternatives to ditching were studied. For example, it could be returned to the L1 point or maneuvered into a heteroclinic connecting orbit that would alternate back and forth between the L1 and L2 points. An interesting option was to put it into a slightly elliptical heliocentric orbit with a 1-year period so that it would effectively remain bound to the Earth in a distant retrograde orbit of about 5 × 10 million km. If so, then it would be the first spacecraft to enter such an orbit. In this extension of its mission, named Exodus since it would follow Genesis, the bus would study the solar wind using its ion and electron monitors. This data, in concert with that from other near-Earth and L1-orbiting missions, would characterize for the first time structures of the interplanetary environment at separations between 0.01 and 0.1 AU.[116,117] The distant retrograde orbit could be attained by a series of maneuvers concluding on 21 September 2005. But Exodus faced two problems. First, the spacecraft had not been designed for such an orbit, and this would severely limit communications. Second, a magnetometer would be needed to properly characterize the interplanetary medium and none was carried. After reflecting, NASA rejected this proposal, in part because of the relatively high costs of $2.5 million for redirection into heliocentric orbit and $1.5 million per annum thereafter.[118] After making one full distant orbit of Earth, the bus fired its thrusters on 6 November 2004 to head back to the L1 point, from which it entered a 0.896 × 0.990-AU solar orbit to be abandoned. The last contact was on 16 December 2004.

Meanwhile, at 15:52 UTC on 8 September the capsule hit the Earth's atmosphere. It was tracked by aircraft which recorded its ultraviolet, infrared and visible-light spectra to calibrate meteor re-entry physics as in the case of Stardust. As the deceleration peaked at 27 g a little over 1 minute later, the heat shield reached 2,500°C. Meanwhile, two Eurocopter Astar 350 helicopters piloted by movie stunt pilots were aloft in the recovery area, waiting. As soon as the capsule was acquired by long-range cameras, it became evident that the drogue chute had not opened. On going subsonic the capsule started to tumble end over end, and at 15:58 hit the ground at a speed of 311 km/h. It was 8.3 km south of the aim point, but still well within the nominal landing ellipse.[119] The recovery team rushed to the impact site to remove and secure the remaining pyrotechnics devices and retrieve the science payload. Their work was complicated by the fact that about 50 per cent of the capsule was buried in the Utah soil, which had been dampened by rain the previous week. It took 8 hours to retrieve the capsule and extract the sample canister. The surrounding soil was meticulously sifted for 2 days using tweezers and trowels to retrieve thousands of fragments. Meanwhile, the shattered sample canister was taken to the clean room and carefully disassembled over a period of a month.[120] It transpired that the base plate of the canister had broken loose from the side walls, metallic foils were completely crumpled and the collector arrays had shattered with many of the fragments less than 5 mm in size. Only one full-hexagon and three half-hexagons were recovered intact. Three out of the four concentrator targets survived intact, largely due to the fact that they were suspended on the arms of the instrument which had absorbed much of the shock. The 15,000 retrieved fragments larger than 3 mm were individually documented in Utah and packaged, many in vials or tissue culture dishes and in some cases on the adhesive of post-it notes. It was all flown to the Johnson Space Center on 4 October, taken to a 'space-exposed hardware' facility and the highest-value samples placed in a dry-nitrogen environment.

The capsule hardware had suffered various degrees of contamination of different types. During 3 years in space, the sample canister had received a micrometeorite hit 0.4-mm across which chipped the paint. Contaminants on the canister and collectors

Genesis' sample-return capsule laying semi-destroyed after its parachuteless landing.

had been polymerized by solar ultraviolet, creating "brown stains". Obvious sources of contamination from the hard landing were Utah dirt and burned carbon fibers off the heat shield. The fragments were cleaned with inert gas, sable brushes, cryogenic gases and several solvents. The results were sufficiently encouraging to prompt the announcement that in spite of the crash, the science objectives would probably be met and "the only effect [would be] delays" in meeting them. Meanwhile, over thirty laboratories around the world, in the US, Switzerland, France, the United Kingdom, Japan and Canada started working on the samples using some of the most advanced analytical instruments.[121,122,123,124,125,126]

The Genesis Mishaps Investigation Board, which was established 2 days after the accident, delivered its report in June 2006 – more than 20 months after the event as a result of the decision to await the outcome of the Stardust mission. The root cause of the failure was the Stardust-heritage g-switches which were to sense the atmospheric deceleration and initiate the sequence of events leading up to the deployment of the drogue chute – they had been installed inverted, precisely as specified by a defective drawing! Because it had been felt that the Stardust-heritage components would be reliable, the drawing error had gone unnoticed through three levels of oversight. In fact, a test intended to verify the g-switches in action was canceled. While the cause of the accident was quite trivial, and was identified and announced quite early in the investigation, the report criticized NASA's approach to systems engineering, as this had given rise to a number of failures of 'faster, cheaper, better' missions in addition to Genesis.[127,128]

Despite NASA's optimistic call, only a handful of scientific results from Genesis have been published at the time of writing this book, and those results apply mainly to the few collectors which were recovered more or less intact. Of course, the nature and complexity of the analyses was always unlikely to produce the 'instant science' that planetary missions often yield. Nevertheless, the study of a bulk metallic glass target that was recovered essentially intact solved a decades-old lunar mystery. Dust recovered by the Apollo missions had seemed to contain neon and other noble gases from two distinct sources and different isotopic ratios. Whilst one of the sources was the solar wind, the second had posed a puzzle that seemed to require periods of very intense fluxes of energetic solar particles in the past. The Genesis samples showed varying isotopic ratios with depth of implantation, indicating that the differentiation is a depth effect. Consequently, no dramatic variation of solar activity is required to explain the Apollo data.[129] Some results for noble gases from the arrays have been published, and indicated that the isotopic composition of the different types of solar wind were extremely similar in terms of neon, with an enrichment of light isotopes in the slow solar wind in terms of helium. The data also provided a determination of the isotopic and elemental compositions of the different regimes of the solar wind.[130,131,132] Another mystery was the isotopic ratios of oxygen, nitrogen and noble gases at the formation of the solar system. Some meteorites which appeared to be enriched in lighter isotopes relative to terrestrial, lunar and Martian rocks had been thought to have picked up 'alien' matter, most likely from a nearby supernova explosion. However, a study of the samples from the concentrators showed the isotopic ratio of the solar wind (and hence presumably of the original nebula from which the solar system formed) to mimic that of meteorites. It was therefore concluded that Earth must somehow have received a skewed dose of these isotopes or that some process must have depleted the nebula from which the inner planets formed of lighter isotopes prior to the formation of the planets.[133,134,135,136,137]

The second Discovery mission selected in October 1997, the Comet Nucleus Tour (CONTOUR) was another of the original Discovery proposals of 1993. Conceived by APL, which had already submitted the NEAR mission, its objective was to fly by at least two cometary nuclei to characterize them by providing global topographical and compositional maps, detailed imagery of selected areas, and the structure and composition of their comas. To achieve this without a heavy propulsive capability, it was to be put into a heliocentric orbit which would provide regular encounters with Earth to adjust its trajectory for a succession of targets. The baseline mission called for 2P/Encke, 73P/Schwassmann–Wachmann 3 and 6P/d'Arrest. There was the possibility of an extended mission with at least one more flyby. The spacecraft was to be launched during a narrow window in August 2003 and make a flyby of Encke on 12 November at a range of about 100 km and a relative speed of 28.2 km/s, at a point some 0.27 AU from Earth. The spacecraft was then to return to Earth on 14 August 2004 and again on 10 February 2006. During this time, it would be in an orbit inclined at 12 degrees to the ecliptic, the second highest reached by a space probe (the record of course, being the out-of-ecliptic Ulysses mission). Except for 50-day periods centered on the Earth encounters and a 75-day period devoted to Encke activities, it would operate only thermal control and a minimum level of housekeeping activities whilst listening for commands via its low-gain antenna; most of its subsystems and all of its instruments would be off, including propulsion and attitude control.[138] The Schwassmann–Wachmann 3 encounter would be on 19 June 2006 at a low relative speed of 14.0 km/s and at a point 0.33 AU from Earth. This would involve another 75-day period of activity. After that, the spacecraft would follow an orbit that would fly past Earth on 9 February 2007 and on 10 February 2008. The d'Arrest encounter would occur on 16 August 2008 at a relative speed of 11.8 km/s and at a point 0.36 AU from Earth. All of the encounters would take place within 0.4 AU of Earth, ensuring that simultaneous studies could be carried out by ground-based telescopes. After the d'Arrest flyby, the spacecraft could have returned to Earth to set up a second encounter with Encke on 8 October 2013 at approximately the same position as previously. Other alternative extended missions included a flyby of Grigg–Skjellerup in 2013 and Giacobini–Zinner in 2018, or of Tempel 2 in 2015 and Encke in 2023. Alternatively, the extended mission could visit a newly discovered comet. In fact, owing to its repeated Earth flybys the orbital design was extremely flexible, allowing encounters with comets at heliocentric distances in the range 0.8 to 1.4 AU. Had CONTOUR been in space in 1995, for example, it could have been diverted to encounter the long-period comet C/1995O1 Hale–Bopp on 6 May 1997, which was one of the intrinsically brightest and most active such objects yet discovered.[139]

A search was started for new, long-period comets that may cross CONTOUR's path and could be encountered. One such comet was discovered, C/2001Q4 NEAT by the Near-Earth Asteroid Tracking program that the probe could theoretically fly by when it passed within 0.32 AU of Earth in May 2004, if the Encke encounter was dropped. But the scientific management decided to stick to the original schedule.

Unfortunately, by late 1999 the mission cost had overrun the estimates and the management decided to cancel the encounter with comet d'Arrest, which could be reinstated in a separately funded extended mission, and to purchase a solid rocket motor for departing Earth that had exceeded its 5-year 'shelf life'.[140]

After further consideration of the constraints of the initial phase of the mission, it was decided to advance the launch by 1 year to July 2002. The spacecraft would first be inserted into an eccentric Earth orbit with an apogee of 115,000 km and a period of 1.75 days. After as much as 45 days it would fire its motor to enter a heliocentric orbit with a period of 1 year that would provide an Earth flyby on 15 August 2003, which was the original launch date. This preliminary orbit provided the flexibility to guarantee the conditions required to initiate the complex tour.[141,142]

The first target of CONTOUR, 2P/Encke, was one of the most important and best known periodic comets. It was discovered on 17 January 1786 by Pierre Méchain of the Paris Royal Observatory, but it was not until 1822 that Johann Franz Encke was able to link it to comets seen in 1795, 1805 and 1819 and thus establish it to have a period of just 3.3 years.[143] Furthermore, Encke also discovered that the period was 2.5 hours shorter at each revolution. This remained a mystery until 1950 when Fred L. Whipple, having analyzed almost two centuries of observations of comet Encke, introduced his 'dirty snowball' model of comet nuclei and realized that the loss of matter by the activity at each perihelion passage served to provide a 'rocket effect' which progressively modified the trajectory of the

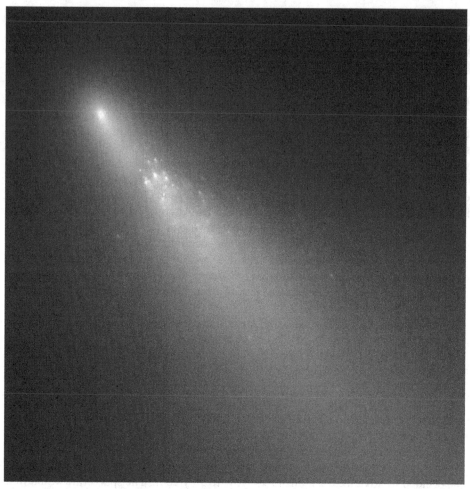

An amazing image of the fragments of comet Schwassmann-Wachmann 3, taken by the Hubble Space Telescope in 2006. (NASA, ESA, H. Weaver (JHU/APL), M. Mutchler and Z. Levay (STScI))

nucleus. Whipple calculated that comet Encke lost about 0.2 per cent of its mass at each perihelion.[144] Nevertheless, with a diameter estimated at 2.4 km, this nucleus was the largest of the CONTOUR mission's targets.

The recent history of 73P/Schwassmann–Wachmann 3 was also interesting. It was discovered on 2 May 1930 by Arnold Schwassmann and Arthur Arno Wachmann of the Hamburg Observatory shortly before it made a very close approach to the Earth at a range of just 0.0617 AU. As the comet is intrinsically faint, it was not seen again until 1979. The 1984 perihelion was not observed, but it has been recovered at each apparition since. When it was recovered in December 1994 it already had a hint of a tail, despite still being 3 AU from the Sun. Images and measurements taken on this occasion enabled astronomers to calculate an upper limit of 1.1 km for the size of its nucleus.[145] Spectra taken when it suddenly brightened 100-fold in September 1995 showed it to be erupting spectacular quantities of water. When large telescopes were aimed at it in December at least three and possibly four condensations were visible in the coma, suggesting that the nucleus had fragmented between the end of October and mid-December.[146,147] By the next apparition in 2001 one fragment had already disappeared. It looked as if another would disappear by the time of the CONTOUR encounter, but the largest fragment would almost certainly be present and scientists were very eager to inspect the freshly exposed surface.

The third candidate, 6P/d'Arrest has a rather bland history in comparison to the other two. It was discovered by Heinrich Ludwig d'Arrest in June 1851, but recent studies have confirmed its link to a comet of 1678.[148] Its nucleus was estimated to be a little over 1.5 km in diameter.[149]

The CONTOUR spacecraft was an 8-sided cylinder 1.8 meters tall and 2.1 meters in diameter. On the front was a 25-cm-thick Whipple shield comprising four layers of Nextel fabric and seven layers of Kevlar that projected 12.5 cm beyond the rim. The sides and the rear were covered with solar cells to assure up to 670 W of power. A low-gain antenna centrally positioned on the rear would provide communications for the Earth-orbit and cruise portions of the mission, and an 18-cm-diameter dish on a fixed mount offset towards the edge of the rear provided a high-gain link for the cometary encounters. As in the case of Giotto, CONTOUR had a 'kick stage' buried inside its main structure. This 503-kg STAR 30BP solid rocket was to provide the 1,922 m/s velocity increment needed to escape from the initial Earth orbit into solar orbit. There was 80 kg of hydrazine for use by two 22-N thrusters in making course corrections and by fourteen 0.9-N attitude control thrusters. The spacecraft had three attitude modes: a fast-spin mode of 15 to 25 rpm for hibernation periods, a slow-spin or 'rôtisserie' mode that would provide both accurate attitude and passive thermal control, and a 3-axis stabilized mode for use during encounters. It had four scientific instruments. The most important was a combined imager/spectrometer which used a 100-mm-diameter, 680-mm-focal-length Ritchey–Chrétien telescope and a 10-slot color filter carousel. It was mounted on one of the side panels, towards the rear, and to protect it from being 'sandblasted' was aimed at the cometary nucleus by a two-sided pivoted mirror. The imager was to provide a maximum resolution of 4 meters per pixel (some 25 times better than had been achieved by Giotto). The geometry of the vehicle, optics and mirror would allow the imager/spectrometer to track a target during the inbound leg of an encounter until just a few seconds before the actual flyby. An imager that viewed through the Whipple shield was for optical navigation and to take wide-field color images to study the evolution of gas and dust jets. This used 60-mm-diameter, 300-mm-focal-length refracting optics and was very sensitive to facilitate identification of a target when it was still very distant and dim. It had an interchangeable 4-position mirror so that if one mirror was ruined by dust it could be replaced by another for the next encounter. The imagers were to be calibrated during the Earth flybys. An impact dust analyzer similar to such instruments on Giotto, the Vegas and Stardust was installed on the front to study cometary dust, and a mass spectrometer based on that of Cassini was to measure the composition of gases and the ratio of hydrogen to deuterium in the

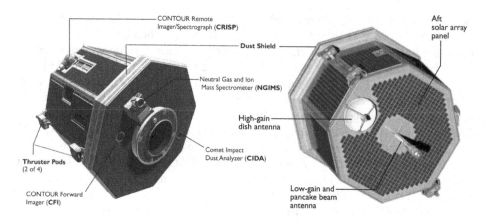

CONTOUR Remote Imager/Spectrograph (**CRISP**)

Aft solar array panel

Dust Shield

Neutral Gas and Ion Mass Spectrometer (**NGIMS**)

High-gain dish antenna

Comet Impact Dust Analyzer (**CIDA**)

Thruster Pods (2 of 4)

CONTOUR Forward Imager (**CFI**)

Low-gain and pancake beam antenna

The ill-fated CONTOUR spacecraft.

coma. Data from each encounter would be stored on two 5-gigabit solid-state recorders before being transmitted to Earth. The cost of the mission, including the launch vehicle, was $159 million.[150],[151]

The spacecraft had an overall mass of 970 kg. It was delivered to Cape Canaveral for launch by a Delta 7425, a form of the Delta II equipped with four solid boosters. The 24-day launch window opened on 1 July 2002, but the launch had to be delayed in order to clean a light dust contamination from the spacecraft. It finally lifted off at 06:47 UTC on 3 July and entered the planned high-apogee orbit some 50 minutes later above Australia, where portable tracking stations had been erected.[152] The Delta slightly underperformed, but the error was easily manageable. No fewer than 23 orbital corrections were executed during 43 days and 25 orbits to ensure that on 15 August the spacecraft would be passing through perigee at the desired altitude within a few minutes of the optimal time for the heliocentric orbit insertion burn. At 08:49 on that date, approaching its perigee at 224 km over the Indian Ocean and out of contact with ground stations, CONTOUR ignited its solid rocket for a burn which was to last 50 seconds. Contact was to be re-established by the Deep Space Network 45 minutes later, but there was no signal. The spacecraft's programming stated that after 4 days without a contact with Earth it should switch antennas in order to try to re-establish a link. Nothing was heard. Nor was anything heard in early December, when the spacecraft's attitude should have been particularly favorable for the high-gain antenna. NASA reluctantly declared the CONTOUR mission a total loss. The first clue as to what had happened came 20 hours after the burn, when the 1.8-meter-diameter Spacewatch telescope at Kitt Peak in Arizona, which was usually used to detect near-Earth asteroids, was turned to inspect the spacecraft's predicted position. Two objects were spotted 1,000 km behind it. Their distance was estimated at about 480,000 km from Earth and some 460 km apart. On the assumption that these were large fragments of the spacecraft, it was calculated that the rocket had delivered an impulse 3 per cent below that expected, suggesting that just before the burn finished the vehicle broke apart and caused these fragments

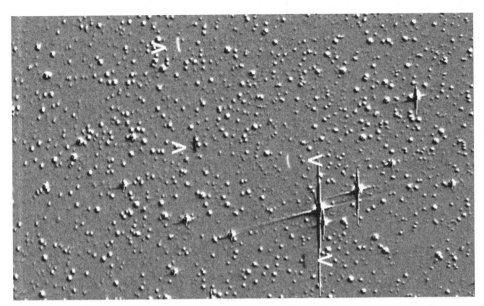

Two items of debris from CONTOUR imaged by the Spacewatch telescope on 16 August 2002. Positive tracks were taken earlier, negative tracks later. (Image copyright 2002 The Spacewatch Project, Lunar and Planetary Laboratory, The University of Arizona)

to separate at a relative speed of at least 6 m/s. Further observation revealed there to be a much fainter object trailing 6,000 km from the other two. The three fragments were tracked for four nights, and modeling of their orbital perturbations indicated that the brightest one accounted for one-third of the original mass of the spacecraft, whilst the mass of the second was no more than 4 kg. Data supplied by "Department of Defense assets" (i.e. early warning satellites) confirmed that the spacecraft had indeed broken up while performing the escape burn.[153,154]

As usual, a Mishap Investigation Board was established. This delivered its final report in May 2003. First of all, the report noted that despite the experience of Mars Observer and Mars Polar Lander, both of which had failed while out of contact with Earth, the CONTOUR mission design did not call for telemetry transmission during such a critical event as the insertion burn into solar orbit, making the cause of the loss difficult to identify. However, it was discovered that the thermal analyses made to validate the embedded engine position had used rather optimistic data, and that independent analyses indicated that the structure surrounding the engine would have endured thermal stresses much higher than predicted. In particular, the degradation of the properties of the material contaminated by the engine's exhaust had not been fully considered. It was therefore concluded that the most probable cause of the loss was a structural failure of the spacecraft during the firing. Less likely explanations included a structural failure of the motor, a micrometeoroid impact and a complete loss of control.[155] APL mission managers maintained, however, that the loss was due to the catastrophic failure of the 'old' solid fuel motor. Given this determination, hopes for an immediate CONTOUR 2, or New CONTOUR, in order to recover the science were dashed. Such a mission could have been launched in 2006 to encounter the same targets as the original mission, or launched in 2007 or 2008 to such targets as Grigg–Skjellerup, Honda–Mrkos–Pajdušáková, Hartley 2, Giacobini–Zinner or Tuttle–Giacobini–Kresak. In this case, it would have been launched directly into solar orbit by a more powerful form of the Delta II and been equipped with a larger hydrazine course correction engine.[156,157] Revised forms of this proposal were submitted for subsequent Discovery selections, but they never made the finalists.[158]

## FROM FIRE TO ICE ... TO NOWHERE

In 1997 NASA, impressed by the successes of the Discovery program and the first Mars Surveyor mission, began the Outer Planets/Solar Probe exploration program of low-cost missions to address some of the top priority but difficult destinations in the solar system. Three JPL-managed missions would constitute the project nicknamed 'Fire and Ice'. These were a Europa Orbiter, the Pluto–Kuiper Express flyby mission and the Solar Probe. The plan was to encourage commonality of spacecraft, core systems, launch and operations so as to cap each mission at $190 million. In particular, there would be common avionics, software, communications and propulsion, and a single small operations team was to manage all three missions. First to be launched would be the Europa Orbiter in 2003. It would be followed by the Pluto–Kuiper Express in 2004 and the Solar Probe in 2007, at about the same time as the Europa Orbiter would be approaching Jupiter.

After the Galileo mission's indication that there could be an ocean of liquid water under the icy surface of the Jovian satellite Europa that could theoretically harbor some kind of life, the planetary science community had been busy formulating new mission proposals to investigate this further. The Europa Orbiter would be the first logical step in this direction. It was to study the surface to interpret its history and evolution, look for signs of recent geological activity and determine once and for all whether there was a subsurface ocean, and if so map its extent. A laser altimeter and radio tracking by the Deep Space Network would map the shape of the moon and measure the amplitude of the tidal deformation as the surface strained and relaxed on a cyclical basis, to determine whether the crust was anchored to the rock below or isolated by a liquid water buffer. If there was a thin shell of ice over an ocean, a triple-frequency radar would attempt to detect echoes from both the surface and the underside of the shell to measure its thickness. The orbiter would also have wide-angle and narrow-angle cameras with resolutions of 300 and 20 meters respectively. The other objectives included determining the surface composition with particular emphasis on mapping the distribution of molecules in the hope of finding prebiotic chemistry, and to characterize the moon's environment for future missions.

For most of its interplanetary cruise, the spacecraft would transmit only a beacon signal capable of being received by antennas 5 to 10 meters in diameter manned by relatively untrained university staff. This was one of the many ways by which JPL hoped to reduce the cost of these missions. After insertion into orbit around Jupiter, the spacecraft would spend several years flying a Galileo-like mini-tour of the three

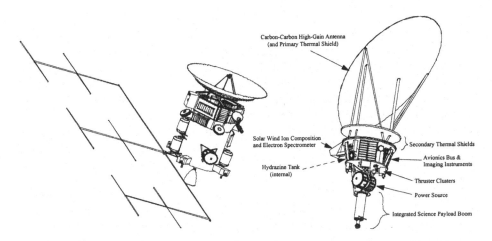

The 'Fire and Ice' Europa Orbiter, with the three element Yagi antenna for the radar (left), and the RTG-powered Solar Probe (right).

outermost Galilean satellites, using a dozen flybys to shed most of its orbital energy. This phase would be followed by a 5-month 'end game' that included the final half-dozen flybys of Europa and maneuvers designed to enable the spacecraft to slip into orbit around its target. A distinguishing feature of this mission was the propellant tanks for the rather large total velocity change of 2,500 m/s involved in Jupiter orbit insertion and the subsequent maneuvering to enter orbit around Europa.[159] The in-system evolution would require better navigational tools than for Galileo, whose tour schedule was less hectic.[160] Owing to the intense radiation at Europa's position in the Jovian magnetosphere, the part of the mission conducted in orbit of that moon would last only 30 days. Nevertheless, the 300 revolutions during this time would be sufficient to make the intended observations. In fact, the activity would start with gravity measurements in a slightly eccentric orbit, followed by global mapping in a high-inclination 200-km circular orbit. Radar soundings of the surface of Europa had been conducted from Earth before, but owing to the short wavelength used they could probe only a few meters. To penetrate to a greater depth, the Europa Orbiter was to deploy a large 3-element Yagi antenna for a 100-W radar transmitting at a wavelength of several meters. Even operating at a low rate, this would generate 100 times more data than could be transmitted in real-time. Onboard data processing and compression would be required. This problem could be partially solved if one of the technologies under investigation for the mission could be implemented: namely an optical communication terminal consisting of a 30-cm telescope 'receiver' and a laser downlink capable of a data rate from Jupiter several tens of times greater than for a conventional radio system. This would ensure that an adequate amount of data could be returned in the 30-day period around Europa.[161] Otherwise, it would have to rely on a 2-meter-diameter high-gain radio antenna. The mechanical design of the Europa Orbiter's bus would resemble Deep Space 1 in using load-bearing panels to mount the electronics and stackable avionics boards without bolts and fasteners etc.

The electronics design would use a 'firewire' architecture in order to enable PCs to be linked for development, tests, simulations etc. For protection against the Jovian radiation, the electronics would be shielded by 25 mm of aluminum.[162]

The plan was to use the Space Shuttle, an Inertial Upper Stage and an additional 'kick stage' during the November 2003 launch window to inject the Europa Orbiter on a direct trajectory that would require 3 to 4 years to reach Jupiter. The reasons for reverting to using the Space Shuttle are not clear, but the mission would have been delayed by the loss of Columbia in the same manner as Galileo had been delayed by the loss of Challenger in 1986. However, the mission was transferred to the Evolved Expendable Launch Vehicle (EELV) that was being developed for the US military and for NASA.

For JPL, the Europa Orbiter was to be the first step in an effort to explore Europa and its oceans and assess its capability to support life. The engineers were already at work on technologies to enable a lander to probe through the icy crust into the liquid water ocean. The most important of these technologies, cryobots, which were robots able to melt a path through several kilometers of ice, had already been tested in the early 1960s to explore the Greenland icecap. There was also the possibility of using this technology on the polar caps of Mars. In the case of Europa, the spear-shaped cryobot would draw 1 kW of power from an RTG and pass through the ice either by passive heating or by jets of warm water, and communicate with its lander stage and thence with Earth using miniature radio transceivers embedded on the icy surface. (Tethers were deemed unsuitable for communication since they would probably be cut as the ice re-formed behind the cryobot.) On approaching the ice/water interface, the cryobot would split. The upper section, housing the control and communication systems, would anchor itself in refrozen ice. The lower section would melt through to the liquid water, where it could release either sensors to analyse the biochemistry or a small 'hydrobot' submarine that would explore the ocean and possibly even the rocky seafloor. Such a probe could be tested in the Antarctic Lake Vostok, a body of liquid water 10,000 $km^2$ in extent that was discovered in 1974 by Soviet scientists of the Vostok station beneath a 4-km-thick crust of ice. This environment has evidently been isolated for millions of years, and may host a unique ecosystem.[163,164] (It was penetrated by Russian scientists in February 2012 using relatively conventional drills.)

The second Fire and Ice mission was the RTG-powered Pluto–Kuiper Express, which has already been described in detail in Part 2 of this series.[165]

The final mission was the Solar Probe. This was to exploit over two decades of studies which started with the Arrow to the Sun of the 1970s and ran through to the multi-billion-dollar proposal in the 1980s for the Shuttle-launched Starprobe.[166] Its objectives were to determine in-situ the mechanism that accelerated the solar wind and trace the source of energy that heated the corona to a temperature exceeding that of the photosphere, to map the structure of the corona and the configuration of the magnetic field at a wide range of latitudes, and to identify the origin of the fast and slow solar winds. In fact, the structure of the solar wind and the associated magnetic field was so difficult to study that tracing single structures back to the surface of the Sun was nearly impossible for heliocentric distances greater than 1

A Europa 'hydrobot' submarine exploring hydrothermal vents on the floor of Europa's ocean. (JPL/Caltech/NASA)

AU. In addition to sampling the low-speed solar wind near the Sun's equator, a trajectory that passed low over the polar regions near the time of solar maximum would permit a study of the well-developed coronal holes which issued the high-speed solar wind. Moreover, measurements could be taken within coronal structures such as streamers and 'open' magnetic field regions at the pole. By flying a trajectory that would make the pole-to-pole passage in as little as 13 hours, the Solar Probe would use remote-sensing instruments to study small structures on and nearby the Sun and in-situ sampling to characterize the corona. The three remote sensing instruments were to be a visible-light magnetograph and helioseismograph, an extreme-ultraviolet camera (or perhaps an X-ray telescope) and an all-sky coronal imager. The five in-situ instruments were an ion and electron spectrometer, a detector suitable for the fast solar wind, a plasma-wave sensor, a magnetometer and an energetic particles spectrometer. The visible and extreme-ultraviolet images would provide our first insight into the polar regions, which are inaccessible for observers in the ecliptic, would provide a full map of the corona at rapidly varying latitudes and would image structures on the photosphere at unprecedented resolution – the smallest features on the photosphere resolvable from Earth are about 100 km in size and the Solar Probe would increase this resolution by at least an order of magnitude.

Unlike the 1980s Starprobe, the Fire and Ice Solar Probe would not be particularly well suited to investigating the gravitational field of the Sun, and this was not even mentioned as a secondary objective. In fact, the orbital design was such that the first perihelion would be at quadrature (when the Earth-Sun-spacecraft angle was close to 90 degrees) in order to ensure that communications would not be disturbed by solar radio noise. Because in this position the gravitational field of the Sun would tend to perturb the vehicle's motion on a direction perpendicular to the line to Earth, there would be little or no measurable Doppler shift in the probe's transmissions. The plan was to launch the Solar Probe in 2007 on a direct trajectory to Jupiter, for a flyby in which it would shed most of its heliocentric velocity and fall towards the Sun. The launch vehicle was to be the Delta III, a heavy-lift form of the Delta II that gained the unenviable record of failing on all its launches and was abandoned. As in the 1980's projects, the Jovian flyby was to establish a final orbit with a perihelion at about 2.8 million km (a mere 4 solar radii) in a polar inclination. The Solar Probe would be traveling at in excess of 300 km/s when it reached perihelion in October 2010, which was expected be in the run up to solar maximum, and if the spacecraft survived for a full orbit it would be able to make a second pass about 4.5 years later, shortly before solar minimum.

Given the extreme environment in which it would operate, the mechanical design of the Solar Probe would need particular attention. It was decided to make the high-gain antenna out of carbon to withstand very high temperatures and use it to double as the main Sun shield, with secondary heat shields added between the antenna and the 8-sided bus. This design would also greatly reduce the mass. A series of ports (or 'soda straws') would have to be provided for instruments which required to face the Sun at perihelion; for example the ion detectors that were to sample particles coming from that direction. However, because the front of the antenna would reach 2000°C, there would be abundant sublimation that could interfere with measurements. Other instruments would be mounted on a boom that would extend radially to place them at the edge of the antenna's shadow cone. The antenna would provide a data rate of up to 88 kbps, although there would be long periods during the perihelion pass when the requirement to maintain the bus in the shadow of the antenna would prevent real-time communication with Earth. The spacecraft would be powered by two sets of solar panels, one for low temperatures and the other for high temperatures during the cruise, with both being retracted for perihelion, when it would run for several days on its batteries. As an alternative, an advanced RTG like that of the Europa Orbiter might be used instead of solar panels. Although there was to be some commonality between the three Fire and Ice missions, for the Solar Probe a small monopropellant engine would suffice, since the maneuvering would be mostly to precisely target the Jupiter slingshot.[167,168]

With the mission design more or less established and its funding in place by 1998, the Fire and Ice program looked set to become a reality. But the schedule was soon threatened. One issue was slippage by the Department of Energy in developing the advanced RTG required for the Europa Orbiter and Pluto–Kuiper Express. Another problem was the slow development of the Evolved Expendable Launch Vehicle – so much so, in fact, that by 2000 it looked as if the Europa Orbiter would serve as a test payload! JPL therefore recommended that this launch be postponed from 2003 to 2006.[169] Unfortunately, after 2000 the entire program rapidly collapsed because the projected cost of each mission had risen far in excess of the $190 million cap per mission which, in hindsight, was a completely unrealistic budget for such complex tasks. Moreover, the 'faster, cheaper, better' concept itself was attracting ever more intense criticism as a result of the failure of both of the Mars Surveyor missions that were launched in 1998 and 1999. The Pluto–Kuiper Express was the first to be canceled, in an effort to give priority to the Europa Orbiter, but when the cost of this mission rose to $1.4 billion in 2002 it was abandoned.[170] Development of the Solar Probe had never advanced very far, and it readily fell by the wayside.

## THE WOUNDED FALCON

As an alternative to pursuing the Stardust-style SOCCER (Sample of Comet Coma Earth Return) joint mission with NASA, Japanese engineers and scientists at ISAS (Institute of Space and Astronautical Sciences) studied a low-cost sample-return to a near-Earth asteroid as a means of demonstrating several new techniques including interplanetary ion engine propulsion, autonomous station-keeping control systems, low-gravity sample collection and high-speed re-entry on a mission that would also provide important scientific results. In particular, they wished to be the first to visit a sub-kilometer member of the most conspicuous population of objects that cross the Earth's orbit. The project was approved in April 1996 with a budget of $170 million. Since it was to be the third technology demonstration of the MU Space Engineering Spacecraft series and followed the experimental MUSES-A (Hiten) lunar orbiter and the MUSES-B (Haruka) radio-telescope, it was named MUSES-C. Fabrication of the spacecraft started in 1999.

One early issue that faced the engineers was the manner in which a few grams of material should be collected in the very low gravity environment of the surface of an asteroid. Several ideas were assessed, including using rotating brushes to collect bits and pieces of regolith and using coring cylinders that would be pushed against the surface. These, however, would be almost useless if the asteroid proved to lack a regolith. On the other hand, traditional drilling would be impractical unless the spacecraft could be anchored in some way.[171] Instead, the engineers devised a 'touch and go' sampling system. First the spacecraft would use laser altimeters to align its axis to the local vertical, then as it descended it would detect the contact of a 1-meter-long telescopic sampling horn by sensing its deformation, at which time the spacecraft would fire a number of 5-g tantalum pellets that would strike the surface at 300 m/s. Even in low gravity, some of the material chipped off the asteroid would find its way through the sampling cone and into the sample canister mounted on one side of the vehicle. The canister was a cylinder 5 cm in diameter and 6 cm in height, with two separate chambers for two sampling runs and a moving aperture that could be positioned over either. As the laser range finders could not measure the horizontal drift relative to the asteroid, just prior to landing the spacecraft was to release a 10-cm optical marker sphere that was a soft bag filled with polymer pellets designed not to bounce, and whose surface was highly reflective. When illuminated by flash-beams, this would easily be spotted by the cameras and provide a reference for canceling the vehicle's horizontal drift. Three markers would be available to allow the sampling to be repeated at different points. As the surface characteristics of the target asteroid could not be precisely determined in advance, sampling tests were conducted at both full Earth gravity and at very low gravity in drop towers, and against a wide variety of materials ranging from bricks to coarse and fine-grained gravel and a simulated powdery lunar regolith.

The MUSES-C spacecraft was built by NEC–Toshiba Space Corporation, which was also responsible for the ion thrusters. The body was a $1.0 \times 1.6 \times 1.1$-meter box. A 1.5-meter-diameter wire-mesh parabolic antenna was on the side opposite to the platform that carried the scientific instruments and sampling horn. A pair of solar panels with a total area of $12 \text{ m}^2$ were capable of generating 2.6 kW at 1 AU. The propulsion system on one of the short ends had four gimbaled xenon microwave-driven thrusters, each of which was throttleable in the range 4.2 to 7.6 mN. The ion engines eliminated electrodes and introduced carbon parts to replace metallic ones, in order to reduce erosion in firing and to maximize the operational life. The sample-return capsule on the opposite end of the bus was 40 cm across and had a mass of just 18 kg. To test the capsule under realistic flight conditions, ISAS built a small technology demonstrator and launched it in 2002 on an H-IIA rocket supplied by the nation's other space agency NASDA (National Space Development Agency). DASH (Demonstrator of Atmospheric re-entry System with Hyperbolic velocity) was to have been recovered after 3 orbits of the Earth, re-entering at a speed only slightly less than that of an interplanetary return, but shortly after launch all contact with it was lost.[172]

The science camera used a five-lens f/8 refracting optic 15 mm in diameter which was mated to a $1024 \times 1024$-pixel CCD capable of 1-meter resolution at a range of 10 km. It had a choice of seven science filters with pass-bands near those of standard asteroid photometry, a magnifying lens for detailed close-ups at the time of landing, and a filter for optical navigation. It also carried an infrared spectrometer and an X-ray spectrometer to measure the composition of the asteroid, and two wide-angle navigation cameras.[173] To the 380-kg dry mass was added 60 kg of xenon for the ion engines and 70 kg of fluids for the 23-N bipropellant attitude control thrusters. The autonomous navigation and sampling maneuvers would exploit a number of sensors, including a wide-angle navigation camera, the science camera, a laser range finder, laser altimeter and fan-beam sensors to detect surface obstacles and the deflection of the sampling horn. Once the spacecraft had reached a 'bounding box' in space close alongside the asteroid, the laser altimeter and navigation cameras were to ensure that it did not stray out of position.[174,175,176,177,178]

It must be noted that at this same time the University of Arkansas was proposing the Hera mission for the Discovery program. It would use ion propulsion to visit up to three near-Earth asteroids and collect samples from each by using a pair of counter-rotating cutters to push material into a vessel at the end of a long boom that would be lowered to the surface. Other collection methods were also considered, including a sticky collector.[179] Asteroid sample-return was a highly competitive sport!

The MUSES-C spacecraft, minus it solar panels, during preparations on the ground. The circular object is the sample-return capsule. (ISAS/JAXA Image)

The MUSES-C mission was supported by NASA in the form of a nano-rover built by JPL that was to be delivered to the asteroid to explore its surface using a camera, an infrared spectrometer and an alpha-particle and X-ray spectrometer derived from Mars Pathfinder hardware. The rover was called MUSES-CN, weighed only 1.3 kg, and was designed to right itself in case it came to rest on a side. Once operating, it would either roll on its four small wheels or hop in the very low gravity.[180] But in November 2000 rising costs obliged NASA to cancel the $21-million rover.[181] The undaunted Japanese devised MINERVA (Micro/Nano Experimental Robot Vehicle for Asteroid) as a replacement. This 16-sided 'hopper' was 12 cm across and 10 cm tall, covered by solar cells and at 600 g was less than half the mass of JPL's design. It was equipped with a suite of six thermometers mounted on pins sticking out of the body to make thermal measurements of the asteroid's surface, a pair of stereoscopic cameras and a short-focus camera capable of a surface resolution of 1 mm. It was to explore autonomously and transmit its data using one of two loop antennas to the mothership over a distance of up to 20 km. Its locomotion system used a turntable and a flywheel to cause it to hop – the turntable would set the direction and the flywheel would supply energy capable of setting the rover flying for as much as 15 minutes in the low gravity.[182]

The chosen launch vehicle was the all-solid M-V, which had already been used to

JPL's MUSES-CN rover. (JPL/Caltech/NASA)

send Nozomi on its mission to Mars (see the next chapter). The original plan called for MUSES-C to set off in January 2002 and encounter near-Earth asteroid (4660) Nereus, which was the original target intended for the NEAR mission. If the launch were delayed the fall-back target was the unnamed asteroid (10302), which was also known as 1989ML. This was estimated at only 600 meters across and its surface to resemble typical black chondrite meteorites. In fact, delays soon promoted 1989ML to the prime target with launch scheduled for July 2002, the encounter in 2003 and sample-return in 2006. However, technical difficulties with the M-V launch vehicle, which failed on its third flight carrying an astronomical satellite in 2000, meant the 2002 launch window for 1989ML could not be achieved. The search for a new target settled on 1998SF36, which was a small body discovered in September 1998 by the automated telescopes of LINEAR (Lincoln Near Earth Asteroid Research). With an orbit ranging from just inside that of Earth out to 1.7 AU at a small inclination to the ecliptic, it was able to make close approaches to both Earth and Mars. In fact, such encounters were so common as to cause its orbit to be chaotic – meaning that small uncertainties in its present position become major differences over a period of only a few centuries. Statistical and dynamical considerations suggested it originated in the inner portion of the main belt, and is at risk of either hitting one of the inner planets or falling into the

MUSES-C's MINERVA hopping rover. The hoops on both ends were the antennas, while spikes sticking out were temperature sensors. (ISAS/JAXA Image)

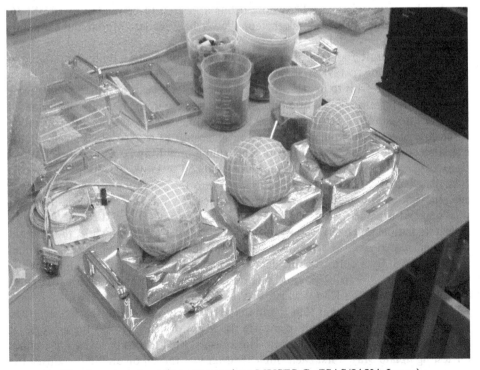

The three target markers mounted on MUSES-C. (ISAS/JAXA Image)

Sun at some time in the next 100 million years.[183] At the moment the energy requirement for an interception is one of the lowest for any body in the solar system apart from the Moon.

On the selection of 1998SF36 as the target for MUSES-C, a worldwide campaign was mounted to determine its rotation period, shape, taxonomic type etc. The most important results were radar observations during two close Earth flybys, the first in March–April 2001 and again in 2004. It proved to be a slightly asymmetrical and flattened ellipsoid of 0.55 × 0.3 × 0.28 km, making it 60 times smaller than Eros, the actual target for NEAR. The radar echoes showed the topography of 1998SF36 to be rather subdued compared to other objects either seen by radar or by flyby spacecraft. It appeared to rotate in just over 12 hours and its taxonomic type was determined to be S, and so related to stony meteorites.[184] If the mission went well then this would become the first asteroid 'imaged' by radar techniques to be visited by a spacecraft, and the radar astronomers were eager for visual observations to evaluate the fidelity of the technique they had introduced in the late 1980s and used to study a number of near-Earth asteroids. Also, being of a size intermediate between the stony meteorites and the already inspected S-class asteroids, observations of 1998SF36 might shed light on the 'stony meteorite paradox' of no known S-class asteroid being a precise match to the spectra of stony meteorites. The proposed solution was that the spectral characteristics of the asteroids had been masked by 'space weathering'. Remarkably, archive surveys identified five meteors photographed between 1953 and 2000 whose orbits appeared to be related to 1998SF36, and it was concluded that a weak meteor shower between late March and early July originated from this asteroid.[185]

As the spacecraft and its propulsion system had already been built and tailored to 1989ML, switching to 1998SF36 imposed some compromises, in particular adding a preliminary solar orbit to provide an Earth-flyby after 1 year, and thereby postpone the sample-return to 2007. Although a defective O-ring in the launch vehicle caused the launch to slip from December 2002 to the spring of 2003, the flexibility afforded by electric propulsion would still allow the spacecraft to reach 1998SF36 essentially as planned. MUSES-C lifted off from Kagoshima on 9 May 2003 on a direct ascent trajectory and was inserted into a solar orbit of 0.860 × 1.138 AU. It then opened its solar panels and extended the sampling horn into the operational position. As was customary for a Japanese mission, once the spacecraft was successfully on its way it was assigned a name: in this case 'Hayabusa' (peregrine falcon) because it was to collect its sample much as a rapacious bird catches its prey. Meanwhile, the asteroid 1998SF36 was dedicated to Hideo Itokawa, the aeronautical engineer who had founded ISAS. By a curious coincidence, he had worked for the Nakajima Aircraft Company from 1939 to 1945, and there designed one of the most successful fighter aircraft of the war, the Ki-43, which was also called Hayabusa.[186,187]

As the late launch meant that the initial orbit would not return the spacecraft to Earth at the time and position to initiate the trajectory to intercept the asteroid, it had to thrust to restore this slingshot. Hayabusa started to power up its ion engines for testing in late May. First each single thruster was turned on and left to thrust for about an hour. They were then fired in pairs but this operation encountered some contamination problem, as had Deep Space 1 just after launch. After heating and baking, tests proved the ability to fire single and pairs of thrusters for long periods of time, and then from July three thrusters simultaneously. Three engines would be thrusting at any given time, with the fourth in standby in case one of the engines had to be shut off because of a malfunction. In fact, one of the thrusters failed soon after launch. Meanwhile, in an overdue rationalization, ISAS, NASDA and the Japanese National Aerospace Laboratory were merged to create the Japanese Aerospace Exploration Agency (JAXA). After a smooth year in space, the spacecraft returned to Earth on 19 May 2004. On the inbound leg it took a series of color images of both Earth and the Moon to calibrate the camera. It flew over Japan heading towards its closest point of approach, 3,725 km above the eastern Pacific, and then departed for Itokawa in a 1.01 × 1.73-AU heliocentric orbit. Following this flyby, Hayabusa was caught in a record solar eruption which severely degraded the performance of its solar cells and thereby reduced the power to the ion engines. As a result, it was announced that the spacecraft's arrival at Itokawa would be delayed from June to September 2005, and its departure with the samples for the return to Earth would be postponed from November to December in order to preserve most of the time allocated to scientific observations and sampling.[188],[189] After solar conjunction in the summer of 2005, the ion engines were switched on again in late July to resume thrusting for Itokawa. On 31 July one of the US-built reaction wheels of the attitude control system failed.

Hayabusa's star trackers first spotted Itokawa shining at about magnitude 7 in late July, and over the next few weeks made 24 measurements of its position against the stars in order to refine its ephemeris. Once the range had reduced to several hundred thousand kilometers, it started to take pictures every 20 to 30 minutes to measure the rotation period and axis. A search of the space around the asteroid established there to be no satellites larger than 1 meter in size. On 28 August, now 5,000 km from its target, Hayabusa deactivated its ion engines. It was then in a 0.95 × 1.70-AU orbit, matching that of Itokawa. The ion engines had fired for a total of 25,800 hours in space, delivering a speed increment of 1.4 km/s for the consumption of only 22 kg of xenon. Of the four engines, one had accumulated a total firing time of 10,400 hours alone. At 01:17 UTC on 12 September 2005, Hayabusa used its bi-propellant thrusters to cancel all relative motion with respect to Itokawa and end the approach phase at the 'gate position' – which was 20 km from the asteroid along the line to Earth and also almost sunward. On the one hand this offered the engineering advantage of not requiring gimbals to be installed for either the high-gain antenna or the solar panels, but on the other it meant the encounter had to be when Itokawa was on the far side of the Sun from Earth, making communications susceptible to solar interference. It was impractical for Hayabusa to enter orbit around Itokawa, because its very weak and irregular gravity would make most close orbits unstable. The plan called for an active control technique by which the spacecraft – whilst remaining in solar orbit and independent of the asteroid – would use its thrusters to station-keep in a 'box' centered on the desired position. However, frequent corrections would be needed since relative speeds of as little as 1 cm/s would soon cause the spacecraft to drift out of position. A first estimate of Itokawa's mass was obtained during this stay at the 'gate position', but its gravity proved so negligible as to be readily exceeded by small perturbations due to solar radiation pressure. On 30 September Hayabusa moved to the 'home position', which was a nominal position for hovering some 7 km from the surface.

About 1,400 images were taken during these early phases of the reconnaissance. Itokawa appeared to be a 'rubble pile' asteroid, and the first such object visited by a spacecraft. It was divided into smooth sections and rough sections strewn with large boulders. High-resolution pictures showed over 500 boulders exceeding 5 meters in size, ranging to tens of meters. The abundance of boulders and the paucity of craters from which they could have come probably implied that the boulders were related to the collision in the main belt that created Itokawa itself, with the rocks cast adrift by that event soon being captured. All these details implied that the age of the surface was less than 100 million years. As in the case of many of the asteroids whose shape is known, Itokawa appeared to be made of two objects in stable contact, in this case resembling a sea otter with a small rounded lump forming the 'head', an elongated 'body' and a smooth area at the join. How these two objects came in contact without disintegrating in the process is not known. Hints of landslides were found at the base of the 'head', where surface slopes were the greatest.

Features on Itokawa acquired informal names drawn from Japanese space history, but the largest crater, which was 150 meters in diameter, was named Little Woomera after the site in the Australian outback where the sample-return capsule was to be recovered. Even including indistinct circular depressions, there were fewer than 100 craters in evidence, and they were mostly buried by coarsely grained regolith shaken by subsequent impacts – which was a characteristic shared with Eros. Three smooth regolith-covered regions existed. The 60 × 100-meter MUSES Sea was between the 'head' and the 'body' and extended to the south pole. The Sagamihara region was at the north pole. In fact, because the poles of such a small elongated body posses the weakest gravity, it was not surprising that the regolith should have pooled there. The third region, called Uchinoura, probably consisted of a cluster of three or four buried craters. In addition to these differences in texture and roughness, unusual variations in color and brightness could indicate either differences in the sizes of the grains in the regolith or differences in mineralogical composition. The infrared spectrometer obtained over 80,000 spectra at varying ground resolutions, showing Itokawa to be similar to some types of chondrite meteorites. This seemed to confirm that Itokawa (or at least its parent body) formed in the warmest part of the main belt, just beyond the orbit of Mars.

Altimeter readings taken at a distance of 7 km and images with a resolution of 70 cm were used to build a 3-D model of Itokawa to refine its size, shape and volume. Its irregular shape and varied terrain complicated the task of selecting landing and sampling sites. The requirement to have the solar panels properly illuminated ruled out areas with large boulders and steep slopes. Most of the MUSES Sea appeared to be sufficiently smooth, but was too steep apart from several small safe patches near the equator. Eventually, landing sites in the MUSES Sea and Little Woomera were chosen.

On 3 October, as Hayabusa was working on mapping Itokawa, the attitude control system lost a second reaction wheel, leaving only one operating. The thrusters took control as programmed, but the science operations were affected because it was no longer possible to accurately aim instruments. Furthermore, the reliance on using the thrusters introduced some doubt as to whether the propellant supply would last until 2007, when the spacecraft was due to return to Earth. During October it repeatedly left the home position to make a series of passes at varying altitudes and solar angles to obtain high-resolution vertical perspectives of the polar regions. On such a pass at an altitude of 3 km between 21 and 22 October, attitude control was switched off in order to collect 'clean' tracking data to improve the estimate of the asteroid's mass.

Hayabusa flew its first landing rehearsal on 4 November. The objectives were to test the laser ranging system, to release the first marker to confirm that the camera was able to detect it on the surface, to release the MINERVA hopper, and to check whether the single remaining reaction wheel could be used in concert with thrusters for precise attitude control. The spacecraft began its descent at an altitude of 3.5 km, but owing to the failed reaction wheels it had difficulty pointing the laser altimeter at Itokawa and received no range readings. At an altitude of 700 meters, still unable to

A view along the longest axis of Itokawa. Note the quantity of boulders on the 'rubble pile' asteroid and the lack of craters. The smooth patch of terrain at bottom is the Sagamihara Sea. (ISAS/JAXA Image)

'feel' the surface, the rehearsal had to be canceled and Hayabusa withdrew. Neither the marker nor the hopper had been released. When the very high resolution pictures obtained during the exercise were studied, they showed the Little Woomera target to be far too rocky for a safe landing. It was decided to undertake the second rehearsal and both sampling runs in the MUSES Sea and to postpone the first run, assigned to 12 November, by one week. An impromptu test was made on 9 November in which the camera was used to control the descent and this time the rehearsal was entirely successful, approaching to within 70 meters, withdrawing and then closing again to 500 meters. On the second descent it released one of the markers, but because it did so at a much greater altitude than originally intended the marker missed the asteroid altogether. However, images showed the marker drifting away with Itokawa in the background, and the fact that it was possible to measure the position of the marker verified the method. A second mass-determination pass was made on 12 November at altitudes ranging from 1,400 down to just 100 meters. Combined with estimates of the asteroid's volume, the mass measurements indicated that at least 40 per cent of it must be void space, which argued the case for it being rubble more porous than Eros and other asteroids and also of terrestrial analogs such as gravel beaches.

Another landing rehearsal was made on 14 November in which Hayabusa came within 55 meters of Itokawa, firing its engine from time to time to hold this range. The slope of the asteroid and the difficulty in controlling its orientation caused the spacecraft to approach much closer than planned. On realizing that Hayabusa was so close to the surface, the controllers on Earth commanded it to release MINERVA. Unfortunately the tracking was in the process of being transferred

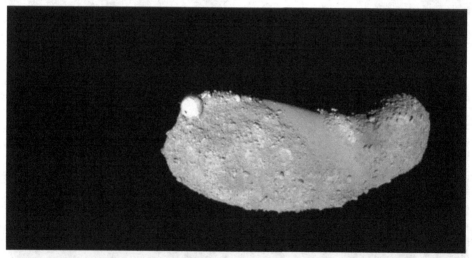

An image of Itokawa. The asteroid spins along an axis almost vertical in this picture, with the south pole on top and the north pole at bottom. The smooth patch of terrain is the MUSES Sea and Hayabusa's first sampling run (and unplanned landing) was made near the center right edge of the sea. The first target marker is visible drifting towards the asteroid. (ISAS/JAXA Image)

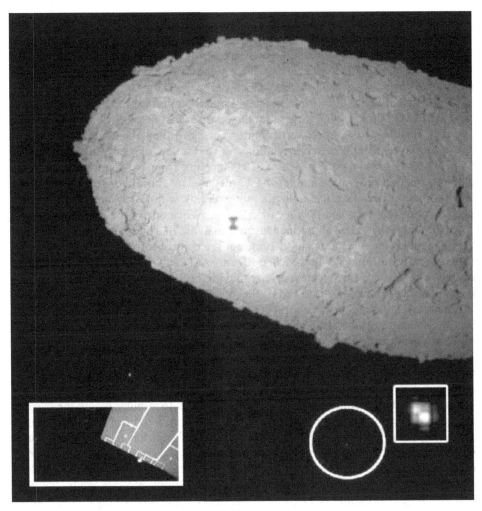

The MINERVA hopper imaged by Hayabusa drifting away and missing Itokawa altogether. Note the shadow of the spacecraft on the surface of the asteroid. In the inset at bottom left, the tip of Hayabusa's solar panels seen by the rover just after its release. (ISAS/JAXA Image)

from a Japanese antenna to a NASA antenna and no information on the vertical speed was available, and the command reached the spacecraft just after it had fired its engines to maintain its height and was moving away at about 15 cm/s, which was faster than the escape speed. Consequently, instead of falling towards the asteroid, MINERVA flew free in solar orbit. As images taken by Hayabusa showed the unlucky hopper and its cover drifting away, the only picture it returned showed a portion of its parent. MINERVA was monitored operating faultlessly until it passed out of communication range after 18 hours and was lost for good.[190]

Despite the fact that issues in controlling Hayabusa in its low-altitude phases had not been fully overcome, it was decided to proceed with the first landing attempt and sampling run on 19 November. Increasingly better X-ray spectra were integrated as the altitude diminished, showing the presence of silicon, magnesium and aluminum. Below 100 meters the images achieved their highest resolution of 6 mm, revealing a gravel-like terrain with small pebbles which were larger, on average, than the grains in the 'ponds' of Eros, and with increasingly smoother grains nearer the center of the 'sea'. The cables holding the target marker were severed at an altitude of 54 meters and then Hayabusa reduced its vertical rate in order to allow the target time to fall to the surface and settle. In a public relations move, the Japan branch of the Planetary Society had arranged for the names of 880,000 people from 149 countries around the globe to be engraved on the marker. At 17 meters, Hayabusa reoriented itself to the local horizontal, severing the high-gain antenna link with Earth and switching to a beacon mode that returned only a carrier signal. When one of the fan-beam sensors reported an obstacle, Hayabusa maneuvered to avoid it. The spacecraft should have executed an automatic emergency ascent, but the control software canceled this and the spacecraft continued to approach at 6.9 cm/s. At 21:09:32 the sampler horn made contact at 6°S, 39°E (the prime meridian having been defined as passing through a prominent black boulder at the end of the 'head') in the middle of the MUSES Sea. It rebounded and remained 'airborne' for 20 minutes, then rebounded again prior to finally coming to rest at 21:41 in the southwestern part of the MUSES Sea, close to the 'head', with either part of its body or the tip of a solar panel in contact with the ground. The sampling mechanism recognized the anomalous situation and correctly did not fire the projectiles. Nevertheless, the faster-than-intended landing could have dislodged material that passed up through the horn into the sample canister. Owing to the large distance from Earth, some 1.93 AU, it was 16 minutes before the manual emergency ascent command could be ordered, and this did not reach Hayabusa until 22:15, at which time the spacecraft dutifully lifted off after having spent 34 minutes on the surface of Itokawa. This was the first-ever launch from a solar system body other than Earth and the Moon. It ascended at high speed to an altitude of 100 km, where, owing to large attitude disturbances, it placed itself into safe mode. Before a new sampling effort could be attempted, engineers would have to regain full control of the spacecraft, and it would have to adopt the correct attitude and move back in to the holding point. It was also necessary to determine whether any of its components had been damaged by the high temperature while on the surface of Itokawa. In fact, scientists were able to use the temperature recorded on the X-ray spectrometer radiator, which had attained thermal equilibrium before the emergency liftoff, to estimate the temperature (about 40°C) and the thermal inertia of the surface. A reconstruction of the descent, landing and rebounds gave a measurement of the bearing strength of the coarsely grained surface. The first sample chamber, open since before launch, was closed, and the second was opened.

At about 13:00 UTC on 25 November Hayabusa started its second sampling run at an altitude of 1 km. Because it successfully located the previously deployed target marker, the final marker was retained. The spacecraft canceled its horizontal rates relative to the marker, and at about 35 meters began to receive and process data from the laser range finder. At 15 meters it once again oriented itself to the local

Two close views of the surface of Itokawa showing a rocky area and a transition area at the rim of a smooth 'mare'. Images taken on 12 and 9 November 2005 respectively.

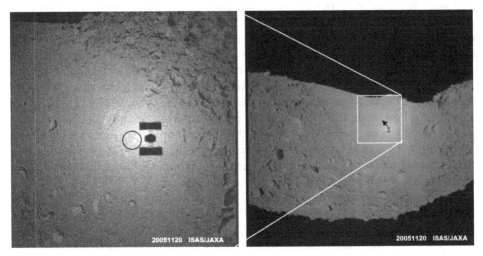

Two pictures taken during the 19 November 2005 sampling run (20 November Japanese time). The arrow in the image at right points to the sampling site. The bright spot on the image at left is the target marker left on the surface. (ISAS/JAXA Image)

vertical, severing the high-gain link. At 22:07, while descending at 10 cm/s, the laser sensor monitoring the horn noted a deformation as this made contact. The software should have fired the two projectiles within an interval of 0.2 second to obtain the sample, and then promptly lifted off. When contact was re-established after 30 minutes, the engineers inferred that the horn of the sampler appeared to be slightly warped as a result of coming into contact with the surface and were confident the computer had sent the firing command. It looked like Hayabusa had fulfilled the most difficult part of its ambitious mission, with the sample safely on board. This time, however, it had not survived contact with the surface in such good shape. Within hours of its liftoff, engineers discovered a serious leak in one of the attitude control propellant lines, which they tried to isolate by closing latches and valves. The redundant lines would not produce enough thrust, possibly because the fuel had frozen. The leak continued. This problem prevented the spacecraft from pointing its high-gain antenna at Earth, and without the full telemetry stored during the landing it was difficult to diagnose the attitude control problem. Worse, because the spacecraft was no longer able to maintain its solar panels facing the Sun, it was short of power and the drain on its the batteries was wreaking havoc on the computer and data management systems.

After several days of only intermittent contact, the recovery operation was begun on 30 November. But it was not yet clear whether Hayabusa could be brought back online in time to restart the ion engine in mid-December to head home. Meanwhile, tests were being conducted to assess the possibility of using xenon, of which there was ample supply, as a backup to the malfunctioning attitude control system. On 1 December the spacecraft again lost attitude control and as it drifted the falling power caused the computer to switch off some systems. With some control re-established, and using the medium-gain antenna, which swept Earth every 6 minutes, Hayabusa was able on 6 December to return the telemetry it had recorded during the sampling run. It was not clear, owing to the recording of events having been corrupted by the power reset, whether the projectiles had been fired or not. The only encouragement was that the temperature readings of the firing mechanism were higher than usual, which suggested that the command had been executed. It was later found, however, that a faulty command sequence sent to the spacecraft prior to the run had activated a self-protection mode that prevented the projectiles from being fired. The problem was attributed to inadequate rehearsal of the whole sampling sequence and to a lack of oversight. There was no possibility of making another landing attempt. Other lost data included imagery that could have helped to determine the attitude at landing and identified the precise location of the sampling site.[191,192,193,194,195,196,197]

As the recovery was progressing, on 8 December tracking stations noted a sudden attitude disturbance and a decrease in the signal strength. The attitude control system had been overwhelmed by the thruster effect of the continuing propellant leak. With Hayabusa tumbling, contact was not possible. Still, the engineers were confident that once all of the leaking propellant had either been vented overboard or had frozen in the pipes the wild spin would slowly stabilize, and they predicted a good chance of re-establishing contact at some time in 2006 or early 2007. If this could be done and the spacecraft was otherwise healthy, then there was the possibility of departing Itokawa's vicinity in the first half of 2007 and reaching Earth in June 2010, some 3 years behind schedule. In fact, the beacon signal was recovered on 23 January 2006 and on 6 February the spacecraft began to respond to commands. During the time that it had been out of contact with Earth its spin axis had tilted almost 90 degrees, the direction had reversed, and it had spun up. At one point it had completely lost power, severely damaging the batteries. The heaters had switched off, and since the frequency of the radio oscillators was controlled by their temperature, this had also changed. Finally, the pressure of the tank which was leaking had dropped to zero. Hayabusa was at that time leading Itokawa by about 13,000 km in their similar orbits around the Sun, and gaining by 3 meters with every passing second. Attitude maneuvers in February and software patches managed to restore some control over pointing, and with communications re-established the next several months were spent 'baking' the spacecraft to ensure that any propellant that might be coating its body was evaporated. The ion engines were checked and tested in late April, although thrusting to Earth would not resume for another year. Two of the engines were functional, one was inoperable and, as previously, the fourth was to be held in reserve. On 17 and 18 January 2007, after the health of the batteries had been assessed and the surviving ones recharged, the lid of the sample catcher was closed and this was stowed and latched into the return capsule to secure any material that might have been collected either during the first 'hard' landing or on the second run. A thermometer recorded the cold canister being pushed into the warm capsule, cooling the latter by several degrees, thereby providing indirect confirmation that the catcher was indeed safe.

After further tests, Hayabusa started its much-delayed return to Earth on 25 April 2007. It maintained thrusting attitude by carefully balancing the torques generated by the ion engines and by solar radiation pressure, and also by parsimonious use of the single available reaction wheel. A complication was that the orientation had to prevent the frozen fuel lines from being exposed to the Sun, especially at perihelion in June, lest leaks restart and disrupt attitude control. After the first phase of thrusting ended on 24 October, the spacecraft was again spun up for stability. By this time, its engines had operated for a total of 31,000 hours. Only one was used at any given time in order to prevent excessive wear and minimize the odds of failure. Meanwhile, another one had failed. Aphelion was in February 2008. At solar conjunction in May, Hayabusa was 2.5 AU from Earth. On 4 February 2009 it began the 13-month thrusting phase that would provide the 400-m/s velocity change required to reach Earth in June 2010. After exactly nine months of continuous firing, on 4 November a voltage surge in the only usable thruster was recorded, indicating its impending failure. This would leave the spacecraft with no reliabily working thruster to complete the final 4 months of firing required to return to Earth. In this case as well, Japanese controllers found an ingenious solution to the problem: by combining the ion source from one engine with the neutralizer from another, thrusting could be resumed. A workaround was also devised just in case the only remaining reaction wheel should fail. The engine was turned off for a week starting on 5 March 2010, by which time the trajectory of Hayabusa would allow a close pass by Earth. A precise orbit determination was performed. The main thrusting phase finished on 27 March, in a 0.983 × 1.654-AU orbit. The final weeks of thrusting had moved the point of closest approach to Earth from the day-side to the night-side, so that it would travel in the same direction as the Earth's rotation and minimize its relative speed at re-entry.

Re-entry operations started 2 months prior to arrival back at Earth. Four course corrections were made by the ion engine during this period. The first was 42 days and 17 million km out, and the second 21 days and 9 million km out, with the latter aiming for a point 630 km above the Earth's surface. This had been increased from the original 200 km to preclude an accidental re-entry if the spacecraft were to fail during the final approach. On 12 May it returned a heavily bloomed picture of Earth and the Moon taken by a star tracker from a range of more than 13 million km; the

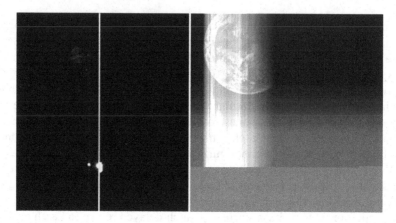

Two images of the Earth captured during the approach phase by star cameras on board Hayabusa. The first (left), taken on 12 May 2010, shows the overexposed Earth as well as the Moon. Hayabusa at the moment was 13.5 million km distant. The second was taken just one hour before re-entry and in fact it had not been entirely transmitted to Earth when contact was lost. (ISAS/JAXA Image)

science camera having been inoperative since the final sampling run. The longest correction started on 3 June and moved the closest point of approach to a collision course over the Australian outback. The fourth and final maneuver 3 days before re-entry fine-tuned the approach trajectory and landing site. This was in the desert part of the Woomera test range, from which Europe attempted to launch a satellite in the 1960s. The spacecraft was imaged during its final approach by some of the world's largest telescopes.

On the last day of the cruise, Hayabusa's attitude control system malfunctioned, but it was successfully reset just in time for re-entry. On 13 June, the capsule was released 3 hours before striking the Earth's atmosphere. At that time, Hayabusa was 40,000 km from Earth and over India. About 2 hours later, Hayabusa was flipped to take a final picture of Earth using its star-tracker camera, but contact was lost after about 75 per cent of the image had been returned to the ground. The spacecraft then slammed into the atmosphere at 13.51 UTC. Both it and its sample capsule were traveling at a speed of 12.2 km/s relative to Earth, the second fastest re-entry ever. Hayabusa started to decelerate at an altitude of about 75 km and then it shed debris until, at 47 km, it had completely burned up. The fireball was monitored by cameras and spectrographs on the ground as well as by an international team flying NASA's instrumented DC-8 aircraft along the track. The re-entry and destruction of the spacecraft were thus monitored in extreme detail. For example, a strong lithium line that appeared in the spectrum at an altitude of 55 km was traced back to the explosion of the lithium-ion batteries.[198,199] The capsule experienced a 25-g deceleration, but was unscathed. Upon going subsonic, the capsule released its aft heat shield and deployed a parachute. At the same time, the spent carbon phenolic forward shield was released to fall clear. The capsule landed only 500 meters from the aim point at 14.12 UTC. A helicopter tracked the beacon as the capsule

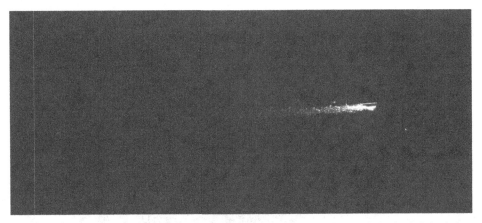

Hayabusa breaking up in the Earth's atmosphere. The sample capsule is the bright dot at lower right. (NASA/ARC-SST/SETI Institute)

descended, and then marked its position with GPS. After dawn, helicopters located the shiny sample canister and its parachute intact. The fore and aft shields were located some 5 km away. Although the ablator material was extensively eroded, the heat shielding was in remarkably good shape. Before the capsule could be taken to a clean room at the Woomera Range Control Center, its unused pyrotechnics were safed. Once in the clean room, the batteries were removed and the capsule was cleaned with nitrogen jets.

On 18 June it arrived at the purpose-built JAXA curation center in Sagamihara, in the outskirts of Tokyo.[200] A CT scan confirmed that the lid had properly sealed, but disappointingly it showed no evidence of any dust particles exceeding 1 mm in size. The inner sample container was then removed and cleaned with carbon dioxide and plasma jets. One of the two sample chambers had been left open at launch as a precaution against a failure of the opening mechanism before reaching Itokawa. This of course created risks of contamination by terrestrial dust, and JAXA had collected samples of dust from the area around the launch pad in order to be able to identify possible contaminants after re-entry. On the other hand, the second chamber would have remained sealed and perfectly clean. On 24 June chamber A of the sample canister, used for the second run was opened first, in near vacuum. Chamber B, which was more likely to have collected particles from Itokawa, was not to be opened until later. Traces of gas were found early on. That was not supposed to be there. It could have come from Itokawa, but it could equally represent spacecraft outgassing or even terrestrial contamination. It was later determined to be air that had leaked inside the container through a double O-ring seal in the days prior to the opening. Minuscule specks of dust were discovered inside the canister that scientists hoped were off the surface of Itokawa, although only detailed chemical and mineralogical analyses would show whether this was the case.

The first attempt to pick out grains using a quartz glass probe was unsuccessful. The team then switched to scraping the walls of the chamber with a Teflon spatula to minimize contamination. This approach was more successful and retrieved 1,534

An explosive expert safes the minuscule Hayabusa sample capsule on the morning after its landing in the Australian outback. (ISAS/JAXA Image)

samples larger than 10 micrometers. Finally a "brute force" approach was used, simply turning the container upside down over a quartz disk and tapping it with of a large screwdriver, thereby recovering an additional 40 particles, including several up to 180 micrometers (0.18 mm) across. Many aluminum flakes were also found, off the container itself. The results of their preliminary scanning electron microscope analyses were reported in November 2010, relating to the four of the largest grains whose elemental abundances had been measured. They were composed of iron-rich olivine and pyroxene, and had mineral phases matching the ordinary chondrite class of meteorites and with Itokawa's surface, whilst having no corresponding terrestrial analog. Moreover, none of the types of rock that could be expected if the sample

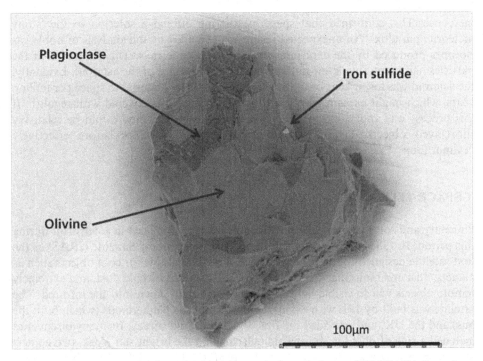

A scanning microscope view of a largish sample from Itokawa. (ISAS/JAXA Image)

chamber had suffered contamination either at the launch site or in Woomera was found. JAXA could then confirm that the sample, minuscule though it was, was indeed from the asteroid; the first sample ever returned from the surface of another celestial body other than our own Moon. Detailed analysis of the microscopic particles would be difficult but, as Stardust's experience had indicated, not impossible.

The B chamber was opened on 7 December, but a visual inspection revealed no large particles. Preliminary investigations started in late January 2011 at a number of Japanese laboratories. These addressed the elemental composition, the presence of carbon-bearing organic molecules, the morphology and petrography of samples, their isotopic ratios and trace element composition, their crystal structure and mineralogy, the internal structure and the isotopic ratio and presence of noble gases from the solar wind. Particles tenths or hundredths of millimeters across could be used to search for basic organics and amino-acids, if they too came from Itokawa. Amino-acids are well known to be present in certain primitive types of meteorites.

Particles were made of a mix of several minerals, with complex three-dimensional textures. Analyses showed that exposure to the solar wind had created opaque "blobs" of iron sulfide, only nanometers across, on the surface that reddened the appearance of the asteroid. Beneath this surficial layer there was a second layer, tens of nanometers thick, entirely made of metallic iron blobs. Together, these layers masked the real spectrum of the asteroid and made it look different from any known meteorite. This confirmed that space weathering offered a solution to the 'stony meteorite paradox'. An unexpected result was provided by the analysis of noble gas isotopes produced by the exposure of grains to cosmic rays. This showed that the particles had spent only a few million years on the surface of the asteroid. Evidently, the asteroid was losing tens of centimeters of superficial material to space per million years, which might explain why so little of its surface was covered with regolith. If this process was applicable to other near-Earth asteroids, they would be relatively short-lived objects, lasting no more than a billion years before effectively "evaporating".[201,202,203,204,205,206,207,208,209,210]

## A SPACE TELESCOPE

Planetary and solar probes were not the only spacecraft placed in solar orbit during this period. In 1983 NASA launched the Infrared Astronomy Satellite (IRAS) as the first satellite designed for astronomy at wavelengths suitable for cool objects such as planets, dust and molecular clouds, hot objects shrouded by cold dust, and extremely remote objects whose visible light has been 'redshifted' down into the infrared. The satellite was built by JPL with extensive input from the Netherlands (which built the bus) and the UK, and provided the first all-sky infrared survey. Its many discoveries included a ring of dust (or an asteroid belt) around the bright star Vega, two comets and a remarkable near-Earth object in the form of asteroid (3200) Phaethon – which could actually be a dormant comet. In order to avoid the detectors at the focal plane of the detector from being blinded by infrared radiation from the telescope itself, the structure was cooled to a few degrees above absolute zero by the act of evaporating liquid helium. The mission ended when the helium supply ran out after 10 months. Building on this experience, the Netherlands pushed the European Space Agency to fund the Infrared Space Observatory (ISO) that was approved in 1986 and launched in 1995.

Astronomers in the US also planned a successor to IRAS which would be able to perform detailed observations of selected objects revealed by the sky survey. In fact, such a mission had first been recommended by the National Academy of Sciences in 1974. The proposed Shuttle Infrared Telescope Facility (SIRTF) was to be operated in a Shuttle's payload bay, make observations for a fortnight and then be returned to Earth for refurbishment and re-flight. This idea built upon NASA's early tenet that Shuttle flights would be cheap and frequent, and that human intervention would be crucial to operating complex hardware in space. In this case, technical reality would have doomed the concept. In particular, the Shuttle's low orbit would have severely limited the observation time of individual targets to tens of minutes at best. Also the cloud of gas from the vehicle's attitude control system would probably have blinded an infrared telescope, and in any case all manner of exhaust and debris would have condensed on the cryogenic optics. In fact, Challenger carried an infrared telescope as part of a Spacelab laboratory in July 1985 and the results showed that the orbiter made a very poor platform for astronomy in this wavelength range.[211] Although the SIRTF gained the support of the scientific community, it was ranked in priority after a large X-ray telescope. Together with the Hubble Space Telescope, this Advanced X-ray Astrophysical Facility, the Gamma-Ray Observatory and the renamed Space Infrared Telescope Facility were to form a group of Shuttle-launched and serviced 'Great Observatories' that would, between them, cover most of the electromagnetic spectrum. By the end of 1985, the SIRTF had been redesigned as a 'free flyer' that would be carried into low orbit by the Shuttle and transported to a more appropriate orbit by an automated 'space tug' named the Orbital Maneuvering Vehicle (OMV). After delivering its payload, the OMV would fly back to the Shuttle to be returned to Earth for refurbishment. When the SIRTF's helium was almost exhausted, another Shuttle would send a tug to collect the telescope and return it to the Shuttle to enable astronauts to replenish the coolant, after which the tug would redeploy the telescope and return to the Shuttle. However, this unrealistic scenario was doomed by the loss of Challenger in January 1986 and by the cancellation of the OMV in 1987 when its budget spiraled out of control.

In an effort to save the SIRTF mission, scientists proposed a low-cost IRAS-based observatory that could be launched by an expendable Titan IV into a highly elliptical orbit having an apogee at about 100,000 km – similar to that in which the European ISO had been operated. The high orbit would allow longer observation times for any given object, would minimize interference from the radiation in the van Allen belts, and by spending most of its time far away from Earth would minimize the thermal loads and the rate of coolant consumption in order to maximize the operating life. Such a telescope would have a mass of 5,700 kg and cost in excess of $2 billion to build, launch and operate.

In 1991, in view both of its shrinking budget for science and of the poor technical performance of two major missions like the Hubble Space Telescope and Galileo, NASA canceled the SIRTF – only to have a change of heart and reinstate it later the same year because it was supported by communities which normally did not share a common interest: astrophysicists and planetary scientists. However, NASA sought to reduce the complexity (and hence the cost) of the mission by applying the 'faster, cheaper, better' approach. The spacecraft was totally redesigned to trim its mass (but not its capabilities) to about 2,500 kg in order to use a less expensive Atlas launcher. The crucial decision was to inject the SIRTF into a solar orbit trailing behind Earth. This had important consequences in terms of mass. In particular, by removing the telescope from the heat radiated by Earth it could either operate for longer using a given amount of helium or require less coolant to operate for a given time. It was, of course, decided to carry less coolant in order to reduce the mass. And flying in solar orbit would require fewer orbital adjustments, which would further trim the mass by reducing the amount of propellant carried. Finally, if the telescope were placed in an orbit that receded at a rate of only about 0.1 AU per year, communications could be readily maintained throughout the planned operating life. Unfortunately, just as the scientists believed they had achieved a viable design which would cost less than $1 billion, the project was canceled again! But all was not lost, and yet another redesign was approved and hardware construction started in 1996. The mass of the spacecraft had been further trimmed by a 'warm launch'. Whereas the IRAS and ISO satellites were launched with their telescopes and detectors already cryogenically cooled, the

SIRTF's telescope would be launched at room temperature, chilled in the shadow of a sunshade for a month, and then lowered to its operating temperature using helium. This would require much less helium – so little, in fact, that the spacecraft would be sufficiently small and lightweight that it would be able to be launched by a relatively cheap Delta II. In fact, the SIRTF should be able to operate for 5 years using a tank of helium with one-sixth the capacity of that of the ISO satellite, which had lasted 28 months. Even when the helium ran out, some of the SIRTF's instruments would be able to operate uncooled for several years if the telescope remained in the shadow of the sunshade. And by then the spacecraft would be over 0.6 AU from Earth and the mission would end because that was too far to efficiently transmit the scientific data. As a result of all this clever innovation, the cost of the mission was reduced to about $720 million.

In the work breakdown, the SIRTF's bus would be built by Lockheed Martin and the cryogenic telescope assembly would be developed by Ball Aerospace, which had considerable experience in this field. The spacecraft was a cylinder which stood 4.55 meters tall and had a diameter of 2.11 meters. The fairly conventional bus provided power, communications, attitude control, etc. It controlled its attitude using reaction wheels and compressed nitrogen jets which drew from a spherical tank located at its center. Including 50 kg (360 liters) of liquid helium and 16 kg of nitrogen, the 923-kg launch mass was one-sixth of the Titan IV concept. The mirror was 0.85 metres in

The Spitzer infrared space telescope.

diameter and made of lightweight beryllium. It was figured for a focal ratio of 1.2, and this was boosted to 12 using Ritchey–Chrétien optics. The ridge of solar panels shaded the entire length of the telescope tube and produced 427 W of power. This side of the spacecraft was to continuously face the Sun. One side of the tube and the dewar of liquid helium were polished to reject heat radiating from the solar panel, while the opposite side, facing deep space, was painted black to radiate all excess heat. Between them, radiative cooling and liquid helium ensured that the optics and instruments were cooled to about 5.5K. Furthermore, to minimize heat conduction to the telescope, the bus and solar panels were mounted on specially designed trusses, with the solar array barely connected to the telescope. Three scientific instruments was installed near the focal plane: a camera, a photometer and a spectrometer.

The spacecraft was delivered to Cape Canaveral in March 2003 for launch by the 7925-Heavy form of the Delta II that used nine strap-on boosters of the type made for the canceled Delta III and which were 40 per cent more powerful than standard. The plan to launch in mid-April was confounded first by a suspected delamination in the carbon composite motor casings of the strap-on boosters, and then by conflicts with missions which had tighter launch windows, including a military GPS satellite and the two Mars Exploration Rovers. When the launch was postponed to no sooner than August, the mission logo was removed from the launch vehicle and replaced by the one for the Mars Exploration Rover that it would now launch. After all the other spacecraft had been successfully launched, the SIRTF followed on 25 August. Once released into the planned 0.996 × 1.019-AU, 369-day solar orbit, it shed the cover that had sealed the telescope and opened the door that sealed the dewar. After status tests, the first stage of the cooling process started. Once the telescope had reached 40K, heat exchangers were enabled to allow helium to take it down to its operating temperature. As was customary for NASA's space observatories, the SIRTF was given a name: in this case the Spitzer Space Observatory, in honor of the astronomer Lyman Spitzer who, in the 1940s, was one of the first to propose that telescopes be placed in space in order to observe free of the effects of the unsteady atmosphere. Its contributions to astronomy during its 5 years of operation have been significant and wide-ranging, but are mostly beyond the scope of this book. To mention only those which are pertinent, it helped to characterize minor bodies such as comets Tempel 1 and Churyumov–Gerasimenko and asteroid Steins, all of which have been selected as targets for other deep-space missions.[212,213]

## REFERENCES

1 Conway-2007
2 Soderblom-2000a
3 NASA-1998a
4 Rayman-1999
5 Conway-2007
6 Casani-1996
7 Kronk-1984a
8 Buratti-2004a
9 Lazzarin-2001
10 NASA-1998a
11 Soderblom-2000a
12 Wang-2000
13 Rayman-1999
14 Soderblom-2000b

15  Desai-2000
16  ST-1999
17  Kerr-1999
18  Riedel-2000
19  Buratti-2004a
20  Richter-2001
21  Conway-2007
22  Kronk-1984b
23  Spaceflight-1992
24  Kronk-1984c
25  Lamy-1998
26  Soderblom-2001
27  Rayman-2000
28  Rayman-2002a
29  Rayman-2002b
30  Soderblom-2002
31  Soderblom-2004a
32  Buratti-2004b
33  Britt-2004
34  Young-2004
35  Tsurutani-2004
36  Soderblom-2004b
37  Farnham-2002
38  Conway-2007
39  Rayman-2002b
40  Racca-1998
41  Scotti-1998
42  For Discovery and 1993 studies see Part 2, pages 349–352
43  For HER, CISR, Giotto 2, CAESAR and SOCCER see Part 2, pages 47 and 103–106
44  Królikowska-2006
45  Carusi-1985
46  For Whipple shields see Part 2, page 31
47  Mitcheltree-1999
48  Tsou-1993
49  Schwochert-1997
50  Hirst-1999
51  Brownlee-2003
52  Brownlee-1996
53  NASA-1999a
54  Taverna-1999a
55  For NEAR camera contamination see Part 2, page 365
56  Krüger-2000
57  Lämmerzahl-2006
58  Domokos-2009
59  Duxbury-2004
60  Newburn-2003
61  Kennedy-2004
62  Brownlee-2004
63  Tsou-2004
64  Sekanina-2004
65  Kissel-2004
66  Tuzzolino-2004
67  Covault-2004b
68  Desai-2008a
69  Desai-2008b
70  Mecham-2006
71  Martínex-Fríaz-2007
72  Brownlee-2006
73  Hörz-2006
74  Sandford-2006
75  McKeegan-2006
76  Keller-2006
77  Flynn-2006
78  Zolensky-2006
79  Burnett-2006a
80  Stephan-2008
81  Nakamura-2008
82  Ishii-2008
83  Matzel-2010
84  Westphal-2010
85  Wolf-2007
86  Veverka-2011
87  Carroll-1997
88  Carroll-1993
89  Suess-1956
90  Carroll-1995
91  Cheng-2000
92  Pieters-2000
93  Mustard-1999
94  Baines-1995
95  For Lagrangian points and halo orbits see Part 1, pages liii, 387; and Part 2, pages 58–60
96  Rapp-1996
97  McNeil Cheatwood-2000
98  Barraclough-2003
99  Neugebauer-2003
100  Stansbery-2001
101  Jurewicz-2003
102  Nordholt-2003
103  Hong-2002
104  Jurewicz-2003
105  NASA-2001
106  Koon-1999

107 Veazey-2004
108 Hong-2002
109 Burnett-2003
110 NASA-2001
111 NASA-2004
112 Smith-2003
113 Wilson-2002
114 Barraclough-2004
115 Reisenfeld-2005
116 Wilson-2002
117 Wilson-2004
118 Steinberg-2003
119 Desai-2008c
120 McNamara-2005
121 Burnett-2005
122 Burnett-2006b
123 Stansbery-2005
124 Allton-2005
125 Lauer-2005
126 Hittle-2006
127 NASA-2005
128 Morring-2006
129 Grimberg-2006
130 Hohenberg-2006
131 Heber-2007
132 Heber-2009
133 Science-2008
134 Burnett-2011
135 Marty-2011
136 McKeegan-2011
137 Clayton-2011
138 Reynolds-2001
139 Farquhar-1999
140 Farquhar-2011
141 Farquhar-1999
142 Veverka-1999
143 Kronk-1999a
144 Whipple-1950
145 Boehnardt-1999
146 Crovisier-1996
147 ST-1996
148 Kronk-1999b
149 Meech-2004
150 NASA-2002
151 Cochran-2002
152 Covault-2002
153 Dunham-2004
154 Morring-2002a
155 NASA-2003
156 Dunham-2004
157 Morring-2002b
158 Farquhar-2011
159 NASA-1999b
160 Staehle-1999
161 Woerner-1998
162 Woerner-1998
163 Zimmerman-2001
164 Carroll-1997
165 For the Pluto–Kuiper Express see Part 2, page 374–379
166 For early solar probe missions see Part 2, pages 125–130
167 NASA-1999c
168 Staehle-1999
169 Reichhardt-2000
170 McFarling-2002
171 Kawaguchi-1995
172 Morita-2003
173 Nakamura-2001
174 Kawaguchi-1996
175 Kawaguchi-1999
176 Kawaguchi-2003a
177 Kubota-2003
178 Sekigawa-2003
179 Sears-2004
180 Wilcox-2000a
181 Flight-2000
182 Kubota-2005
183 Michel-2006
184 Ostro-2004
185 Ohtsuka-2007
186 Harvey-2000
187 Francillon-1970
188 Kawaguchi-2004
189 Clark-2005b
190 Yoshimitsu-2006
191 Yano-2006
192 Saito-2006
193 Fujiwara-2006
194 Demura-2006
195 Abe-2006a
196 Abe-2006b
197 Okada-2006
198 Abe-2011
199 Borovicka-2011
200 Ishii-2003
201 Nakamura-2011
202 Ebihara-2011

# 9

## Mars invaded

### LOSING HOPE

Encouraged by their success with probes to investigate Halley's comet, in the mid-1980s Japanese scientists at ISAS started to plan their second step in solar system exploration. In addition to technological missions that would eventually lead to the Hayabusa asteroid sample-return, they also desired science-oriented missions such as orbiters to study the atmospheres of Venus and Mars. Although Venus would be easier to reach, and a spinning orbiter based on the successful Halley probes was designed, it was decided that Mars was a more worthy scientific target. Go-ahead for this Planet-B mission was authorised by the government in the early 1990s and detailed work started in 1992. Venus exploration was left to the Planet-C mission that was launched in 2010. Planet-A had been the Suisei probe sent to Halley's comet.

Planet-B was to pick up the program of the ill-fated Soviet Fobos orbiters in studying the Martian atmosphere. Specifically, the aim was to gain insight into the process by which much of the atmosphere was lost to space, including possibly its water vapor. The Fobos 2 investigations had been hampered by the fact that this spacecraft had never approached the planet closer than 850 km. Planet-B would be inserted into an orbit better suited to such studies. It would be almost equatorial, but retrograde, and have a very low periapsis at 150 km which radio-occultation studies had shown to be the altitude at which the electron density of the ionosphere was greatest; and with an apoapsis at 15 Mars radii, well beyond the orbit of the outer moonlet Deimos, in order to study escaping ions. The low periapsis would also have enabled Planet-B to investigate the patterns of magnetism on the surface that would soon be discovered by NASA's Mars Global Surveyor.[1] In addition, Planet-B would perform moderate-resolution imaging of the planet and its duo of moonlets, and seek evidence of tenuous rings formed by dust escaping from, yet confined by, the moonlets. The nominal mission duration was one Martian year, which lasts 687 Earth days, but it could be extended for a second local year.

The launch vehicle would be ISAS's new M-V (sometime called the Mu-5) that had replaced the Mu-3SII used to dispatch Suisei and Sakigake to Halley's comet. The three-stage M-V, which was the world's largest 'all-solid' space launcher, had twice the lifting capacity of the Mu-3 and, when fitted with a fourth 'kick' stage, could dispatch several hundred kilograms on an Earth-departure trajectory.[2] The mission was to start in 1996, in the same window as Mars 96, Mars Pathfinder and Mars Global Surveyor, with arrival at the planet in September 1997, but delays in the development of the M-V meant it had to be slipped to the next window, with arrival on 11 October 1999. However, the 1998 window was not as effective as that of 1996 in terms of energy, and when it was realised that even after engineers had gone to great lengths to reduce the mass of the spacecraft the M-V could not dispatch it directly to Mars, ISAS investigated using the gravity of the Moon to overcome the energy shortfall. Although lunar gravity-assists would not provide sufficient boost to reach Mars, a solution was found in the solar perturbations near the boundary of the Earth's sphere of gravitational influence and solar-dominated space – the Weak Stability Boundaries (WSB). The first flyby of the Moon was to place Planet-B in an Earth orbit with a distant apogee. On returning to perigee, the orbit would have been so perturbed by the Sun that its direction of travel would be retrograde. This would result in a much higher relative velocity during the second lunar flyby. Although this would boost it just sufficiently to leave the vicinity of Earth, on departing the Moon its trajectory would be towards Earth. Nevertheless, its energy would not yet be sufficient to reach Mars. To achieve this, two further maneuvers would be required. The first would be an Earth flyby several days after the lunar encounter. In addition to giving the spacecraft an additional boost, this would redirect it towards Mars. And then, as Planet-B was at its closest point of approach to Earth, it would fire its engine for a 420-m/s kick that would enable it to reach Mars. Powered flybys such as this were first pioneered by Voyager 2 at Jupiter. The energy derived equated to 24 kg of propellant, and this saving in mass brought the mission within the capability of the M-V launcher. Fortunately, most of the complex maneuvers needed for this and other ISAS missions, including the Geotail Earth satellite and the Lunar-A orbiter, had been tested in 1990 by ISAS's small technology demonstrator Hiten, with a series of lunar flybys prior to entering orbit around the Moon.

Having inherited some of its systems and structure from the EXOS-C scientific satellite, Planet-B was a fairly conventional spacecraft. It consisted of an octagonal bus with alternating short and long sides, 1.6 meters across and 58 cm tall. There was a 1.6-meter wire-mesh high-gain antenna on top, capable of a data rate in the S/X-Bands of 4 kbits per second. Two low-gain antennas, one mounted on the feed of the primary parabolic antenna and the other on the opposite side, could receive commands and transmit low-rate engineering data. The second antenna was offset 1 meter from the spin axis in order to allow the spin rate to be determined from the Doppler shift modulations arising from this geometry. The propulsion module on the underside had two tanks each for hydrazine and nitrogen tetroxide, the associated plumbing, valves and pressurization hardware, and a truss with the bi-propellant main engine that generated a thrust of 500 N. It was about 2.5 meters from the top of the low-gain antenna on the feed of the dish antenna to the far end of the antenna on the underside. Power was provided by two panels with a total span of

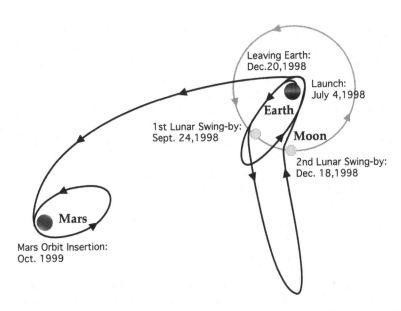

The journey of Planet-B 'Nozomi' from Earth to Mars via the Moon as initially planned. (ISAS/JAXA)

6.22 meters and a solar collector area of 4.6 square meters. Nickel metal-hydride batteries provided backup during launch and eclipses by Mars, Earth and the Moon. The spacecraft was normally to be spin-stabilized at 7.5 rpm, but faster rates would be employed during engine burns at Earth departure and orbit insertion. It would determine its attitude using a suite of sensors that included a lightweight star tracker. Ten 2.3-N hydrazine thrusters arranged in clusters on the short sides of the bus would provide attitude control to hold Earth in the high-gain antenna beam, and also make small trajectory corrections. A small aluminum plaque had been etched with the names of 270,000 project supporters. Planet-B had a launch mass of 540 kg, including 282 kg of propellant and a scientific payload of almost 33 kg. At 18.6 billion yen or $166 million the mission was relatively cheap, even in comparison to NASA's low-cost missions.

The payload of fifteen instruments included six which were either provided by foreign nations or by international cooperation. A neutral mass spectrometer from the United States, and ion spectrometers provided by Canada and Sweden were to make in-situ measurements of ions escaping from the atmosphere. These were also to be monitored remotely by an ultraviolet imaging photometer. The latter would also measure the hydrogen-to-deuterium ratio, the value of which would provide information about the history and evolution of the atmosphere. Conditions in the ionosphere were to be monitored using a plasma-wave sensor which had two pairs of wire antennas spanning 52 meters from tip to tip that ran out from the corners of the bus, as well as by an electron temperature probe. The former could also be used as a radio altimeter to measure the altitude around periapsis and as a radar to determine the roughness and texture of the surface and structure of the subsurface, including the possible presence of water ice within the uppermost 100 meters.[3,4] A very-low and extremely-low frequency (VLF and ELF) radio-wave receiver was to detect plasma waves. A 5-meter mast which had a 3-axis fluxgate magnetometer at its far end was balanced on the opposite side of the spinning bus by the Canadian spectrometer on a 1-meter boom. The German dust counter was a lighter version of an instrument flown on the Hiten lunar orbiter. It was to measure the mass and speed of impacting dust. Atmospheric ions picked up and accelerated by the solar wind would be investigated by instruments for high-energy particles and energetic ions, as well as by the Swedish spectrometer. Complementary observations would be made by an energetic-electron detector. The camera consisted of a small push-broom CCD telescope that would be capable of resolving details on Mars as small as about 60 meters at periapsis. It would also be used for low-resolution full-disk imaging near apoapsis for meteorological studies, during a number of encounter seasons with Phobos and Deimos, and to search for faint dust rings by forward-scattered sunlight. The camera included image compression hardware provided by the French Space Agency CNES (Centre National d'Etudes Spatiales). This would enable 10 images to be collected and returned on each orbit.[5] Solar observations would be made during the cruise to Mars, with an extreme-ultraviolet spectrometer meant to search for helium ions in the Martian atmosphere being used to measure helium in the heliosphere. Finally, a US-provided ultrastable oscillator was to be used to precisely track the spacecraft and to sound the atmosphere of Mars during radio-occultations. Scientific data would be stored in a 16-Mbyte recorder, being collected at different

A model of the Planet-B probe in its final form. (Courtesy Brian Harvey)

speeds depending upon the altitude over the planet, and later transmitted to Earth using the high-gain antenna.[6]

While the Planet-B spacecraft and its instruments were under development, the M-V succeeded on its first launch and placed the MUSES-B space radio-telescope into orbit in 1997. Although Planet-B was to be the third payload, ongoing delays with the Lunar-A mission led to this being canceled, advancing Planet-B to second in line. It was launched from ISAS's Kagoshima range on the southern tip of the island of Kyushu on 3 July 1998. After several minutes in a low parking orbit, the fourth and final stage fired to achieve a 703 × 489,382-km orbit with its apogee at lunar distance and a period of about 15 days. On successfully deploying its solar panels, Planet-B was renamed 'Nozomi' (Hope) in line with the Japanese tradition of not assigning a spacecraft a proper name until its mission was underway. (It was also referred to as SS-18 because it was the 18th Japanese scientific satellite.) Its historical significance was that it was the first non-Russian, non-US Mars mission to be launched.

During the first days and weeks of the flight, instruments were activated one by one for checking and calibration. The dust counter was activated on 10 July and it recorded its first impact several days later. When near an early apogee, the camera snapped pictures of the half-lit Earth some 300,000 km away as well as the distant Moon. After looping around Earth 6.5 times, Nozomi passed within 5,000 km of the Moon on 24 September, with the resulting gravity-assist raising its apogee to 1.7 million km. Observations of Earth and the Moon were made by the camera and the ultraviolet imaging spectrometer for calibration purposes.[7] Particularly amazing were some close-ups of the lunar surface, including the first Japanese pictures of the far-side. After reaching apogee on 18 November the spacecraft encountered the Leonid meteors a day after the stream had put on an excellent show for terrestrial observers, and it was decided to deactivate the instruments to protect them from dust motes impacting at a relative velocity in excess of 70 km/s. Nevertheless, before being deactivated the German dust counter detected two hits by *bona fide* Leonid particles.[8] On 18 December, heading back to Earth, Nozomi again encountered the Moon close to the orbital position of the flyby made 3 months earlier, this time at a range of 2,809 km. The closest point of approach of the final flyby of Earth on 20 December was 1,003 km above the Pacific Ocean. Nozomi was to spin itself up for stabilization, fire its engine, then spin down again. As the spacecraft would be far from any land at the time of the maneuver, ISAS made a naïve planning error of not using a mobile station to obtain real-time tracking. Given the malfunctions of spacecraft by other nations while making similar maneuvers, ISAS ought to have known better. In Nozomi's case, a latching valve in the pressurization system of the oxidizer suffered a malfunction that caused the engine to burn a non-optimal mixture of more fuel and less oxidizer than required, and to produce less thrust than intended. For safety, the burn was automatically terminated after 397.5 seconds. As a result, Nozomi was left with a velocity shortfall of about 100 m/s. If ISAS had been in communication with the spacecraft, it could have ordered an immediate follow-on burn to compensate, but contact was not established until 12 hours later and by that time the engine could not recover the trajectory and still fly a nominal mission. It was decided to perform a maneuver of 340 m/s in order to enable Nozomi to reach the vicinity of Mars on schedule, even though this jeopardized the planned orbit insertion maneuver.

Undeterred, Japanese engineers began a mission redesign exercise which had to be finished by early January. They identified five options. Although the remaining propellant was indeed insufficient to enter orbit of Mars in October 1999, on that date Nozomi could fire its engine during a flyby to revise its heliocentric orbit and set up another encounter with the planet either on the opposite side of the Sun after half an orbit in August 2000, or after a full orbit in July 2002, at which times the spacecraft could try to enter orbit; but in both cases the propellant supply would be marginal. A non-powered flyby would also yield another encounter, but not until July 2006. Whilst there would be more propellant available to make the insertion burn, this would occur so far into the future as to be beyond the nominal operating life of the onboard systems. However, Nozomi happened to be in a 16-month orbit that ranged between about 1 and 1.5 AU and would return it to the Earth's vicinity in December 2002. This and a second Earth encounter 6 months later would enable the spacecraft to arrive at Mars in late 2003 or early 2004 and then enter orbit with sufficient propellant for a meaningful mission. As this was the only sensible option for rescuing the mission, a maneuver was performed on 15 June 1999 to refine the Earth flyby. One concern arising from the protracted interplanetary cruise was the deployment of the booms and antennas. Ground tests implied these should deploy successfully despite spending so long in their stowed configuration, but they seem never to have been deployed.[9]

For a brief period just after the start of the program, studies were carried out on the possibility of using *aerocapture* to insert the spacecraft into Mars orbit.[10,11] In redesigning the mission, the possibility was investigated of using *aerobraking* to adjust the parameters of the orbit obtained by firing the engine to those of the final operating orbit, but this option was not pursued.[12,13,14]

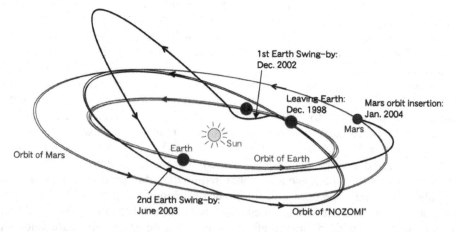

The circuitous journey of Nozomi from the Earth to Mars via two Earth flybys to compensate for an engine malfunction. (ISAS/JAXA)

In July 1999 Nozomi suffered another setback by losing the use of the amplifier of its S-Band transmitter. This would not only compromise the atmospheric radio-occultation experiment once the spacecraft was in orbit of its target, it also made getting there more difficult because the much narrower X-Band beam necessitated attitude adjustments which perturbed the interplanetary trajectory. Furthermore, at some point between the two Earth flybys it would become necessary to point the high-gain antenna away from home in order to enable the solar panels to produce adequate power, and this would make navigating and pinpointing the spacecraft's position impossible for a period of 2 months. For this reason, and to obtain better positional data during other periods, the Japanese engineers decided to employ the technique used for the balloons released into the atmosphere of Venus by the Vega missions, and precisely located Nozomi on the sky by reference to distant quasars. Meanwhile, on 7 September 1999, around aphelion, Nozomi flew within 4 million km of its objective.[15] In late December 2000 and early 2001 it went through solar conjunction, but since it had insufficient memory to store their data all of its instruments were switched off except the dust counter. And since contact was lost for fully 3 weeks, radio soundings of the corona were possible only for relatively large angular separations.

Some interesting scientific data was collected during 1999, 2000, 2001 and the first third of 2002. In particular, the ultraviolet and extreme-ultraviolet instruments obtained all-sky maps of interstellar hydrogen and helium penetrating the solar system.[16] During this 3-year cruise the dust counter detected over 100 particles, including several which could have been interstellar dust specks similar to those detected by Ulysses and collected by Stardust.[17] Solar studies were carried out in concert with the Advanced Composition Explorer (ACE) which was in a halo orbit of the L1 Lagrangian point, in particular mapping interplanetary magnetic fields, solar wind structures and the extent of coronal-mass ejection shocks. Nozomi also observed the 'far-side' of the Sun when it was near solar conjunction and antipodal to Earth before its instruments had to be shut off. Furthermore, Nozomi and ACE provided correlation for the solar flare of 21 April 2002 which would significantly influence the remainder of the Japanese Mars mission.[18]

At the end of April 2002, Nozomi ceased to issue telemetry, apparently after a power supply malfunction. The cause was never determined, but could have had something to do with the massive solar flare that completely saturated the solar proton monitors. Engineers soon found that they could not perform attitude maneuvers or course corrections. It was inferred that the loss of the power supply must have also switched off the hydrazine tank heaters, causing the fuel to freeze. For some time only the radio beacon signal was received, but this too was lost on 15 May. At that point it looked as if the mission was doomed. After 2 months, however, it proved possible to restore the beacon. Japanese controllers then employed a clever trick in which they reprogrammed the on/off status of the beacon as a means of producing the telemetry required to gain some understanding of the health of the spacecraft. Finally, at the start of September, approaching perihelion, the hydrazine thawed in the plumbing and both attitude and trajectory control were regained, but not the main engine. For the first time in months it became possible to determine the spacecraft's position in space. In order to avoid reorienting the vehicle during this period, the engineers used a 'side lobe' 90 degrees off the high-gain antenna beam for communication, telemetry and orbit determination.[19] Three small engine firings made between September and November precisely targeted the Earth flyby. On 21 December 2002 Nozomi flew by Earth at a range of 29,510 km and a relative speed of 13.8 km/s. Bad weather prevented observatories in Japan from imaging the vehicle. The flyby lowered the inclination of the spacecraft's orbit and cut its period to 1 year, in an orbit that would cause it to cross the Earth's path again 6 months later. During this interval, Nozomi would remain within about 15 million km of Earth, and in the northern sky. The second flyby occurred on 19 June 2003 at a range of 11,023 km and the comparatively low relative speed of 3 km/s. By design, this was right in the middle of a launch window for Mars. Japanese engineers managed to obtain precise positional data for a period of more than 30 days using nine antennas scattered all over Japan and one in Canada. And on this occasion astronomers at the Kuma Kogen observatory in Japan took two pictures of the probe streaking across a star field, confirming the positional fix obtained by radio methods. This flyby not only stretched Nozomi's aphelion back out to the orbit of Mars, it also rotated the aphelion some 90 degrees around the Sun relative to the 1999 encounter in order to set up the arrival in late 2003. However, engineers had still to determine how to regain use of the power supply, as this would be essential for orbit insertion and for scientific operations. The fact that frozen fuel remained in the pipes meant that the use of the main engine for orbital insertion would be difficult, if not impossible. With Nozomi finally heading for its target, disaster struck again with contact being lost on 8 July 2003 and never re-established. The reason for this failure was never determined.

The failure came as the new JAXA space agency was poised to start operations in October. On 13 December 2003 Nozomi silently flew within 1,000 km of Mars; JAXA gave an altitude of 894 km. In November a Japanese newspaper 'revealed' that the Earth flyby could conceivably have put the probe on a course that would cause it to collide with the planet and contaminate it with the micro-organisms that its unsterilized hardware had protected for years in space. To avoid this possibility, on 10 December controllers sent commands 'in the blind' to modify the trajectory by 100 km and thereby reduce the chances of impact, believed to be of the order of 1 in 100. Whether the spacecraft ever received the command, let alone executed it, will never be known. Presuming that there was no collision, Nozomi would have been perturbed into a heliocentric orbit similar to that of its objective. All mission support was discontinued after 19 December.[20,21,22]

Although Nozomi provided useful scientific data on the Sun and interplanetary space, the fact that it failed to return any data from its intended target means that it was a failure. Japanese scientists have expressed an interest in flying a mission to recover some of the lost science, but this is unlikely until later in the 2010s.

## DINNER FOR THE 'COSMIC GHOUL'

When the Mars Global Surveyor orbiter was conceived, it was as the first mission of the Mars Surveyor program. This was expected to run at least into the 2010s, with each mission being designed, managed and 'flown' by JPL. However, after a competitive selection a 10-year agreement was reached in February 1995 between JPL and Lockheed Martin Astronautics of Denver, Colorado. The company would provide the spacecraft hardware and establish an operations center in Denver. Only Mars Pathfinder and possible 'micro-missions' were excluded from the agreement. The idea was to provide a natural evolutionary path for the missions and maximize commonality of hardware in order to minimize costs. The program envisaged the launch of no fewer than four spacecraft in 1998: an orbiter to continue the re-flight of instruments which had been developed for Mars Observer and not assigned to Mars Global Surveyor, a 'conventional' lander, and a pair of penetrator probes, all of which were to be designed and built in just 37 months. In fact, only two of the Mars Observer instruments remained. Sometime after the program was started, the pressure modulation infrared radiometer was selected for the 1998 orbiter, with the gamma-ray spectrometer following in 2001.

The aim of the Mars Climate Orbiter (MCO) mission, which would be launched in 1998, was to function as a Martian meteorological satellite by providing regular observations of the weather, measuring temperatures, dust and water content and clouds from the surface up to an altitude of 80 km. Moreover, twin color cameras were to provide horizon-to-horizon global weather images and medium-resolution monitoring of seasonal surface features. The 42-kg radiometer was built by JPL with the assistance of Oxford University in England and the Russian IKI (Institut Kosmicheskikh Isledovanii; Institute for Cosmic Research), and utilized methods tested in Earth orbit and applied by the Pioneer Venus Orbiter. Specifically, it was to map the thermal structure of the Martian atmosphere, the global atmospheric dust loading, and the distribution of water vapor in the lower atmosphere; distinguish between water vapor and carbon dioxide; monitor how the surface pressure varied seasonally; investigate the energy balance of the polar regions and characterize the interaction of the atmosphere with the ground. The instrument was to observe in nine channels ranging from visible to infrared using special filters and sealed gas-filled cells.[23] The 1.1-kg imager consisted of a visible and ultraviolet wide-angle camera that provided kilometer-scale resolution, and a medium-angle camera with a maximum ground resolution of 40 meters over a 40-km-wide area. Both cameras used 1,000 × 1,000-pixel CCDs with identical electronics and power supplies, the only difference was the optics.[24]

The basic structure of the spacecraft was a 2.1 × 1.6 × 2-meter bus of carbon composite and aluminum honeycomb. This housed the equipment deck and tanks for a total of 291 kg of hydrazine and nitrogen tetroxide propellants. Like Mars Global Surveyor, Mars Climate Orbiter would not use a scan platform; instead, all of its instruments would be on a Mars-facing science deck. On the opposite side of the science deck was the main 640-N orbit insertion engine. The spacecraft was 3-axis stabilized. Star, solar sensors and gyroscopes provided attitude determination and thrusters and reaction wheels provided attitude control. There were four pulsed 22-N and four 0.9-N monopropellant thrusters operating in two sets for attitude control and trajectory corrections. A 1.3-meter high-gain antenna on an articulated boom and a medium-gain antenna provided communications. Like its predecessor, Mars Climate Orbiter had a radio relay to communicate with the sister Mars Polar Lander and future missions. On one side of the bus was mounted a 5.5-meter solar array consisting of three deployable panels capable of generating a total of 500 W of power at Mars.

Mars Climate Orbiter was to insert itself into a high Mars orbit with a period of 12 to 17 hours. This would then be circularized and lowered using a series of burns and aerobraking passes. There were small drag flaps on the solar panels to assist in aerobraking maneuvers. Since the insertion orbit would have a lower apoapsis than that of Mars Global Surveyor, the circularization process was to be completed in a shorter time. Indeed, within about 2 months of arrival the spacecraft was to reach a 373 × 437-km polar orbit that was Sun-synchronous and crossed the equator in the local late afternoon and then again before dawn half a revolution later. The first 3 months in this orbit (to the end of February 2000) were to be spent supporting the Mars Polar Lander by serving as its main radio relay to Earth. Only then would the orbiter initiate its own science mission, which would last a full local year.[25]

In mid-1994 NASA briefly considered joining forces with the Russian Space Agency for the 1998 mission. In addition to integrating Mars 96 hardware into the US orbiter, this Mars Together proposal would involve a Russian rover, a French balloon, a Sojourner-like US rover, a Fregat propulsion stage, and possibly a pair of Mars 96 penetrators. Other countries, including Germany and Italy, considered joining in, with the latter proposing a rover-mounted drill. The payload would be launched by a Proton rocket. The Fregat would then perform the escape maneuver, course corrections, insertion into Mars orbit and, after separating from the orbiter, deorbit the descent module. Although no formal estimates of Mars Together were provided, the combined mission was to cost less than the US Mars Surveyor 98 and the Russian Mars 96 combined. The launch mass of the spacecraft stack would be near 8,000 kg, with some 700 kg allocated to the descent capsule and less than 1,300 kg to the orbiter.[26,27,28] This was one of several joint US–Russian planetary missions under consideration at that time, the others being the Proton-launched US Pluto-Kuiper Express equipped with Russian 'Drop Zonds' (which the Russians called Lyod; Ice), and a joint Solar Probe (Plamya; Flame).[29] In the end, NASA decided to launch the 629-kg Mars Climate Orbiter using one of its own Delta II rockets.

Mars Polar Lander (MPL) was the surface portion of the 1998 Mars Surveyor program. It was designed to be the first lander to explore a high-latitude region of the planet. Its primary tasks were to search for ice in the soil and to investigate the polar cycles of water and carbon dioxide. Despite the recent interest in a landing system which used airbags to cushion a hard landing, in this case JPL reverted to the Viking-era architecture of a 3-legged spacecraft that would use parachutes and retrorockets. It was built around a central aluminum frame which housed the main computer, batteries, radio equipment and other hardware. A separate compartment housed other electronics and the radar altimeter that would be used to control the descent. The radar was a modification of the altimeter of the F-16 fighter jet, with a capability to sense vertical and horizontal velocities. The graphite-epoxy science deck on top of the bus stood 1.06 meters above the landing pads. On the sides were four solar panels which spanned 3.6 meters when deployed and were to provide up to

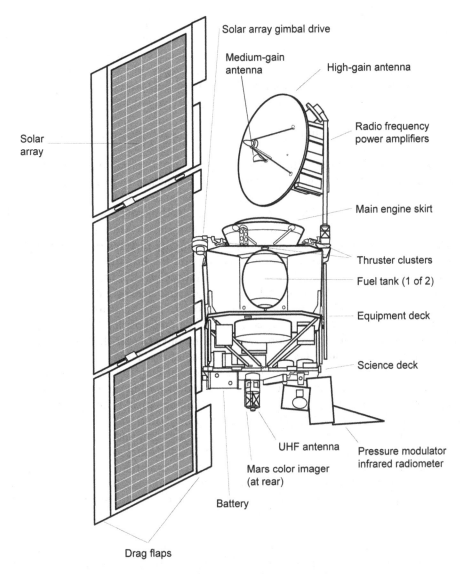

Solar array gimbal drive

Medium-gain antenna

High-gain antenna

Solar array

Radio frequency power amplifiers

Main engine skirt

Thruster clusters

Fuel tank (1 of 2)

Equipment deck

Science deck

UHF antenna

Pressure modulator infrared radiometer

Mars color imager (at rear)

Battery

Drag flaps

A drawing of the Mars Climate Orbiter, carrying the reflight of the Mars Observer infrared radiometer.

200 W of power immediately after landing. Two smaller auxiliary solar panels were mounted on the body of the spacecraft. Under the bus were three aluminum legs, each supported by a pair of struts and deployed from its stowed position by a spring, with crushable honeycomb inserts to absorb the shock of contact with the surface. The bus had a dozen 266-N descent engines, the hydrazine fuel for which was carried in a pair of spherical tanks installed under the solar arrays. Mars Polar Lander was to communicate directly with Earth using either a low-gain antenna or a dish at a

maximum rate of 5,700 bits per second, or by relaying through Mars Global Surveyor or Mars Climate Orbiter using a separate radio system capable of 128 kilobits per second. All the lander's components had been thoroughly cleaned by ethyl alcohol, and large components such as the mylar blankets used for passive thermal control or the 11.8-meter parachute canopy were sterilized by being held at 110°C for 50 hours. It was calculated that the lander would contain a maximum of 300 microbiological 'spores' per square meter.

The science payload comprised a stereo camera, a robotic arm, a meteorological station and a gas analyzer.

As a clone of the Mars Pathfinder imager, the coiled spring-loaded mast would place the stereo camera some 1.8 meters above the surface to provide stereoscopic images of the surface out to about 3 meters from the lander, aid with the operation of the robotic arm, and provide panoramas of the landing site for correlation with orbital imagery obtained by the two Viking orbiters and by Mars Global Surveyor. Moreover, imaging sessions would be devoted to taking pictures of the atmosphere in order to measure its dust loading and to measure the brightness of the Sun in a spectral band that would measure absorption by vapor water. One interesting idea was to capture views of Martian parhelia, sundogs and halos created by crystals of carbon dioxide ice. Such observations would provide 'ground truth' to the confirm that clouds of carbon dioxide ice occur. The shape and position of the halos with respect to the Sun would place constraints on the physics and thermodynamics of Martian cloud formation.[30] The camera had a resolution of 256 × 256 pixels and a 12-position carousel with filters tailored to geological and atmospheric studies. As usual, there were magnets mounted on the science deck within sight of the camera to study particles of magnetized dust.[31] The 5-kg graphite-epoxy robotic arm had four degrees of freedom (azimuth, shoulder, elbow and wrist) and was capable of reaching 2.2 meters from the lander. It had a backhoe configuration and drew an average of 10 W in digging trenches to a depth of 50 cm. It could acquire samples using a small scoop, position them for inspection by the stereo camera, position a forearm-mounted camera close to the ground to view details as small as 0.025 mm, and insert a 175-mm spike temperature probe into the ground. By design, the arm could position its camera to observe the underside of the lander and the soil which had been disturbed by its retrorockets. As the arm articulated, a thermometer on its elbow joint could measure temperatures over a wide range of heights. The arm was also a scientific instrument in its own right, reading actuator currents, torques and forces to determine the mechanical and physical properties of the surface.[32,33] The meteorological station used a 1.2-meter mast that pointed upward and a 0.9-meter mast that pointed downward on which were installed sensors to indicate the speed and direction of the wind.

The most important instrument was the Thermal and Evolved Gas Analyzer (TEGA) that would be fed by the robotic arm. It comprised eight single-use ovens, each a quartz vial capable of heating a 0.038-ml sample of soil to temperatures up to 950°C in order to measure water in abundances as small as 8 parts per million, peroxides, carbon dioxide and carbon isotopes. Furthermore, a record of the heat absorbed by the sample with respect to the applied temperature would provide an

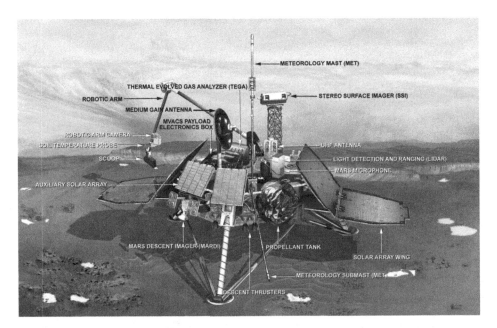

Mars Polar Lander was the first spacecraft to attempt a soft landing near one of the poles of Mars.

indication of phase changes in the constituents. Each oven had a 35 × 100-mm loading port which had spring-loaded doors to prevent material from one sample finding its way into the adjacent ovens.[34]

The payload was completed by a descent imager and a laser radar (lidar). The descent imager had a 1,000 × 1,000-pixel frame and was to snap up to 30 pictures in the interval between heat shield ejection and touchdown, to provide context for the images taken on the surface. The IKI-built lidar would characterize hazes, ice and dust in the lower atmosphere. It was the first Russian instrument to be carried on a US planetary spacecraft. There was a 50-gram microphone in the electronics box of the lidar sponsored by the Planetary Society to capture the sounds of Mars. It was claimed to be the first such instrument ever carried on a planetary probe, but there were acoustic sensors on the Soviet Venera landers and on the ESA Huygens probe which the Cassini spacecraft was at that time ferrying to Titan.[35]

Mars Polar Lander was to be protected during its interplanetary cruise by a 2.4-meter-diameter aeroshell and heat shield with similar geometrical characteristics to the somewhat larger Mars Pathfinder shell. As then, a simple solar powered cruise stage was mounted on top of the aeroshell for trajectory control, power generation and attitude control during the 11-month flight.[36] In this case, however, the cruise stage also carried two spacecraft funded by the New Millennium Program. These Deep Space 2 penetrators were the first such devices flown by NASA. It was their addition which took the US probe count for the 1998 Mars window to four. These microprobes were extremely small, massing just 2.4 kg each, and were designed to

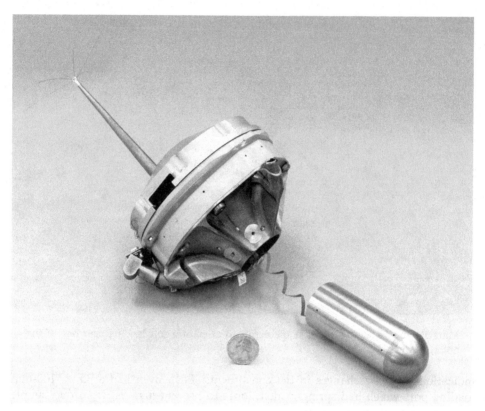

The Deep Space 2 penetrator microprobe included an aftbody (left) and a tethered forebody (right). Its mass was 2.4 kg.

pass through the atmosphere and survive the deceleration of high-speed impact with the surface. Complete with aeroshell, cruise stage and the microprobes, the Mars Polar Lander payload had a total launch mass of 576 kg.

The penetrators were enclosed in melon-sized aeroshells 35 cm in diameter and 27.5 cm high, with a forward angle of 90 degrees. These were designed to provide sufficient aerodynamic drag to slow the rate of descent to just below Mach 1 and then maintain in-line stability through to impact within 10 degrees of the vertical. No parachute, braking rockets or airbags were to be used to cushion the impact at a speed of between 180 and 200 m/s. Tests of the aeroshell were carried out in the supersonic wind tunnel at JPL at Mach numbers up to 3, and in the hypersonic wind tunnel at Mach 5. Transonic tests were carried out at research facilities of the US Air Force, and in the wind tunnel of the Russian Central Scientific Research Institute of Machine Building, TsNIIMASH, at Kaliningrad to replicate the unique character-istics of the rarefied Martian atmosphere.[37] The aeroshell was designed to shatter upon experiencing a deceleration of 30,000 g at the moment of impact. At the same time, the penetrator was to split into an aftbody that would remain on the surface and a bullet-shaped forebody that would bury itself to a depth of 0.6 meter in unconsolidated soil. The aftbody was 14 cm in diameter and 12 cm tall, and had a communications system with which to transmit data to an orbiter at 7 kbits per second by using 'whisker' antennas mounted on a short boom. The forebody was a steel cylinder with a rounded head, 10 cm in length and 39 mm in diameter. It was housed in a recess in the aftbody in flight. The two parts would remain connected by a 1-meter umbilical. Heavy tungsten was used for the penetrator's head so as to position the center of mass of the probe as far forward as possible for maximum stability at impact. Each penetrator had four heavily miniaturized instruments. An accelerometer was to record the deceleration history during atmospheric entry in order to permit atmospheric profiles over the poles to be reconstructed. Another accelerometer of a completely different design was to record 30 milliseconds of data on impact to measure the hardness and possible stratification of the soil. Two temperature sensors mounted on the forebody measured the thermal conductivity of the soil. And a 0.9-W electric drill pushed by springs on the side of the forebody was to collect soil with a volume equivalent to approximately half that of a typical soft-drink can during a period of 5 minutes. This would be heated to 10°C and the evolved gases scanned by a laser spectrometer to detect traces of water. Three series of development tests were performed utilizing air guns. The first series of 61 shots validated the mechanical design of the microprobe by firing it at speeds in excess of 600 km per hour against various types of target ranging from dry sand to clay to wet sand chilled by liquid nitrogen. The second series of tests demonstrated that the impact accelerometer could resolve layers in the target, and the third series of tests gave calibration data for dry sand, clay and damp sand.[38] Sample acquisition was also demonstrated.

The main purpose of the Deep Space 2 mission was to validate technologies intended for missions in which a single carrier spacecraft would disperse dozens of such probes across Mars.[39] After launch, the microprobes were named in honor of the leaders of the two expeditions which were first to reach the Earth's south pole: Roald Amundsen of Norway and Robert Falcon Scott of England.

The total cost of the two Mars Surveyor 1998 missions was estimated at $327.6 million including two Delta rockets. The microprobes were not financed by the Mars Surveyor program, and added another $28.2 million. Excluding the launch vehicles, the overall cost was a mere $193 million, which was only a little more than the single (already lost-cost) Mars Pathfinder and actually less than that in terms of constant dollars. This raised the possibility that NASA was stretching the 'faster, cheaper, better' management style too far, and an internal JPL study warned that "some missions may be headed toward a cliff on this".[40]

Mars Climate Orbiter was to be launched on 10 December 1998, but had to be postponed by one day due to an issue involving the software which controlled the electrical system of the spacecraft. It lifted off at 18:45 UTC on the second day of its 2-week-long window and entered a type-2 transfer orbit that would enable it to reach Mars in 9.5 months. A 19.1-m/s course correction on 21 December removed the trajectory dispersions inherited from the launch vehicle and adjusted the arrival parameters. Several days later, the radiator door of the radiometer was opened for the first time. A course correction of 0.86 m/s was made on 4 March 1999, and the instruments were calibrated during a week-long session later in the month. During this time, the camera performed scans of selected bright stars. By the third course correction on 25 July, which was 60 days prior to Mars orbit insertion, the flight team had begun to suspect something amiss with the design and execution of the maneuvers. Trajectory data showed the spacecraft to be outside the nominal ellipse and aimed closer to Mars. However, the team's attention was distracted by the spacecraft entering safe mode due to a problem in the solar array articulation. It took a full 6 weeks to troubleshoot this anomaly and prepare a contingency plan in case it were to reoccur during aerobraking and circularization. Meanwhile, on 7 September the camera took calibration images of Mars from a range of 4.5 million km, with the planet spanning only a few pixels. No further images were to be taken before the primary mission started in February 2000. On 15 September 1999 the spacecraft performed its fourth correction maneuver. Tracking showed the periapsis altitude at arrival to be 173 km instead of the planned 210 km. Since the low pass would be survivable, it was decided not to make a fifth maneuver on 20 September. On the morning of 23 September the spacecraft stowed its solar arrays and reconfigured its communications system to transmit only a carrier tone using the medium-gain antenna. Meanwhile, JPL navigators were starting to report a periapsis altitude as low as 110 km. This would still be survivable, but a major burn would be required at the first apoapsis to raise the subsequent periapsis altitude. Pressurization of the propellant lines occurred as planned, and at 09:49 UTC the engine was ignited for the 16-minute 23-second insertion burn.

Some 5 minutes after ignition, the spacecraft disappeared behind the limb of the planet 39 seconds earlier than expected for an occultation that was to last about 20 minutes; it never re-emerged. Analysis of the trajectory data immediately revealed that the altitude at closest approach had been a mere 57 km, which was deep in the atmosphere. At 98 km the temperature of some components would have exceeded their design limits. Then at 85 km aerodynamic torques would have overwhelmed the attitude control system. The solar array probably detached early on. If the hull of the spacecraft did not rupture, it could have been captured in an eccentric orbit only to burn up on the next pass. Attempts to establish contact were discontinued on 25 September, by which time it was evident that the spacecraft must have been destroyed.[41] Having been subjected to less stringent planetary contamination and sterilization requirements than a lander, Mars Climate Orbiter was potentially the worst case of contamination of another planet by microbiological material from Earth. It is possible that small pieces reached the surface, but hopefully they were sterilized by the brief exposure to temperatures exceeding 500°C.

Investigation teams were created at JPL and NASA to find the cause of the loss. The main investigation delivered its conclusions as early as 10 November 1999, having identified an amazingly trivial and embarrassing 'college-level' root cause. This consisted of an error in how the files used to model 'small forces' acting on the spacecraft during its interplanetary cruise were handled. Small forces were all external forces like solar radiation pressure, attitude control thruster imbalances, etc, which imparted minor perturbations on the path. Of course, while these forces were minor, their effect over a prolonged time could be considerable. Small forces files were generated by Lockheed Martin and used by JPL in designing trajectory corrections. The problem was that engineers at Lockheed Martin, as is the habit of US technicians, generated the files in imperial units of pounds-force, while JPL, as is standard for NASA, assumed that these were in metric units of Newtons. Within a week of the loss of the spacecraft, it was realized that the small forces had been consistently underestimated by a factor of 4.45, this being the conversion factor between the two units. A contributing cause was the fact that the solar panels were mounted in a strongly asymmetric configuration and that solar radiation pressure had imparted sufficient torques to necessitate frequent attitude corrections. These maneuvers were made every 17 hours on average, some 10 to 14 times more often than expected based on prior experience. As a result of this, and the fact that these perturbations were modeled in imperial units and assumed to be in metric units,

Mars Polar Lander is attached to the third stage of its Delta launcher. The shell of one of the microprobes is visible beside a folded cruise stage solar panel.

small trajectory errors were introduced which produced a substantial error by the time of the arrival at Mars. Unfortunately, the trajectory errors were aligned in the direction of the planet (the only direction in which they could be catastrophic) and lowered the initial periapsis into the atmosphere.[42] Organizational issues were also identified, with engineering teams not consulting spacecraft design experts at JPL; with the existence of two separate navigation teams for the prelaunch development phase and for the actual flight; and with a geographically dispersed organization that made communication difficult at best. Moreover, these communications were informal contacts such as personal emails, rather than formal reports. There had been ample indications of the small forces error in the form of the consistent, non-random discrepancies during the interplanetary maneuvers, but the understaffed navigation team never raised the issue for peer review. Ultimately, therefore, the loss of the mission was attributed to poor program management. As if the problem were not sufficiently embarrassing in itself, it was all the more so for JPL, which took particular pride in its deep-space navigation.[43,44]

Meanwhile, another Delta II rocket had lifted off on 3 January 1999 with Mars Polar Lander on the first day of its launch window and placed the spacecraft on a type-2 transfer orbit that would reach Mars in exactly 11 months. As it was not to perform any observations whatsoever during the cruise phase, the only activities were course corrections and routine monitoring of the hardware. The first course correction of 16 m/s was made on 21 January. The second was on 15 March. The third maneuver on 1 September fine-tuned the aim for a point 800 km from the south pole. This area at 76°S, 195°W, was expected to be on layered terrain made of stacked beds of dust and ice deposited within the last several tens of millions of years, where rocks were expected to cover no more than 10 per cent of the ground. The landing was to occur in the Martian southern summer so that the Sun would remain above the horizon throughout the months during which the lander was to operate. Meanwhile, the microprobes would land about 10 km apart some 60 km from the lander, within a 200 × 20-km ellipse centered at 75.3°S, 195.9°W, where the winter carbon dioxide frost cover would have retreated only 1 month earlier. Their batteries would support a surface mission lasting up to 1 week.[45]

While NASA reviewed the design and operation of Mars Polar Lander after the loss of Mars Climate Orbiter, the spacecraft performed its fourth course correction on 30 October. The investigation identified a potential problem: the performance of the retrorockets could be affected by the cold temperatures of Mars during the descent. It was therefore decided to run the heaters of the propellant tanks in the hope that this would warm the adjacent thrusters before they were to be called into action. However, owing to the substantial thermal protection that existed between the tanks and engines the results of this operation would be hopeful at best. A fifth maneuver was carried out on 20 November to refine the speed of the vehicle by a mere 0.06 m/s. A final correction was made 6.5 hours before entry on 3 December 1999. At 19:49 UTC Mars Polar Lander adopted the attitude for atmospheric entry, thereby severing its link with Earth. Five minutes prior to entry the aeroshell that cocooned the lander separated from the cruise stage. The cruise stage released the two microprobes after waiting an additional 18 seconds to ensure that there was no collision between the four separate objects. Amundsen and Scott could have been released in any attitude because the cruise stage was no longer stabilized, but their aerodynamics would ensure entry at the correct angle.

Mars Polar Lander was to fly a narrow 10 × 40-km corridor into the atmosphere at a speed of 6.9 km/s, experience a deceleration peak of 12 g and slow to 430 m/s within several tens of seconds. At an altitude of about 7.3 km the parachute was to deploy, the heat shield would be jettisoned and the descent camera would begin to snap pictures. Then the landing legs would be deployed. The altimeter would be activated at a height of 2.5 km. After jettisoning the backshell and the parachute at a vertical speed of 75 m/s, the retrorockets would be activated for the most critical phase – slowing the vehicle to a survivable landing speed of 2.4 m/s. At a height of about 40 meters the altimeter would be switched off and the spacecraft, continuing to fire its engines, would wait for sensors to report the deflection of the legs upon contact with the ground.[46] After allowing 5 minutes for the dust to settle, the solar panels were deploy, and 3 minutes after that the parabolic antenna would begin to search for Earth to resume contact. The camera was to take a preliminary 5-color scan prior to being raised on its mast.[47] Contact was expected within 20 minutes of touchdown, but the team waited in vain. Surprisingly, nothing was heard from the microprobes either.

Communications windows opened and closed without response for several days in succession. By 17 January 2000 engineers had investigated all plausible failure modes. After the 45-meter Stanford University radio-telescope was asked to assist and reported detecting weak signals on 18 December and 4 January, the services of large radio-telescopes at Westerbork in the Netherlands, Jodrell Bank in England and Medicina in Italy were enlisted. Unfortunately, no obvious transmissions were detected. Further analysis established that the signals reported by Stanford were noise generated in the receiver of the radio-telescope. In addition, on 16 December 1999 Mars Global Surveyor began a campaign of scanning more than 300 square kilometers centered on the target at a resolution as fine as 1.5 meters per pixel, but no indications were evident of either the lander (spanning a couple pixels) or the parachute (spanning about four pixels).

An investigation was made to find the cause of the failure of the trio of probes. Unfortunately, early during the design phase it had been decided to save money by ceasing communications during the descent, with the result that no telemetry was available to assist the 'post mortem'. A key recommendation of the investigation for future missions was to transmit telemetry during all critical events. This should have been obvious after the loss of Mars Observer in 1993, but it was evidently not so. Owing to the lack of telemetry, the investigation could not identify a definitive cause for the loss, it could only formulate hypotheses for the most likely scenarios. Plausible ones included a terrain that was too rough; a heat shield structural failure due to manufacturing defects; an oversimplified model used for the descent control system; an excessive center of mass shift caused by the migration of propellants; and a catastrophic collision between the lander and its backshell. A plausible culprit was the landing engines, which had been inherited from the warhead delivery bus of an intercontinental ballistic missile. Accepted on the basis of (fortunately) limited previous flight experience, they had not been properly tested in their new role. But the *most likely* cause was a trivial logic error in the descent engine software that caused a premature shutdown. Each leg of the spacecraft was fitted with a bending sensor to detect contact with the ground and shut off the engines, but it was found that the sensors could also be triggered by the jolt as the legs deployed and locked into place. In that event, when the software began to interrogate the sensors at an altitude of 40 meters it would have found them already tripped and so would have shut off the engines, leaving the vehicle to fall freely and be disabled or destroyed upon impact at a speed of about 80 km per hour. This sensor behavior was known from tests prior to launch, and although the software was required to ignore such events they were not specifically described in the requirements and therefore were not properly accounted for. A simple solution would have been a single software command to reset a 'tripped' flag to 'false' prior to polling the sensors for the first time.

As for the Deep Space 2 microprobes, the investigation identified four possible failure modes. The probes were not required by design to penetrate ice that could lie a few centimeters beneath the surface, and so could have bounced off on impact because ice at that temperature is as hard as rock. The electronics, and in particular the radio system and batteries, which were not properly tested, could have failed at impact, or the 'whisker' antennas could have electrically discharged in the thin air in the same manner as is believed to have disabled Mars 3 almost 30 years earlier. Other plausible scenarios included the probes not landing in a vertical orientation, being damaged by the shattered heat shield, and a broken antenna.[48] As a whole, the microprobes were unequivocally deemed "not adequately tested, and [...] not ready to fly". Moreover, the whole Mars Surveyor program was found to have been underfunded by about 30 per cent from the outset.[49,50]

Following the review of these disastrous failures, NASA Administrator Daniel Goldin went to JPL to apologize for pushing too far on mission costs. Fortunately, instead of abandoning 'faster, cheaper, better', the agency decided that from now on such programs would receive adequate support in money and personnel. But as Goldin stressed, "there won't be a return to the days of big, expensive spacecraft". While this may not have been true in the long run, the decision was instrumental in saving the Mars and Discovery programs. Despite some evident shortcomings and the occasional lost mission, these programs had delivered valuable results in terms of science and public appreciation of solar system exploration at a sensitive price. Improving the management to be able to continue these 'low-cost' programs was important, since there would be no increase in the overall budget for solar system exploration.[51] It is rarely recognized that the 1999 Mars failures which drew to an end an *annus miserabilis* for JPL that began with the low-cost Wide-field Infrared Explorer (WIRE) space telescope leaking coolant soon after launch and remaining 'blind', highlighted a major difference between the open, civilian, scientific space program run by NASA and the vastly larger but classified military program. While NASA and JPL were ridiculed for their losses, at the same time the US military suffered several launch failures and satellites stranded in wrong orbits for which, despite representing losses of several billion dollars, no-one was ever publicly held to account.

Mars Polar Lander again hit the news after the landing of the Mars Exploration Rovers in 2004. It was possible for the first time to pinpoint the landing sites of the rovers and for Mars Global Surveyor to image them. Then, finally knowing what a lander looked like on the surface, the December 1999 coverage of the Mars Polar Lander area was re-examined, revealing what appeared to be a parachute, with the lander 200 meters away as a bright spot in the middle of a dark splotch that could have been produced by thruster exhaust. However, when the area was revisited in high-resolution in September 2005 the 'parachute' turned out to be the illuminated slope of a small hill and the spot was revealed for what it actually was: electronic noise.[52],[53] So the fate of Mars Polar Lander remains a mystery.

## CANCELED SURVEYORS

The disastrous loss of the 1998 Mars Surveyor missions led to yet another redesign of the US approach to exploring Mars. In the meantime, all of the Mars Surveyor missions under study and under development were halted and, with one exception, eventually canceled. Only the 2001 orbiter was permitted to continue in a modified configuration. The 2001 lander was canceled in February 2000, with the spacecraft 4 months into its final assembly phase and a little more than 1 year from its launch window. Hopes of flying it in 2003 in modified form were undermined by the fact that the reason for Mars Polar Lander's loss could not be definitively established.

The plan had been to launch this lander in April 2001 on a trajectory to arrive early in 2002. The Mars Polar Lander-based spacecraft would use thrusters early in the descent while still traveling at hypersonic speeds in order to reduce the size of the landing ellipse to about 10 km, which for Mars was a pinpoint landing. Such accuracy would be vital for the sample-return missions that were to follow. After landing, the spacecraft would deploy two circular fan-like solar panels that would provide power for the nominal mission of 100 Martian days (or 'sols'). The Mars Surveyor 2001 missions were to carry several instruments relevant to the future human exploration of the planet provided by a NASA enterprise called the Human Exploration and Development of Science, which was funded by the Johnson Space Center. These included radiation monitors on both lander and orbiter to assess the radiation exposure on the interplanetary cruise and in the Martian environment, an experiment to assess the health hazards and toxicity of Martian soil to humans, and an experiment to evaluate the production of propellants from the atmosphere. This research was believed by some to be essential to facilitating a human mission. The lander would have the same robotic arm as Mars Polar Lander, but this time one of its tasks would be to deploy onto the surface a rover initially stowed on top of the lander's deck.

The rovers for the 2001 and 2003 Mars Surveyors were to employ technologies evaluated by the FIDO (Field Integrated, Design and Operations) prototype. At 1 meter in length and 0.5 meter in height, the 6-wheeled vehicle was several times the size of the Sojourner rover delivered by Mars Pathfinder, but shared the same 'rocker-bogie' design and could negotiate obstacles up to 30 cm high. It carried a deployable and tiltable mast on which science and navigation cameras were set at eye-level for an adult human.[54] FIDO also carried a robotic arm of its own, to be used to place instruments on rocks and other targets. To obtain samples, it had a miniature drill system beneath its belly that was monitored by a mini-camera. The prototype was extensively trialed in 1999 in the Mojave Desert and in 2000 near the Lunar Crater Volcanic Field in Nevada, successfully undertaking traverses of 600 meters.[55,56] The Mars version would carry an integrated payload called Athena (born of a separate Discovery proposal) that included 14 cameras for science and navigation, an alpha-proton X-ray spectrometer, a Mossbauer spectrometer and an infrared thermal emission spectrometer. As the plan was to demonstrate collection techniques for a sample-return mission, it was equipped to select, collect and cache rocks in a container having a capacity of 92 samples.[57] But budget realities caused the first large rover to slip to 2003. The 2001 lander was redesigned to deploy the refurbished Marie Curie rover, the Sojourner test item. This severely diminished scientific interest in the project because it was not clear that Marie Curie would be able to perform a useful mission. The rover would be lifted by the robotic arm and gently lowered to a suitable position on the surface. The arm would then be used to collect samples for onboard instruments, dig trenches and perform soil mechanics experiments. A Mossbauer spectrometer like the one envisaged for Athena was to be placed in contact with the soil. Following the cancellation of the big rover, two of its instruments, the panoramic camera and the thermal emission spectrometer, were to be mounted on the lander's deck.[58] The payload would be completed by a

The canceled Mars Surveyor 2001 lander together with the Marie Curie rover.

downward-looking descent camera. When the mission was canceled, the target had yet to be decided but it would have been within the latitude range 3°N to 12°S to satisfy solar power constraints.[59,60]

The total cost of the 2001 orbiter and lander missions was projected at about $311 million.

The original Mars Surveyor plan for the 2003 launch window envisaged using a rover to collect a diverse suite of rocks for a return mission. The orbiter would be a small relay satellite which could either be launched together with the lander or by a small rocket. The Italian Space Agency, ASI, would provide the communications payload. Alternatively, the relay role could be assigned to the Mars Express orbiter that was being developed by ESA. A low-cost Mars Sample Return mission would then be flown in 2005. This would involve launching a lander carrying a rover and a Mars Ascent Vehicle (MAV), and an orbiter carrying the vehicle which would enter the Earth's atmosphere with the samples. The orbiter would be launched first and enter a highly eccentric orbit around Mars in August 2006, then aerobrake to circularize its orbit while waiting for the arrival of the lander in November 2006. The lander was to make a pinpoint descent by homing on a beacon from either the 2001 or 2003 rovers. A 'fetch' rover would then drive to its predecessor to retrieve its cache. Additional samples could be collected either by this rover or indeed by the lander itself. Sample collection would last between 30 and 180 sols, then the ascent vehicle would liftoff with its payload and enter a 300-km parking orbit. The ascent vehicle was a 270-kg two-stage design using hypergolic propellants which could safely be stored for a long period. This rocket had an extremely squat shape with a diameter of 1.7 meters and a total height of only 1.33 meters. The hermetic sample canister would be carried in a 'stinger' installed backward in a hole inside the first stage. After rendezvousing, the orbiter would transfer the canister into its return capsule, jettison the ascent vehicle and depart Mars orbit on a trajectory to reach Earth in April 2008. As it had been decided not to employ a parachute, the 1-meter entry vehicle was to fall freely through the atmosphere to a recovery site in the Australian desert. A thick layer of shock-absorbing material was to protect the sample canister from the impact. A metal-on-metal knife-edge seal would ensure that Martian 'contaminants' did not leak out of the canister. So, some 3 years after launch, scientists would have a 1-kg cache of Martian rocks, dust and atmosphere to study in their laboratories.[61,62]

However, the design of the liquid-fueled ascent vehicle soon ran into serious difficulties because it required to be more sophisticated than initially expected and would need much new technology development. Finally, its cost alone had soared to at least $120 million, which was impractical for a mission whose total cost had been capped at $200 million.

Otherwise, the Mars program looked to be progressing well, so much so that NASA decided to update the planning and tasked JPL to revise the architecture. In doing so, JPL not only included the scientific community in the planning but also representatives of human spaceflight and the space agencies of Europe, France and Italy. Moreover, following the hotly debated discovery of Martian fossils within a meteorite, NASA decided to move its first Mars Sample Return mission forward to 2003.

The redesigned mission would use a completely different mini-MAV based on the late-1950s launcher for the classified NOTSNIK (Naval Ordnance Test Station – with a 'nik' to rhyme with Sputnik) project to use air-launched unguided rockets to place extremely small satellites into orbit.[63] The solid-fueled mini-MAV would use a spinning table for stabilization at launch and a NOTSNIK-era trick for orbit insertion. The third stage would be installed with its nozzle facing forward, and the first two stages would place it on a suborbital trajectory such that after completing a half-revolution of Mars its nozzle would be facing aft, at which time it would be fired to enter a circular orbit.[64] This approach greatly simplified the design of the vehicle, theoretically requiring no active stabilization or guidance systems. It was calculated that a 45-kg mini-MAV could deliver a 200-gram sample to Mars orbit. However, because an unguided vehicle could not assure that the resulting orbit would be accessible to an orbiter, a more complex guided two-stage solid-fueled rocket was designed. This 150-kg vehicle was 1.75 meters tall and included an 'igloo' for thermal protection while on the surface. It was able to deliver 0.5 kg of samples to orbit.[65]

The adoption of a narrower rocket (its diameter was just 45 cm) released space on the lander base to accommodate new instruments and an autonomous sampling arm. This was to be developed by the Italian Space Agency, would possess four degrees of freedom and would have an overall reach of several meters. It would have on its tip a large box containing a sampling drill (called the DeeDri for 'Deep Driller') and a sample management system. The drill was capable of penetrating to a depth of 50 cm using just 20 to 35 W of power depending on the hardness of the soil. Follow-on versions were expected to be equipped with additional drill rods to penetrate several meters. Based on technology developed for the Rosetta cometary lander, the hollow auger employed a drill bit of polycrystalline diamonds that was capable of coring into rocks as hard as basalt. The DeeDri was to provide samples to several instruments mounted on the lander for in-situ analyses, and to the ascent vehicle. One of these instruments was to crush the sample and seek key organic molecules such as amino acids, amines and hydrocarbons in the interiors of rocks.[66] Having its own microscopic camera and thermocouples, DeeDri would function as an independent scientific instrument measuring the hardness of the soil and other parameters. There was scope for carrying instruments inside the drill box to collect data on the electrical characteristics of the Martian soil, etc.[67,68] Also carried on the lander would be a 12-kg self-contained laboratory with its own thermal control and electronics for instruments supplied by Italy to perform various chemical analyses of the surface, to study the properties of Martian dust, and to monitor the radiation during the interplanetary cruise and on the surface of the planet in order to provide data for planning human missions.[69,70,71,72] In addition to the samples collected by DeeDri, a large rover would explore the landing site and collect surface samples.

At 1,700 kg, the Mars Sample Return lander was to be the heaviest surface craft yet sent to the Red Planet. It would be launched in June 2003 on either a Delta III or Atlas III rocket on a trajectory that would reach Mars in December of that year. For power generation purposes, the landing site had to be within a 20-degree band centered on 5°S, and a mission requirement was that there be a small probability of operations being endangered by boulders larger than about 35 cm in size.[73] After 90

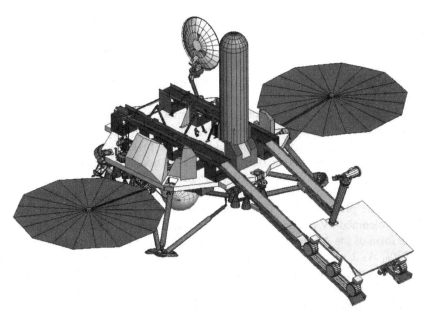

A rendition of the 2003 Mars Sample Return. The solid-fueled Mars Ascent Vehicle is shown in the launch position and the large rover is on the ramps. Not shown in the image is the Italian robot arm and sampler drill.

days on the surface, the ascent vehicle was to lift off and attain a circular orbit at an altitude of 600 km with the rather large uncertainty of plus or minus 100 km. The payload was a spherical canister only 16 cm in diameter with a mass of 3.6 kg which, in addition to the samples, had a radio beacon.[74]

In 2005 a 2,700-kg spacecraft consisting of a French orbiter, four small French NetLanders and NASA's OSCAR (Orbiting Sample Capture and Return) payload would be dispatched by an Ariane 5 rocket, together with a second sample-return lander. After deploying the NetLanders, the spacecraft, protected by a large heat shield that would provide lift and trajectory control, was to perform an aerocapture maneuver by dipping deep into the Martian atmosphere. After this risky maneuver and jettisoning the spent heat shield, the orbiter would start to search for the radio beacon of the sample canister, taking up to 6 months to complete the rendezvous. When the orbiter was within 2 km of its target, a system using laser radars would facilitate an approach in which a capture cone 'swallowed' the canister and guided it into a return capsule. Meanwhile, the second vehicle would have set down on the planet and used the advanced DeeDri to obtain samples from a depth of several meters. Its ascent vehicle would be injected into the vicinity of the orbiter and the capture process repeated, with the cone steering the second canister into a separate return capsule. The orbiter would then adopt a highly elliptical orbit in readiness for a burn in July 2007 to escape on a trajectory which would reach Earth in April 2008. Because celestial mechanics dictated that only landing sites in the southern hemisphere would be accessible on this date, the plan envisaged an Earth flyby to set up an encounter 6 months later that would permit a landing in the continental United States.[75],[76]

After these two sample-return missions, the Mars Surveyor program could have flown the same architecture in the launch windows of 2007 and 2009 and returned the samples to Earth in 2012, taking the total number of sites sampled to four, but there were no firm plans.

NASA hoped to launch these sample-return mission on a relatively tight budget of $500 million. In contrast, all previous sample-return studies had been priced at several billion dollars. French involvement would amount to an additional $400 to $500 million. But a re-evaluation after the loss of the 1998 missions showed that at least five key technologies required substantial development work, and until then it would not be possible to confidently estimate the price of the mission. At first, it appeared that the start of the project would simply slip to a later launch window, a delay which would be welcomed by some people because it would provide time to develop a more powerful form of the Ariane 5, but the first mission was deferred to no sooner than the early 2010s. As a point of historical interest, this was the first and only time that a Mars sample-return mission got within a few years of a scheduled launch date.

In parallel with these 'flagship' concepts, JPL, NASA and the French Space Agency studied $20–30 million Mars Micro-missions for payloads up to 20 kg that could be launched in large numbers either piggybacked on commercial launches of the Ariane 5 or on dedicated low-cost rockets. On being released in a geostationary transfer orbit with an apogee of 36,000 km, the spacecraft would use its engine and a combination of lunar and Earth flybys to depart for Mars in the same manner as Nozomi. The plan envisaged the first two missions arriving at Mars in December 2003 with an aircraft and a communications relay orbiter. The aircraft was dubbed 'Kitty Hawk' after the town in North Carolina where the Wright brothers flew the first 'heavier than the air' machine exactly 100 years earlier on 17 December 1903. Studies of aircraft for Mars had been undertaken in the 1970s as Viking follow-up missions and exploited work on very-high-altitude sampling drones configured for low 'Reynolds numbers' (i.e. the ratio of inertial forces to viscous forces, in this case for very rarefied air or very low speeds).[77] Theoretical work continued during the 1980s and early 1990s. In particular, NASA's Glenn center built a terrestrial prototype of a long-duration solar-powered aircraft. Martian airplanes were also offered at the landmark workshop in 1993 that effectively kicked off the Discovery program. NASA's Ames center, for example, suggested flying an aircraft low over Gusev crater, which in orbital imagery appeared to be an ancient lakebed. During these years, research continued into human propulsion and solar-powered flight for low Reynolds numbers. In 1988 the MIT Daedalus aircraft flew from Crete to the island of Santorini in the Aegean powered solely by an Olympian cyclist pilot. Work on Martian airplanes continued at NASA's Ames and Langley centers in the 1990s. The former experimented in 1996 with a glider that could be carried inside a heat shield by using a folding wing with the fewest possible number of hinges. Alternative concepts included roll-up wings, bat wings and parafoils. After being released from the heat shield the aircraft would deploy its wings and fuselage, and then dive to start straight and level flight at low altitude. Exiting the dive would be a particularly critical maneuver, because it would have to occur before the falling aircraft

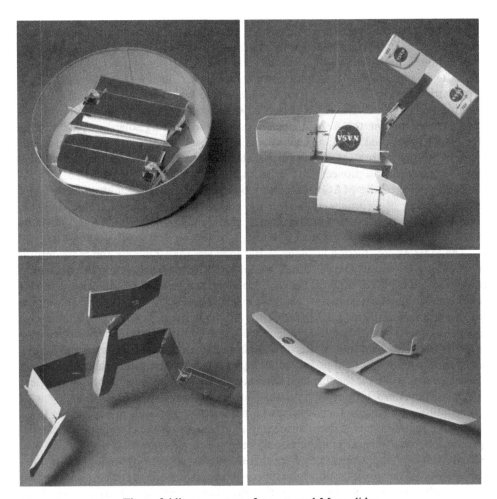

The unfolding sequence of a proposed Mars glider.

could approach the speed of sound, which in such rarefied air could pose a variety of aerodynamic problems. A propeller-driven version of the Ames glider was proposed as a Discovery mission in 1996 as the Aircraft for Mars Exploration. The Kitty Hawk airplane was then proposed to the Discovery program in 1998 as a joint endeavor by Ames, Langley, the Naval Research Laboratory, Malin Space Science Systems and Orbital Sciences. This Mars Airborne Geophysical Explorer (MAGE) was to mass 135 kg, have a wing span of slightly less than 10 meters, and cost $246 million.

The most attractive MAGE mission was a low-altitude flight down a canyon of Valles Marineris to map it at the highest possible resolution. In addition to still and video cameras, the payload would have gravity, magnetic and electric field sensors for lightning detection, an infrared spectrometer and a laser altimeter. The cruise stage would release Kitty Hawk in its heat shield several hours before atmospheric entry, and would then perform a deflection maneuver to make a flyby. The camera package would document the deployment of the wing at an altitude of about 2,000 meters. Over the next 3 hours, the aircraft would travel 1,800 km along one of the canyons. As the aircraft was not designed to communicate directly with Earth or to survive landing, it would transmit its data to the cruise stage whose trajectory was designed to pass overhead at this time. The departing cruise stage would replay the data to Earth.

As an alternative, the AeroVironment company (building solar-powered aircraft for NASA) proposed flying a flock of small gliders. With JPL it flight-tested a full-scale prototype with a span of just 1.5 meters. This aircraft was dubbed Otto after the German aviation pioneer Otto Lilienthal.

NASA rejected the MAGE Discovery proposal in late 1998, but revived it as a Micro-mission in February 1999 with the endorsement of Daniel Goldin himself, who found the opportunity of flying an airplane on Mars on the Wright brothers' Centenary irresistible.*

The Mars Micro-mission airplane would span less than 2 meters and mass only 20 kg, entering the atmosphere in a heat shield just 75 cm in diameter. The cost would be of the order of $40 million. As an alternative, the mission would carry a trio of even smaller gliders. The fuel would sustain a flight of up to 20 minutes and a range of the order of 200 km. Like MAGE, the Micro-mission would require the cruise stage to undertake a deflection maneuver after releasing its payload, in this case to time its flyby for 15 minutes after the airplane had entered the atmosphere. Unlike MAGE, the mission would be for technology demonstration rather than for science. Development work on the airplane was initially divided between NASA's Langley, Dryden and Ames centers, but in the end it was concentrated at Langley with contributions by CNES. Langley issued a request for proposals in September 1999, listing the requirements without specifying anything about the configuration, shape, propulsion, etc. Several configurations were investigated, including straight and swept wings, flying wings, propeller-driven versus rocket-powered, etc. But the complexity of the project soon grew overwhelming, and the launch was slipped to 2005 at the earliest, thereby missing the attraction of flying a Martian aircraft in December 2003. It was quietly canceled in November 1999, just before the loss of Mars Polar Lander.[78,79,80,81,82]

An alternative to an airplane was a Mars Aerobot Mission that would deliver a long-duration balloon. Scientists envisaged swarms of such balloons roaming the Martian atmosphere, returning tens of thousands of centimeter-resolution images and meteorological measurements. In fact, due to their endurance, balloons would be better than aircraft for exploring Mars – although of course, the latter would be more likely to capture the attention of the public.[83]

A key problem for the Mars program was that of communicating with a variety of

---

\*   Note, however, that an aircraft would not be the first object to fly in the Martian atmosphere, because that honor must go to the Viking landers which flew a 'lifting' trajectory during their descent.

static landers, rovers, balloons, penetrators, airplanes, ascent vehicles and their sample canisters. JPL therefore designed a two-element Mars communications and navigation infrastructure for relaying high-resolution imagery and streaming video from long-range rovers, aerobots, etc. The first 'layer' would be low-altitude relay satellites which could be piggybacked onto commercial Ariane 5 launches. These micro-satellites would have a mass of 220 kg, including 140 kg of propellants and a payload of 6 kg, and would aerobrake into a low operating orbit. The possibility was also examined of deploying an inflatable 'ballute' (balloon-parachute) trailing the small spacecraft at arrival for aerocapture by the planet's atmosphere. A low-cost 3-axis-stabilized bus was designed by Ball Aerospace and Aerojet that could be used as an orbiter to conduct science as well as relay communications. It had a peculiar curved shape to allow it to fit into the Ariane 5 shroud. Alternatively, this bus could be used to deliver small probes to the surface of the planet; possibly, for example, four pairs of Deep Space 2-like mini-penetrators. In fact, one plan was to use this for one of the airplane projects. The first $50 million demonstrator micro-satellite was to be launched in 2003 and be followed by two more on successive launch windows. Six satellites in near-equatorial and high-inclination orbits would provide connectivity at

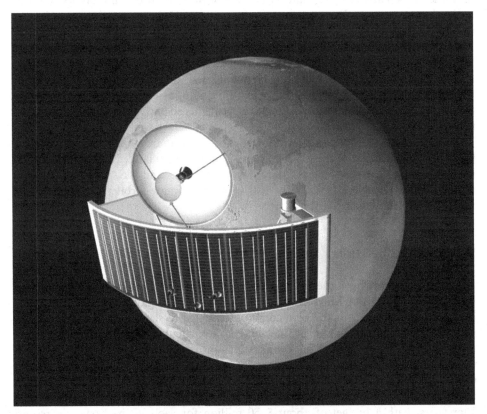

The 2003 Mars Micro-mission relay satellite.

all latitudes and GPS-like navigation to localize a payload on the surface to within about 5 meters. The second 'layer' of this infrastructure would comprise a small number of areostationary orbiters in distant high-altitude equatorial orbits to transmit to Earth as much as 100 Gbytes of data per sol. Unlike the low-altitude orbiters, the areostationary ones would be large spacecraft with a pair of high-gain antennas, one for receiving data from the surface of Mars and the other for transmitting to Earth. Because these spacecraft would consume a lot of propellant in maneuvering into their operating positions, they would not be able to be launched as readily as the piggybacked satellites. The first areostationary relay satellite was to be launched in 2007 at the earliest.

In parallel with building up this architecture around Mars, JPL was to upgrade the Deep Space Network, expanding its capabilities with tracking assets of other nations. Finally, new systems and software would be put in place to transform this architecture into a veritable interplanetary Internet.[84,85,86]

Other possible Micro-missions were a small Dynamo orbiter to map the fossil magnetic fields discovered by Mars Global Surveyor, a network of Deep Space 2 penetrators, the Mimas 9.7-kg microwave sounding orbiter, the US–French Pascal network of 24 meteorological stations, a pair of orbiters to use mutual occultations to develop vertical profiles of the atmosphere at a wide range of latitudes and times of day, and a variety of other aircraft, balloons, etc.[87,88]

Most scientists endorsed this program, but there were significant critics. Gerald Wasserburg, a geochemist at the California Institute of Technology, for example, said that it lacked focus. In a decade, he maintained, "we could end up with a lot of small triumphs but a number of big scientific questions still unresolved". And the inclusion of human exploration-oriented experiments so early in the program was, he insisted, "absurd".[89]

Nevertheless, the 2003 launch window was intended to be a busy one, seeing in the best case the departure of the US-led Mars Sample Return lander and rover, the first communications Micro-satellite, the Mars Airplane, and ESA's Mars Express orbiter with its Beagle 2 lander. The loss of Mars Climate Orbiter and Mars Polar Lander completely changed this.


## WATER EVERYWHERE!

After a thorough review, the US approach to exploring Mars was reorganized and rationalized. Much like Mars Surveyor, the new Mars Exploration program would have four main themes linked by a common strategy: to search for past and present life; to understand the history of the climate, in particular of atmospheric volatiles; to understand the geology of the surface and subsurface; and to inventory resources which could be exploited by human exploration missions. Key to all these themes was the search for water and water-modified geological features. The new program called for several well-defined missions in its initial years and for more flexibility in later years, with a budget of about $450 million for each of the first 5 years and significant contributions from international partners.

Orbiters and landers were to be launched on alternating windows, so that if one were to fail there would be several years available in which to work out what went wrong and make the necessary modifications prior to sending either a replacement or a successor. The program would start with the Mars Surveyor 2001 orbiter. This would be followed in 2003 by long-duration rovers, and in 2005 by a 'spy satellite-class' Mars Reconnaissance Orbiter which would build on the experience of Mars Global Surveyor with imagery of an even higher surface resolution. In parallel with the main program, starting in 2007, NASA planned a new class of missions. Called Mars Scout, this was similar to the Discovery program in that it was to consist of competitively-selected, relatively inexpensive missions costing about $300 million each that might involve innovative technologies such as aircraft, balloons, etc. It would provide the flexibility required to investigate newly discovered features and phenomena. The main 2009 mission would be the Mars Smart Lander to deliver a payload, most likely a large rover, to a precise spot. The sample-return mission was slipped from 2003 to 2011, possibly 2014, and a detailed definition study deferred. The French space agency was to undertake an aerocapture mission to demonstrate this for a sample-return mission, in this case with the orbiter establishing a network of landers. The Italian Space Agency would supply a communications orbiter. The head of space science at NASA, Edward Weiler, stressed that his agency had "not abandoned 'faster, cheaper, better'". However future experience, in particular with the 2009 mission, would prove otherwise.[90]

The Mars Surveyor 2001 orbiter was initially meant to complete the reflight of Mars Observer instruments, of which only the gamma-ray spectrometer remained. It was to be launched inside a Viking and Mars Pathfinder-derived heat shield and perform the first aerocapture experiment in which the atmosphere would be used to achieve the capture orbit prior to aerobraking to settle into the operating orbit. Although using aerocapture would permit the propellant load to be reduced, the technique was contingent upon precise navigation and trajectory control, and its effectiveness would strongly depend on knowledge of the atmospheric parameters, which varied. Too shallow a trajectory would result in too little braking effect and in too high an orbit, possibly even a missed insertion. Too deep a trajectory would expose the spacecraft to unplanned dynamic pressure, possibly sufficient to cause damage (as in the *inadvertent* aerocapture of Mars Climate Orbiter) or even strike the surface. It was a matter of trading off the mass of propellant for an insertion burn against the heat shield for aerocapture in an effort to devote more mass to the scientific payload. Other innovative approaches were planned for the mission. Due to the relatively high inclination of the Earth parking orbit for the 2001 window, JPL planned to launch the orbiter from Vandenberg Air Force Base on the coast of . California, which gave better access to highly inclined orbits than Cape Canaveral in Florida. This would make it only the second deep-space mission to depart from there; the first being the Clementine lunar orbiter developed by the Department of Defense. It was about to enter the assembly phase when Mars Climate Orbiter was lost. Fortunately, the error that doomed the 1998 mission was so trivial that it had no schedule impact on the reformulated Mars Exploration program, allowing the 2001 mission to be confirmed for that window.

A 'Red Team' headed by a former Cassini program manager was created at JPL to thoroughly scrutinize the plan and to oversee the development of the spacecraft at Lockheed Martin. The 'major action' list intended to assure success resulted in almost 200 modifications. These were covered by about $20 million of additional budget. The risky aerocapture maneuver was discarded early on, and the spacecraft reverted to a more conventional design. The modifications included specifying all documentation in both imperial and metric units to preclude misinterpretation, and improving navigation on making the initial approach to Mars by measuring against the background of quasars. The navigation team was also increased to six full-time staff (there was only one person assigned in the case of Mars Climate Orbiter). As an additional safety factor, the planned altitude at arrival was raised by 50 km. The principal hardware change was the installation of check valves between the lines carrying fuel and oxidizer, to enable the propulsion system to be pressurized early. It was decided to retain the Cape Canaveral launch site, and since the inclination of the parking orbit would make an unprecedented pass over western Europe it was decided to enlist the help of European tracking facilities, including a military space communications center in the UK and the Fucino station in Italy. US Navy telemetry aircraft operating from a base on the island of Crete would be on station over the Mediterranean. The US Air Force would have a mobile tracking station in Oman to acquire telemetry just after the final stage of the launch vehicle burned out. The unusual departure trajectory also meant that during the first month of its interplanetary cruise the spacecraft would be so far south of the celestial equator that the only Deep Space Network station which would be able to track it would be at Canberra in Australia. For this reason, and to provide additional coverage in the early portion of the flight, the assistance of a radio-telescope of the University of Chile was enlisted.[91,92] Finally, the mission was renamed Mars Odyssey in tribute to the novel by Arthur C. Clarke and the movie by Stanley Kubrick, both of which were set in the year 2001.

The mission profile for Mars Odyssey was to be similar to that of Mars Global Surveyor, with a 6-month interplanetary cruise ending with a propulsive insertion into a long-period capture orbit. In the case of Mars Global Surveyor a faulty solar panel had ruled out rapid aerobraking, resulting in this phase of the mission being drawn out for almost 12 months. This issue had been remedied. Mars Odyssey was therefore to lower its periapsis to about 100 km to generate aerodynamic drag and lower its apoapsis to about 400 km over an interval of 2.5 months, at which point adjustments would produce the final circular orbit at that altitude.

The Mars Odyssey bus was based on the structure of the Mars Climate Orbiter, which played no part in its own loss. It was 2.2 meters long, 1.7 meters tall and 2.6 meters wide, and was divided into a propulsion module and an equipment module that housed avionics, instruments, etc. It had a single solar panel on a gimbal with 2 degrees of freedom. When the three gallium-arsenide arrays were unfolded, they spanned 5.7 meters and provided up to 750 W of power at Mars. For periods when the orbiter was in the planet's shadow, or when otherwise needed, nickel-hydrogen batteries were available. A 1.3-meter high-gain antenna on a gimbal with 2 degrees of freedom was mounted at the end of a short boom. As in the case of Mars Global

Surveyor, this would remain in its stowed configuration until the aerobraking was complete. There were redundant 15-W amplifiers and X-Band transmitters with a maximum data rate of 124.4 kbits per second. There was also a receive-only low-gain antenna for commands sent from Earth. Like Mars Global Surveyor, the new orbiter was capable of relaying for surface missions, but in this case it was also capable of sending commands. It had a transmit-only medium-gain antenna and a UHF antenna, and was able to support data rates from the surface of 256 kbits per second. The propulsion module carried two hydrazine tanks, one nitrogen tetroxide tank and a helium pressurant system. These tanks fed the main 640-N bipropellant engine located at the center of the module. This would be used only once, for orbit insertion. Four 22-N hydrazine thrusters were to be used for low-thrust corrections in the interplanetary cruise and thereafter for trimming the orbit. For simplicity, accelerometers were not to be used to determine the duration of the orbit insertion burn. Instead, the main engine would burn to exhaustion. As a result, the apoapsis of the capture orbit could not be precisely predicted. The vehicle had star mappers, Sun sensors and an inertial platform for attitude determination, and four reaction wheels and four 0.9-N monopropellant thrusters for attitude correction.

Most of the scientific instruments for Mars Odyssey were on the nadir-viewing side of the equipment module, which would remain facing Mars. The mission was to map the elemental and mineralogical composition of the surface, the abundance of hydrogen in the near-subsurface as an indicator of water, and yield insight into the structure of the surface. It would also provide data on the radiation-related risk to hardware and humans. The Thermal Emission Imaging System (THEMIS) used uncooled 'multispectral' infrared technology developed for military satellites. By providing a 30-fold increase in spatial resolution across 10 spectral bands, it would enhance the results of Mars Global Surveyor's thermal emission spectrometer by

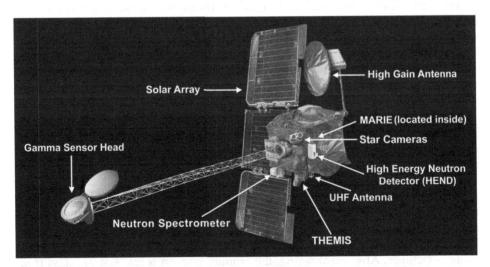

The Mars Odyssey orbiter completed the reflight of the Mars Observer instruments, carrying the gamma-ray spectrometer.

distinguishing between carbonates, silicates, sulfates, etc. It was also to correlate the mineralogy of Mars with its surface features. Working as a thermal imager, the instrument could identify 'hot spots' such as hydrothermal springs to assist in the search for indigenous life. In addition to 18-meter-resolution visible-light imagery to fill a gap between the moderate-resolution images of the Viking orbiters and the high-resolution images by Mars Global Surveyor, the Mars Odyssey camera would undertake multispectral mapping of the entire planet at a resolution of 100 meters to produce a survey comparable to the Landsat coverage of the Earth's surface. About 15,000 images would be collected during the entire duration of the primary mission. A gamma-ray spectrometer derived from that lost on Mars Observer was to collect 300-km-resolution maps of the mostly unknown composition of Martian rocks. The only gamma-ray data available was that provided by Soviet orbiters and landers in the 1970s, so data from this new instrument was eagerly awaited. The spectrometer was also to detect solar flares and participate in the triangulation of celestial gamma-ray bursts. In order to minimize interference from gamma rays issued by the vehicle itself, the instrument was mounted at the end of a 6.2-meter-long boom that would be deployed after aerobraking. As part of the upgrade after Mars Observer, the instrument now incorporated a high-energy neutron detector and a neutron spectrometer supplied by the Russian IKI. These were on the science deck to detect hydrogen present within the uppermost meter of the planet's surface as an indicator of water ice. In particular, Mars Odyssey was to investigate the sites where Mars Global Surveyor data on topography and geology suggested the presence of water ice. Long-term studies could characterize whether the distribution of water ice was seasonal. The neutron and gamma-ray spectrometers could also be used together to measure the mass of atmospheric carbon dioxide over the poles, and estimate the quantity of gas that was exchanged between the atmosphere and surface deposits.

The science suite was rounded out by a spectrometer to monitor radiation levels and energetic particles from the Sun, as well as cosmic rays, in the interplanetary cruise and in orbit around Mars. Space radiation is one of the main issues facing a human mission to Mars, since over the duration of such a mission the crew will accumulate dosages many hundreds of times greater than that of an average person on Earth. Similar instruments had been flown before, but none outside the Earth's magnetosphere.

Mars Odyssey had a non-redundant 1-Gbyte memory for storing data from the instruments, pending transmission to Earth. Unlike most other planetary flyby or orbiter missions, no scientific measurements were to be extracted from the radio tracking.

Mars Odyssey was flown to Florida early in January 2001 for final testing and mating with instruments, solar panels, and then the third stage of the Delta II 7925 rocket. The window opened on 7 April 2001 and ran for 20 days, with two launch opportunities per day. Of course, after two failures in a row, most of the future US Mars exploration hinged on a successful launch. The spacecraft was to arrive at Mars in October. After aerobraking, the primary mission would last 917 days from February 2002 to August 2004, during which it would also provide relay services to the US and European landers that were to arrive in 2003. It had a launch mass of about 730 kg, including 225 kg of hydrazine, 122 kg of nitrogen tetroxide (to be consumed by the single firing at orbit insertion) and 44.2 kg of scientific payload. Including launch and operations during the primary mission, the overall cost was estimated at $297 million. It lifted off on the first day of the window, traveled up the coast, being tracked by stations in New Hampshire, and then entered a 52-degree inclination parking orbit just off Nova Scotia. After a brief coast, the second stage fired over Europe. A video camera on the second stage observed the phases of the ascent with stunning clarity. The third stage made the escape burn over the Middle East and released the spacecraft into a 0.982 × 1.384-AU solar orbit for a type-1 trajectory that would reach Mars after sweeping less than 180 degrees around the Sun.[93] The launch vehicle had been programmed to place the spacecraft in an orbit that would miss Mars by about 450,000 km to prevent the unsterilized third stage from contaminating the planet, but the trajectory was such that the 'miss distance' was only about 70,000 km. This was a bonus for the mission, since the propellant saved in maneuvering during the interplanetary cruise would become available for an extension beyond the primary phase of the orbital activities. Twelve days after launch, the camera was activated to take visible-light and infrared shots of the faint crescents of Earth and the Moon from a range in excess of 3.5 million km, both for calibration purposes and to verify the accuracy of the attitude control system. As Mars Odyssey was at that time near the perpendicular to the Earth-Moon line, this view showed these bodies at almost their true separation. Infrared scans measured the temperature of the Earth's night-side over Antarctica and Australia.

The first of four trajectory corrections had been scheduled for 1 week into the mission, but it was deferred until 23 May. Owing to the beneficial inaccuracy in the departure trajectory, a total of 18 kg of propellant would be saved during the interplanetary cruise. The radiation monitor collected data for more than 4 months. When ordered on 13 August to download its data there was no acknowledgement and even the 'heartbeat' signal of the instrument was absent, possibly because its electronics had been hit by an energetic particle. Automatic resets and emergency commands failed to resolve the issue, so the instrument was turned off pending an investigation. Hopefully, the instrument would be able to be reactivated after the critical orbit insertion and aerobraking maneuvers.[94] The data obtained prior to its being shut down included two showers of solar energetic particles. The only other hardware problem appeared to be the star camera for attitude determination, which was letting in too much sunlight. The gamma-ray and neutron spectrometers were operated for 1,000 hours during the cruise, and returned good data which included 25 gamma-ray bursts between early May and late September.

Mars Odyssey started to prepare for orbit insertion 10 days from the planet. The final course correction was not required. With 9 minutes to go, the propellant lines were pressurized. Two minutes later, the spacecraft slewed to the burn attitude and maintained contact only by transmitting an unmodulated carrier over the medium-gain antenna. The engine ignited at 02:18 UTC on 24 October 2001. Ten minutes into the burn, the vehicle was eclipsed by the planet from both the point of view of the Sun and the Earth. It reached periapsis at an altitude of 300 km. After firing for 1,219 seconds, the engine stopped with only about 0.5 kg of oxidizer remaining in its tank, having delivered a 1,433-m/s velocity change. After an occultation lasting about 19 minutes, the spacecraft reappeared and resumed communication. It was in a 272 × 26,818-km orbit with a period of 18.6 hours, well within the acceptable range, and inclined at 93 degrees to the planet's equator. If the orbital period had exceeded 20 hours a reduction maneuver would have been performed at the third periapsis. The neutron detector and spectrometer were reactivated the next day, and were to operate for most of the aerobraking phase in order to calibrate them by recording the background radiation attributable to the body of the vehicle. During the fourth periapsis, passing over the north pole, the neutron detector was used to determine its ability to detect and resolve particles originating from the planet. Although this observation was cursory, it was consistent with a polar terrain rich in hydrogen and therefore probably also rich in water. Four days after orbit insertion, the aerobraking walk-in phase began, lowering the periapsis to 111 km in stages over 8 orbits. The limiting factor for the atmospheric passes was the thermal stress on the solar panels. The camera was used sparingly during the aerobraking phase of the mission, but on 30 October it took its first infrared images of the planet. At that time it was at an altitude of about 22,000 km, at apoapsis on the ninth orbit. The view was of the cold late spring south polar ice cap and the Argyre basin. Two days later it recorded the variation in temperature across the south polar cap. In all, 600 atmospheric profiles were obtained across altitudes in the range 95 to 170 km. Unlike Mars Global Surveyor with its broken solar panel, Mars Odyssey suffered no particular anomaly during this phase, which produced a more or less complete spatial coverage of the atmosphere. As then, the density of the upper atmosphere was variable. As the air density was believed to be capable of doubling between passes, the initial passes were conservative. As confidence grew, the periapsis was allowed to diminish, and eventually dipped as low as 95 km.

Throughout the aerobraking, the thermal emission spectrometer on Mars Global Surveyor monitored the planet's atmosphere, in particular to provide a warning of the occurrence of storms which could send heat-absorbing dust into the uppermost reaches of the atmosphere and increase its temperature and density. Mars Odyssey flew through the 'uncharted' northern polar upper atmosphere and provided insight into its dynamics and structure. The temperatures at high altitude were found to be twice the values predicted by atmospheric models. Precise orbit tracking provided an independent measurement of the atmospheric drag, and indirectly measured the average density. Wind-shears of 200 m/s were detected, and wind speeds within the polar vortex were measured in some detail.[95,96,97]

After more than 2 months of aerobraking, the orbit was so low that if no action were taken the spacecraft would crash within 24 hours, so the periapsis was raised above the atmosphere. A 20-m/s burn on 11 January 2002 completed the walk-out phase by placing the periapsis at 201 km. A total of 332 passes through the upper atmosphere had achieved a total velocity change of just over 1 km/s. If only engine burns had been used, the propellant requirements would have been increased by about 200 kg.[98] Over the ensuing days, the orbit was modified to attain the correct inclination, orientation and timing required by the scientific mission, and then the 400-km circular orbit was established on 30 January, about 18 km higher than that

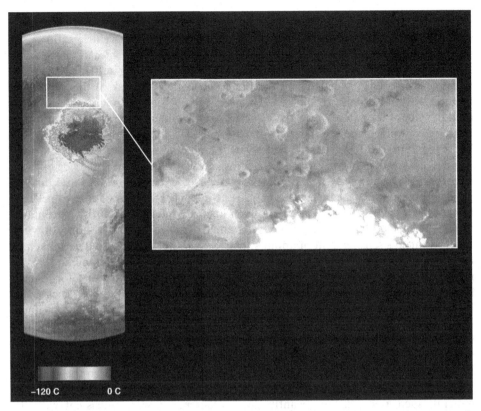

The first Mars image taken by the camera on Mars Odyssey at apoapsis on its 13th orbit on 2 November 2001. The thermal image at left shows the temperature field of the southern hemisphere, with the minimum over the south pole, and the visible-light image shows detail at the edge of the polar cap. (NASA/JPL/Arizona State University)

of Mars Global Surveyor at the time, and with a period of just under 2 hours. For most of its first Martian year, Mars Odyssey's equator crossing would be allowed to gradually drift from 3:45 p.m. towards 5 p.m. local time, providing acceptable illumination for high-quality imagery, and then a burn in late 2003 would 'lock' it at 5 p.m. This was a compromise between the requirements of the infrared camera and the gamma-ray spectrometer, with the former requiring an equator crossing in the mid-afternoon (or earlier) and the latter requiring it at dusk. However, the late-afternoon crossing would make the orbiter a relatively poor relay satellite, because the orbit would provide communication opportunities for a lander only in the early morning and late evening, at times when a lander would be required to draw upon its batteries.[99] The high-gain antenna was deployed on 4 February, and the camera started routine imaging two weeks later.

The gamma-ray spectrometer reopened its door on 20 February. After warming and cooling periods, it started to collect data in late March. Troubleshooting of the radiation monitor enabled it to resume operating on 6 March.[100] Deployment of the gamma-ray spectrometer's boom was delayed to allow engineers time to develop software patches for use in the extremely unlikely event of the vehicle safing itself during deployment at the same time as an attitude control system hardware failure striking in the orbit where the solar panels were not providing sufficient power to recharge the battery. Management did not want to take the risk of this occurring, because it would be catastrophic.[101] However, even before the boom was extended (and in spite of the 'noise' from the vehicle itself) the first month of observations in March and April made the mission's most important discovery. The abundance of hydrogen appeared to vary quite widely, with high concentrations southward of 60 degrees of latitude in the southern hemisphere indicating the presence of vast reserves of water ice near the surface, probably as a layer of mixed ice and dirt. At polar latitudes, water ice seemed to be present in concentrations of several per cent by weight in the uppermost few tens of centimeters, and in substantial percentages (up to 30 per cent by weight, equivalent to more than 50 per cent by volume) down to the instrument's sensing limit of about 1 meter; it was not possible to determine how far down it extended. This ice could be within reach of a Mars Polar Lander-style mission, and the discovery would also have important implications for human exploration because it would be possible to produce air, water and even rocket fuel by electrolyzing it into oxygen and hydrogen. At the time of these observations it was not possible to confirm the presence of water ice at the north pole, where it was winter and the surface was coated with a thin a cap of frozen carbon dioxide. Surprisingly, patches of hydrogen-rich terrain were present near the equator. These latitudes were too warm for water ice to be stable near the surface, so the signature was theorized to be either ice which had been buried or chemically-bound water in minerals such as clays. Significantly, the measurements were consistent with the amount of chemically-bound water that had been measured in-situ by the Viking 1 lander. Although only preliminary, these observations convinced most planetary scientists that Mars possessed a vast reserve of water.[102,103,104,105] The gamma-ray spectrometer boom was successfully deployed on 4 June.

By mid-October the carbon dioxide cap at the north pole had sublimated, giving the neutron spectrometer a view of the exposed surface. It proved to contain an even greater percentage of water ice than the southern pole. 'Permafrost' deposits at northern latitudes appeared to be even nearer the surface than those at southern latitudes. And synergy between Mars Odyssey measuring the flux of neutrons and Mars Global Surveyor using its altimeter to measure how the thickness of the polar cap varied revealed how densely the ice was packed at high northern latitudes. The ice appeared to have the density of freshly laid fluffy snow.[106,107] Low-latitude and temperate-latitude deposits also attracted interest. It was found that the amount of water ice there was too great to be currently in equilibrium with the atmosphere. This suggested that Mars was just emerging from a colder period or 'ice age' that could have ended as recently as 400,000 years ago, with ice deposits still adapting to the new climate.[108,109] Simulations also showed the spin axis of the planet to be unstable, and that during periods when the axis is steeply inclined the polar caps will sublimate and ice will migrate to form glaciers several kilometers thick on the slopes of the highest mountains; namely, Olympus Mons and the volcanoes of the Tharsis bulge.[110] Long-term observations of the polar regions revealed the seasonal variation of the ice caps as carbon dioxide sublimated and water ice was exposed. It was found that the neutron flux increased during spring, then stabilized after all of the carbon dioxide had sublimated to leave only pure water ice. The gamma-ray spectrometer results provided the best evidence to-date of a possible ancient ocean in the northern lowlands in the form of potassium, iron and other elements which could have been 'leached' from the southern highlands, transported and deposited as sediment on the ocean's floor.

From the low circular science orbit, the camera returned the first orbital night-time images of Mars, and could discern warmer rock fields from sand and dust by their different temperatures and thermal inertia. A plateau was thereby identified in images of a part of Valles Marineris where valleys, buried under a layer of loose sand-like material, had all of the characteristics of terrestrial rain drainage features. As the valleys appeared to be about 3 billion years old, this could indicate that the climate remained 'warm' for long enough to support liquid water until at least that time. Moreover, this discovery could indicate that the Valles Marineris complex and its tributary canyons were formed by the action of flowing surface water rather than as a result of crustal tectonics as many geologists had concluded.

Vast areas of bare bedrock were discovered. This unexpected result showed that processes must be at work to clear away the otherwise omnipresent dust. Piles of bare rocks accumulating on hillsides also showed that they must have formed quite recently. The same technique was able to reveal the blockiness of crater ejecta and

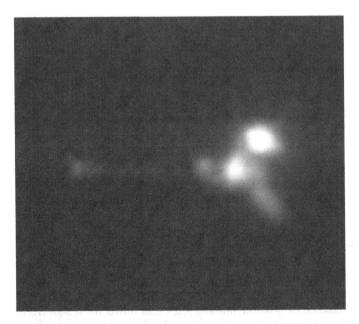

Mars Odyssey photographed from a distance of 90 km by Mars Global Surveyor. (NASA/JPL/MSSS)

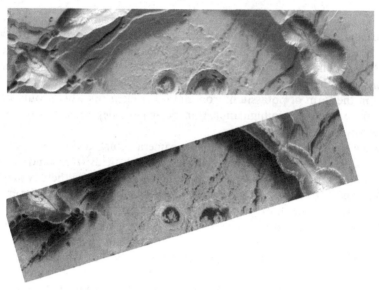

This pair of images shows the capabilities of the thermal camera on Mars Odyssey on a portion of Noctis Labyrinthus seen in visible-light during the day (top) and in infrared during the night (bottom). The night-time image allowed scientists to recognize the different thermal inertia characteristics of the terrain, and to discriminate between rocks, dust and sand. North is to the right. (NASA/JPL/Arizona State University)

A portion of Valles Marineris at 19-meter resolution, showing landslides and gullies on the canyon walls. North is to the right. (NASA/JPL/Arizona State University)

avalanches on the walls of canyons. Conversely, Tharsis, the flanks of Olympus Mons and ancient cratered terrains were found to be covered with thick blankets of fine dust with few if any rocky ridges exposed. The ability to use thermal inertia to distinguish between rocky and sandy terrains was used to evaluate landing sites, in particular those being considered for the two Mars Exploration Rover missions. In fact, reconnaissance of possible landing sites accounted for a significant fraction of the observations during the primary mission.[111] The caldera in Nili Patera proved to be solid rock, but there were dunes of coarse dust superimposed in some places.

Other such outcrops were discovered, including one near the Mars Pathfinder site. But there was absolutely no sign of endogenous heat sources such as hydrothermal springs or fresh lava flows; even young-looking volcanic deposits did not give off any residual heat.

The Mars Odyssey camera found areas at high southern latitudes that appeared cooler than adjacent regions. Observations in concert with the thermal emission spectrometer on Mars Global Surveyor over a period of 2 local years indicated a thermal inertia that could be explained as exposed deposits of a lag of water ice grains mixed with dust. A search of over 25 years' worth of imagery revealed that these regions corresponded to patches of brighter terrain in Viking pictures. Mars Odyssey established such patches of exposed ice at high latitudes to be extremely common after the carbon dioxide caps had retreated. This enabled the mapping of

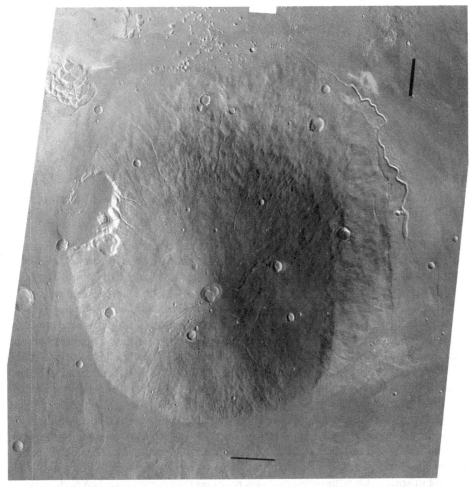

A mosaic of infrared images of the volcano Hecates Tholus. (NASA/JPL/Arizona State University)

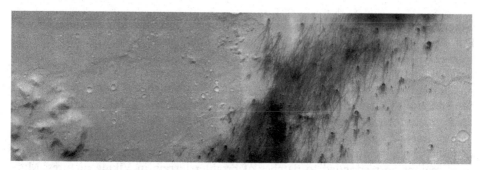

A portion of the landing ellipse of the Spirit Mars Exploration Rover inside Gusev crater seen by Mars Odyssey. Dark streaks represent areas cleaned of bright dust by the wind or dust devils. North is to the right. (NASA/JPL/Arizona State University)

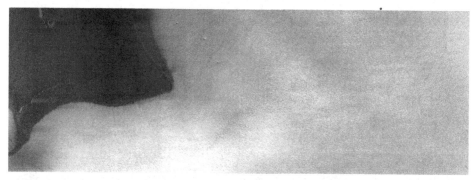

This featureless terrain includes part of the Mars Polar Lander landing ellipse. North is to the right. (NASA/JPL/Arizona State University)

surface water ice deposits at a scale of less than a kilometer; a considerably finer spatial resolution than was possible using the gamma-ray spectrometer.[112,113] Other landforms, including smooth deposits, polygonal terrains similar to polar features on Earth, 'gullies', dust-covered glaciers, etc, indicated the direct emplacement on the surface of a mix of water ice (probably snow) mixed with dirt and dust.[114] The sudden melting of water ice in small deposits at temperate latitudes where they had accumulated during the last 'ice age' offered a plausible mechanism to explain the intriguing 'flows' discovered by Mars Global Surveyor. These gullies were one of the priority targets for the Mars Odyssey camera, and its mid-resolution imagery showed a relationship between gullies and 'pasted-on' terrain laden with snow and protected by dust. As already discovered by Mars Global Surveyor, such snow deposits survived only on shadowed slopes, since they would rapidly melt in direct sunlight.[115] In the polar regions the Mars Odyssey camera was able to distinguish between carbon dioxide and water frost, and to characterize temperature patterns, convection cells in the atmosphere, the deposition of ices in the autumn and winter and their thawing in spring. Day-time observations revealed that meter-scale layers discovered years before by Mars Global Surveyor in some cases corresponded to different physical and compositional characteristics, suggesting that the process of deposition had changed over time.[116]

Operating in its multispectral mode, the camera, in conjunction with the thermal emission spectrometer on Mars Global Surveyor, established that, just as on Earth, lavas evolved over time as the interior of the planet differentiated and cooled. For example, the lava flows in the Nili Patera part of Syrtis Major that appear to be billions of years old began as silicon-poor basaltic and later became glassy silicon-rich lavas as the source evolved. In contrast, the walls of the 4.5-km-deep Ganges Chasma cut through olivine-rich basalt kilometers below the surface. Olivine is an iron-magnesium silicate that is so easily 'weathered' by water that its presence is normally used as an indicator of a dry environment. Finally, quartz-bearing rocks which indicated very highly evolved magmas were identified for the first time on Mars on the flanks of volcanic structures in Syrtis Major.[117] As already suggested by Mars Global Surveyor, large areas of the planet were confirmed to be covered with olivine basalt. Small regions exposed to ground water were discovered by the hundreds in the southern highlands, consisting of small but widespread irregular patches several kilometers in size showing spectral characteristics compatible with the chlorides and chlorine-bearing minerals that typically imply the evaporation of bodies of standing water. The irregular outlines of these deposits proved to trace the outlines of sinuous channels and the floors of small craters. In higher resolution imagery obtained by other orbiters, these areas were brighter than their surroundings, and in some cases displayed polygonal patterns typical (on Earth) of desiccated mud.[118]

One of the most astonishing discoveries by the Mars Odyssey camera was the presence of black circular spots on the flanks of Arsia Mons, the southernmost of the Tharsis volcanoes. Analysis of multispectral images revealed that these were not darker patches of surface, nor impact craters, but very likely the 'skylights' of caves. They were typically several hundred meters across, and the caves were at least as deep. This discovery was interesting not only for geology, but also because if life developed early on in the planet's history, caves could be an environment in which micro-organisms could have survived to this day. Furthermore, caves could provide shelter for human explorers. Unfortunately, caves will be quite difficult to explore by

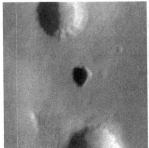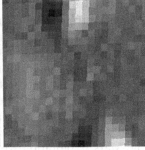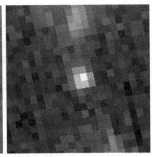

One of the cave skylights discovered by Mars Odyssey seen in visible light and in two infrared bands. (NASA/JPL/Arizona State University)

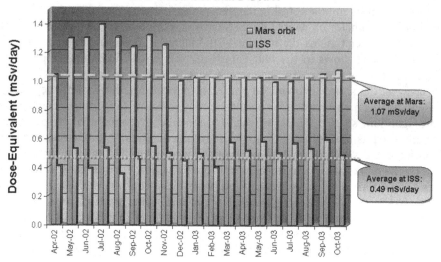

**Radiation Dose-Equivalent at Skin Level (mSv/day): ISS Orbit vs. Mars Orbit**

A graphic of the radiation dose (in milliSieverts per day) measured in Mars orbit compared to that measured in Earth orbit on the International Space Station (ISS). The accumulated dose on Mars was on average twice that in Earth orbit. (NASA/JPL/ Johnson Space Center)

robotic landers, not only because their skylights are much smaller than a typical landing ellipse, but also because, being on the flank of a volcano that sits atop a ridge, the air is too thin for conventional braking during the initial phase of atmospheric entry.[119]

The only serious instrument glitch of Mars Odyssey's primary mission occurred on 28 October 2003. In a period of increased solar activity, the radiation monitor ceased to function properly and remained silent despite efforts to revive it. During its 18 months of operation in Mars orbit, it reported the total accumulated radiation dosage to be on average twice that in low Earth orbit, but significant solar activity imposed short-term irradiation well above that average.

The lander relay also acted as an instrument by issuing an unmodulated carrier signal for a 'specular reflection' bistatic radar experiment when the planet was in opposition in August 2003 (at its closest to Earth for 60,000 years) while terrestrial antennas sought the reflections. At the next opposition in 2005, a radio-telescope 'illuminated' Mars and the spacecraft sought the reflections.[120,121]

A $35-million supplement to its budget enabled Mars Odyssey to start the first phase of its extended mission in late August 2004; this was to last for a Martian year. In addition to monitoring seasonal climate changes, polar ice cycles, etc, it was to be the primary orbital relay for the Mars Exploration Rovers. It would also evaluate the north polar terrain for the 2008 Phoenix mission which, it was hoped, would recover the science lost by Mars Polar Lander and the canceled 2001 Mars Surveyor lander. Long-term observations by the gamma-ray spectrometer provided data on the composition and circulation of the atmosphere, revealing details of the complex carbon dioxide cycle in which the atmosphere and polar caps both played a role, with the results revealing just how oversimplified the models of the polar weather had been. In particular, it provided an estimate of the total mass of carbon dioxide that condensed at the poles, and revealed that atmospheric argon increased at high latitudes during autumn and then dissipated during the winter and spring. Indeed, Mars Odyssey revealed that the very composition of the atmosphere varied with the seasons. Gases such as argon, nitrogen and oxygen that cannot condense contribute on average 5 per cent, but after the carbon dioxide had frozen onto the ground their overall proportion reached 30 per cent of the winter polar atmosphere and stratified the polar air in a similar manner to that of water of different salinity in the terrestrial oceans. Furthermore, this stratification and the meteorology of the Martian polar regions were intriguingly similar to the 'ozone holes' which develop above the terrestrial poles. Also, whereas the model in vogue since the mid-1960s predicted that carbon dioxide would condense at the poles during late autumn and winter, then sublimate in spring, infrared and neutron data showed that frost was already condensing by the late summer.[122,123,124,125]

As in the interplanetary cruise, the gamma-ray spectrometer continued to yield astrophysical data from its vantage point in orbit around Mars, in particular joining other spacecraft of the third international network of gamma-ray burst detectors in observing the 'soft gamma-ray repeater' flare of 27 December 2004.[126]

During its extended mission, Mars Odyssey had opportunity to investigate the dark spots, typically several tens of meters in size, which characterize the southern polar terrain in the spring, often in association with small networks of 'spidery' radial channels. The camera took hundreds of images during different seasons and found the spots to be the same temperature as the surrounding carbon dioxide ice, thus showing them to be a thin veneer of dust rather than warmer patches of bare soil as some scientist had thought. As the Sun rose over the polar cap, the thawing carbon dioxide that was mixed in with darker and hence warmer dust would tend to create jets of gas that carried dust and sand to create patches of darker materials. The jets would last several months (at most) after sunrise, and then cease, only to reappear the next spring.[127]

In 2006 the Mars Odyssey mission was granted a second extension and then on 30 September 2008, at the end of this extension, the spacecraft fired its thrusters to depart the Sun-synchronous orbit that it had flown for the last 5 years and began to drift in a manner calculated to move the afternoon equator crossing from 5 p.m. to 3:45 p.m. to improve the infrared mapping and obtain additional high-quality data for the entire planet. The penalty paid was that the gamma-ray spectrometer would soon overheat in this orbit and have to be deactivated. Its final contribution was gamma-ray spectra of a dozen different volcanic regions that revealed trends in the composition of the ancient lava flows compatible with the evolution of the Martian mantle, its cooling and the thickening of the lithosphere. Older volcanoes appeared to have spewed warmer lava than younger ones.[128]

Starting in 2009, the camera began to operate by looking sideways instead of at the nadir. This allowed stereo imaging of a few selected areas, as well as mapping of some polar latitudes that would never be directly overflown. On 9 June 2009 the vehicle fired its engine again to become Sun-synchronous with the plane of the orbit in the desired orientation. Meanwhile, the spacecraft underwent some maintenance. Back in 2007 a power distribution unit had failed and the computer had switched over to the backup unit. If a problem were to develop with the backup, this would end the mission. An investigation had concluded that rebooting the computer might revive the primary unit. On command from Earth the computer was restarted for the first time in flight, regaining the lost level of redundancy. The vehicle, meanwhile, was showing signs of aging – entering safe mode when the primary encoder of the solar panel gimbal malfunctioned by erroneously detecting a mechanical problem in the actuator.

By mid-2010, Mars Odyssey's infrared camera had mapped the entire planet at low-to-medium-resolution, having taken over 21,000 images to enable scientists to build the most complete, 100-meter-resolution map of the entire planet, and on 15 December this spacecraft surpassed Mars Global Surveyor as the longest-lived orbiter, having spent 3,340 days in situ. It is estimated to have sufficient propellant to provide attitude control and orbit changes through to 2015 at least, and probably into the 2020s, barring any incapacitating mechanical problem. In 2012, it will be called to support the landing of the Mars Science Laboratory 'Curiosity' and then, it will probably be approved for a fifth extended mission, starting in September 2012. By then, the 11-year mission will have cost a total of $508 million.[129] A maneuver is to be made at the end of the mission to raise the orbit to minimize the risk of microbiological contamination by preventing the spacecraft from crashing onto the planet for at least 50 years.


## EUROPE GOES LOW-COST

European scientists had developed a suite of instruments for the Russian Mars 96 mission, and its loss left them frustrated. The state of the Russian economy meant that it would be in no position to mount another mission any time soon, and there was little chance of the US adding so many foreign instruments to its missions.[130] As a result, in late 1996 the French Space Agency, CNES, began a study into the feasibility of modifying its Proteus small-satellite platform to undertake an orbital Mars mission carrying a 100-kg scientific payload which would include the French visual and infrared imaging spectrometer and occultation experiment, the German high-resolution camera, the Swedish ion and neutral particle spectrometer, and the Italian Fourier spectrometer; the conclusion was that this should be feasible.

This Mars 2001 mission was initially proposed as a bi-national French-German project, but it was soon integrated into ESA's scientific program.[131] Spurred by the apparent success of the 'faster, cheaper, better' approach of the US Discovery and Mars Surveyor missions, ESA had decided to establish its own series of 'flexi' (i.e. flexible) missions which would employ similar management principles in order to keep costs below a stringent ceiling. Crucial aspects of this approach were to be a 'lean' (streamlined) management, early selection of instruments and 'freezing' of the payload, reuse of technologies developed for other missions and greater industrial responsibility and oversight. A flexi-mission was cost-capped at 175 million euros. However, when Mars was selected for the first mission, a cap of 150 million euros was imposed because the instruments were already available. Mars Express, as it was called, would finally – two decades after early studies such as Kepler – mark the European debut in Mars exploration.[132] As initially envisaged, the mission was to carry the five instruments from Mars 96 and a pair of 150-kg landers, but one of the landers was deleted to accommodate an Italian-American ice-seeking subsurface radar. The Russian Academy of Sciences was at the time reviewing plans for a small mission to recover some of the data expected from Mars 96. The difficulty in funding this Molniya-launched mission made a Russian contribution to Mars Express a more sensible option.[133]

In early 1997 the Space Science Advisory Committee of ESA supported such a Mars orbiter. After a mission definition study, Mars Express progressed smoothly and the Scientific Program Committee confirmed its support in November 1998 so long as the funding requirements did not impair already approved projects. After a competitive industrial study phase, a contract was issued to Matra Marconi Space (now Astrium) for the spacecraft at a cost of 60 million euros. Work got underway early in the new year. As for the lander, in December 1997 ESA had invited bids. It received three proposals: one from the UK, one from Russia and Germany, and a proposal from France and Finland to fly a prototype NetLander. By the summer of 1998 the UK proposal, called Beagle 2 in homage to Charles Darwin's ship and its contribution to our knowledge of biology, was the preferred option. Mars Express was given full approval on 19 May 1999. Development would have to proceed apace, because it was to be launched during the window of the 'great opposition' of 2003, for which the energy requirements at departure from Earth would be most favorable. In fact, this would mark the closest that the two planets had approached each other for thousands of years! As Europe did not at that time possess a cheap, medium-performance launch vehicle, a Russian launcher would be used. ESA had already selected the Soyuz-Fregat to re-fly its 4-satellite Cluster mission, the first version of which was lost on the inaugural flight of the Ariane 5 in June 1996. But in order to provide flexibility, the structural design and interfaces of Mars Express were made compatible with the US Delta II. The Japanese H-II was also briefly considered. Tracking was to be performed by European facilities, in particular the deep-space antenna built at New Norcia in Australia for the Rosetta mission.

After launch in June 2003, the spacecraft would reach its objective towards the end of the year. The spacecraft would deploy the British lander prior to entering a polar orbit with a periapsis at 250 km and an apoapsis in excess of 11,000 km that would allow the instruments to cover the entire planet during the nominal primary mission timescale of 1 Martian year. It was to provide global topographic mapping at high resolution, a mineralogical survey at high resolution, study the permafrost and the subsurface, study how the surface, subsurface and atmosphere interacted, study the circulation and composition of the atmosphere, and study how the upper atmosphere interacted with the interplanetary environment.

The objective of ESA was to incorporate the orbiter into the larger context of international Mars exploration. When the mission was approved, this still included the US Mars Surveyor sample-return, whose sample canister would be tracked by Mars Express; joint studies with Nozomi, which was to enter a similarly eccentric but equatorial orbit; and support to the French-led NetLanders that were expected to be launched in 2005 or 2007.

Various management techniques were applied to ensure that the cost cap would not be exceeded. These included a small ESA project team of about 10 persons. In the flexible approach, payload specifications were frozen early on (the fact that the instruments were carry-overs from Mars 96 made this relatively straightforward to achieve) and direct interfaces were established between the payload scientists and the makers of the spacecraft hardware, with ESA retaining full control of payload performance issues. Furthermore, the fact that about 80 per cent of systems and hardware were common to Rosetta, with both projects running in parallel, enabled key personnel to be shared.[134] Several non-ESA states participated to some degree in either Mars Express or Beagle 2. These included the US, Russia, Poland, Japan and China.

The orbiter had a low-cost modular bus comprising an aluminum honeycomb 1.7 × 1.7 × 1.4-meter box connected to a cylindrical launch vehicle adapter and to an internal frame. The bus carried a pair of solar panels, a fixed high-gain antenna, the instruments, the Beagle 2 lander and its relay antennas, space radiators, and the main engine and the launcher interface. The bipropellant propulsion system was inherited from the Eurostar communications satellite. It consisted of a 416-N main orbit insertion engine and two quartets of 10-N thrusters for attitude control, all of which were mounted on the face opposite to the instruments and drew propellants from two tanks of monomethyl hydrazine and mixed nitrogen oxides. If the main engine were unavailable for the orbit insertion maneuver, the 10-N thrusters could achieve a highly eccentric capture orbit whose apoapsis would be reduced by mild aerobraking. The solar panels were derived from those of the Globalstar low-Earth orbit communications satellites. They had a total area of 11.2 square meters, which was sufficient to provide up to 660 W at Mars, had 1 degree of freedom and gave the vehicle a total span of 12 meters. As the orbiter would experience 1,400 eclipses during its primary mission while passing through the shadow cone of the planet out near apoapsis, each lasting potentially up to 95 minutes, it was provided with three rechargeable batteries for use in darkness. The communications system had a 65-W X/S-Band transmitter and a 1.65-meter high-gain antenna capable of a peak transfer rate of 230 kbits per second. Attitude determination was by a number of sensors, including laser gyroscopes, Sun sensors and two wide-field star trackers on the side of the bus opposite from the high-gain antenna. Attitude control was by reaction wheels and thrusters.

Mars Express was designed from the start to be highly autonomous and capable of looking after itself, because the single European deep-space antenna (a second was planned, but would not be commissioned until the middle of the next decade) would be able to provide support for at most several hours per day. Hence 12-Gbit memories were carried for storing housekeeping and scientific data. NASA's Deep Space Network would provide support during the early phases of the mission, and thereafter on request. ESA would receive data from Beagle 2 and then pass it on to

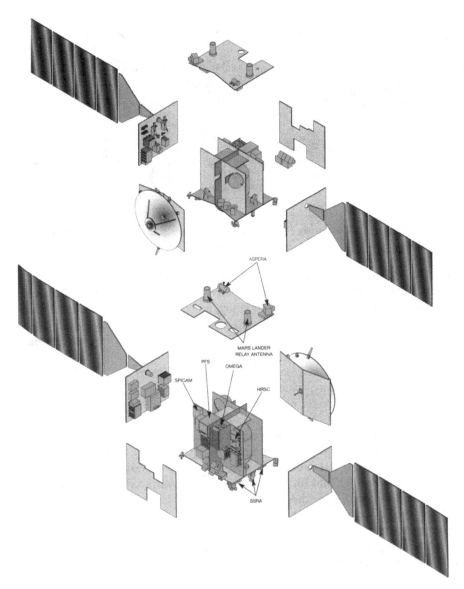

Mars Express, the first European Mars orbiter. (ESA)

an operations facility at the British National Space Centre in Leicester, England. ESA issued a tender for the scientific payload in November 1997. In addition to Beagle 2 it selected 7 instruments with a total payload mass of 113 kg. The most amazing instrument, at least from a public awareness point of view, was the high-resolution stereoscopic camera developed by Germany for Mars 96, upgraded with a super-resolution channel. This was a 9-CCD push-broom camera that was able to provide a surface resolution of 10 meters. Although its resolution was not as fine as

The German High Resolution Stereo Camera on Mars Express was a re-flight of the instrument lost on the Russian Mars 96 mission. (ESA)

that of Mars Global Surveyor, it would be able to scan wider and longer swaths, and over time would build up the first global stereoscopic color coverage of the planet to characterize its geology, climate, topography, morphology and geological evolution. Imaging at about 2 meters per pixel was also possible by using motion-compensation techniques, and was planned for several specific areas including the Beagle 2 landing site.[135]

The Observatoire pour la Minéralogie, l'Eau, les Glaces et l'Activité (OMEGA; observatory for mineralogy, water, ices and activity) was a French-led visual and infrared imaging spectrometer derived from instruments on Fobos and Mars 96. Its resolution would vary between 300 meters and 5 km depending upon altitude, and it would make a spectrum in 352 adjacent channels for each pixel. It was to provide global maps of the composition and mineralogy of the surface, and characterize frosts, ices, atmospheric dust, OH-radical-carrying hydrated minerals and carbonates.

An infrared Fourier and an ultraviolet and near-infrared spectrometer were to study the atmosphere, providing vertical profiles of temperature, pressure, carbon oxide and dioxide, ozone, hydrogen, water vapor, etc. The infrared spectrometer was also to seek trace gases such as methane. The instruments would also monitor the optical properties of aerosols and profile dust in order to study atmospheric circulation patterns. Although the ultraviolet spectrometer was boresighted with the other instruments in the nadir direction, it was equipped with a small mirror to enable it to scan the limb of the planet during solar and stellar occultations so as to detect chemicals in the atmosphere at altitudes up to 150 km.[136]

A low-frequency radar was to isolate reflections from features buried within the Martian crust to a depth of several kilometers, with a vertical resolution of around 150 meters. In addition to detecting water ice and, if present, liquid water, it could measure the roughness of the surface and provide altimetry and ionospheric data. It used a pair of 20-meter-long dipole antennas and a 4-meter monopole antenna. The monopole would be aligned vertically, and the dipoles perpendicular both to it and the direction of flight. The antennas were to be deployed several months after arrival.

Radar data would be collected only when the vehicle was within 850 km of the surface and primarily over the night-side, when other instruments would not be operating. Each radar pass would involve 5 minutes of ionospheric sounding, 26 minutes of subsurface sounding, and a further 5 minutes of ionospheric sounding. Supplied by the University of Rome in cooperation with JPL, it was the first deep-space subsurface radar sounder since Apollo 17 in December 1972.

The Analyzer of Space Plasmas and Energetic Atoms (ASPERA) was a reflight of a Swedish instrument developed for the Fobos and Mars 96 missions to detect atoms and ions escaping the atmosphere, and to investigate the interactions of the solar wind with the planet's ionosphere in conjunction with a similar instrument on Nozomi. As the only 'particles and fields' package on Mars Express, it comprised a pair of energetic neutral atom sensors, one electron spectrometer, and one ion spectrometer.[137]

The final experiment on the orbiter was a radio system to sound the atmosphere and ionosphere during radio-occultations, perform bistatic radar observations and map the gravity fields of the planet and its moonlets. It would also be used at solar conjunction for coronal studies. Other hardware included the Beagle 2 data relay system and a small 640 × 480-pixel wide-angle webcam-style camera to verify the lander's release and monitor its separation.[138,139]

The idea of a small British lander cushioned by airbags was born in the wake of the Mars Pathfinder mission that marked the US return to the Red Planet and the participation of scientists from the Open University in the analysis of the meteorite ALH84001 that led to the 'fossils on Mars' debate. Beagle 2 was to characterize the morphology, geology, chemistry and mineralogy of its landing site in order to determine whether conditions had ever been conducive to the development of life. Specifically, it would investigate the oxidation state of the rocks, soil and dirt, look for water and water-modified materials such as carbonates, search for organics and characterize their isotopic ratios, and analyze the 'air' for trace gases which could indicate currently extant life. In short, it would be the most comprehensive search for life since the Vikings and would benefit from recent discoveries of how micro-organisms can exist in extreme environments. The original 100-kg lander was to be pyramidal with four solar panel-covered leaves deploying from its sides, and a miniature rover for sampling. When this design proved to be too heavy, the rover was replaced by a robot arm and an ingenious 'mole' that could fulfill most of the same functions. Dubbed the Planetary Underground Tool (PLUTO), this was a tethered sampler 28 cm long and 2 cm wide, cylindrical with rounded ends. Massing only 0.5 kg, the mole was carried in a 'launch tube' on the lander. It could either dive straight into the ground or crawl across the surface to a rock and sample the soil beneath. An internal magnetically-actuated hammer that imparted one shock every 5 seconds enabled it to burrow into sand to a depth of 1.5 meters, measuring the temperature along the way. Two claws on its front side could open to collect and store 0.2 cubic centimeters of soil. Once the mole had a sample, the tether would reel it back into its tube so that the sample could be analyzed. This 'puppy' version of Beagle 2 just met the stated mass constraints. However, the mass available to the lander then shrank to 60 kg. To accommodate this constraint, the lander was

A full-scale model of the British Beagle 2 Mars lander. The instruments and tools on the 'end effector' of the robotic arm are (counterclockwise from left): mole funnel, wind sensor/wide-angle mirror/sampling 'spoon', left camera, rock corer and grinder, Mossbauer spectrometer, microscope, right camera, X-ray spectrometer.

redesigned as a 66-cm-diameter flat disk which would be cushioned on impact by a segmented airbag and operate on the surface for 180 sols.

The main structure was a shell of carbon fiber and aluminum honeycomb with an external protective layer of Kevlar and internal layers of foam to absorb shocks. In addition to the core systems the base unit housed the Anthropomorphic Robotic Manipulator (ARM), at the end of which was the Position Adjustable Workbench (PAW) incorporating the mole and a panoply of instruments. The clamshell lid had solar cells on its interior, and would hinge a quartet of rounded panels out onto the adjacent surface. The overall area of about 1 square meter would generate 650 W. The power system included rechargeable lithium-ion batteries. The lander had a non-redundant UHF link, and was to use an aerial on its lid to communicate with Mars Express or Mars Odyssey at up to 128 kbits per second. There was no facility for communicating directly with Earth. Thermal control was ensured by multiple layers of insulation, with gold foil absorbing solar heat during the day to keep the electronics warm.

Although Beagle 2's scientific payload was only 10 kg, it carried an impressive suite of miniaturized systems. At 5.74 kg, the gas analysis package was heaviest. Its mass spectrometer was to measure the abundances and isotopic ratios of carbon dioxide, hydrogen, nitrogen, oxygen, neon, argon and xenon, as well as trace gases such as methane. There were 12 miniature ovens on a carousel for heating soil samples to 1,000°C to study rock-trapped gases, the geochemistry of Mars, water-related processes, formation processes and organic chemistry including complex organic salts and acids. In particular, it would look for the relative abundances of the carbon isotopes because terrestrial organisms prefer to feed on carbon-12 over carbon-13; and by comparing the ratio in the soil to that in the atmosphere it would be possible to determine whether Earth-like organic processes were active. Being able to detect mere nanograms of carbon in a sample of 50 to 100 milligrams, this instrument was far more sensitive than the mass spectrometer used by the Viking landers.

The ARM was 109 cm when straight, but its maximum reach across the surface was 70 cm. Its 5 degrees of freedom made it more versatile than a human arm. The end-effector, the PAW, had a variety of tools. An X-ray fluorescence spectrometer was loosely based on the instrument carried on Sojourner, and was to be used for determining the elemental composition of rocks and soils by irradiating them with X-rays from four radioisotope sources. In conjunction with the base-mounted mass spectrometer, it would perform a rough rock-age determination, a type of analysis never before made on another celestial body. The Mossbauer spectrometer would measure the oxidation state and the nature of iron-bearing minerals as a means of investigating the history of water on Mars. A pair of 1,024 × 1,024-pixel cameras were mounted 19.5 cm apart to take stereoscopic pictures. One was equipped with a wide-angle mirror to enable it to take a 360-degree panorama with the ARM still stowed. Each camera had a 12-position filter wheel, in one case with a lens which magnified 64 times. The equipment of each camera also included a wiper blade to clean dust off the optics, and LED white-light 'torches' to take images at night that would reveal the true color of the Martian material without the day-time red light produced by dust in the atmosphere scattering sunlight.[140] The imaging calibration target comprised 16 brightly colored dots on an aluminum plate; the first example of extraterrestrial art, as the 'spot painting' had been prepared by the artist Damien Hirst. Moreover, to further raise public awareness on the mission, Beagle 2 was to initiate its transmissions from Mars with a 9-note tune especially composed by the British pop group Blur. A third camera was part of a microscope with a resolution of 4 micrometers to search for microfossils and textures of biological origin. Color information would be obtained by illuminating the target with four LEDs installed around the camera. A small corer/grinder was provided by a Hong Kong dentist. It massed only 370 grams and drew a mere 2 W, but was capable of penetrating up to 1 cm into rock to obtain powdered samples for the mass spectrometer and prepare dust-free flat surfaces for the microscope and the spectrometers. A larger prototype was tested aboard the Russian Mir space station.[141] As a backup in case neither the mole nor the corer/grinder were able to deliver samples, the PAW carried a small 'spoon'. A set of environmental sensors were distributed on the lander and its arm. The consensus derived from the Viking biology experiments was that the Martian surface contains oxidants inimical to life. One sensor was to test this hypothesis by seeking oxidants such as hydrogen peroxide and ozone. The environmental sensors included six detectors to monitor the flux of cosmic rays and ultraviolet in order to understand the sterilizing mechanism of the planet's surface, a dust impact sensor, the angle of repose of dust, air pressure and temperature, wind speed and direction. Overall, they massed a mere 156 grams. (The Fourier spectrometer on the orbiter was to scan the atmosphere above the landing site to calibrate the environmental sensors.) In addition, during atmospheric entry and descent a 3-axis accelerometer would measure the density and pressure profiles and the speed of the winds at low altitude.

As with US landers, the parachute/airbag system proved difficult to develop. In mid-2001 Martin Baker Aircraft, which supplied the system for the Huygens probe and was contracted to provide the entry, descent and landing system for Beagle 2, withdrew from the consortium; its role was taken by the prime contractor Astrium. In May 2002, after the lander's configuration was 'frozen', and a mere 13 months prior to launch, tests demonstrated that the airbags would not survive an impact at the predicted speed. It was necessary to completely redesign either the airbags or parachute. A larger parachute was made in just 3 months and validated by tests in August and September. Beagle 2 would be protected during atmospheric entry by a carbon-fiber heat shield that spanned 92.4 cm, had a 120-degree front angle and was covered by tiles of Norcoat Liège, a mix of powdered cork and phenolic resin. The backshell of carbon-fiber and titanium used a milder thermal protection. The two-part aeroshell would isolate the lander from biological contamination during preparations for launch. The probe would penetrate the atmosphere at 5.5 km/s at an altitude of 120 km and an angle of about 16.5 degrees. The deceleration would peak at 14 g. At an altitude of 7 km, having slowed to Mach 1.5, accelerometers would trigger a mortar that would deploy the drogue parachute that would slow the rate of descent to about Mach 0.5, then the aeroshell would open and the drogue would ensure that the backshell was drawn clear. As the heat shield was jettisoned, the 10-meter main parachute would slow Beagle 2 to the final vertical velocity. A radar altimeter was to detect the surface at an altitude of 200 meters and command inflation of the airbag. Built by the same US company that supplied the airbags for Mars Pathfinder and the Mars Exploration Rovers, this consisted of three segments that were laced together like cloves of garlic to create an almost spherical envelope some 1.9 meters in diameter. However, owing to mass constraints the design had fewer layers of fabric for protection against ripping than in the case of US landers. Beagle 2 was expected to strike the surface at about 17 m/s. The parachute was to be retained until after accelerometers reported the jolt of touchdown, to ensure that the lander was oriented the 'right way up' for the first rebound. Like other airbag-cushioned landers, it would bounce many times before coming to a halt. Once set on the surface, the laces holding the airbag segments would release and Beagle 2 would fall the final meter onto the surface. Like Mars Pathfinder, it would be able to right itself irrespective of how it came to rest. Finally, the lander was to open its lid, spread its solar panels and await first contact from an orbital relay.

All lander and aeroshell components were sterilized and cleaned with hydrogen peroxide, plasma and high-temperature treatments prior to being assembled in an aseptic chamber purpose-built for the project. The total mass of Beagle 2 at launch was 68.8 kg, of which the lander proper contributed 33 kg, the remainder being the heat shield, parachute, airbags and other landing devices.[142,143,144,145,146]

Three potential landing sites were selected from Mars Global Surveyor images: the Maja Vallis outflow channel on Chryse Planitia, Tritonis Lacus on the margin of Elysium Planitia, and Isidis Planitia. All three were sedimentary deposits where Beagle 2 could seek hints of past life. The site selected was Isidis Planitia, an old impact basin filled with debris believed to have washed down from the southern highlands. In order to gain the greatest braking by the parachutes, the specific site was over 3 km below the mean 'sea level' of the planet. (Although Mars does not have any seas, the planetary datum was set at the elevation where the atmospheric pressure is 6.2 hPa, the 'triple point' of water, and it is commonly referred to as 'sea level'.) There were pitted ridges, a number of small craters and small light-toned ripples and dunes, and it seemed not so rocky as to threaten a safe landing yet still sufficiently rocky to be scientifically interesting.

At its inception, it was expected that Beagle 2 would be funded by academic and industry money, in addition to external sponsorships and partnerships, with the British government being only marginally involved. The cost of the mission was estimated at £25 million, with up to one-third being by sponsorship managed by a well-known London-based publicity agency. Sponsorship was expected to build on the half-million 'hits' by the Mars Pathfinder Internet site in 1997. However, when sponsorship stalled, the Open University approached several government programs for funds. When ESA accepted the lander as part of the project, the Department of Trade and Industry in Britain contributed £5 million and the Particle Physics and Astronomy Research Council, whose remit was to fund astronomy and planetary exploration, allocated £2.77 million.[147] An independent review of the lander was undertaken in 2000 by a team of experts led by a former JPL project manager. This said that although the mission was challenging, it was "eminently doable". But the management structure was complex and "fragile", risk management was virtually non-existent, and other areas required particular attention. As a result, the Beagle 2 team asked ESA for help in minimizing the technical risk and regaining financial control of the project. This included £17 million to assist development and testing and to reinforce the bonds between contributors to the program, which was mostly at "gentlemen's agreement" level. Two-thirds of that, however, would have to be reimbursed by the UK to ESA. The final cost of Beagle 2 exceeded £42.5 million. In addition to ESA, this sum included about £25 million provided by the British government, funds by Astrium, by the Open University and by the National Space Centre.[148] Despite the financial and technical problems, engineers at Astrium and scientists at the Open University managed to finish the lander in January 2003 and the next month shipped it to Astrium's facility in Toulouse, France, for integration with the orbiter.

Astrium gave some thought to how this technology might be reused on future missions. The most amazing suggestion was a Beagle-class sample-return mission. An airbag-cushioned mini-lander carrying a two-stage ascent rocket weighing just 90 kg would land and use a mole to collect samples of the subsurface. The rocket would rendezvous with the orbiter that had released the lander. The samples would be returned to Earth using ion propulsion and a Hayabusa-style entry capsule. This was estimated to be capable of returning 200 grams of surface and subsurface fines using a spacecraft with a launch mass of just 1,200 kg.[149]

Meanwhile, in Toulouse the Mars Express orbiter was finished. As a publicity stunt, several drops of 'Rosso Corsa' (racing red) paint from the Italian carmaker Ferrari were installed on the vehicle in a small glass sphere.[150] Scientists felt such stunts illustrated the inability of ESA to properly communicate to the public (and taxpayers) why space missions were being flown.[151]

In March, Mars Express and Beagle 2 and their support apparatus were flown to Baikonur in Kazakhstan on separate Russian heavy transports. Only a few minor problems were encountered during the 12-week 'launch campaign'. A fault in one of the electronics modules that took some time to fix slipped the launch into early June. The final preparation and integration of Mars Express with its launcher was undertaken in an unusual setting. This was a western-standard clean room built for operators of satellites intended to be launched on Soyuz rockets, located in a 'safe' corner of the huge horizontal integration hall erected in the 1960s to assemble the N-1, the Soviet equivalent of the US Saturn V. The roof of the main building had collapsed in 2002, killing eight workers and destroying the prototype of the Buran space shuttle together with the only flightworthy Energiya heavy lifter. By the end of April Mars Express was complete, its thermal blankets were in place and it was ready for fueling prior to integration with first the Fregat stage and then the Soyuz booster. With its instruments, the Beagle 2 lander and a full load of propellants the spacecraft had a launch mass of 1,223 kg.

The launch window extended from 23 May to 21 June, but could be extended to 28 June at the expense of a slight redesign of the mission profile. The fact that this was a 'great opposition' meant the spacecraft would reach Mars in late December. When it lifted off at 17:45 UTC on 2 June it kicked off a busy window because, if all went well, by early 2004 there would be three landers operating simultaneously on the Red Planet: Beagle 2 and a pair US Mars Exploration Rovers.

After a 2-burn ascent profile, the Fregat released Mars Express. The spacecraft then deployed its solar panels and oriented them to the Sun, and established radio contact with the New Norcia station. The first course correction was made 3 days later. Next, the clamps that had held Beagle 2 firmly in place during launch were released. This was a particularly critical operation, since if the clamps had failed then the lander would not have been able to be released and its retention would have jeopardized the Mars orbit insertion maneuver.[152] On 6 June the spacecraft was slewed to enable the 'webcam' to take a picture of Earth with Beagle 2 in the foreground. The checkout of the instruments began 1 week into the interplanetary cruise. On 3 July the science camera was aimed back towards home at a range of 8 million km and snapped calibration images of the Moon and the half-lit Earth, which spanned 180 pixels, showing the Pacific Ocean and clouds over equatorial regions. A few days later Beagle 2 was powered up to carry out the first of several in-flight checks. The Mars Express glitches during the first few months included a faulty

Beagle 2 is mated to the Mars Express orbiter in a clean room at the Baikonur launch site. Various instruments can be identified on this, the Mars-facing side of the probe. The box at the upper left corner is the ASPERA analyzer, the circular ports under Beagle 2 is the French OMEGA spectrometer, and the two ports to the right of that, one rectangular, one circular belong to the high resolution camera. (ESA)

solid-state memory and a malfunction in one of the connections between the solar panels and the bus. Although the latter reduced the power supply by 28 per cent, it would still be possible to operate the spacecraft as intended for at least 85 per cent of the orbits of the primary mission, and thereafter schedule the instruments to fit the available power; software patches were developed to ensure that enough power would be available during both routine operations and possible emergencies. The very energetic solar flares in November were the most serious problem during the cruise because they saturated the star trackers, preventing them from determining the orientation of the spacecraft for a full day. But various procedures were put in place, the attitude remained remarkably stable and the computer did not even enter safe mode. The next check of Beagle 2 showed that it had not been affected by the

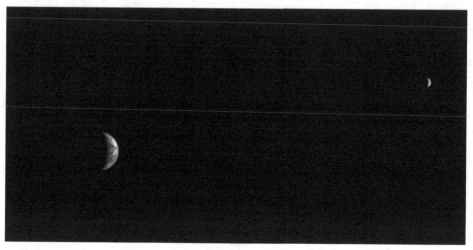

The Earth and the Moon seen by Mars Express from a distance of 8 million km one month after launch. (ESA)

flares. Later in the month the main engine was fired for 2 seconds to calibrate its performance and measure the attitude imbalances that it imparted.

On 1 December, with preparations for the release of Beagle 2 in full swing, the camera on the spacecraft took images of Mars from a range of 5.5 million km. The release of the lander was scheduled to occur 6 days prior to orbit insertion; as late as possible so as to arrange its trajectory with the greatest possible precision. The Deep Space Network was called on to ensure very accurate navigation during this critical phase. In fact, the approach trajectory was so well defined that the landing ellipse for Beagle 2 was able to be cut to an estimated 6 × 30 km. On 19 December the lander was spun up at 14.2 rpm, Mars Express adopted the required separation attitude, which severed real-time contact with Earth, and at 08:31 UTC springs in a lightweight separation mechanism pushed the two spacecraft apart at 0.3 m/s. The resulting change in the velocity of the orbiter was apparent in Doppler tracking as soon as the signal was re-established, and it served to confirm that the lander had been released. Furthermore, images taken by the 'webcam' at 50-second intervals showed the aeroshell of Beagle 2 against the background of the dark sky. On the clearest image, taken 67 seconds after deployment, the separation was 20.5 meters, and 150 seconds later it was over 65 meters. Beagle 2 was deployed 5 million km from Mars. It was the first Mars lander to make its approach completely without propulsion. Beagle 2 was passive except for a timer that would expire 60 minutes prior to arrival and initiate the descent and landing sequence. If everything went to plan, nothing would be heard from the lander until it was sitting on the surface. Its trajectory would have intersected the atmosphere at 02:47 UTC on 25 December and the schedule envisaged it landing about 7 minutes later near 11°N, 269.7°W in Isidis Planitia. As celestial mechanics prevented Mars Express from gaining a line of sight with Beagle 2 for the first 10 days of the surface mission, Mars Odyssey was to relay its signal. This was in position to listen 2.5 hours after landing, but it heard nothing.

Later that day the radio-telescope at Jodrell Bank in England tried to eavesdrop on the lander's transmission, but without success. Radio-telescopes at Stanford University in the US and Westerbork in the Netherlands were called on for assistance, but in vain. Commands were sent 'in the blind' in an effort to reset the lander's clock, just in case this had stuck, but again without result.

After releasing Beagle 2, Mars Express had to perform a deflection maneuver to move off a collision course with the planet, in order to pursue a trajectory suitable for the orbit insertion burn. At 01:31 UTC on Christmas Day the vehicle adopted the desired attitude. The engine was fired at 02:47 and shut down 32 minutes later, having reduced the velocity by 800 m/s. The ensuing telemetry confirmed that the spacecraft had entered a 260 × 187,500-km capture orbit with a period of 10 days, making Europe the third 'space power' to successfully insert a spacecraft into orbit around Mars. At apoapsis 5 days later Mars Express made a 4-minute burn which refined the plane of its orbit to the desired almost-polar orientation of 86 degrees. In January 2004 it commenced a series of apoapsis lowering maneuvers, first using the main engine and then the 10-N thrusters, in order to achieve an orbit in which it would make 3.6 circuits of the planet per sol. Moreover, maneuvers performed between 4 and 6 January placed the periapsis above the intended Beagle 2 landing site. Its first attempt to communicate with the lander was made on 7 January. Since this was the primary relay link, the Beagle 2 team were optimistic but there was no signal. Attempts continued with ever feebler hopes of establishing contact until 12 March.

This is the clearest image of the sequence documenting the deployment of the Beagle 2 lander. Several bright objects can be seen in the original image (one is at bottom left) that are believed to be small debris from the spacecraft. There is a mysterious bright spot between 2 and 3 o'clock on the rear of the shell cocooning Beagle 2 that, to some, was interpreted as a damaged area of the heat shield. (ESA)

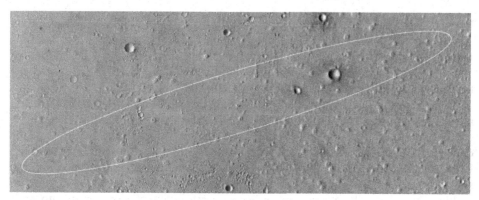

The Beagle 2 landing ellipse imaged by the camera on Mars Odyssey. (NASA/JPL/ Arizona State University)

In the meantime, on 6 February ESA declared Beagle 2 to be officially lost. The joint ESA and British National Space Centre inquiry reported in late May but the need to protect commercial interests meant that, at least initially, only a summary was published. The investigation was hampered by the fact that the designers had fallen victim to the same naivety of many of their US and Russian colleagues in not equipping their vehicle with any system to return telemetry in real-time during the critical entry and descent phases.

In the absence of data, the report could only develop hypotheses and rank them in terms of likelihood, and issue technological and managerial recommendations. In fact, the agencies put most of the blame on management shortcomings and lack of industrial oversight. In terms of ESA management, Beagle 2 was treated as just another payload, providing its managers and scientists considerable autonomy. Of course, this also meant treating the lander's funding in the same way as the other instruments, which were paid for by national agencies rather than directly by ESA. The lander should have been regarded as a fully independent spacecraft, and been managed centrally. The science director at ESA went so far as to declare that "the creative approach used with Beagle 2 was probably a step too far", and the lander should not have been allowed to fly. On the other hand, during its development the fact that academics were bearing the entire responsibility for the lander had been hailed as "an incredible achievement".[153] From the technical point of view, it was recognized that owing to lack of time and funds it had not been possible to test the lander adequately, with even some "mandatory" tests being omitted. One key test that was not performed was the deployment of the parachute by releasing a lander from a high-altitude balloon, which would have been expensive. Another omission was to shock test the aeroshell separation pyrotechnics. The redesigned parachute, which was validated just months before Beagle 2 was shipped for integration with the orbiter, was considered particularly critical. Some form of parachute or airbag failure was deemed the most likely cause of the loss of Beagle 2. In particular, the investigation concluded that the design of the heat shield separation system could have caused the jettisoned shield to recontact the lander before the parachute could fully deploy and draw the lander clear. Several other possible failure modes were identified, including the airbags getting wrapped in the parachute canopy after the first rebound. The laces holding the airbags would have severed 130 seconds after the initial impact, and if Beagle 2 was still mobile at this time it could have been so damaged upon striking the ground as to preclude it opening its lid. Analyses of the 'webcam' pictures taken as Beagle 2 moved away from Mars Express showed an unidentified spot several centimeters across on the backshell in one image and objects apparently receding from it in other frames. These objects could be either fragments of the aeroshell insulation blanket or chips of ice that had formed in the shadow of the lander. However, none of these objects could be associated with any of the failure scenarios.[154,155]

In August the Open University Beagle 2 consortium released its own report. It blamed loose management on ESA and the technical failure on "bad luck". It was noted that the design had been obliged to accommodate a shrinking mass budget in the development phase. Responding to criticism that the lander was not capable of providing telemetry during the descent, the consortium pointed out that a telemetry system had been designed but could not be installed because ESA had advised that no receiver would be available for it. The report observed that both of the US Mars Exploration Rovers, which landed just days after Beagle 2's arrival, had found the atmosphere to be significantly thinner than expected as a result of recent regional dust storms. Early scans of the atmosphere by the occultation experiment on Mars Express confirmed this. A 15 per cent reduction in atmospheric density could have caused Beagle 2 to deploy its parachute or airbag too late, and perhaps never. Even if it deployed, it was possible that the airbag had burst or ripped open on striking the ground at a higher vertical speed than intended. The Open University report re-analyzed the Beagle separation pictures and found one possible failure scenario: perhaps one of the carbon-composite doors on the backshell that provided access to the heat shield separation mechanism had become unglued; if so, then hot gases could have penetrated the shell during atmospheric entry.[156,157,158]

Shortly after the publication of the ESA report the British House of Commons began its own inquiry. The report released in November noted the problems faced in securing funds and criticized the management of the project on both the UK and ESA sides. It said ESA could have stopped Beagle 2, or at least made known its objections, but took the attitude that so long as the lander did not interfere with the orbiter's mission it should be allowed to proceed. This was also motivated by the political implications of cancellation, in particular after so much money had been spent. The House of Commons report described its own government's contribution as too-little-and-too-late. The short development time and uncertain funding only made an already high-risk venture even riskier. Worse, the 'British-only' flavor of the project discouraged other ESA countries from joining in. And given the nature of the design, there was not much heritage hardware to pass on to missions such as the French NetLanders.[159]

Some years later, Australian researchers studied the aerodynamics of Beagle 2 during the first few seconds of its entry and published a possible reason for its loss. They showed that for a realistic range of angles of attack, the probe's flight would have been unstable as it passed through the uppermost fringe of the atmosphere; if its attitude had in some manner become perturbed, then it would not have been able to restore its angle of attack.[160]

Undeterred, the team that created Beagle 2 sought ways to recover some of the intended science. One possibility was to launch a replica as early as 2007, but this presupposed that there would be an interplanetary bus to deliver it. The team then devised an evolved proposal in which a computer-controlled vented toroidal airbag would absorb all of the kinetic energy of impact and avoid rebounds. This way, the lander would be able to be built without mechanisms to right itself after landing, making more mass available to other systems. Next was the BeagleNET. Seen as a precursor to ESA's ExoMars rover, this would carry life-detection equipment and release a mini-rover that would deploy seismometers. The team envisaged it being delivered by Russia's Fobos-Grunt sample-return mission.[161,162] Other (probably more realistic) proposals involved flying Beagle's instruments on a European mission, possibly even on the ExoMars rover.

Beagle 2 was occasionally sought by US orbiters. A low-resolution image taken by Mars Global Surveyor only 18 minutes after the expected landing showed that despite a storm that had raised dust over a large part of the planet, the weather was clear at the landing site. This image also showed a 1-km crater within the landing ellipse and a blanket of rocky ejecta spanning several kilometers. The likelihood of the lander descending there was small, but if it had then it could very easily have been damaged – recall that one of the design criteria for the airbag system was that the site be free of large rocks. In response to a request by ESA, high-resolution images of the eastern part of the landing ellipse were taken in January through to April by Mars Global Surveyor. Only one candidate for Beagle 2 was suggested. A young 20-meter-diameter crater was identified in 0.5-meter-resolution imagery of the rim of the 1-km crater. There was a darker patch on the northern wall of this smaller crater and three brighter spots on its floor. With some imagination, these could be viewed as Beagle's airbags, if not the lander itself with its solar panels deployed. However, when this was imaged at higher resolution some years later by Mars Reconnaissance Orbiter there was nothing unusual, and certainly no trace of the lander.

Up to mid-January 2004 media attention was focused on the fate of Beagle 2, then, with Mars Express returning a lot of data, some of which was unprecedented, scientists presented their first scientific results at ESA's European Space Operation Center in Darmstadt on 23 January. Although the performance of the orbiter more than offset the loss of Beagle 2 in terms of the overall objectives of the mission, it went largely unrecognized by the public.[163,164] While the orbit was being refined, the high-resolution camera demonstrated its worth. On 14 January it took a 1,700 km-long swath of Valles Marineris from an altitude of 275 km showing the canyon with a resolution of 12 meters per pixel in places, in color and in 3 dimensions. Other early pictures included Reull Vallis, Acheron Fossae and the summit caldera of the 4.5-km-tall Albor Tholus. During the first month in orbit a solar occultation was observed, and occultations of stars indicated ozone in the upper atmosphere. Bistatic radar and other experiments using the radio carrier signal had been carried out when

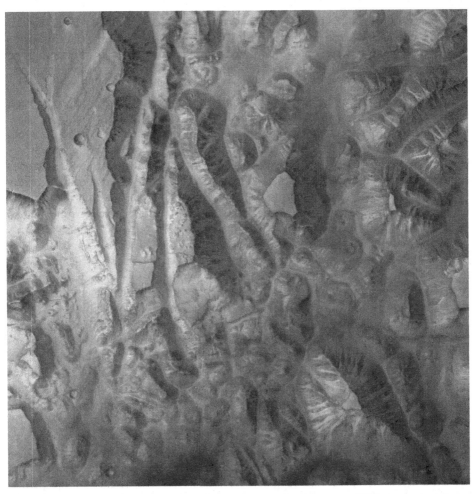

This image of a portion of Valles Marineris, taken on 14 January 2004, was one of the first returned by Mars Express. (ESA)

it was impossible to aim other instruments at the planet during the period of maneuvering. But the mineralogy spectrometer provided the most fascinating results. Beginning in the second half of January, scientists used it to survey a large fraction of the south polar region shortly before the planet's autumn equinox. The south polar cap had been expected to consist of carbon dioxide ice, although this had not been confirmed by observation. The tens of thousands of spectra obtained by Mars Express during this time confirmed that carbon dioxide ice was present in some areas, but revealed it to be a layer that was nowhere more than 10 meters thick and was resting on a base of water ice several kilometers thick. Three distinct areas were identified where water ice was concentrated: the bright polar cap itself, where water and carbon dioxide ices were mixed; scarps around the residual cap, where dust and water ice pooled; and zones free of carbon dioxide which extended downslope in stratified terrain tens of kilometers from the perennial cap. And the 'Swiss cheese' terrain observed by Mars Global Surveyor was found to be predominantly carbon dioxide ice.[165] These observations confirmed that a major portion of the south polar cap was water ice, and that this probably constituted the single largest reservoir on the planet.[166]

Only a few observations were possible in February and March while the orbiter underwent a 'season' of solar eclipses which included a 95-minute occultation on 3 March. The Fourier spectrometer and the mineralogy instrument observed the south polar cap. In early May the vehicle achieved its operational orbit, ranging in altitude between 300 and 10,110 km in a plane inclined at 86.3 degrees to the planet's equator, and with a period of 6 hours 43 minutes that enabled it to revisit sites every 11 orbits.[167] Owing to the orientation of the orbit the periapsis would remain over the night-side for a portion of the summer of 2004, so spectroscopic instruments had priority. The commissioning of the instruments was completed on 3 June. The only remaining activity was to release the radar booms. This had been planned for May but in April JPL, which supplied the booms, had asked for this to be deferred. The tubular booms, 3.8 cm in diameter, were made of fiberglass and Kevlar and were stowed folded up like an accordion along 13 hinge points. Finite-element simulations raised the concern that upon being released the booms might violently whip about and damage the solar panels. This required further analysis, and the expectation was that the radar would not be able to be deployed any earlier than March 2005.[168,169]

The first solar conjunction occurred in August and most of September. Also in August, the vehicle's orbit led to a first series of relatively close flybys of Phobos, culminating on 22 August with a pass at 149 km. Although the orbital periods of the two objects were similar, their orbits were almost perpendicular and there were multiple close passes over consecutive orbits and then periods without encounters. By measuring the position of Phobos against background stars during this and later 'encounter seasons', it was possible to refine the orbit of the moonlet, revealing it to be several kilometers ahead of its predicted position. This was confirmed by the Mars Exploration Rovers observing eclipses of Phobos from the surface of the planet. Images by Mars Express at resolutions of several tens of meters covered some regions of the moonlet which had previously been documented only at a poorer resolution. Whereas the orbit of Mars Express intersected that of Phobos, the orbit of Deimos was further from the planet and close encounters were impossible. Nevertheless, a 10,931-km flyby of Deimos on 22 October 2004 provided a number of observations.[170]

Expectations were quite low for Mars Express in the US, where the feeling was that the European orbiter was merely repeating observations made by US missions. Because this was ESA's first planetary orbiter, European planetary scientists were relatively inexperienced but the results proved that the spacecraft and its managers and scientists were able to produce first-class science to complement that from the US orbiters.

Averaging data collected by the infrared Fourier spectrometer over many orbits revealed five unidentified spectral features, one of which could only be methane. Remarkably, the discovery of methane (and ammonia) on Mars had already been announced just two days after the Mariner 7 flyby in 1969, from infrared spectra of

An early multispectral image of the southern polar cap taken on 18 January 2004 by the mineralogy spectrometer on Mars Express. The image at right shows the visible-light channel, the image at center shows the signal from carbon dioxide and the image at left that from water ice. (ESA/OMEGA)

the southern polar cap. The announcement was later retracted, however, as the spectra were recognized to show carbon dioxide lines. More recently, methane was reported in 2003 by US teams using Earth-based near-infrared high-resolution spectroscopy. Methane seemed to account for about 10 parts per billion of the already tenuous atmosphere on average, but one of the Earth-based teams had reported detecting concentrations tens of times greater over Valles Marineris and the highlands north of Hellas. (For comparison, methane is about 2 parts per million in the Earth's atmosphere and it constitutes 5 per cent of Titan's atmosphere.) Its abundance varied greatly from one orbit to the next, and it was not uniformly distributed, tending to concentrate over Arabia Terra, Elysium Planum and Arcadia-Memnonia. Interestingly, these were three of the equatorial regions where Mars Odyssey had detected water ice at shallow depth. Longer-term observations revealed seasonal variations, with the mixing ratios slowly decreasing from northern spring to southern summer. Methane must be short-lived in such a highly oxidizing atmosphere (with a molecule having a lifetime of several hundred years) therefore in order for it to persist it must be being replenished. Over 90 per cent of the methane in the Earth's atmosphere is produced by micro-organisms; the remainder is of geochemical and volcanic origin. The presence of methane in the Martian atmosphere therefore had two potential implications: either there was some form of life or volcanism was currently active. The Viking results appeared to have ruled out biology at the surface, so if there were micro-organisms they must be in 'sheltered' environments beneath the surface. They could exist in hydrothermal springs, but an infrared search by Mars Odyssey for such sites had been futile. Other possible methods of production included the alteration of basalt by water as occurs on the floor of the Earth's oceans, but there was no evidence of such activity. Alternatively, methane might be only a temporary constituent of the Martian atmosphere; it occurs in cometary spectra, and may have been introduced to Mars by the impact of a cometary nucleus 100 meters in size.[171,172] The Fourier spectrometer also detected formaldehyde, which is a more complex hydrocarbon, in concentrations 10 times greater than methane. This shorter-lived molecule could originate from the oxidation of methane by iron oxide, which is abundant on Mars, but in that case it ought to be present in *lesser* concentrations than methane. Other minor gases such as hydrogen fluoride and bromide were also detected.

The Fourier spectrometer also made simultaneous observations with the thermal emission spectrometers on the Mars Exploration Rovers on the surface, providing overlapping profiles of the lowest several kilometers of the atmosphere.[173] Starting in April 2005 the Fourier spectrometer suffered a hardware mishap which meant it was unable to produce usable data, but it was recovered in November by switching it to a more powerful backup motor.

When the magnetometer on Mars Global Surveyor discovered a 'patchy' fossil magnetic field on Mars, it was speculated that the miniature magnetospheres could generate the Martian equivalent of auroras by interacting with charged particles in the solar wind. To investigate this, the field of view of the ultraviolet spectrometer on Mars Express was allowed to drift over the night-side limb of the planet. On 11 August 2004 the instrument detected an emission peak clearly due to the excitation of gas molecules by a flux of solar wind electrons which had been concentrated by a localized magnetic field. This was the first evidence of auroral emissions in the upper atmosphere. The line of sight of this observation passed over the strongest of these crustal magnetic fields.[174] The signature of the small magnetospheres was also detected by the only particle instrument on board.[175] The ultraviolet emission from nitric oxide over the night-side caused by the recombination of nitrogen and oxygen was also detected. These ions were formed by the breakdown of nitrogen, oxygen and carbon dioxide molecules on the day-side by solar radiation, and they were carried to the night-side by atmospheric circulation. It was possible to exploit this 'night-glow' to measure the differences in atmospheric circulation between the two hemispheres and monitor their seasonal changes.[176] Ozone was detected using the instrument operating in several modes, including stellar and solar occultations. Simultaneous observations of ozone were carried out from Earth and from orbit. As the ozone molecule is easily destroyed by water, it serves as an important tracer of photochemical reactions and cloud activity. Coordinated studies of atmospheric ozone continued through to at least 2009 and, with Earth-based observations, gave a database spanning more than two decades.[177,178] The ultraviolet channel was also used during the Phobos encounter season in mid-2004 to obtain spectra of sunlight scattered by the moonlet. An anomalous absorption feature was detected that may indicate a coating of organic materials.[179] Operating in occultation mode the instrument complemented Viking measurements of the planet's atmosphere by providing the first complete density and temperature profiles for carbon dioxide, and by measuring the abundance of molecular oxygen at altitudes between 60 and 120 km.

During its first Martian year, the high-resolution camera covered one-quarter of the planet's surface at a resolution of 20 meters per pixel (or better) in color and in stereo, and more than half of it at 50 meters. It concentrated on volcanic features, ice

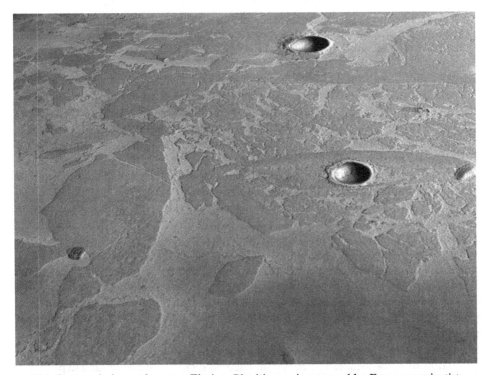

This flat terrain in northeastern Elysium Planitia was interpreted by European scientists as a slab of pack ice covered by a thick layer of volcanic ash. The interpretation is not unanimous, and high-resolution images by US orbiters proved it to be a solidified smooth lava flow. (ESA/DLR/FU - G. Neukum)

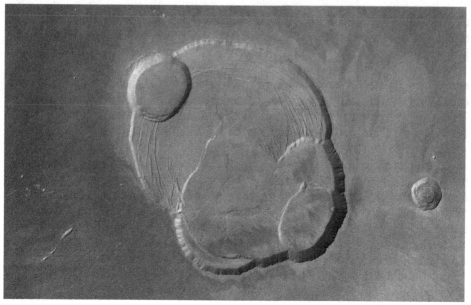

The summit caldera of Olympus Mons photographed in high resolution by Mars Express. The relative lack of craters indicates that the volcano must have been active until relatively recently. North is to the right. (ESA/DLR/FU – G. Neukum)

deposits and glaciers, and the action of water. The observations of volcanic and ice-related features sparked some controversy between US and European planetary scientists. An area in northeastern Elysium Planitia that was imaged in 1998 by Mars Global Surveyor had been interpreted by US scientists as slabs of basalt that once floated on a major lava flow which was apparently erupted from the Cerberus Fossae fissures as recently as 5 million years ago. But the Mars Express pictures challenged this interpretation: the floor appeared to be younger than the slabs and its surface was flatter than solidified lava. To European scientists the region 800 × 900 km in size resembled terrestrial pack ice, with ridges forming where ice had been compressed, and with indications of pieces having broken apart and the fragments rotated relative to one another. The ice had since been covered and protected from sublimation by layers of volcanic ash. The slab of ice, if still present, could be as much as 45 meters thick.[180] American scientists, of course, were very skeptical of the European 'discovery' and countered that the region actually showed signs of a cooling lava flow and pointed out that the impact craters had characteristics which suggested they formed in solid rock rather than ice. Higher resolution images would prove them correct.

High-resolution images by Mars Express enabled crater counts on the calderas and flanks of volcanoes. Using commonly accepted cratering ratios, lava deposits on the slopes of Olympus Mons and Hecates Tholus were dated between 3.8 billion years ago and just 100 million years ago. The build up of volcanic edifices over such protracted intervals confirmed that the Martian crust was immobile. In the case of

Earth, the process of plate tectonics causes mantle 'hot spots' to produce chains of volcanoes, with the Hawaiian Islands being an excellent example – the upwelling plume in the mantle remains fixed and the lithosphere passes over it. On Mars, the process produces enormous volcanoes above the plume. In this regard, it is perhaps significant that Olympus Mons and the three large volcanoes atop the Tharsis ridge all summit at the same altitude, as if the mantle pressure was unable to force magma any higher. The present-day calderas of Olympus Mons, Ascraeus Mons, Arsia Mons, Albor Tholus and Hecates Tholus all appear to have been formed during a fairly narrow interval about 150 million years ago. Intriguingly, this corresponds to when most of the known Martian meteorites (with the notable exception of ALH84001) were formed. Furthermore, Mars Express and Mars Global Surveyor together facilitated the identification of three lava flows on the lower flanks of Olympus Mons that could be dated between 115 million years ago and 2.4 million years ago, which is 'yesterday' in geological terms. If the cratering ratios used to derive these ages are correct then there is no reason to believe that Mars has ceased to be volcanically active.

Yet dust-buried glacial deposits and mud flows on the flanks of Olympus Mons, situated 20 degrees from the (current) equator, appear to be as young as 4 million years in some places. The fact that other small tongue-like deposits could not be reliably dated by crater counting was an indication of their extreme youth.[181] On 19 January 2004 Mars Express imaged a 10-km caldera on the northwestern flank of the shield volcano Hecates Tholus which apparently formed 350 million years ago and showed evidence of explosive volcanism. Most volcanism on Mars was of the effusive type. Explosive volcanism was believed to have occurred only in very ancient eruptions, but this discovery proved otherwise. Moreover, glacial deposits both near the caldera and inside it were extremely young, at most 24 million years old.[182] Other 'recent' features of glacial or snow-accumulation origin were evident at mid-latitudes in the eastern Hellas region. These included peculiar 'hourglass-shaped' flows and debris aprons in craters. All these observations confirmed that an 'ice age' had recently occurred during which glaciers several kilometers thick formed on the flanks of the tallest mountains on the planet.[183,184]

Echus Chasma, the source region of Kasei Valles, the greatest outflow channel, clearly showed that liquid water was present on the surface billions of years ago, that gigantic waterfalls poured over the 4-km-tall cliffs to feed a lake that formed on its floor, and that glaciers developed when the climate chilled and carved Kasei Valles.

A number of other observations were carried out by the camera. For example, on 2 February 2005 Mars Express took one of its most iconic images, capturing a 35-km crater in the northern Vastitas Borealis with a bright white patch of water ice covering a dune field on its floor. Images of the edge of the northern polar cap showed fields of small, young volcanic cones with material piling up on one side. This terrain led to a chasm and thereafter to a ring of flat ridges. This seemed to be evidence of a recent catastrophic flood, perhaps only 20,000 years ago, triggered as polar ice was melted by a volcanic eruption which left a tuya, a rock feature that forms when lava erupts into ice.[185,186]

The OMEGA imaging spectrometer mapped over 90 per cent of the planet at a

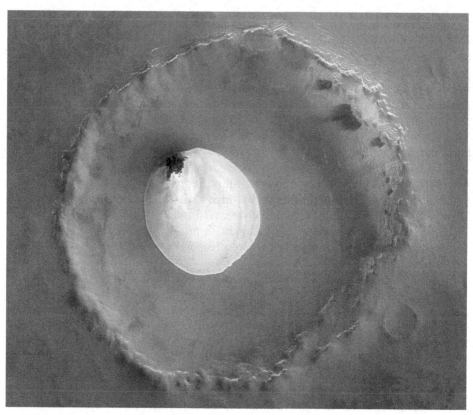

This image of melting ice inside a small 35-km crater in the northern Vastitas Borealis, taken in February 2005, is one of the most iconic images of the European Mars Express mission. (ESA/DLR/FU – G. Neukum)

resolution of 1 km during the first local year of operations. The most interesting results concerned water-altered minerals, namely phyllosilicates (essentially clays) and sulfate salts probably formed in the presence of water on or near the surface. The minerals included gypsum, a calcium sulfate; kieserite, a magnesium sulfate; jarosite, an iron sulfate hydrate known to form only in the presence of relatively acidic water; and hematite. The sulfates were concentrated in Valles Marineris, Margaritifer Sinus and Meridiani Terra (where 'ground truth' was provided by the Opportunity rover). The Meridiani sulfate layer was overlain by a hematite layer. Cross-correlation between Mars Express spectrometry and Mars Global Surveyor images established that the sulfates formed light-toned layers in the walls of mesas and hills in Valles Marineris. These layers could have originated by the alteration by groundwater of volcanic ash, or they could be windblown deposits. Light-toned patches of sulfate salts were also identified in chaotic terrain within Margaritifer Terra. The extent of such deposits argued for the existence of large quantities of rather acidic water on the surface at some point in the planet's history. However, whilst kieserite forms in the presence of water, it is easily modified by prolonged exposure. So this seemed to indicate a transitory, sporadic and possibly unstable presence of water in the past. Moreover, salt deposits would not strictly imply the presence of liquid water on the surface, as they could also be created by groundwater, snow or frost. And the common presence of the mineral olivine, a silicate that readily 'weathers' in a warm and wet environment, implied the surface had been dry for a very long time.

A surprising discovery was the presence of sulfates, and in particular gypsum in a 60 × 200-km patch surrounding the north pole after the carbon dioxide cap had retreated in the spring and summer. This feature was found to match fields of dark dunes in Olympia Planitia, the largest 'sea of sand' on the planet. The origin of the polar gypsum is disputed, but some scientists have argued that it indicates recent geological activity. It could result from the alteration of a substrate by sulfur-rich gases leaking from the interior of the planet.

Analyses of Viking samples and Martian meteorites had suggested the presence of clays. These commonly appeared to be associated with dark deposits in Arabia Terra, northern Syrtis Major, Meridiani Terra, Xante Terra and Lunae Planum, in some outcrops eroded from the ancient heavily cratered regions of Syrtis Major, Nili Fossae, in Mawrth Vallis, and in several spots in Isidis, Hellas and Meridiani. In some cases, the imaging spectrometer discovered ancient clay deposits which had been buried by olivine-rich lavas and later unearthed by impacts, implying that the water that was present when the clays were formed had disappeared when the lava was erupted. In some locations, clays were traced as they were eroded by channels and concentrated in river deltas. This was particularly evident in an ancient lakebed in the northern hemisphere crater Jezero. It was deemed critical to determine how the composition of the clay deposits varied from one deposit to the other, so as to infer the conditions in which they formed. On Earth, clays are formed when rocks are exposed for long periods of time to warm water, such as in hydrothermal springs. The different exposure times to water produce different types of clay containing increasingly insoluble elements up to aluminum, and by identifying the elements in a clay it is possible to say how long the surface was in contact with water. Martian clays mainly included magnesium and iron, but in some cases aluminum was detected in various concentrations. Analysis of the spectra showed deposits in Nili Fossae probably resulted from hydrothermal sources, whilst those in Mawrth Vallis, which has the largest abundance of hydrated minerals on the whole planet, may well be sedimentary in form. These results were particularly important for the planning of future surface missions, because clay deposits could represent some of the best places to look for evidence of past life. Moreover, unlike sulfates, which form in highly acidic water, clays form in watery conditions that are conducive to (terrestrial) life.[187]

Owing to the different environmental conditions required for their formation, the presence of both sulfates and clays on Mars might represent two different water-related episodes, separated in time. It could perhaps also attest to a change in the chemistry of Martian water from neutral or alkaline when clays formed, to acidic when sulfates formed. This change may be related to the Tharsis volcanism, which would have injected vast amounts of sulfur into the atmosphere.[188,189] Because all the water-related deposits appear to be billions of years old, the surface must have been dry for most of the planet's history. Furthermore, no water-altered minerals were detected on the younger-looking northern plains. Instead, the spectral data implied that they were covered by glassy volcanic materials. If there was once an ocean filling these low-lying regions, then any sediments that it laid down must be masked by volcanic deposits.

Although the signature of hydrated minerals was not detected in locales where gullies had been observed, this could be a result of the instrument's relatively low spatial resolution. Despite being capable of detecting carbonates in concentrations greater than 1 per cent, the instrument did not find any of the deposits that ought to have formed as atmospheric carbon dioxide reacted with water, so it was not able to confirm the discovery of carbonates in airborne dust inferred from the thermal emission spectrometer on Mars Global Surveyor.

Mars Express paid particular attention to the polar regions, to monitor seasonal effects and to investigate the nature and distribution of the different terrains. It monitored the changes of the northern polar cap during spring and early summer, observing the sublimation of carbon dioxide frost and the retreat of the residual cap of water ice. Observations from late 2004 to 2005 were devoted to the mystery of the 'cryptic' regions, dark patches of ice seen during the early to mid-spring by Mars Global Surveyor whose low temperatures had suggested the presence of a slab of almost pure carbon dioxide ice. Only a very weak signal of carbon dioxide was found, however, and the regions were established to be covered every year by carbon dioxide heavily contaminated with dust, possibly because of some aspect of the atmospheric circulation. Unfortunately, links between this contamination and dark spots, fans, 'spiders' and vents were not simple to establish.[190] Data collected over several Martian years facilitated a detailed reconstruction of the sublimation processes occurring in the south polar cap during spring and summer.[191]

The landing site of Mars Pathfinder in Ares Vallis was revisited and rocks there were confirmed to be mostly basalts unaltered by water, as Sojourner analyses had suggested. Of course, this was hard to reconcile with the appearance of that area as an ancient flood plain.[192,193,194,195,196,197,198,199,200,201]

The mineralogy instrument also made atmospheric observations. For example, in combination with the Fourier spectrometer it studied how the concentration of water vapor above Hellas changed over a full local year. An abrupt increase over a mere 3 days has yet to be explained.[202]

The plasma instrument started to obtain data during the commissioning phase. Plasma from the solar wind and ionospheric ions accelerated by it were observed down to the 270-km periapsis, verifying that the solar wind directly influences the upper atmosphere.[203] The instrument's low-energy neutral particle imager was also used to map populations of interplanetary particles and it saw streams coming from a variety of sources, some of which are not fully understood.[204]

A large number of radio-occultation experiments were performed to 'sound' the Martian atmosphere. In contrast to the US orbiters in low Sun-synchronous orbits, the elliptical orbit of Mars Express provided occultations at a wide variety of local times and latitudes. It revealed a previously undetected sporadic ionospheric layer some 80 to 90 km high formed, amongst other things, by ions carried by meteors.

While the orbiter was operating routinely, the dynamics of deploying the radar booms were investigated by separate ESA and JPL teams and then independently reviewed. Tests were made on the boom tubing material, and a full-scale model of the antenna was deployed in a vacuum chamber. It was eventually concluded that although there was a fair chance of the boom recontacting the vehicle, this was not likely to cause any damage and any residual vibrations after deployment would be able to be damped by the attitude control system. The first boom was deployed on 4 May 2005, shortly before the vehicle made its 2,000th orbit of the planet. Soon afterwards, however, it was discovered that the boom was not straight as intended. Apparently the tenth hinge, two-thirds of the way to the tip, was stuck at an angle of 40 degrees. Exposing one side of the hinge to the Sun softened it, then it slowly straightened out and locked in place. After further analyses the second dipole was released on 14 June, this time without incident. Three days later, the monopole was deployed.[205,206] Commissioning and calibration began immediately, with the first radar echoes being received on 19 June. The instrument would not be declared fully commissioned, calibrated and operational until early 2006, but deployment was timed to occur when periapsis fell on the night-side of the planet in order that preliminary scientific observations could start immediately. The data-taking began on 26 June, sounding the northern hemisphere and in particular the polar layered terrains. This continued to mid-August, when the periapsis shifted to the day-side and imaging took precedence. After that, ionospheric data would be collected on about 20 per cent of the orbits until periapsis returned to the night-side and shifted toward the south pole.

During two early orbits the radar received a large, parabola-shaped subsurface echo from the northern mid-latitude Chryse lowlands which was interpreted as a buried impact basin some 250 km in diameter and 2.5 km deep, apparently totally filled with a thick layer of material that could be water-rich. A total of 11 buried circular depressions ranging from 130 to 470 km in diameter were found beneath the smooth northern plains during the first radar campaign. Hints of these features had been apparent in Mars Global Surveyor laser altimetry as enigmatic circular depressions. Interestingly, the size and distribution of these filled basins allowed the buried crust of the northern hemisphere to be dated at more than 4 billion years old, making it similar to the ancient southern hemisphere.[207,208] When over the polar cap, the radar echo split into an upper trace that represented surface deposits and a lower trace that represented the boundary between the ice and the underlying terrain. The northern polar cap itself appeared to consist of a deposit of radar-dark fairly pure water ice as much as 1.8 km thick.

Ionospheric studies performed after the northern hemisphere campaign clearly detected echoes corresponding to areas of fossil magnetic field. This demonstrated the ability of the ionospheric sounder mode of the radar to measure the planet's weak magnetic field. Between November 2005 and April 2006 the radar sounded the southern polar regions during more than 300 orbits in local winter. Analysis of this data revealed very fine layering and confirmed the detection of pure water ice by the mineralogy instrument. The southern cap was found to be layered to a great depth and to comprise deposits with different dust-mixing ratios. There is probably carbon

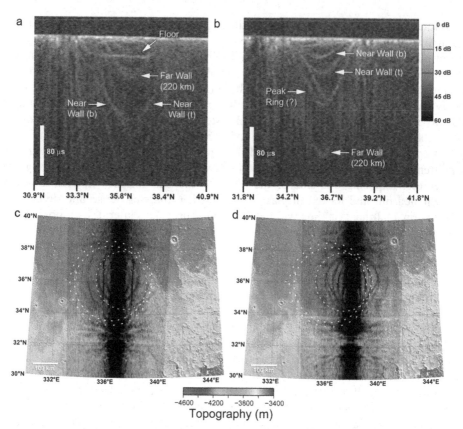

Radar echos collected over Chryse Planitia in July 2005 uncovered the presence of buried structures probably belonging to ancient impact basins. The dashed white circles on the topographic maps are approximate fits to the radar echo arcs. (ESA/ASI/NASA/University of Rome/JPL/Smithsonian)

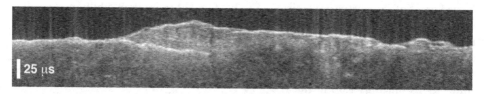

A 'radargram' cutting through the southern polar cap. Note that the trace splits into two on the left side of the image. The upper trace represents the surface of the polar layer deposits, while the lower trace is the echo from the boundary between the lower surface of the polar cap and the underlying material. From the characteristics of the echo, the material between the two traces was interpreted as made of mostly water ice. (ESA/ASI/NASA/University of Rome/JPL/Smithsonian)

dioxide ice present but the radar was unable to distinguish this specifically. The southern polar cap seemed to be much thicker than in the north, sitting on the crust at a depth of 3.7 km. Large quantities of ice were found buried under Dorsa Argentea, a plain near the south pole with an area of 3 million square kilometers. Although previously thought to be of volcanic origin, it is more likely to be a sheet of ice as much as 1 km thick covered by dust. The radar data yielded an estimate for the total volume of water ice present in the southern polar regions: sufficient to cover a smooth globe of an equivalent size to a depth of 10 meters. Nevertheless, this was still one or two orders of magnitude less than that required to explain the observed surface erosion.[209] The Swedish plasma instrument collected data on ions escaping from the atmosphere over an entire Martian year. The loss rates were a full two orders of magnitude less than were measured 15 years earlier by a similar but inferior instrument on Fobos 2, indicating that solar wind interaction alone had removed the equivalent of only a few hectopascals of carbon dioxide and several centimeters of water from the atmosphere over the last 3.5 billion years. If the atmosphere was much thicker early in the planet's history, then some other erosion process must have been responsible for its loss.[210]

In parallel with the southern radar sounding, Mars Express had a second season of Phobos encounters in November and December 2005. At the nominal end of its mission, it received an extension to enable it to monitor the planet for evidence of long-term trends. During most of the summer of 2006 the vehicle was in 'survival mode' because with Mars at aphelion the solar panels were generating less power and the plane of the spacecraft's orbit produced long eclipses. At one time, it was spending up to 75 minutes in the planet's shadow, preventing full recharging of its batteries. When necessary, systems were switched off and the vehicle was turned with its high-gain antenna facing away from Earth in order to better aim the solar panels at the Sun. To make things even more complicated, during this period Mars and the vehicles on and around it went through solar conjunction and reached the maximum distance from Earth. Scientific observations were essentially suspended for a period of almost 10 weeks except for radio soundings of the solar corona and several coordinated studies with the US rovers and Mars Reconnaissance Orbiter. Nevertheless, on 22 July Mars Express acquired 14-meter-resolution pictures of the infamous 'Face on Mars' in the Cydonia region. After conjunction, imaging and surface mineralogy observations were impaired by storms which injected fine dust into the atmosphere. Also during this period, a 2-km-thick layer of haze was detected in Valles Marineris. Meanwhile, the occultation spectrometer discovered the highest clouds over any planetary surface: fleeting clouds of carbon dioxide ice crystals hovering at altitudes of 80 to 100 km.

At the end of 2006 ESA evaluated the possibility of reviving the small camera that had monitored the deployment of Beagle 2, this time to provide wide-angle 'webcam-like' images that would be released to the public on the Internet within hours of their being received. After 3 years of remaining off, the camera was tested later in the year by experimenting with a wide range of exposures. The results were extremely encouraging, although only the fastest exposures could be used to avoid saturating the sensor. But using the camera posed a number of challenges: despite using a

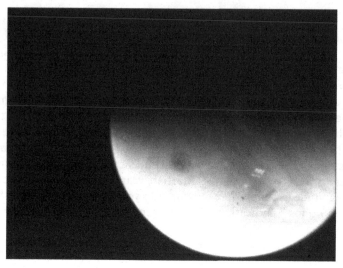

An image of the half-lit Mars and Olympus Mons captured by the Beagle 2-monitoring 'webcam' on 9 November 2007. (ESA)

negligible fraction of the power available and producing a relatively small amount of data, it was not boresighted with the other instruments and its data bus prevented the use of the rest of the payload. For these reasons, this camera could be operated only for about 1 hour when the orbiter was at apoapsis and slewing with all of its other instruments off. Although these constraints limited it to distant, full-disk images of Mars it successfully raised the mission's public profile. In the end, the camera's website became one of the most visited of the space agency's Internet pages. Furthermore, the images proved to have scientific value, with scientists using them to identify large-scale atmospheric phenomena which could then be investigated in detail by the other instruments. During May 2010 the camera was operated for a complete orbit in order to create for the first time a movie as the vehicle traveled around the planet. In the longer term, it is also planned to use 'webcam' images of the night-side to detect meteors at times when the planet is passing through major showers.[211]

For the second mission extension a slightly modified orbit was agreed between engineers and scientists to increase the number of periapsis passes occurring over the day-side, in order to improve observing for imaging instruments. Of course, observations by the non-imaging instruments were not neglected.

During 2006 and 2007 the radar was used to assist in solving a long-standing mystery, namely the nature of thick sedimentary deposits in equatorial regions. In the 1970s Mariner 9 identified one such feature in Medusa Fossae. Judging by the degraded appearance of superimposed craters, it was inferred to be made of easily eroded material. But what? Some scientists believed it to be a blanket of volcanic ash. However, the presence of thick circular 'pedestals' around craters suggested the presence of water ice, drawing similarities to the ice-rich deposits of the polar layered terrains. Its age was disputed. Although the paucity of craters suggested it was young, their absence was also explicable if the terrain was easily eroded. The radar on Mars Express showed Medusa Fossae to be made of low-density material. This could be water ice, but the evidence was not conclusive. The echoes resembled those of the polar terrains but with layers richer in dust and sand. Interestingly, the configuration of the fossil magnetic field of Mars implied that the magnetic pole (and possibly also the axial pole) was situated near Medusa Fossae in the remote past.[212,213]

Operations by all instruments except the imagers had to be reduced while Mars Express endured another eclipse season from June to August 2007. As it flew its 5,000th orbit on 23 November it was undertaking another series of encounters with Phobos. Over several such seasons, the orbiter had passed within 3,000 km of this object a total of 46 times and had supplied over 230 images.[214]

In January 2008, Mars was well positioned to enable simultaneous studies of the impact of the solar wind on the upper atmospheres of Earth and Mars. A joint campaign was carried out using ESA's four identical Cluster magnetospheric satellites and Mars Express, comparing the fluxes of oxygen ions under similar solar wind conditions. It was thus found that the flux leaving Mars was one order of magnitude higher than that of Earth. The influence of the different distant from the Sun on atmospheric erosion rates was also studied. The intention of scientists was to integrate observations by Mars Express' twin, Venus Express, also, in order to have a more complete understanding of how planets with different atmospheres, and in the case of Earth with a magnetosphere also, reacted to 'space weather'. In May, Mars Express assisted in the arrival of NASA's Phoenix lander (of which more in Part 4). It adjusted its orbit to be able to provide a 13-minute backup relay capability over the high-northern-latitude landing site. It was an additional orbiting reference for precisely determining the approach trajectory, and reported upon the state of the atmosphere at the intended entry site. After helping Phoenix, a sizable portion of Mars Express's observations targeted Phobos in support of the Russian Fobos-Grunt sample-return mission. In July and August it had yet another series of close encounters with this moon, culminating on 23 July with a 97-km flyby which provided 3.7-meter-resolution images of its grooved surface, although unfortunately the two main candidate landing sites for the Russian mission were then in darkness.[215] By monitoring the Doppler shift on the orbiter's radio signal during these encounters, it was possible to refine the mass of the moonlet and (given its volume) its density. The radar was also used to 'sound' the internal structure of Phobos. Later in the year it was discovered that Mars Express would make a dangerously close pass of Phobos and the spacecraft had to maneuver to open the range. On Mars itself, images confirmed Iani Chaos and other 'light-toned terrains' to have been formed by groundwater erupting from the surface. Another eclipse season occurred in early 2009, culminating with a 49-minute pass through the planet's shadow on 9 March. Later the spacecraft's orbit was adjusted in order to prevent the periapsis from drifting over the night-side. On 5 November 2009 Mars Express had a unique opportunity to image both Phobos and Deimos sailing together through the field of view of its camera.

There were a dozen flybys of Phobos over a period of 6 weeks in February and March 2010. The closest point of approach on 3 March would have been 50 km, but

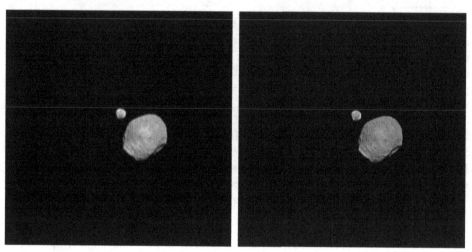

Mars Express was the first orbiter to capture a conjunction of Phobos and Deimos, in 2009. (ESA/DLR/FU - G. Neukum)

the spacecraft performed a maneuver to open the range to 67 km to prevent an occultation by Phobos from interrupting the tracking data. Nevertheless, this was still the closest flyby of Phobos by a spacecraft, beating the 3-decade old record of 80 km established by the Viking 1 orbiter. This was an excellent opportunity to gain insight into the interior of the moonlet. The flyby was tracked by NASA's Deep Space Network as well as by a number of European radio-telescopes. The carrier-only signal of the orbiter was powerful enough to be received by radio-amateurs. The closest encounter occurred over the night-side of the moonlet, preventing high-resolution imaging. However, later encounters in the sequence had favorable lighting and provided detailed photographs of the Fobos-Grunt landing site. On the seventh flyby of the series, on 7 March, thousands of radar echoes were received from Phobos down to a range of 175 km and then the camera sprang into action as the orbiter passed 107 km from the surface producing images with a resolution of better than 5 meters. Tracking confirmed the moonlet to have a very porous interior, with up to one-quarter of it being empty space. Moreover, seen by the Fourier spectrometer, the surface material appeared to be made of iron- and magnesium-rich minerals as well as, surprisingly, of clays. Clays were found in particular around Stickney crater. As this composition appeared quite similar to that of Mars, it suggested that Phobos was formed by the accretion of debris hurled from the surface of the planet by an impact.[216]

Meanwhile, a series of orbital adjustments in February and March increased the period of the spacecraft's orbit to 7 hours to improve the illumination of the planet at periapsis. Further Phobos flyby seasons occurred in August 2010, and between December 2010 and January 2011. The latter culminated, on 9 January, with a flyby that took the European orbiter within 111 km of the center of the moonlet, its third closest flyby. This time the encounter was over the dayside, and allowed the cameras to take 3-meter resolution and stereoscopic images of the poorly covered southern

Part of a 4-meter resolution image of Phobos captured during one of the February 2010 flybys. (ESA/DLR/FU – G. Neukum)

hemisphere and of the intended landing site for Fobos-Grunt. The other instruments also collected data. The radar collected 50 seconds of data at ranges less than 230 km. On 1 June the orbiter captured a sequence of images of Phobos gliding through the camera field of view with distant Jupiter in the background.

An important result for meteorology and the history of water on the planet was obtained by long-term observations by the occultation spectrometer. This provided vertical concentration profiles of water vapor over several more Martian years. Data collected during northern spring and summer, when Mars was at aphelion and the atmosphere was clean of dust, provided the first evidence of an unusual 'super-saturation' state when water vapor could not find enough dust particles on which to condense into droplets and therefore remained in the atmosphere as gas, resulting in a relative humidity in excess of 100 per cent and giving rise to tens of times more water than predicted by atmospheric models. This discovery could explain discrepancies in predicting Martian weather, as models had assumed that vapor would turn to ice whenever saturation was reached. Super-saturation occurred in the middle atmosphere, at altitudes in excess of 50 km, where water molecules are readily dissociated by solar radiation. This in turn implied that the quantity of water that had escaped from Mars' atmosphere over the ages must have been much greater than expected.[217],[218]

In mid-August 2011, Mars Express had to halt all observations when it entered safe mode after encountering difficulties in reading and writing to the solid-state memory used to store data. It was instructed to use the backup memory. However, it again safed itself in September and October. While a work-around was studied, these problems not only showed that hardware was aging, but they wasted much fuel when the probe had to be reoriented to locate the Sun at the end of each safing event. Mars Express slowly restored scientific observations in November, in time for its 10,000th Martian orbit. New procedures were developed involving a complex redesign of the way the orbiter was commanded, without making use of the solid-state memory. This allowed Mars Express to complete a 600-orbit, 5-month campaign to map the subsurface of the entire northern hemisphere from the pole to 45 degrees of latitude. Analysis of this, and other data collected over the northern lowlands for several years by the radar, showed the plains to be covered by a low-density porous material that had a low dielectric constant typical of a sediment, perhaps mixed with ice. But the areas covered with this material extended to low latitudes where near-surface ice would not be stable, and they closely matched topographic lows. This was the first evidence, beside topography, to indicate that the northern hemisphere may previously have been covered by an ocean. But the whole subject of the northern ocean remains controversial.

Extensions to the Mars Express mission are currently financed to 2014, the spacecraft being in a remarkably good state despite a number of minor instrument and hardware glitches. Only about 250 grams of fuel are being consumed annually, so the supply could last for as long as 14 more years. During 2012, it was to make a 2-month methane monitoring campaign in support of observations by the world's largest telescopes, as well as new observations of the upper atmosphere and its response to the solar wind when Earth and Mars were in opposition in February and March. It would then be made available to support the arrival and early operations of the large NASA Mars Science Laboratory rover. During the next years, it can be expected to support this and other US probes and to pave the way for future European missions.[219]

## A LANDER NETWORK

The next logical step for Europe after Mars Express would be a mission based on the MARSNET and INTERMARSNET studies of the 1990s. It would attempt to deliver a set of identical landers to the surface of the planet.[220] However, in the late 1990s ESA was thinking more in terms of flying a second Mars Express mission in 2005. How this would contribute to science was questionable. In addition, it would have required the Europeans to fund the Russians in developing an uprated Soyuz-Fregat.[221]

A NetLander deploying its solar panels after landing on Mars. (CNES)

In 1998 French scientists recommended that the national space agency, CNES, undertake an ambitious program of Mars exploration. First proposed in response to a 1997 call for small scientific missions, this would include not only cooperation with ESA on Mars Express and with NASA on the sample-return mission, but also a French PREMIER mission (Programme de Retour d'Echantillons Martiens et Installation d'Expériences en Reseau; Mars sample-return and network experiment establishment program). PREMIER was to be launched in 2005 or 2007, and in addition to the OSCAR (Orbiting Sample Capture and Return) payload for NASA it would carry small landers. NetLander (for Network of Landers) was to be a joint project by CNES, the Finnish Meteorological Institute, the German Institute for Planetology and several other institutions in Europe and the US, reusing hardware developed for Huygens, for the Russian Mars 96 small surface stations and for the Rosetta lander. On approaching Mars, four semi-hard landers would be released to use parachutes and airbags for landing. This idea was well received by both the Science Program Committee of CNES and by French authorities. A demonstrator network was proposed to ESA as the lander for Mars Express but the reduction in mass allocated to this payload ruled out NetLander. When the loss of the 1998 Mars Surveyor missions resulted in the sample-return mission being postponed to 2011 at the earliest the role of PREMIER was revised from an operational sample-return orbiter to an aerocapture, sample rendezvous and docking demonstration. It would be launched in 2007 and (as envisaged for the sample-return orbiter) would consist of a cruise stage to provide power during the flight to Mars, and, as well as attachments for the NetLanders, the aerocapture stage with a heat shield sized to fit in the shroud of the Ariane 5 launcher and accommodate the orbiter proper. With a $400 million share of the mission, France would be responsible for supplying and operating the orbiter.

Having deployed the landers, the orbiter would use aerocapture to enter a 50 × 1,400-km orbit, and then it would use thrusters to circularize at low level. It would release a grapefruit-sized subsatellite to simulate a sample canister in autonomous rendezvous and docking tests that would be performed using hardware supplied by NASA. The subsatellite would be jettisoned after these tests, and the orbiter would maneuver into a circular orbit at 250 km to perform its primary scientific mission and demonstrate long-range tracking of the subsatellite. The nominal duration of the orbital mission was 3 years.[222,223] The NetLanders were to investigate the surface, subsurface and internal structure of Mars, and its meteorology and climate over a period of one local year. The four landers would be individually deployed, starting 20 days out. Three landing sites were in Lycus Sulci, Memnonia and Tempe Terra, spaced some 30 degrees from each other around the Tharsis volcanic region. The fourth lander was to be in Hellas, antipodal to this triangle, to detect seismic waves which passed through the planet's core. The total mass of a NetLander at entry would be about 66 kg, reducing to 22 kg after landing, of which a mere 5 kg would be the payload. During entry, it would be protected by a 900-mm-wide heat shield incorporating carbon composite and aluminum structures able to accommodate a variety of entry trajectories. It was to be slowed by a parachute system based on that of Huygens, and the landing would be cushioned by the two-lobed airbags of the Mars 96 landers. Atmospheric parameters would be measured during entry and descent using 3-axis accelerometers. All the systems of the lander proper occupied the thermally-controlled half of a cylindrical can 58 cm in diameter. As many as five circular 'petal' solar panels would be deployed, along with booms and masts for instruments. The payload was to comprise atmospheric sensors deployed on a boom very similar to that of Mars Pathfinder, electric field sensors to measure the conductivity of the atmosphere and its electrical activity, a low-frequency ground-penetrating radar with its associated antennas to analyze structures down to a depth of 2.5 km, a magnetometer, a seismometer sensitive enough to record crustal tides arising from Phobos passing over, and a pair of 1,024 × 1,024-pixel CCD cameras on a deployable scissor-boom for stereoscopic panoramas. A penetrometer for soil temperature and mechanical strength measurements was added at a later date. The radio system would be used to measure the nutational motions of the planet (which it was hoped would confirm the presence of a liquid core) and the electron density of the ionosphere. The four NetLander stations were not expected to communicate directly with Earth; they would relay through orbiters.[224,225,226]

The first major modification of the mission plan occurred when CNES canceled the aerocapture experiment. As a result, the launch mass of the PREMIER orbiter increased to 3,000 kg in order to carry the propellants for all of the required orbit changes.[227] Finally, CNES deleted the entire orbiter in 2002. It tried to keep the landers alive as piggybacked payloads on the Mars Telecommunications Orbiter planned by NASA. Russia offered a ride on a Soyuz rocket. After NASA withdrew from the project in April 2003 owing to financial constraints CNES made one last effort to save the landers by proposing that ESA integrate them into its ExoMars mission. Shortly thereafter, however, CNES decided not to fund the $88 million to complete the development of the project, and announced that henceforth it would pursue robotic exploration of Mars only within the context of ESA initiatives.[228]

## BLUEBERRIES ON MARS

When the Mars Exploration program was started in 2000, several options were considered to recover some of the science of the failed 1998 Mars Climate Orbiter and Mars Polar Lander missions and of the canceled 2001 lander. One possibility was to refit the 2001 lander with the instruments of Mars Polar Lander and fly it at the earliest possible opportunity, which would have been to launch in 2002 and use a flyby of Venus to reach Mars in 2003. Another proposal was to create a network of a dozen small landers.[229] However, two more attractive options emerged and were retained for further study. Serious thought was given to flying the long-range rover initially envisaged for the 2001 lander, but equipped with the Athena suite of instruments. It was to operate for at least 30 days, and would explore a wider area than Sojourner by driving up to 100 meters per day. But the issue was how to land the rover. A modified Mars Surveyor 2001 lander was the obvious choice, but it was no longer clear that this would be viable. Instead, JPL opted for the tetrahedral airbag proved by Mars Pathfinder, and tried to squeeze the Athena rover within its limited volume. Another possibility was a Mars Science Orbiter (MSO) to recover some of the science of Mars Climate Orbiter and seek signs of water. In addition to the infrared sounder of that mission, it would have uprated versions of instruments carried by Mars Global Surveyor and an ultraviolet spectrometer. The cost of such an orbiter was estimated at about $220 million, excluding launch, as against $260 million for the rover. Slightly favoring the latter from the programmatic point of view was the fact that, after Mars Pathfinder, NASA had no successful experience in landing, whereas the successful Mars Global Surveyor would hopefully soon be followed by Mars Odyssey.[230,231]

In July 2000 NASA announced the choice of the rover over the science orbiter. Moreover, at the suggestion of the Administrator himself, the agency would return to the strategy of the Mariners, Vikings and Voyagers by launching two identical rovers for redundancy. The only obstacle was that Congress refused the additional $250 million required to complete the second rover, leaving NASA to reassign funds from other projects.[232] Naming the mission Mars Exploration Rover (MER) marked the start a grueling period of development aimed at launching in May and June 2003. This window simply had to be achieved because, as mentioned above, it happened to coincide with a 'great opposition' of Mars which minimized energy requirements and enabled a given rocket to send a larger payload to the planet. For the mission as a whole to be considered a success, at least one rover would have to land safely and function on the surface for at least 90 days, during which it would drive at least 600 meters.

The golf-cart sized Mars Exploration Rover was about seven times heavier than

An early rendition of a Mars Exploration Rover after landing. (NASA/JPL/Caltech)

the Sojourner rover delivered by Mars Pathfinder. Across the wheelbase it was 1.4 meters long and 1.2 meters wide. With its solar panels unfolded, it was 2.25 meters wide and 1.7 meters long. It had a ground clearance of about 30 cm, and the mast of the panoramic camera stood 1.5 meters above ground level. The rover was built around a rectangular 'warm electronics box'. This was a composite honeycomb structure insulated by panes of aerogel and gold foil, designed to retain most of the heat from the apparatus it contained and to be as insensitive as possible to external conditions. Operations would be possible as long as the internal temperature was above –40°C. The components themselves would survive a chill of –55°C. Eight small radioisotope pellets containing a few grams of plutonium oxide were placed inside to provide residual heat. The box contained the main computer, which was a radiation-hardened PowerPC chip running a VxWorks real-time operating system, the communications system, interfaces to the instruments, and an inertial platform and three identical fiber-optic gyroscopes to determine the heading and orientation of the rover while driving on the surface. Power would be provided by six solar panels. Although the total area of 1.3 square meters would yield 140 W at landing, the power was expected to decline as dust accumulated on the panels. During the night, or whenever required, two rechargeable lithium-ion batteries were available. A central triangular solar panel was fixed on the main body of the rover, atop the warm electronic box. To this were attached a rear folding panel and two additional double-fold panels which gave the rover an hexagonal shape, and there were two 'winglets' extending to the rear. An omnidirectional and a 'hockey puck' high-gain antenna would provide direct communications with Earth in the X-Band at up to 28.4 kbits per second. However, the preferred means of communication would be to relay via either Mars Global Surveyor or Mars Odyssey using a UHF close-range antenna that could transmit data at between 128 and 256 kbits per second. As Mars Odyssey was capable of transmitting at 124.4 kbits per second, this was a much faster way of returning data to Earth than by the direct link. If necessary, the European Mars Express could also act as a relay.[233]

The rover used the same six-wheeled 'rocker-bogie' architecture as Sojourner, which was a small bogie at the end of a master bogie, with an internal differential connecting the two sides of the suspension. All of the links were made of welded titanium box-beams. This distributed loads quite evenly on all six wheels, ensuring that slippage of one wheel was overcome by the remaining five. It also enabled the vehicle to roll over rocks larger than its wheels. One advantage of the rocker-bogie that was not particularly evident on the small Sojourner, was that it readily scaled up to provide a good shock absorber on a more massive vehicle. The four corner wheels were given independent steering to enable the rover, when necessary, to turn in place. A drive motor was placed in each wheel hub together with a 1,500:1 gearbox and a magnetic brake to hold the wheel even against quite a steep slope. Although the rover would not overturn on slopes up to 45 degrees, self-protection software would intervene at 30 degrees. Each wheel was 26 cm across and 16 cm wide, with an external black anodized coating to prevent metal burrs from sticking to the airbag fabric of the landing base. Each wheel was machined as a single piece of aluminum, incorporating a spiral pattern to act as a first-point-of-contact shock absorber to lessen the load on the rocker-bogie suspension system. The rover had a top speed of 4.6 cm/s on hard, flat ground, but its computer would be programmed to halt every few seconds to assess its position, giving it a typical average speed of less than 1 cm/s.[234]

A small robotic arm with 5 degrees of freedom was mounted on the front of the rover. Having articulated shoulder azimuth and elevation, elbow, wrist and turret rotation this was the most dexterous manipulator yet flown on a lunar or planetary mission, but it had a maximum reach of just 0.75 meter. The arm was made mostly of titanium, but a small cable cover on each rover was made of scrap aluminum salvaged from the World Trade Center in New York. While the rover was driving, the arm would be stowed loosely in front by a pin and socket mechanism and by a hook capturing the elbow.

The Athena scientific payload was provided by Cornell University of Ithaca, New York. It was designed to make each rover act as a robotic field geologist. The rotatable 'end-effector' of the arm carried a 2-kg payload of heavily miniaturized instruments. Mars Pathfinder scientists had been frustrated by the quantity of dust adhering to rocks, making it impossible determine whether the analyses referred to rocks or to the encrusting dust. The Rock Abrasion Tool (RAT) of the new rovers would use its two diamond-tipped grinding heads to brush or scrape a circular spot some 45 mm in diameter several millimeters into a rock to enable the arm to place instruments directly onto unweathered material. There was a brush on the forearm to enable the grinding tool to be cleaned of any fragments of rocks remaining in its teeth. Although the tool was capable of abrading hard volcanic rock, the degree to which it was itself worn down would depend on the frangibility of the rocks that it cleaned. The microscopic imager was a 1,024 × 1,024-pixel camera. This relied on the arm positioning for focus. It provided a resolution of 30 microns over an area of 31 × 31 mm with a depth of field of 3 mm. Its detailed imaging of rock textures would provide context for the other instruments. The imager was protected from dust by a retractable transparent plastic cover.

German researchers provided chemical and mineralogical analysis instruments. The alpha-particle and X-ray spectrometer was a greatly improved version of the

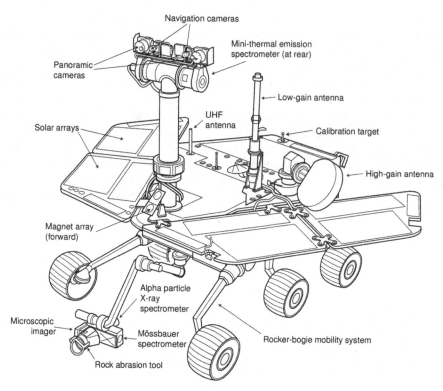

A diagram of a Mars Exploration Rover. (NASA/JPL/Caltech)

instrument used by Sojourner to analyze the rocks and soil. Target areas would be irradiated by six curium-244 sources, and the spectrum of backscattered radiation would be recorded by sensors in the instrument itself. Although it did not have the proton capability of the Sojourner instrument, this was covered by improvements in the X-ray sensors and by better correction for the atmospheric carbon dioxide which complicated previous analyses. A Mossbauer spectrometer used two sources of radioactive cobalt-57 (which had a half-life of a mere 270 days) to measure the abundance of, and distinguish between, the various oxidation states of iron-bearing minerals; correlating them to water presence, formation temperature, etc. The only instrument of the original Athena package that could not be retained was a Raman spectrometer designed not only to assess the mineralogy of the rocks but also, and most importantly, to look for organic molecules.[235] Each instrument on the robotic arm had a sensor to halt the arm's motion when the instrument established contact with its target.[236]

The panoramic camera was to provide high-resolution color images. It used two separate imagers set 30 cm apart on a mast that provided a vantage point similar to human eye-level. Each camera used a 1,024 × 1,024-pixel CCD and an 8-position filter wheel for color imaging, direct Sun imaging to measure the dust content and opacity of the atmosphere, and multispectral mineralogical imaging. The CCD had a

One of the Mars Exploration Rovers during ground mobility tests. (NASA/JPL/Caltech)

42-mm-focal-length optical system optimized for focusing at a range of 3 meters but able to extend out to infinity with a resolution similar to that of the human eye. A full-color, full-resolution panorama would generate 100 to 500 Mbits of data.[237] The rover also had a miniaturized version of the thermal emission spectrometer on Mars Global Surveyor, an infrared instrument to identify carbonates, silicates, and minerals created or modified by the action of water. If pointed skyward, it could measure the abundance of water vapor in the atmosphere and obtain temperature profiles. It was mounted at the base of the mast and viewed through a periscope in the mast that used a mirror on the opposite side from the two panoramic cameras. Once the mast had articulated from its stowed horizontal position and locked itself vertically after landing, its 3 degrees of freedom permitted full azimuth rotation, camera elevation from straight up to straight down, and elevation of the mirror for the spectrometer.

Four black-and-white wide-angle hazard cameras were mounted near ground level, two on the front and two on the rear of the rover. The overlapping fields of view of each pair of camera facilitated a stereoscopic reconstruction of the terrain within 3 meters of the rover. Those at the front assisted in operating the robot arm. Another pair of black-and-white cameras were mounted alongside the panoramic cameras to provide images for navigation purposes.

As on the Viking and Mars Pathfinder landers, the payload included three sets of magnets to study fine-grained iron-bearing particles. One was mounted on the abrasion tool, a second on the front of the rover within reach of the instruments on the arm, and a third on the solar panel deck within view of the panoramic camera. A secondary role was to assist in assessing how much reddish dust had settled on the color calibration target of the panoramic camera. As on Mars Pathfinder, the magnets were provided by Danish scientists.

What was lacking, compared to previous landers, were dedicated atmospheric instruments. But the cameras were to regularly survey the sky to measure the dust and water ice clouds, as well as to monitor the local weather. The thermal emission spectrometer could also measure dust, ice particles and water vapor, and provide profiles of atmospheric temperature from altitudes of 20 meters to 2 km, the range where the atmosphere directly interacts with the surface. Orbital observations such as radio-occultations did not have the resolution to profile the lowest kilometers of the atmosphere.[238,239]

For landing, the rover was mounted on a base made of four triangular graphite-epoxy petals with bonded titanium fittings that folded to form a pyramid. The base would completely enclose the rover during cruise and landing, and then provide a stable platform for the initial surface activity. As with Mars Pathfinder, each petal was equipped with a deployment motor that was powerful enough to right the base after coming to rest. Flexible ramps (dubbed 'batwings') of Vectran cloth were set between each pair of petals, and would be emplaced as the petals opened. These were to provide routes for the rover to drive off its base. Each rover would have available a choice of three ramps, rather than just two as in the case of Sojourner. The base also carried the devices that the rovers themselves would not need, such as a radar altimeter, a UHF antenna for communications during the entry, descent and landing, the airbags and their inflation and retraction systems. A 25-cm lifting device was on the main petal of the base to elevate the rover while its bogie was unfolded and locked into place, and while the front wheels made 'reverse origami' motions into their operating positions.

During the interplanetary cruise the folded base was enclosed in a 2.65-meter aeroshell that had a 140-degree forward angle inherited from Mars Pathfinder and Mars Polar Lander. There was a circular cruise stage on top of the shell, with solar cells able to provide 600 W at Earth and 300 W at Mars. A propulsion system with two clusters of four 4.5-N thrusters would provide trajectory corrections as well as attitude and spin control. In nominal circumstances, the spacecraft would cruise in a 2-rpm spin using Sun sensors and a star scanner to determine its attitude. The rover's computer was to command the entire vehicle in flight. To keep this cool, a pumped freon cooling system using pipes and rim-mounted radiators was added to the cruise stage. The rover's communications system was directly linked to both a low-gain and a medium-gain antenna on the cruise stage.

The landing system resembled that of Mars Pathfinder, with a 2-part aeroshell and a 15-meter-diameter parachute. The aeroshell consisted of the front heat shield and backshell made of aluminum honeycomb sandwiched between two face sheets of graphite-epoxy. The two shells were then covered with the same phenolic resin and

The Mars Exploration Rover cruise stage undergoing thermal tests. (NASA/JPL/Caltech)

One of the Mars Exploration Rovers folded up and about to be enclosed in its landing base. (NASA/JPL/Caltech)

powdered cork ablator used for all US landers since the Vikings. As the heat shield was jettisoned, a 20-meter-long plastic bridle would lower the lander from the backshell, six-lobed airbags on each side of the pyramid would inflate and, at a height of about 15 meters, a solid-fuel retrorocket would zero the vertical descent rate. To make a landing survivable in a scientifically interesting site such as Gusev crater, where relatively strong low-altitude winds were to be expected, a system of three lateral solid-fuel rockets spaced at intervals of 120 degrees was installed to reduce the horizontal velocity. To measure the horizontal velocity, engineers came up with the idea of a camera taking a sequence of three images during the descent to enable onboard software to measure the displacement of features. Although this system was added relatively late in the development, the rover's design provided interfaces for ten cameras, including a canceled upward-looking one, so a slot was readily available for the descent camera. The system was qualified using helicopter drop tests. In addition to providing lateral control, other modifications were made to the landing system in order to accommodate the fact that the landed mass would be about 50 per cent greater than Mars Pathfinder: primarily increasing the size of the parachute by 40 per cent and strengthening the airbags. The 5-meter-diameter airbags had six layers to provide abrasion protection and gas containment, making them 24 kg heavier than previously. These modifications were made necessary by an early test in which the airbags of Mars Pathfinder were ripped open upon first impact when carrying a model of the new rover. The maximum survivable landing speed had to be cut from 30 m/s to just 25 m/s. It was also felt that the mass of the new rover was about the most that airbags could safely land on Mars, so missions to deliver heavier payloads would require the development of some other means of landing.[240] The parachute was also a cause for concern. It would clearly have to be much larger than that of Mars Pathfinder, and the one designed failed several tests by shredding or improperly deploying. A workable modification was demonstrated only 8 months before launch.[241] Fortunately, due to the 'great opposition' window the entry velocity of 5.7 km/s would be about one-third less than Mars Pathfinder and the flight path would be shallower, thereby reducing the aerodynamic heating and deceleration levels and allowing the parachute to open at a higher altitude than Mars Pathfinder. Having learned from the loss of Mars Polar Lander, a radio link would be maintained throughout. A 1-bit-per-second beacon sent directly to Earth through the rover's low-gain antenna would indicate the health of the vehicle until the unreeling of the payload from the backshell allowed an 8-kbit-per-second UHF link with Mars Global Surveyor.

Affixed to each base platform was a mini-DVD bearing over 4 million names of persons who had expressed their support. The disk, sponsored by the Danish Lego toy-maker, was held by simulated 'Lego bricks' and included a color calibration chip and a magnet to attract dust particles. The rovers themselves had microchips etched with the scanned signatures of over 30,000 visitors to the JPL facility where the rovers had been assembled. A symbolic sundial on the top deck was the main color calibration for the cameras. The 7.6-cm-square sundial, bearing the message "Two Worlds, One Sun" as well as the name of the planet in 24 languages, was to have been carried by the 2001 Mars Surveyor lander.[242] As usual, the hardware for the

Martian surface was sterilized. All components were cleaned in assembly using alcohol, the electronics compartments were sealed to retain any micro-organisms, and items like the parachute, which would not be damaged by high temperatures, were heat-sterilized.

By the time development was complete, each MER had a launch mass of about 1,062 kg, with only a few kilograms of difference between them. The lander and its mechanisms (including airbags) was 365 kg, the backshell and parachutes were 198 kg, the forward heat shield was 90 kg, the cruise stage was 183 kg with a load of 52 kg of hydrazine propellant; and the rover itself accounted for only 174 kg.

Earlier cost estimates had put the twin rovers at $688 million, but in an effort to achieve the tight schedule of less than 36 months from approval to launch this was allowed to rise to $804 million. As the 1998 experience had shown, it was possible to meet an ambitious schedule if sensible budgets were provided. But independent analyses raised doubts as to whether the mission, judged to be almost as complex as Galileo or Cassini, could be developed in such a short time, less than half that of the two 'flagships'; and many feared that JPL was racing to yet another embarrassing failure.[243,244]

Launch in the June 2003 window would ensure that the two rovers would reach Mars in January 2004, in the late summer for the southern hemisphere. Hence the landing sites had to be between 15°S and 10°N for adequate solar power. They had also to be low-lying to provide sufficient atmosphere for the parachutes to slow the descent. In addition, the surface winds at that time must be mild. The abundance of rocks had to be low, but the dust had to be sufficiently coarse to provide reliable radar echoes in the final phase of the descent. And, of course, the surface had to be one likely to be conducive to driving. A total of 155 candidates satisfied all of the engineering constraints. Of these, scientists selected seven high-priority targets for inspection by Mars Global Surveyor and Mars Odyssey. Gale crater was deleted as it would not fit the landing ellipse, and the remaining candidates were divided into four primary sites (Meridiani Planum, Melas Chasma, Gusev crater and Athabasca Valles) and two backups (Isidis Planitia and Eos Chasma). However, some sites had to be rejected after wind models showed that an early afternoon landing might not be survivable, or after radar observations raised doubts about their smoothness. One of these lost sites was Melas Chasma, a portion of Valles Marineris that offered the most dramatic scenery. The final selection was made in January 2003. Gusev and Meridiani Planum were the clear winners because they showed evidence of water. Awaiting more tests to confirm that the airbags should survive a landing in Gusev, the team chose Elysium Planitia as a 'site of last resort'; its only merit was that it was low-lying and smooth, but because it lacked indications of the action of water it was of little scientific interest. As with the Viking landers, the sites were on opposite sides of Mars so that at any given moment a direct transmission from Earth would be received by only one of the rovers.

Meridiani Planum lies in the region where Giovanni Schiaparelli, observing the planet during a 'great opposition' in the late 19th century, decided to position the 'prime meridian', dubbing it 'Meridiani Sinus' (Gulf of the Meridian). It was the first of Schiaparelli's dark 'seas' selected for an in-situ investigation. The evidence for the past action of water was mineralogical: the thermal emission spectrometer on Mars Global Surveyor found at least one-tenth of the surface to be a deposit of gray coarse-grained crystalline hematite ($Fe_2O_3$), an oxide that forms either from high-temperature volcanism or from rocks which are immersed in water, usually lakes and hydrothermal springs. Although 1.3 km below 'sea level', it was nevertheless the highest site yet selected for a landing, and as such the 85 × 11-km target ellipse was challenging.

The 160-km Gusev crater was named after the 19th century Russian astronomer Matvei Gusev. Interestingly, there is a character named Gusev in the influential science fiction novel *Aelita* by Alexei Tolstoy. It lies at 15°S, in the highlands near the boundary with the low-lying northern smooth plains, some 1.6 km below 'sea level'. Accurate interplanetary navigation with respect to distant quasars would be required to ensure that the rover would come down inside the 78 × 10-km landing ellipse. In this case the evidence for the past action of water was morphological: Ma'adim Vallis, at 900 km in length one of the largest branching channels on the planet, appeared to have breached the crater's southeastern wall and formed a lake on the crater's floor. However, the channel and lake evidently dried up more than 3 billion years ago. Consequently, water-related features would probably be buried by meters of volcanic and/or aeolian sediments. Nevertheless the lakebed could have been unearthed by small impact craters, and hills protruding from the sediments might preserve traces of this watery past.[245]

The first MER mission would be launched on a 7925-variant Delta II during the main window running from 30 May to 16 June, with two opportunities per day less than 1 hour apart. For the second mission NASA would use a window that ran to 15 July, but because this would require an energy boost from the launcher almost double that of the first, the more powerful Delta II-Heavy would be used; a vehicle which had also been chosen for the Spitzer space telescope and the MESSENGER orbiter for Mercury. In any case, the back-to-back launches would impose a strain on operations at Cape Canaveral. The number of spacecraft and the complexity of their systems was unheard of since the Viking launch campaign in 1975, with two teams working in parallel around the clock. The navigators had to plan trajectories which would see the spacecraft arrive at Mars on 4 January and 25 January 2004 respectively, regardless of the actual launch dates. This 3-week phasing of arrival dates was chosen to provide time to conduct engineering tests of the first rover and perform some early scientific observations before attention switched to doing the same for the second rover. The phasing would also provide some time to address a problem which either impaired or caused the loss of the first rover.

The MER-1 rover was subjected to a longer series of assembly and integration tests than its twin. As a result, when the time came to decide which rover to launch first, hardware for MER-2 was more readily available and it was selected. Hence MER-2 became MER-A, to be launched first with the name Spirit. MER-1 then became MER-B, Opportunity. A competition had invited suggestions for names, and the winner was a 9-year-old girl from Arizona.

A truck delivered Spirit to Florida in late February 2003 for its final integration. Owing to the need for further tests and reviews, and issues with its electronics and

A mosaic of Mars Odyssey images of crater Gusev, where Spirit was targeted to land.
Note the Ma'adim Vallis canyon at bottom which appears to discharge into the crater.
Its presence made scientists believe that they would find an ancient lake environment
inside the crater. (USGS)

hardware, Spirit would not lift off before 8 June at the earliest. By removing some
ballast from the launch vehicle, engineers were confident that the window could be
extended several days. After thunderstorms passed in the vicinity of the flight path
on the 8th and 9th, Spirit finally lifted off on 10 June 2003. It was the second use of
this window, as ESA's Mars Express had been launched on 2 June. Telemetry was
collected by a US military system aboard a commercial ship off the southwest coast
of Africa and another in Botswana. Slightly more than 30 minutes later, the
spacecraft was released by the spent third stage. The Canberra station of the Deep

Space Network received confirmation of successful separation. Spirit then reduced its spin rate to the cruising norm of 2 rpm, with its star camera acquiring targets to enable it to determine its attitude.[246] A course correction 10 days later provided the 14.3-m/s delta-V to remove the deliberate targeting offset designed to prevent the unsterilized third stage from hitting Mars. The target was still undecided between Gusev and Elysium but Spirit could maneuver for either, depending on the results of further airbag tests and on the successful launch of Opportunity.

The launch of Opportunity was a different and more complicated affair. First, a solid-fuel booster was found to be defective during assembly of the launch vehicle on the pad; the damaged booster had to be replaced. After liquid oxygen had been loaded into the tanks of the main vehicle in preparation, a routine inspection on 21 June found that some of the insulation of the first stage was peeling off. The rocket had a 7-mm-thick blanket of cork bonded near the forward end of the boosters to ensure that at supersonic speeds the aerodynamic heating caused by the proximity of their nose caps would not degrade the mechanical properties of the tank walls. It had probably been soaked with water during the 4 months the rocket had spent on the pad and become debonded. The loose insulation was removed and a new layer of cork applied in time for a first launch attempt two days into the window. But on that day, 28 June, a boat strayed into the zone where the first stage would impact, and in any case the winds at high altitude were excessive. When the oxygen was drained, it was found that a 50 × 50-cm portion of cork had debonded again, either because the glue had not been allowed to cure properly or because the draining had distorted the metallic skin to which the insulation was attached. It is possible that if the mission had launched, this loose insulator could have caused a failure. While NASA and Boeing, the supplier of the rocket, investigated the issue, the insulation was replaced yet again and several days were reserved for the adhesive to cure. As a result a launch could not be attempted before 5 July, which was perilously close to the end of the window. A new mishap then occurred when one of the batteries of the launch vehicle's self-destruct system was found to have discharged. With time running out, NASA and Boeing studied whether the performance of the Delta II-Heavy was capable of buying several more days of window, possibly to 18 July, but as NASA Administrator Sean O'Keefe put it, "that would [be] a sporty event".

Everything was ready on 7 July. The first attempt was aborted only 8 seconds before liftoff by a faulty oxygen pressurization valve; but the second chance of the day, 43 minutes later, was successful. The rocket flew a slightly revised trajectory and achieved parking orbit. An hour later, approaching the coast of California, the second and third stages performed the escape burn. Next, 83 minutes after liftoff, Opportunity was released.[247,248] A 16.2-m/s course correction on 18 July timed its arrival for 25 January, targeted at Meridiani Planum. Only then was Spirit targeted at Gusev.[249] If all went well, in 2004 there would be seven spacecraft involving no fewer than 17 countries operating on or around Mars.

A total of four course corrections were performed by Spirit, but only three by Opportunity. During the interplanetary cruise, the cocooned rover had little to do except calibrate its instruments. An anomaly was discovered in Spirit's Mossbauer spectrometer, but its team were confident of being able to adjust the instrument in time for operations on Mars. Starting in July, arrival operations were rehearsed to ensure that the whole staff would be ready. Twice in November, plasma issued by record-breaking solar flares washed over the vehicles temporarily disabling their star trackers. Spirit made a final trajectory correction on 26 December to delay its arrival by 2 seconds and thereby move the impact point northeast by 54 km. Two further trim maneuvers were canceled. Less than a month before the first rover was due to reach Mars, a large dust storm erupted over Ares Vallis and this affected the atmosphere on a global basis. In particular, dust caused the atmosphere to absorb more heat from the Sun, thereby changing its density. JPL and Goddard used data from Mars Global Surveyor to monitor the state of the atmosphere near the landing site and, on the day before its arrival, Spirit's computer was instructed to deploy its parachute several seconds earlier than previously planned in order to cope with the altered conditions. In fact, software updates were being uploaded as late as 5 hours before arrival at the planet on 4 January 2004.

As Spirit was to land at about 2 p.m. local time, it would have only a few hours in which to perform its preliminary activities before shutting down for the night. During this time, it would have 1 hour of Earth visibility and a single relay pass by Mars Odyssey. The first and second sols were to be dedicated to deploying some devices and engineering work. On Sol 3, Spirit was to deploy its wheels and stand up off the base plate. Preliminary observations, including panoramas and scans by the thermal imaging spectrometer, would start on Sol 4. The umbilical between the lander and rover was to be cut around Sol 7, preparatory to Spirit driving onto the planetary surface.[250] A normal day would provide an opportunity for an uplink and downlink direct to Earth early in the morning, and two communications windows with Mars Odyssey or Mars Global Surveyor in the morning or afternoon. For the surface operations, each rover was assigned five groups of scientists and engineers who would recommend which instruments to use and which observations to make on any given sol in order to pursue the geological, mineralogical and atmospheric science objectives.

The cruise stage carrying Spirit assumed the correct attitude for entry 1 hour 25 minutes out, and then vented its Freon coolant. When it separated 15 minutes prior to entry there was a 2-second loss of radio contact, but the aeroshell continued its stabilized flight. Spirit started to feel the atmosphere at an altitude of 125 km, and the peak deceleration of 5.6 g was endured 2 minutes later with the heat shield at a temperature of 1,600°C. Within several minutes of penetrating the atmosphere, the relative speed had been reduced from 5.63 km/s to just 400 m/s. The deceleration history of the entry and descent provided the first in-situ profiles of temperature, density and pressure in the dust-laden atmosphere. Below 30 km, the temperature was almost constant.[251] At an external pressure indicative of an altitude between 6 and 7.5 km, a mortar deployed the parachute. Thirty seconds later explosive bolts were detonated and springs pushed the heat shield clear. Ten seconds after that, the payload was unreeled on its bridle. Because the atmospheric density was 8 per cent lower than predicted the parachute actually opened a little later than expected, and hence closer to the surface.[252] The radar altimeter locked onto the ground at 2.5 km with a descent rate of 70 m/s. After shedding the heat shield, Spirit sent telemetry at

8-kbit-per-second to Mars Global Surveyor (which recorded it for later replay to Earth), in addition to continuing to transmit the very-low-rate beacon to Earth. The descent imager snapped pictures at altitudes of 1,983, 1,706 and 1,433 meters, and the software calculated the horizontal velocity. This, together with the orientation of the lander, determined whether, when, and which horizontal rockets to fire. The airbags were then inflated. The lander was struck by a sideways gust of wind at an altitude of about 300 meters. The retrorockets and one of the horizontal rockets were fired until the lander was about 8.5 meters from the ground, at which time the bridle was cut and the residual thrust of the retrorockets drew the backshell clear.

The airbag-cocooned Spirit hit the surface of Mars at 14 m/s, which included a horizontal component of 11.5 m/s, then bounced to a height of almost 8.5 meters. Although direct contact with Earth was lost midway through the many rebounds, Mars Global Surveyor continued to receive data and its report of this fact to Earth was a welcome confirmation that Spirit remained healthy. In fact, Spirit bounced 28 times before rolling to a stop 250 to 300 meters from its first point of impact. It was precisely determined by radio tracking to lie at 14.5692°S, 175.4729°E, about 8 km from the center of the ellipse and 13.4 km from the intended point. This was only the fourth completely successful landing on the planet and (for the statistics) the first successful southern hemisphere landing in five attempts. Fifteen minutes after direct contact was lost, the Canberra Deep Space Network antenna reported receiving a strong signal: as with Mars Pathfinder, Spirit had settled on its base petal and the 'patch antenna' on one of its petals chanced to be facing Earthward. It took more than 1.5 hours to complete the key post-landing operations, including retracting the deflated airbags, opening the petals and deploying the solar arrays. Mars Odyssey passed over the site early on 4 January (which was some 3 hours after landing) and collected 24 Mbits of data from the lander, which it relayed to Earth. Although Spirit had a preprogrammed list of tasks to perform, how far it progressed would depend on its state of health. In particular, the team were eager to find out whether this first data included some pictures. Engineering data indicated that Spirit was tilted at only 2 degrees, which was excellent. Owing to the dust still lingering in the atmosphere after the late-December storm, the solar panels were producing 83 per cent of their predicted power.

And there were images! The black-and-white views from the navigation camera in its stowed configuration dramatically revealed the site to be a flat plain strewn with small rocks. All of the rocks within 20 meters of the lander were smaller than 50 cm. Trails of debris were visible downwind of the largest rocks. A pair of small pitted pebbles just in front of Spirit were named Sushi and Sashimi. Remarkably, the rocks appeared to be essentially free of the 'rind' of dust that was omnipresent at the Mars Pathfinder site. As the flattest and least rocky of four locations visited by landers thus far, it was obvious that the rover would have no trouble in driving around. When color images were received, the ground was only slightly darker in hue than at the other sites. An area in the immediate vicinity of the lander appeared slightly bowl-shaped and could be either a dust-filled secondary crater or a feature eroded by the wind. It was nicknamed the Sleepy Hollow. Dark features near the lander looked like they might mark where the airbags had made contact during the bounces. Closer

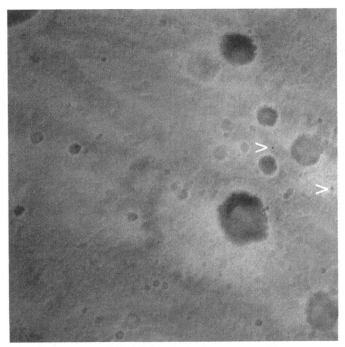

This is the last of the three images taken by Spirit during its descent, at a height of 1,433 meters. The largest crater would later be named Bonneville. The image also included the falling heat shield (arrowed), as well as the shadow of the parachute (arrowed, at the right edge of the picture). (NASA/JPL/Caltech)

An early panorama of Spirit's landing site in Gusev, from images collected on Sol 5. The flat area at left was named Sleepy Hollow. (NASA/JPL/Caltech)

in, there were marks in the soil made by the airbags as these were being retracted. One, dubbed the Magic Carpet, would be examined in some detail by the cameras in the ensuing sols.

The rover was pointing more or less south and a cluster of hills were visible on the horizon, several kilometers away. They were near the 20-km crater Thira (one of the

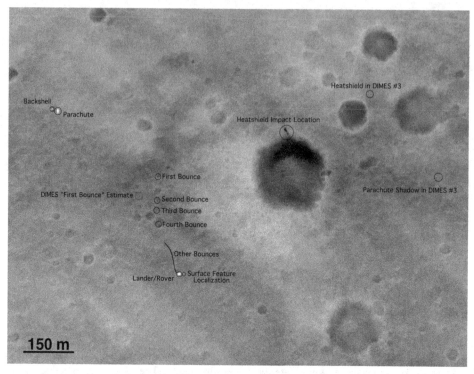

A high-resolution image of Spirit's landing site taken by Mars Global Surveyor, showing not only the lander, but also discarded hardware as well as bounce marks. (MSSS)

candidate sites for the Mars Surveyor 2001 lander) nested inside Gusev, and a case could be made that the hills were a side-effect of that impact. There were hints of layering in orbital imagery which might be water-eroded features. A peak some 7.5 km to the southwest was visible in the distance, as well as isolated mesas and knobs up to 26 km in the approximate direction of the mouth of Ma'adim Vallis. To the north was the slightly brighter rim of a crater about 300 meters distant. This was later named Bonneville. It could potentially expose the stratigraphy of the plain. The crater should be within reach, but the hills, although alluring, were well beyond the expected radius of operations. Reconstruction of the descent revealed that if Spirit had not fired a lateral rocket, its first impact would have occurred on the southern wall of Bonneville with a barely survivable horizontal speed in excess of 20 m/s.

Overhead imagery from Mars Global Surveyor and from the descent camera, as well as surface panoramas, were combined to identify Spirit's position. By chance, orbital images showed dark streaks where dust devils removed a layer of brighter dust and exposed the darker rock beneath. Being a reddish 'pavement' of pebbles, small rocks and coarse dust and sand, the landing site was significantly less dusty than predicted.

NASA named the landing site the 'Columbia Memorial Station' in honor of the seven astronauts who died in the loss of the Space Shuttle mission on 1 February 2003, several weeks before Spirit was sent to Florida. It was also revealed that the rear of both of the rovers' high-gain antennas carried a circular memorial plaque. However, this antenna would be unusable as long as the rover was parked, because for much of the time Earth was masked by the camera mast. The various hills were named for lost astronauts. The cluster of hills were named the Columbia Hills. The tallest, a little over 100 meters high and 3.1 km away, became Husband Hill after Rick Husband, the commander of the mission. The nearest was Brown Hill 2.9 km away and the farthest was Ramon Hill 4.4 km away. Isolated hills were named in honor of the Apollo 1 crew lost in a fire during a ground test on 27 January 1967. Grissom Hill was 7.5 km away, White 11.2 km and Chaffee 14.3 km.

The first color panoramas did not show anything resembling an ancient lakebed of fine-grained sediments. In particular, there were no clearly sedimentary rocks. All of the rocks in the vicinity seemed to be of volcanic origin, fairly angular with pits and vesicles. On Sol 5 the first thermal emission spectra indicated the site to share the characteristics of an average Martian soil: "Not exactly a geologically thrilling locale." The infrared spectra showed signs of carbonates, but these were probably deposited by airborne dust. If the site had been modified by water, it was now up to Spirit to prove it.

The hazard cameras on the front and rear of the rover returned their first images from a point only several centimeters above the ground. No rocks were blocking Spirit's exit but, as was the case for Sojourner, the issue was that the planned route off the lander (namely driving forward) was partially obstructed by deflated airbag material that could snag the solar arrays. Several attempts to withdraw the airbags by slightly raising a side petal and reactivating the retraction system were fruitless, so alternative exits were studied.[253] On Sol 6 the front wheels were deployed, then the rover was lowered to verify that the suspension was properly locked into place. With the deployment of its rear wheels on the next sol, Spirit was ready to roll and the umbilical was cut. It was decided to turn the rover around to exit in a different direction. This was accomplished in two stages, with Spirit first reversing 25 cm and turning 45 degrees clockwise on Sol 10 then completing a turn of 115 degrees on Sol 11 to face in the chosen direction. After advancing 3 meters on Sol 12, the rover halted about 80 cm beyond the exit ramp. Wheel marks showed the soil to be a cohesive fine-grained sand. Some hours later, Spirit aimed its infrared instrument skyward and observed 'bubbles' of warm air forming at the surface and then rising several hundred meters.

Next, the robotic arm which carried the spectrometers, microscope and abrader was deployed. Its first task was to use its microscope to examine a patch of fine-grained soil. During the next few sols the abrader was exercised, the Mossbauer spectrometer obtained its first spectrum of the soil, and then the camera examined the marks that the instrument had made when pressed onto the ground – in fact, the barely visible 'footprint' hinted at the presence of a cohesive 'crust'. Mossbauer analyses showed the soil to contain olivine, a mineral that is readily weathered by water, but again this could have been delivered as airborne dust. After these early

This cluster of hills, located southeast of Spirit's landing site was named after the deceased crew of the Space Shuttle Columbia. The rover would spend its extended mission first reaching the hills and then exploring them. (NASA/JPL/Caltech)

Spirit took this mosaic of its spent landing platform several days after driving off of it. (NASA/JPL/Caltech)

activities, Spirit was to begin its traverse. The plan was to start with an inspection of rocks in the vicinity of the lander. An early idea to drive to Sleepy Hollow to look for bedrock had been rejected with the realization that this depression was too shallow to have excavated through what seemed to be a layer of lava masking the putative lakebed. Instead, Spirit would drive over to the crater Bonneville, which stood a better chance of having punched through the lava. If all went well and the mission was granted an extension, Spirit would then set off in the direction of the Columbia Hills; even if it did not reach the hills, the nearer it approached the better would be its view of any layering in their flanks.

On Sol 15 the arm was retracted and Spirit made a 2.85-meter drive. Although the vehicle was rolling for just 2 minutes, the remainder of this 30-minute interval was devoted to navigation imaging and path computation. The drive brought it to a rock named Adirondack that had been selected for its flat and apparently dust-free surfaces. The plan was to spend several sols examining this rock. While scientists were discussing how best to analyze Adirondack, the rover took pictures of its now useless landing platform. Meanwhile, on 19 January Mars Global Surveyor passed over Gusev and took a 0.7-meter-resolution image that showed not only the lander but also the backshell and parachute, plus a dark feature on the rim of Bonneville identified as the heat shield. A number of circular patches were interpreted as the airbag bounce marks. Spirit inspected Adirondack with the microscope first. After the alpha spectrometer had examined the rock, the Mossbauer spectrometer was put in place to record a spectrum of its weathered surface. Controllers then planned to put Spirit in standby for several sols while everyone prepared for the imminent arrival of Opportunity.

But a malfunction on Sol 18 meant Spirit retained center-stage. Eleven minutes into a communications window the downlink indicated a problem, then it became erratic and 5 minutes after that the signal disappeared. This occurred at the end of a direct downlink by Canberra, where it happened to be raining. After a brief sense of relief when Spirit acknowledged receiving subsequent commands, no telemetry was received either by Earth or by Mars Odyssey, indicating that the problem was not specific to a given antenna or communications system. It was clear that Spirit had suffered "a very serious anomaly".[254] On the next sol, incoherent data was transmitted to Mars Global Surveyor in a shorter than intended relay window. This indicated that although the rover had awakened, it was unable to route valid data to its radio transmitter. It was evident that this was a software issue. As no recorded data was being sent, the prime suspect was a problem in gaining access to the flash memory. This raised the prospect that Spirit was beyond recovery. Telemetry was finally received on Sol 20, and it indicated that the rover had run on battery power for two consecutive nights, resetting its computer every 15 to 20 minutes, severely draining the batteries; it appeared unable to shut itself off to conserve power. Spirit remained uncontrollable for three full sols until it responded to a command which used a software patch to cause a computer restart without trying to use the flash memory. This produced a degraded but stable configuration in which the problem could be diagnosed. While the specialists pursued this investigation, everyone else turned their attention to the second spacecraft, which had flown a slightly faster interplanetary cruise.[255,256]

Spirit's grinder in front of the rock Adirondack. It was while analyzing Adirondack that the rover's computer suffered a major anomaly which took days to diagnose and solve. (NASA/JPL.Caltech)

On 16 January Opportunity made its final course correction, using 120 grams of hydrazine to change its velocity by a mere 0.1 m/s. This burn shifted the landing site some 380 km, placing it about one-third of the way down the target ellipse and north of the centerline. It also slightly adjusted the time of arrival. As in the case of Spirit, measurements by Mars Global Surveyor of the dust in the atmosphere had led JPL to command that the parachutes be opened 2 seconds earlier than planned in order to obtain a small additional margin for the retrorockets and airbags. It flew directly over the flank of Olympus Mons and Valles Marineris prior to striking the atmosphere at 5.7 km/s. Entry was a little rougher than for Spirit, with a maximum deceleration of 6.3 g. As then, the actual density proved lower than predicted, this time by 12 per cent, with the result that even taking into account the advance in the schedule the parachute opened later in the descent and hence nearer to the surface at 7,520 meters with a velocity of 434 m/s.[257] The radar altimeter locked onto the ground at a height of 5.4 km and the descent camera snapped its three images. The first image, taken at 1,986 meters, included the jettisoned heat shield as a black dot some 700 meters below. The other images were taken at 1,690 and 1,404 meters. The results confirmed the prediction of only light winds. Opportunity was slowly drifting to the north and slightly to the west, but the lateral rockets would not be required. The images showed a crater roughly 150 metres in size and several small ones on an

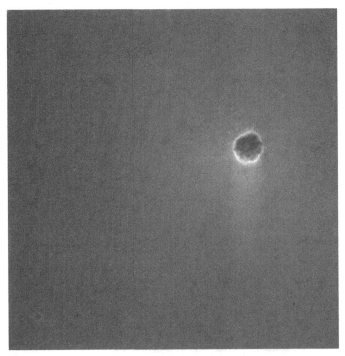

The first descent image from Opportunity. The Meridiani landing site was almost featureless, except for a few craters. The largest was later named Endurance, and would be Opportunity's target during the second half of 2004. (NASA/JPL/Caltech)

otherwise featureless plain. At 121 meters the main retrorocket fired to halt the descent and then, its bridle cut, Opportunity hit the surface at a velocity of almost 14 m/s including a horizontal component of 9 m/s. In bouncing 26 times it traveled a total distance of about 300 meters. However, after the first few rebounds it entered a shallow crater and then rolled up and down its walls prior to coming to rest near the center of the floor. It was precisely determined by radio tracking to lie at 1.9483°S, 354.47417°E, some 14.9 km downrange of the intended point. The site was named the 'Challenger Memorial Station' in honor of the seven astronauts who died in the loss of the Space Shuttle mission on 28 January 1986.

Unlike its predecessors, Opportunity settled on one of its side petals and so had to right itself. A few hours after landing, the first pictures from Meridiani Planum were received by way of Mars Odyssey, starting with the hazard cameras. The site was immediately seen to be more alien than anything yet visited. It was extremely smooth with no rocks resting on the surface and the material was extremely dark, evidently a very fine basaltic sand. Next was a navigation camera image with the mast in its stowed position. This showed a bright rocky outcrop, possibly bedrock, almost in front of the rover and less than 10 meters away. The exposure was less than 50 cm in thickness, but spanned an arc that extended about 180 degrees to the west and north. Significantly, it appeared to be made of layers, each only several centimeters thick.

This was one of the first pictures returned by Opportunity, with the camera mast still folded down for landing. To the surprise of scientists, it showed bedrock only a few meters away. (NASA/JPL/Caltech)

Sediments deposited over the ages and then exhumed provided a feature that scientists had dreamed of finding, as it would enable them to directly 'read' the history of the processes where they occurred, rather than having to infer events from scattered clues. Although the exposure in front of Opportunity looked in these preliminary images to be sedimentary, the layering appeared too fine for a series of lava flows; this left a build up of volcanic ash and sedimentation in standing water. And Opportunity was equipped to perform precisely the detailed analysis to resolve this question.

As more images were received over the ensuing sols, it became evident that the horizon was only a few meters away in all directions because Opportunity was in a small crater, and the bedrock exposure was in its interior wall. The crater was later named Eagle after Apollo 11's lunar module and also, in view of where it came to rest, the connotation of a 'hole in one' in golfing terminology. A 3-dimensional map developed from the rover's imagery showed Eagle to be 22 meters across and 3 meters deep. Its bright rim was easily identifiable in the images taken during the descent. However, the disadvantage of being in a crater was that little, if anything, was visible of the surrounding plain. Close by was a small patch of very fine sand that featured ripples of dark and bright material. Small pieces of airbag fabric were seen swaying in the wind. The powdery soil disturbed by the airbags clearly bore the

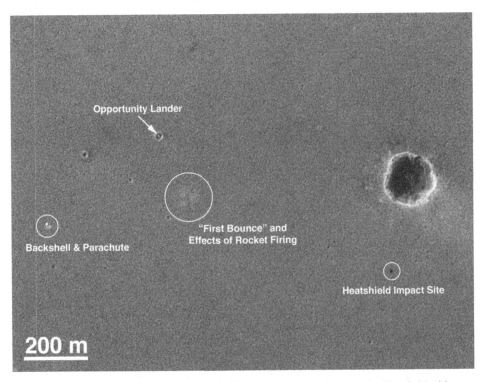

Opportunity's landing site seen by Mars Global Surveyor. The rover had landed inside a small crater, named Eagle after Apollo 11's lunar module. (MSSS)

imprint of the stitching. Extensive scuffing showed where the airbags had been dragged across the surface during retraction. Nevertheless, this was the least dusty landing site to date; so much so in fact that the airbags had remained almost clean. The rover's thermal emission spectrometer mapped the concentration of hematite across the crater. It was greatest in the northwestern portion, and increased towards the rim. Interestingly, it was virtually absent in the imprints created by the airbags. The only technical issue appeared to be the heater for the shoulder joint of the arm, which was to prevent lubricant from freezing during the night. This could not be commanded and could only be turned off by the thermostat installed to prevent it from overheating. On a brighter note, the airbags did not pose an obstacle to Opportunity driving off its lander and straight towards the outcrop. The preparations went smoothly and, just 7 sols after landing, it set off to explore Eagle crater.

Opportunity deployed its arm on Sol 9 to examine a patch of soil using its full suite of instruments. The scientists were particularly interested in rounded pebbles up to 1 cm in size which were scattered all across the site. One hypothesis was that the pebbles were rounded by tumbling in water, like beach gravel. Alternatively, they could be solidified droplets of molten rock ejected from a volcano, melt from a meteor impact, spherical glass beads, or small aggregates of volcanic ash such as buried the town of Pompeii in 79 AD. A problem for a volcanic interpretation was

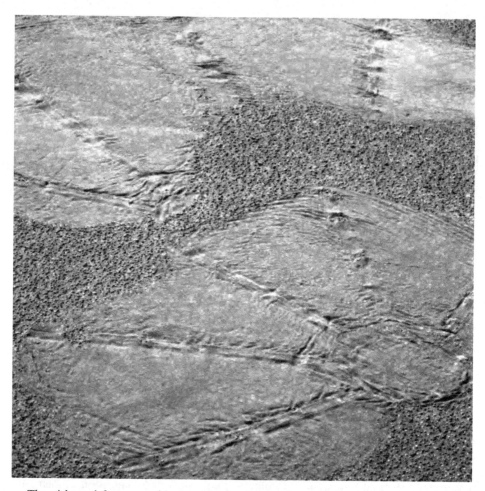

The airbags left some evident marks in the terrain at Opportunity's landing site. (NASA/JPL/Caltech)

the absence of any obvious volcanic vents nearby. The first soil sample analyzed, dubbed Tarmac, comprised olivine-rich fine-grained dark basaltic sand mixed with hematite, and as such was similar to soils in Gusev and the floor of Ares Vallis. It had been hoped that the 'field of view' of the Mossbauer spectrometer would have included some of the spherules, but this was not so; microscope images indicated that placing the instruments on the ground had pressed the spherules into the sand, making them disappear! This implied that they had something to do with hematite, whose signature was absent from thermal emission spectra of the bounce marks of the airbags where the spherules were absent. The fact that the spherules could be buried so easily meant the soil was loosely packed.[258,259] It was decided to forgo a trenching experiment, and instead order Opportunity to advance 3.5 meters to examine the side of the crater which showed the greatest concentration of hematite.

The plan was to conduct a right-to-left pass along the rock exposure. The wheel slippage in driving up the 13-degree slope was consistent with a surface covered with dry and loose sand.

About now, the exact position of Opportunity was pinpointed by correlating the few features evident in the descent imagery with scans by Mars Global Surveyor. The lander was visible as a bright dot at the center of a small crater and the rover was a dark spot near the rim. The backshell and parachute were to the west of the crater. Some of the airbag bounce marks were also clearly visible. A dark spot just to the south of a much larger crater some 750 meters from Eagle marked where the heat shield had impacted. As the rover ascended the interior wall of its crater, its mast-mounted camera was able to peer over the rim and reveal a part of the dark, flat and rock-free plain beyond. Clearly visible 450 meters away was the parachute and backshell.

On Sol 14 Opportunity reached Stone Mountain, an exposure about 16 cm high and 35 cm wide in the wall of the crater. In order to determine what this section of the outcrop was comprised of, it applied its suite instrument to a portion of Stone Mountain named Robert E. The first alpha spectrometer data showed a relatively high sulfur content. The presence of such highly soluble elements as bromine and chlorine was strong evidence of water action. This analysis was sufficient to show that at least 30 per cent of this particular rock consisted of salts that could be made as standing water evaporated. A brief integration by the Mossbauer spectrometer suggested that the sulfur was present as jarosite, a mineral which implied exposure to acidic water. There were also features in the spectrum that could be attributed to hematite, but as yet this identification was not conclusive. Remarkably, no olivine was detected in the outcrop. Microscopic images taken on Sol 15 revealed that the grayish spherules that littered the floor of the crater were embedded in the outcrop "like blueberries in a muffin" and fell out as the surrounding rock was eroded. It was

A mosaic of microscope image of 'blueberries' near Stone Mountain. (NASA/JPL/ Caltech)

A microscope view, 3 cm wide, taken by Opportunity of a blueberry still embedded in the rocky target Robert E on Sol 15. (NASA/JPL/Caltech)

deemed essential to analyze these blueberries (as they came to be known) but this would be difficult because they were so small.

It would have been nice to obtain longer integration spectra of this first rock, but it was decided to push on, since no-one knew how long the rover would last.[260] Over the next few sols Opportunity drove around the crater in a counterclockwise direction, obtaining panoramic images of the outcrop which would enable the team to draw up a sampling plan. Three waypoints along the outcrop were reached from Sol 16 to 21. It then performed its first U-turn and headed for the Hematite Slope, so named due to the high concentration of crystalline hematite seen by the thermal emission spectrometer. There on Sol 23 (17 February) it used its right-front wheel to dig a 50-cm-long, 16-cm-wide and 9-cm-deep trench named the Big Dig. This trench, the first scraped by either rover, exposed a much brighter soil. Whilst this comprised

A navigation camera image of the ledge El Capitan, showing circular abrasion signs in Guadalupe (upper layer) and McKittrick (lower layer). (NASA/JPL/Caltech)

grains which the microscope could not resolve, some unidentified agent enabled them to form into clods. When the bottom of the trench was pressed by the head of the spectrometer, the behavior of the material implied the presence of very fine clay-like particles. When the Mossbauer spectrometer inspected the bottom of the trench only a weak signal of iron was recorded.

The consensus among the scientists was that Opportunity should inspect a ledge in the middle of the exposure. Named El Capitan, this was a band of rock 20 to 25 cm thick which seemed to be made of two distinct geological features. The rover started the 15-meter drive to El Capitan on Sol 26.[261] Seen by the microscope, this ledge showed cross-bedding, a form of layering which arises when the fluid from which the individual particles are deposited (usually water or air) is not motionless but flowing. The result on the Martian rock was that some of the layers were tilted relative to the others. The thinness of the layers indicated that they formed over a very short timescale. Two targets were specifically investigated on El Capitan until Sol 36. The upper one was dubbed Guadalupe and the lower one McKittrick. Both were abraded using the RAT in order to better analyze the unweathered interior of the rock. The material proved to be surprisingly soft. On applying the microscope another startling discovery was made: the exposed surface was crisscrossed by tiny "chicken scratches" that were interpreted as vugs produced when salt crystals were dissolved in water, leaving behind irregular holes. While abrading McKittrick the drill bit had to cease operations when it encountered two blueberries. This showed just how hard the spherules were relative to the enclosing matrix. Guadalupe had a high concentration of sulfur, apparently in the form of salts of magnesium or iron sulfate. In fact, it appeared to comprise up to one-quarter sulfur oxides. To test the past action of water suggested by the presence of salts and vugs, the Mossbauer spectrometer was tuned for the narrow jarosite signal and placed on Guadalupe for a 24-hour integration to obtain a very high signal-to-noise ratio in the spectrum. It clearly identified jarosite, confirming beyond any doubt that Meridiani had once been soaked in water. The concentration of bromine was highly variable, not only with respect to other locations along the outcrop but even between Guadalupe and McKittrick, less than a meter apart. Because on Earth bromine salts form as water evaporates, this could mean the rock was a Martian 'evaporite'. However, as yet it was not possible to state whether the rocks had been at the bottom of a sea or lake, or had been soaked by subsurface water. As the rover withdrew, high-resolution color images of the two holes drilled into El Capitan showed a reddish halo which probably resulted from finely ground iron oxide.

The next task for Opportunity was to inspect nearby rocks named Last Chance and The Dells, both of which were to be documented by 'blankets' of microscope images to study their layering and confirm the cross-bedding process. Sols 37 and 38 were spent driving to Last Chance. The microscope took 120 pictures of Last Chance, which was the size of a football. Opportunity then drove to The Dells on Sol 41 and imaged that. The results revealed extensive small-scale cross-bedding which convinced the scientists that the flowing liquid must have been water acting on loose sediments. The rover then analyzed and abraded Flat Rock.[262,263,264]

The next target had been spotted during the preliminary reconnaissance of the outcrop, when scientists noticed a small depression in which loose berries seemed to have collected. On Sol 46 Opportunity drove to examine this Berry Bowl. This enabled the arm-mounted instruments to be brought into direct contact with the enigmatic spherules to determine their composition. Imaging the berries using the microscope put particular demands on the robotic arm, as they were smaller than its nominal positioning accuracy. Nevertheless, double or even triple berries were imaged. Whilst broken spherules potentially offered a 'window' into their interior, they appeared featureless to the microscope. One half of the feature, named Berry Bowl Empty, was a flat face which was free of spherules, so this was analyzed as a 'control'. Using microscope images as a guide, the arm was commanded to put the Mossbauer spectrometer directly into contact with some spherules. A strong signal was recorded that indicated coarsely crystalline gray-hematite. The spherules were composed of at least 50 per cent hematite. It was concluded that the berries were spheroidal concretions formed as mineral-rich water seeped through sedimentary rock. Similar concretions known as 'moqui marbles' are known to have formed in this way in the Jurassic Navajo Sandstone of the southern Utah desert but, unlike the Martian blueberries, they are at most 30 per cent hematite.[265]

It was evident that all of the characteristics of the outcrops could be explained if the bedrock formed in the presence of water at ambient temperatures and in acidic conditions – the latter most likely due to the presence of volcanic sulfuric acid in the water. The rocks gave the appearance of 'dirty' evaporites formed in shallow, muddy water. Meridiani could thus have been a salt lake that was occasionally wet but spent most of its time in a dry state. The blueberries 'grew' in-situ as mineral-rich water passed through the sediments at some later time. It was not possible to say when these processes occurred but there was nothing to suggest that it was in recent times. By comparing surface measurements by Opportunity with data from the infrared spectrometer on Mars Global Surveyor it became evident that this once-wet environment had spanned thousands of square kilometers. Remarkably, the mineralogy instrument on ESA's Mars Express revealed that the close spatial association between sulfates and iron oxides was not unique to Meridiani Planum, it was also present both in Valles Marineris and Margaritifer Terra.[266]

Although it was difficult to say whether this once-wet environment would have been hospitable to life, only the most 'extremophile' of terrestrial micro-organisms can live in waters of high acidity and salinity. Nevertheless, the confirmation of an ancient 'wet' environment makes Meridiani a prime site for a future life-seeking or sample-return mission.[267,268]

The discovery of rocks formed in sulfur-acidic conditions could also put to rest one of the longest-standing puzzles of Martian geology: the absence of carbonate rock. As carbonates dissolve in acidic conditions, they would not have been able to form until the atmosphere was almost free of sulfur. But by then the atmosphere itself was probably too thin and arid for water to react with carbon dioxide to form carbonate.[269]

While exploring its crater, Opportunity took time to make other observations. In particular, on Sol 39 (4 March) it observed for the first time Deimos transiting the disk of the Sun as a dark feature less than 2 pixels wide. Six sols later it captured a transit by Phobos. Being somewhat larger and orbiting at a lower altitude Phobos appeared as a dark ellipse spanning about half the diameter of the Sun, and it took only half a minute to complete the transit. Other transits, including grazing ones in which the moonlet barely 'bit' the solar disk were observed by both Opportunity and Spirit, enabling the ephemerides of the moonlets to be refined. In particular, Phobos was found to be several kilometers off its predicted position.[270] The Sun itself was often imaged using a filter designed to evaluate the transparency of the planet's atmosphere and the amount of airborne dust. Another very effective (and visually spectacular) means of measuring atmospheric dust was to make time-lapse sunset observations. Direct imaging of the Sun also provided a way to measure the orientation of the rover.

On finishing its study of the Berry Bowl on Sol 48, Opportunity drove to the

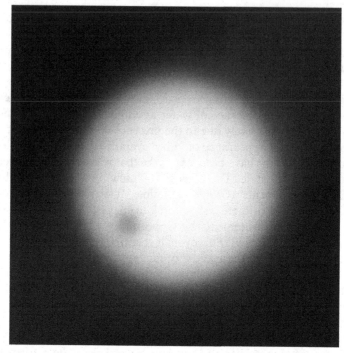

The farthest Martian moon Deimos transiting the Sun on Sol 39. (NASA/JPL/Caltech)

Phobos eclipsing the Sun on Sol 45. (NASA/JPL/Caltech)

Shoemaker Patio at the southwestern end of the outcrop. The cross-bedding there seemed to be airborne rather than waterborne dust. It then drove to a rock named the Shark's Tooth. On Sol 51 it halted and spun its front-left wheel to abrade the rock named Carousel. It then examined five possible routes out of Eagle crater. On the next sol it inspected a small rock dubbed Scoop on which there was small-scale cross-bedding. On Sol 56 it was to drive 12 meters east on a line calculated to take it up the 17-degree wall of the crater and across the rim, after which it was to turn and finish with a short drive around the crater. However, the slope was so slippery that when the rover turned for the second portion of its commanded journey it was still inside the crater and ended up climbing obliquely on the rim. It continued this drive on the next sol and finally emerged onto the surrounding plain. It then drove 6 meters back toward Eagle in order to investigate some bright material on the rim. In that position, Opportunity was commanded to take a multi-sol, high-resolution, full-color 600-Mbit panorama which was nicknamed the Lion King. This featured not only Eagle and its outcrops but also the abandoned lander and the tracks made by the rover during its explorations. The plain was a featureless desert reminiscent of a flat pebbly beach, with a multiplicity of small reddish windblown ripples only a few centimeters tall oriented predominantly northwest-southeast. There was little or no relief visible and the horizon was bland.[271] The view was so spectacular that the scientists felt as if Opportunity had landed for the second time, 2 months after first doing so.

Opportunity provided scientists with their first look at one of the 'tails' seen in orbital images to extend downwind of craters or other obstacles. The images taken during the descent clearly showed a bright tail extending several tens of meters to the southeast of Eagle crater. To the rover's cameras this appeared to be made of aligned ripples of brighter sand which, when analyzed by the alpha spectrometer, proved to be unrelated to the material within the crater, indicating that the tail was windblown dust.[272] On Sol 62 the microscope investigated the impression made by the Mossbauer spectrometer when analyzing a soil site named Munter. It found the uppermost millimeter or so to have cracked, suggesting the presence of a thin crust of dust mixed with salts. Meanwhile, the rover was suffering some minor glitches including a memory problem that was easily fixed; the failure of the doors of the alpha spectrometer to open; and the failure of the contact sensor of the Mossbauer spectrometer to report when the arm was retracted – the latter problem causing the rover to think that it had collided with something.

Opportunity's first stop was a 30-cm rock several tens of meters from Eagle. It was named Bounce because it stood right in the middle of an airbag imprint mark. Remarkably, this was one of the largest rocks visible. The rover paused at Bounce between Sol 64 and 69. After being drilled and analyzed, this rock proved to have a basaltic composition. Not only was it unlike the other rocks at Meridiani, it also differed from the dark sand. In fact it differed from the volcanic rocks in Gusev. It proved to have a composition resembling Martian meteorites found on Earth. The rock looked hematite-free just like Shergotty, a meteorite that fell in India in 1865, and its ratios of magnesium, iron, calcium and aluminum closely matched those of

Opportunity took this mosaic from the rim of Eagle crater, showing the spent landing platform as well as the tracks it left during its exploration of bedrock exposures in the crater walls. (NASA/JPL/Caltech)

EETA79001 which was collected in Antarctica and was the first Martian meteorite to be recognized. This was particularly interesting, since orbiting instruments and spectrometers with kilometer-scale resolution had never been able to find spectral matches to the meteorites which appeared to have come from Mars. Its discovery supported their having originated on Mars. Bounce could represent the volcanic material excavated from a depth of several kilometers by the impact which created the 25-km-wide crater situated 75 km to the southwest, whose rays of ejecta cross Meridiani.

After finishing with Bounce, Opportunity set off for the crater some 740 meters southeast of Eagle in the expectation of reaching it in early May. In orbital images this 150-meter-wide 20-meter-deep crater showed a bright rim that looked as if it might be bedrock. It had been named Endurance, after one of Ernest Shackleton's Antarctic expedition ships. There was a peculiar etched terrain several kilometers to the south which looked hummocky and rough. Scientists suspected that it might be sedimentary rock eroded by the wind. They would dearly love to examine such terrain, but it was well beyond the rover's nominal operating radius. Shortly after setting off for Endurance, Opportunity happened upon Anatolia, a shallow fracture which exposed an outcrop remarkably similar to the bedrock of Eagle crater. This hinted that the plain might be horizontal bedrock covered by a layer of loose dust and dirt several tens of centimeters in thickness. Owing to safety concerns and the need to make haste, the Anatolia outcrop was not studied in detail. A little further on, Opportunity paused and used its wheels to scrape another trench, this time on one of the small sandy ripples. The flat terrain enabled the rover to drive in straight lines. On Sol 82 it established a record for roving on Mars by driving a distance of 140.9 meters, and in so doing nudged its odometer over the 600-meter mark. From Sol 84 to 87 it inspected a shallow crater with a diameter of about 8 metres named Fram after Fridtjof Nansen's Arctic expedition ship. A light-toned outcrop similar to that in Eagle was present in Fram, but it was more eroded, perhaps because the interior of this smaller crater was more exposed to the wind. In places the erosion had left the 'solid' blueberries protruding on long 'sticks' of bright rock. A small rock named Pilbara on the rim of the crater was abraded, analyzed and confirmed to be virtually identical to those in Eagle. The rock named Hamersley on the inner slope was found to have the highest abundance of silicates yet observed.

On Sol 90 (25 April), the nominal end of the primary mission, Opportunity was 200 meters from the rim of Endurance. It had driven 811.57 meters, more than 30 per cent further than the mission-success distance. The panoramic camera alone had taken over 8,900 pictures during this time, including two complete, full-color panoramas, one inside Eagle crater and the other the Lion King taken after leaving the crater, and four high-resolution color mosaics of the outcrop in Eagle. Despite specific searches no ice-crystal clouds were seen in the sky, nor any dust devils on the plain. Many temperature profiles of the lowermost 2 km of the atmosphere had been made by the thermal emission spectrometer. Data obtained at different times of the day facilitated a detailed analysis of the heat exchanges between the surface and the atmosphere, and how this varied during the day and in the longer term. In particular, observations in which the instrument stared in a given direction showed 'parcels' of

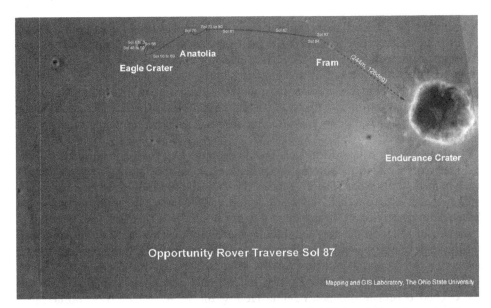

A map of the traverse of Opportunity from landing to Sol 87, just before the end of its primary mission, as it headed to the small crater Endurance.

cold and warm air moving through its field of view. Except for the arm heater and occasional minor glitches, the rover was in a remarkably good shape. In particular, although the current generated by the solar panels had decreased fairly rapidly in the first 25 days as aeolian dust accumulated, the rate of accumulation had then slowed somewhat and the panels were still generated sufficient power to continue operations.[273,274,275,276,277,278,279,280,281,282,283,284,285]

Meanwhile on the other side of the planet, full control had been regained over Spirit. The high-gain antenna was successfully aligned on Sol 25 and images from the hazard avoidance camera showed the arm still resting against Adirondack. The batteries were recharged and the computer restored to health. The fault was traced to how the file system structure represented and managed deleted files in the flash memory, which resulted in out-of-memory errors. This was fixed by modifying the flight software. Each rover had a 256-Mbyte flash memory that could be retained when power was off, and two smaller memories: one that was lost on powering down and a programmable read-only memory that was retained during power-off. The contents of the flash were deleted and the hardware was checked to verify that the issue was not hardware-related. Then the flash memory was formatted. Before the contents of the flash were deleted, however, observations made on Sol 13 in cooperation with Mars Express were recovered, along with microscope images and Mossbauer spectrometer data on Adirondack. The same fixes were implemented on Opportunity several sols after it landed to ensure that it did not suffer the same problem. On Sol 30, after 11 sols of diagnostics, Spirit was judged ready to resume scientific operations.[286,287]

The microscope images (the first ever taken of an extraterrestrial rock) showed

Adirondack as a crystalline rock. Spirit was commanded to abrade its surface and position the microscope to inspect the interior. Remarkably, the team had selected Adirondack for study because its surface had appeared dust-free but the abrasion revealed there to be a lot of dust. The cleaned area was distinctly darker. Although the rock was basaltic it was softer than its terrestrial counterparts, readily enabling the tool to grind into it to a depth of several millimeters. Minerals identified by the Mossbauer analysis prior to and after the grinding included olivine, pyroxene and magnetite.[288]

Driving directly over Adirondack, Spirit began the trek to Bonneville on Sol 36. A 6.4-meter drive tested its automatic navigation capabilities by precisely reaching a rock named White Boat. The 21.2-meter drive on the next sol broke Sojourner's 7-meter record. After another 24.4 meters it approached the cluster of rocks named the Stone Council on Sol 39. It examined a drift on Sol 41. It then analyzed the 'flaky' rock named Mimi on the next sol, finding it to be another basalt. Scientists had been intrigued by Mimi, which had a distinctly layered appearance. Spirit then drove 27.5 meters on Sol 43 and 21.6 meters Sol 44. This brought its total distance traveled to 108 meters, exceeding that of Sojourner. After briefly inspecting Halo, Spirit reached the circular depression named Laguna Hollow on Sol 45, which was another area clear of rocks. There, some 200 meters from the rim of Bonneville, it used its left-front wheel to scrape a 6-cm-deep trench, then it spent the next several sols examining the material that this had exposed. The results indicated that the shallow depression had been filled with basaltic particles ranging from sand to fine dust carried by the wind, topped by the omnipresent reddish dust. After some more driving, extensive analyses were made, starting from Sol 54, of the imposing rock Humphrey. Although it was clearly basaltic, an anomalous bromine concentration indicated that the rock had been in contact with water at some time of its history. Spirit ground more than 2 mm into its surface, exposing small vesicles suggestive of water-deposited minerals.

Astronomical observations were being made in darkness, including imaging the constellation of Orion. One hour before dawn on Sol 63, Spirit captured the first image of Earth ever taken from the surface of another planet. As part of this same imaging sequence it caught a streak of light crossing the sky. Although this could have been a meteor from comet Wiseman-Skiff, whose orbit Mars crossed 4 days later, its speed and direction matched the long-silent Viking 2 orbiter.[289] Spirit also made its first contact with Mars Express by sending data and receiving commands in an engineering test.

On Sol 64 (8 March) Spirit exceeded the 300-meter total traveled distance mark set as a minimum mission-success requirement. By the next sol it was sufficiently near the rim of Bonneville for a panoramic image. As it entered the ejecta blanket of the crater the terrain became rougher and rockier. The rocks also increased in size, with the largest now dwarfing the rover. During the drive to Bonneville there had been times when the onboard navigation software was overwhelmed. On other occasions Spirit had to negotiate steep slopes, suffering substantial wheel-slip. When slippage in climbing the raised rim of the crater on Sol 64 scraped a small trench, scientists named this Serendipity and told the rover to examine the exposed material. Spirit reached the crest of the rim, several meters above the surrounding plain, on Sol 66. Bonneville proved to be relatively shallow: it was just 10 meters deep on average, having been filled with sand. Its walls exposed the same basaltic rock as on the plain. It seemed to have been formed when a meteorite hit a target made of unconsolidated rubble. A bright spot on the far rim had been identified in orbital images as the discarded heat shield, and this was verified by Spirit's high-resolution pictures.[290] Drifts of lighter-toned sand were present nearby on the rim, and the rover spent part of Sol 72 using its wheels to dig through the one dubbed Serpent and finding it to consist of basaltic pea-sized pebbles on top of finer sand. All the soil sites investigated along the rim were unmistakably of volcanic origin. If some of the rocks had been carried to their present positions from upstream in Ma'adim Vallis then some variety could have been expected, but there was none. The source of the basalt was a mystery, since there were no volcanoes or fissures evident on the floor of Gusev. Bonneville did not appear sufficiently scientifically interesting to warrant trying to examine the interior of the crater, so it was decided that the rover should make its way counterclockwise around the rim, starting at the point where it arrived on the southwestern edge, and then depart from the south by crossing a drift field.

Scientists were eager to analyze one of the lighter-toned 'white rocks' scattered all across the plain, in the hope of finding an example of the long-sought Martian carbonates. From Sol 77 almost to the end of its primary mission Spirit was busy analyzing Mazatzal, a 'white rock' on the rim of Bonneville which was marked by wind-abraded incisions called ventifacts. Mazatzal was examined in four phases: in its undisturbed state and then again after three progressively deeper abrasions. It

A portion of the mosaic of Bonneville crater, the closest to the landing site of Spirit. Scientists hoped the impact that dug the crater would have unearthed the ancient floor of Gusev, but it proved too shallow. (NASA/JPL/Caltech)

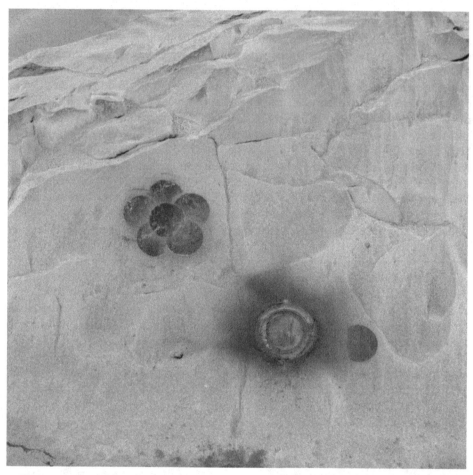

Daisy patterns of brushing and a drill pattern on the rock Mazatzal. Analyzing this rock, Spirit found its first hints of the ancient presence of water, albeit in very small quantities. (NASA/JPL/Caltech)

appeared quite different from those previously encountered, proving in particular much softer; so much so that the tool ground over 8 mm into it. A 'daisy' pattern of six spots was brushed to provide the thermal emission spectrometer with a large viewing area. When inspected by the microscope the brushed and abraded areas showed small cavities and veins filled with bright material. But the analyses by the spectrometers showed this (and other lighter-toned rocks) to be basalts brightened by dust; they were not carbonates. Nevertheless, Mazatzal showed enrichment in sulfur, chlorine, bromine and iron oxides, probably from olivine altered by water. And, interestingly, the exterior of the rock seemed to comprise at least three water-modified layers. Whilst this suggested that Mazatzal had been exposed to ground water or perhaps ice several times during periods when the planet's variable axial tilt placed this location at high latitude, only very small quantities of water would have been involved.[291] While Spirit was parked alongside Mazatzal, Mars Global Surveyor took a 50-cm-resolution image that showed not only the rover itself as a dark spot, but also the tracks it had made in driving from the lander to Bonneville and then around the crater's rim.

On leaving Mazatzal, Spirit was to head southeast towards the Columbia Hills, some 2.5 km away. The elevated terrain seemed to sport older material, as well as hints of layering. It was possible that the hills were an island in the putative lake. If all went well and Spirit managed to drive 60 meters per sol, pausing occasionally to make scientific observations, it should reach the base of the hills around Sol 160 in mid-June. This trek began on Sol 87 with a 36.5-meter drive. On 5 April, Spirit reached the 90-sol milestone that marked the nominal end of its primary mission. By then, its wheels had spun for a total of 637 meters, considerably more than the mission-success requirement. The panoramic camera had taken over 9,300 images, with an overall output exceeding 750 Mbytes, while the microscope had taken 537 images of seven rocks, soil, trenches, etc.[292,293,294,295,296,297,298,299,300]

## THE ROBOTS THAT DID NOT WANT TO DIE

For the extended mission the staff of the Mars Exploration Rovers was cut back to 60 per cent of the primary mission level and the controllers adopted a less grueling 'Earth time' routine than the two-shift 'Martian time' continuous cycles of the first 90 sols. To overcome problems when the communications sessions would occur at night on Earth, the activities of the rovers for these 'restricted sols' were planned several sols in advance using the latest available data. This meant not knowing just where the rover stood, or where its arm had been left by the most recent activity. It was hoped, however, that new onboard navigation software would make driving easier, even on these restricted sols. Activity was managed in cycles of 3 or 4 sols. On the first sol the rover would move to a given point, obtain pictures and thermal emission spectra using its mast-mounted cameras, and then either deploy its arm to investigate selected targets or observe the state of the atmosphere. The second sol would be spent in analyzing the data collected on the first sol, and planning for the third sol. In this way, defining a particular cycle meant specifying the destination for the drive and the targets to be investigated. It was less hectic and more flexible than the earlier ad-hoc process. An engineering analysis estimated that the rovers would probably operate for thrice the 90-day design life, meaning that they would continue until September 2004. The biggest problem would be the accumulation of dust on their solar panels, diminishing their power output. But the decline in power of 0.2 per cent per day was less than expected. By the time the power was down to 20 per cent of the nominal value, the situation was expected to stabilize. Moreover, illumination would reduce as the season changed and the noon-time temperatures would remain chilly.[301]

On 8 April Spirit was commanded to track Earth across the Martian sky for a period of 6 hours and accept new software to enhance its extended mission. Then the computer was rebooted on Sol 98. Opportunity did likewise between 11 and 14 April;

Sol 75 to 78 in terms of its mission. The revised software made mission and power management less conservative, and included new autonomous navigation and self-diagnosis capabilities. The new navigation routines were to make driving on rough terrain such as in Bonneville's ejecta faster. On Opportunity, the software included a 'Deep Sleep' option to offset the power consumption from the defective heater in its arm. This disconnected the battery overnight in order to prevent the heater from drawing power, even if it remained on. When light fell on the solar panels at dawn, this would operate the switching to place the battery back online. But there were significant downsides to Deep Sleep. It precluded operating instruments at night, which was precisely when the arm-mounted spectrometers were supposed to make long integrations. Moreover, night-time relay passes would no longer be possible, thereby reducing the overall data return. Finally, if a critical optical component of the thermal emission spectrometer were to be damaged due to the chill, this would effectively kill that instrument.[302]

Spirit had to drive less than 2 meters to reach its first extended mission target, a 'white rock' named Route 66. It was extensively brushed and the results confirmed the conclusions drawn from Mazatzal. On Sol 100 the rover set off on its southerly trek, beginning with a record drive of 64 meters. The new software made the rover more autonomous in negotiating obstacles and less dependent on intervention from Earth. The transparency of the atmosphere improved as airborne dust settled. This allowed Spirit to see such distant targets as details of Ma'adim Vallis, the snaking channel which breached the rim of Gusev in such a manner as to suggest the crater had once held a lake. Moreover, 'hard' processing of pictures from the panoramic camera revealed the rim of Gusev, some 80 km distant. On 28 April Spirit drove a record 88.5 meters on much rougher terrain than that on which Opportunity was at that time driving longer routes. It then skimmed the rim of Missoula, an older and more degraded crater than Bonneville, 120 meters across. On Sol 113 Spirit dug a 9-cm-deep trench named Big Hole which, being in a topographic low and an area of low thermal inertia more than 500 meters from Bonneville, was expected to be representative of the intercrater plain. It then arrived at the rim of the 75-meter crater Lahontan, but, like Missoula, it was only to be studied from the safety of its rim. The final leg of this trip was halted by the navigation software, which noted a drop as the rover passed the rim.[303] On Sol 125 Spirit traveled for 123.7 meters in a single day, setting a new record. But the time devoted to making long drives was at the expense of in-situ studies, and on these sols the science usually consisted of taking panoramic images and monitoring the atmosphere. From time to time the rover paused long enough for the arm-mounted instruments to investigate soils and rocks in order to gain a sense for how these varied across the plain. Less than 700 meters from the foot of the hills, Spirit scraped an 11-cm-deep trench named Borough and then spent the next several sols thoroughly analyzing it. The soil was rich in the readily altered mineral olivine. Just like Big Hole, this showed sulfur, magnesium and bromine just below the surface, indicating water modification. In both cases a low water-to-rock mixing ratio was inferred. But as Spirit approached the Columbia Hills its trench samples started to show traces of salts which would form in the presence of water. Moreover, the Mossbauer spectrometer found more oxidized iron

The shadow of Spirit near the foot of the Columbia Hills, on Sol 153. (NASA/JPL/ Caltech)

(i.e. 'rust') than was present at the surface itself. About 1 per cent of the soil appeared to be nickel, probably from meteorites.[304] Highly vesicular rocks started to show up as Spirit closed in on its objective, but because the vehicle was well beyond them by the time the images were downloaded to Earth, none of these rocks were examined in detail.

Panoramic images confirmed that ancient bedrock was exposed on the slopes of the Columbia Hills, starting only a few tens of meters above the level of the plain. Even if the bedrock proved to be basalt like the plain, this presented an excellent opportunity to study rocks in the locations where they formed. On Sol 150 Spirit made its final sampling stop on the plain and analyzed the rock named Joshua. To the team's delight, it reached the foot of the hills on Sol 156, several sols ahead of schedule. However, before it started to climb and make its way to the West Spur, it was to spend several weeks performing a detailed study of a number of rocks in a sloping area dubbed End-of-the-Rainbow. Interesting targets included the Rotting Rocks, a cluster of small rocks which resembled rotten slices of bread, and Pot of Gold, a spiky rock barely larger than 10 cm on which there were dozens of nodules and appendages like 'tentacles' projecting in every direction. This odd shape could be explained if it represented the hardest remains of a rock whose softer parts had eroded away. This was made Spirit's next destination. It was difficult to focus the microscope on such an irregular surface, but when the images were obtained they revealed the lumps to be solidly embedded in the rock and to somewhat resemble the blueberries that Opportunity had found at Meridiani. Owing to the soft ground, Spirit took many sols to maneuver into a position from which it could drill into Pot of Gold and in the process the wheels churned up intriguing material that could be debris eroded from that rock. On Sol 169 the RAT was finally in position. When it was placed against its objective, the rock, which was only slightly larger than the drill head, moved, indicating that it was not firmly embedded in the ground. But at least some of its surface had been abraded and so an overnight alpha- and X-ray spectrum was obtained. This indicated the rock to be rich in sulfur, chlorine and phosphorus – unlike any other Gusev rock. Mossbauer spectra showed Pot of Gold to be the first

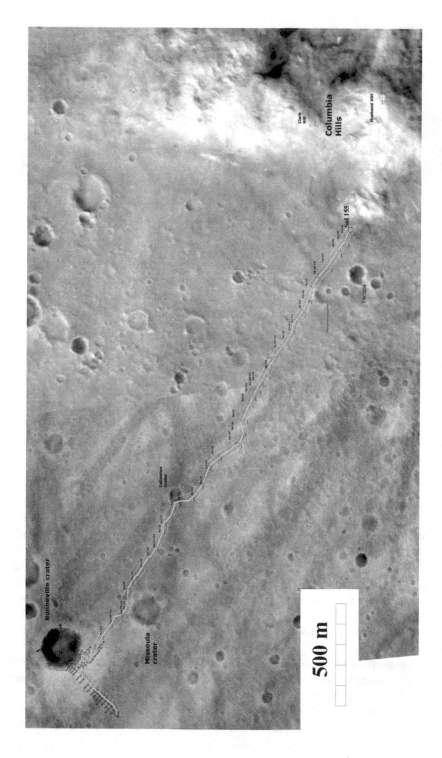

The road traveled by Spirit from landing to the foot of the Columbia Hills. (NASA/JPL/Caltech)

The bizarre, spiked rock Pot of Gold (at right) was examined by Spirit and found to contain hematite. (NASA/JPL/Caltech)

Gusev rock to be free of olivine. Intriguingly, it had an abundance of hematite. Although not direct evidence of the action of water, these analyses were good hints of it. No large standing bodies of water were required, however. Small, transient quantities would have been sufficient; perhaps ground-ice or atmospheric humidity. The important point, though, was that even such small amounts of water implied a warmer, wetter climate than exists today.

As Spirit had operated for twice its design lifetime, it was given a '3,000 meter tune-up' prior to being sent to explore uphill. The motor of the right-front wheel was the most critical hardware concern. It had recently started to draw excessive current, evidently because it was experiencing substantial friction. Controllers had found that on level ground they could drive the vehicle reasonably well in reverse on just five wheels, with the sixth wheel dragging. Climbing on the hills would be a different story. The motor heater was run at its maximum temperature, then the wheel was spun in both directions in an effort to spread the lubricant evenly in the gearbox. Although this did not solve the problem, it apparently prevented it from worsening.

The solstice on 21 September meant the approach of the Martian winter. As the planet's orbit was rather eccentric and winter in the southern hemisphere occurred at aphelion, only 70 per cent of the sunlight available in the summer would fall on Spirit's solar panels. By parking the rover on a slope which would make the best use of the available sunlight, minimizing its activity during the day, and implementing the equivalent of 'Deep Sleep', there was an excellent chance that it would survive the winter and be able to resume operations in the spring by climbing the West Spur in early 2005.[305] Being nearer the equator, Opportunity would be less affected by the winter.

On Sol 188 Spirit started driving in reverse as its primary means of travel. After examining several rocks in the Wooly Patch outcrop, on Sol 200 it set off for the Clovis outcrop almost 10 meters above the plain. The slope was relatively steep but offered the advantage of tilting the rover to the north, which better exposed its solar panels to the Sun. The climb presented a number of problems and challenges, with the rover slipping and finding itself attempting to pursue inaccessible routes. Several sols were spent ejecting a rock lodged in the left-rear wheel. At one point the vehicle was tilted 34 degrees, and on another occasion wheel slippage caused it to slide downhill whilst trying to climb. Finally, on Sol 207 a route was found that avoided the steepest slopes. When Spirit reached Clovis it exploited its elevated vantage point to take panoramas showing the floor of Gusev below, as well as the rim of the crater. Clovis was the first outcrop where rocks could be studied in the very place where they had formed. They were clearly different from those on the plain. Some appeared layered, and microscope images showed grains several millimeters across. Most were gray and clean; possibly unaltered bedrock. Others were lighter in color and crumbly. In fact, Clovis proved much softer than other Gusev rocks and the abrader dug almost 9 mm into it. Alpha spectra showed bromine, chlorine, sulfur, phosphorus and their salts in such high concentrations as to indicate a lengthy exposure to large amounts of water. Moreover, the Mossbauer spectrometer found goethite, an iron-bearing mineral that could only have formed by the alteration of precursor minerals in water.[306]

After 20 sols investigating Clovis, Spirit climbed further upslope in search of a favorable winter parking position, pausing for a dozen sols at Ebenezer and Tikal. There was a hiatus from Sol 244 to 255 for solar conjunction. Whilst it was unable to communicate with Earth, it stood in place near Tikal and undertook pre-loaded observations, mainly atmospheric, but including daily Mossbauer analyses of the dust on the magnets. After contact was re-established, Spirit observed the effects of wind on several previously imaged soil targets. Exploiting the slope to keep its solar panels facing the Sun, it then drove to a rock named Tetl, which proved to be the first truly layered rock seen in Gusev. The arm-mounted instruments showed it to consist of multiple narrow layers of volcanic material, with possible evidence of modification by interaction with water. After spending a week at Tetl, Spirit drove over to Coffee and then to another layered rock named Uchben. When intermittent

The rocky outcrop Clovis seen by Spirit's navigation camera on Sol 205, before it was brushed, drilled and analyzed. (NASA/JPL/Caltech)

malfunctions were reported by the brakes on some of the steering actuators, these faults were diagnosed as sensor problems and the rover was told to ignore them. A more serious problem developed when a short-circuit appeared between the power bus and the chassis, where a fuse had blown during tests on Earth and had not been replaced. Spirit was not in immediate danger, but it had lost its protection against a wire losing its insulation and shorting to the chassis. On Sol 281 Spirit was finally able to start an examination of Uchben that lasted 11 sols.[307,308] Uchben proved to be relatively soft. The microscope showed small sandy particles of various shapes; the angular ones could have been of volcanic origin, but the rounded ones seemed to have been worn by the action of water. By this point, Spirit had survived more than thrice its expected life, and the short half-life of the radioactive sources in the Mossbauer spectrometer required ever-longer integration times in order to produce an acceptable signal-to-noise ratio.

On Sol 300 Spirit was driving to Machu Picchu, 40 meters above the level of the plain. Beyond that it entered a 200-meter flat saddle leading to Husband Hill. Its science in the second half of November was limited to remotely sensing rocks on the way to a ridge named Larry's Lookout, set on the flank of the tallest hill in the cluster. It halted for several sols to inspect the rock Wishstone, which was brushed and then studied by the microscope. This appeared to be a collection of particles of diverse sizes, suggesting that it formed in a high-energy catastrophic event such as an impact, explosion or volcanic eruption. The arm-mounted spectrometers found the abundance of phosphorus to be the highest yet measured anywhere on Mars. It was evident that the rocks from which Wishstone was derived were some kind of water-related phosphate deposits. This was puzzling, as phosphates would form in watery environments different from those in which the more frequent sulfates were formed. Perhaps there had been a change of water chemistry over time, or different watery environments on the planet.

After several locomotion difficulties owing to a potato-shaped rock that became wedged in Spirit's right-rear wheel, the last target for 2004, starting with a drive on Sol 354, was Champagne, so-named because this happened on new year's eve. Since leaving its lander, Spirit had driven a total distance that was only a few meters short of 4 km.[309]

Having completed its 90-day primary mission, Opportunity was continuing its exploration at Meridiani. On Sol 95 (30 April 2004) it was commanded to drive 20 meters to complete its approach to the western rim of Endurance but it halted in-place after 17 meters, having sensed a hazard. When they saw the pictures which it sent, the team were shocked to find that the vehicle was perched on a 5-degree up-slope and within a meter of the crest of the rim! Not only was Endurance deep, its opposite wall was very steep and in some places almost vertical. Although clearly dangerous for the rover, the crater had just what the scientists had been hoping to find: exposed layers of bedrock with a 'stratigraphic column' much deeper than at Eagle crater. The initial impression was that the top layers might well be the same as at the smaller crater, complete with blueberries, but there was an additional rock unit beneath which Eagle had not exposed. On the northern and southern walls, the layering was displayed on a much gentler slope where the rover might possibly be able to descend into the crater. The largest cliff on the far side of Endurance was christened the Burns Cliff, after the late MIT scientist Roger Burns who predicted a significant sulfur chemistry on the planet. In addition to extensive layering, there was cross-bedding at its base on such a macroscopic scale as to be reminiscent of that in 'petrified' sand dunes on Earth. If this interpretation were correct, then the shallow water which wetted the bedrock exposed by Eagle had been preceded by a prolonged period when Meridiani was a dune field. Remote sensing indicated that sulfates and hematite formed the upper part of the cliff, with basalt at its base. There were other layers with a basaltic signature, but these were so thin as to make it unlikely they were lava flows; they were more likely to be deposits of airborne volcanic dust and ash.

Scientists debated whether it would be possible to enter, and more importantly to exit from the crater. The point at which Opportunity had reached the rim, named

A mosaic of crater Endurance taken from its southeastern rim. (NASA/JPL/Caltech)

Larry's Leap, sloped at about 20 degrees, meaning that if the ground was anything like that at Eagle there would be substantial slippage. Nevertheless, this would be a negotiable route in. A very-high-resolution panorama of the interior of the crater was obtained in early May, accounting for several hundred megabits of data. Then stereoscopic images were used to measure the actual depth of the crater and slopes of its interior in order to undertake studies, simulations and tests designed to decide whether it was safe to send Opportunity into the crater. A second promising area to drive into the crater was discovered on the southeastern rim, and named Karatepe after an archaeological site in Turkey. In contrast to Larry's Leap, this entry path would enable Opportunity's solar panel to face the Sun during the descent even as winter was approaching. Although the first few meters of the Karatepe slope were a jumble of rocks, pebbles and blueberries, it led to an orderly stratigraphy which could be directly related to the Burns Cliff. The objective was to drive at least one-quarter of the way down into the crater, to a rocky layer that seemed to be basaltic rather than water-deposited. Beyond this, the sandy surface appeared too soft for the vehicle. The slope started off at about 17 degrees, steepened to 23 and then to 25 degrees. Opportunity would not venture onto the steepest terrain unless the path looked really safe. If the rover were unable to exit, this would preclude interesting activities such as surveying the heat shield whose wreckage was some 250 meters away. Such an inspection would not only enable engineers to assess the margins of tolerance incorporated into the design of the heat shield, if its impact had dug into the surface to a depth greater than the rover's wheels could scrape then this would serve as a bonus trench. Furthermore, the rover had passed by several interesting targets on its way to Endurance. But if Opportunity were to become trapped inside Endurance, that in itself would not be disastrous because, as Principal Investigator Steven Squyres said, "we'd be stuck in a candy store". In preparation, Opportunity was to make a counterclockwise drive around an arc of the rim over the period of a month, during which it would pause to take at least two complete panoramas.[310]

Overnight between Sol 101 and 102, Opportunity tested Deep Sleep for the first time. It suffered no particular problem, but the thermal emission spectrometer was chilled almost to its minimum survivable temperature. As a precaution against the instrument's loss, it took comprehensive scans of the interior of the crater over the next few sols. Starting on Sol 107 a rock some 30-cm in size lying on the rim, and possibly ejected from the crater, was abraded and inspected. Subtle mineralogical differences indicated that this rock, named Lion Stone, may have been formed at a different time and in a different environment than the rocks exposed in the wall of Eagle. Meanwhile, the descent into Endurance was being rehearsed in JPL's Mars Yard using a tilt table, rocks, sand and metal pellets to simulate blueberries, with a lightweight rover simulating Martian gravity, which is only 40 per cent of that on Earth. These tests confirmed that on Mars a rover of this design could negotiate a slope of 30 degrees and would be stable up to 45 degrees. Finally, after a month of reconnaissance on Mars and tests on Earth, engineers and scientists jointly decided to send Opportunity into Endurance at Karatepe. It would advance downslope for a short distance and then reverse slightly to verify that it was capable of ascending, before continuing. By imaging its wheel tracks on the way down and during the short reverses, the rover would provide data on the actual soil characteristics along the way and improve predictions for conditions downslope. In preparation for the entrance, controllers drove Opportunity backward and rubbed its wheels hard on a rocky surface in order to remove their slippery anodized coating and improve their grip. On Sol 133 (7 June) Opportunity eased its two front wheels just over the rim, and on the next sol advanced all six wheels into the crater. It promptly drove back out again. There was negligible wheel slippage and the tilt of the vehicle was as predicted, indicating that it could safely enter and exit. This occurred within a few sols of Spirit reaching the base of the Columbia Hills.[311]

The first target inside the crater was Tennessee, an exposure about 36 cm wide which was abraded to the record depth of 8 mm. An analysis by the instruments on the arm showed a series of parallel layers made of evaporites, mixed with darker thin layers probably deposited by the wind. Close-up observations of the contacts between the layers showed that they were surprisingly complex.[312,313] The pace of activities on the slope was remarkable, with the rover grinding and analyzing no fewer than seven targets between Sol 138 and 161. It used its abrader to drill into the rock several times for each meter of depth. At an exposure named Millstone it ground into a layer that preserved distinct water ripples. Some of the rocky slabs examined had rows of tiny 'razorback' ridges several centimeters tall that could be veins of hard minerals resisting the wind that had eroded the adjacent rock. From its new vantage point, Opportunity gave amazing images of the fine dune fields on floor of the crater. Unfortunately, the rover would not be able to drive on such soft ground. However, scientists briefly considered driving right to the edge of a dune in order to take measurements using the arm-mounted instruments, while the rover remained on firmer ground.[314] Still, maneuvering within the crater was not always simple. For example, on Sol 150 the rover ended up parked with one wheel in the air while driving backwards on a 28-degree slope; as a precaution, it was moved until all of its wheels were firmly grounded.

As in Eagle crater, the Karatepe rock exposures appeared rich in water-formed sulfates as well as jarosite. However, in Endurance the abundance of magnesium sulfate and hematite decreased with depth whilst there was a several-fold increase in chlorine and chlorides, as if the water that produced the deeper sediments was less acidic than that which left the later sediments. Such data hinted at a succession of very different watery environments, with long, wet periods followed by brief, dry periods during which the wind laid down the darker layers. Furthermore, the

The mysterious pointy 'razorbacks' in Endurance crater. (NASA/JPL/Caltech)

hematite berries showed morphological changes further down the Karatepe slope, becoming larger, redder and more angular, and acquiring a hard-looking external shell. After recharging its batteries, Opportunity resumed its descent on Sol 169. Overcoming slippage, it was able to spend some time around Sol 180 examining an exposure named Diamond Jeness, where it abraded more holes. Continuing into the crater despite substantial slippage, it headed east in the general direction of the Burns Cliff. On reaching an outcrop named Axel Heiberg on Sol 192 it began its routine of arm operations, but the abrasion tool stalled several sols later when a piece of debris became trapped in its mechanism. By Sol 210 it was usable again – evidently the debris had fallen out. While the tool had been unusable, the rover had made a number of short, circuitous drives until, on Sol 206, it ended up parked by chance in front of an interesting rock named Escher which the scientists decided to spend several sols investigating. This unusual and revealing rock had a polygonal pattern on its surface similar to 'mud cracks'. If these were indeed mud cracks, then this suggested that water may once have ponded in Endurance crater.

As with Spirit on the other side of Mars, Opportunity was mostly inactive from Sol 222 to 238 owing to solar conjunction. It then resumed work by examining the rock Ellesmere. By this time Opportunity had reached the deepest point that it was to venture, some 10 meters below the rim. It was then sent to investigate Wopmay, a rock named after the first 'bush pilot' to fly geologists into the Canadian Arctic. This

1-meter light-colored rock had a surface that displayed smooth-looking lumps which gave it the appearance of either a brain coral or a toppled 'Michelin man'. To reach it the rover drove almost 20 meters, which was one of its longest drives within the crater. Because of slippage it ended up less than 2 meters away from its target, which was rather closer than planned. And slippage made closing to a range suitable for arm work difficult, especially because engineers avoided approaching from uphill lest slippage cause the vehicle to bump against the rock. It took almost 20 meters of driving over a period of 8 sols to achieve a satisfactory location! The strange-looking rock was examined in detail between Sol 258 and 264.[315] By that time the rover had abraded 21 targets within the crater and obtained 95 spectra using its arm instruments. The plan was to drive uphill from Wopmay and then turn towards the Burns Cliff but the rover slipped so much that after 3.5 meters of an assigned 21-meter drive it stopped, its wheels having dug deeply into the soil. On Sol 268 it was unable to advance more than 40 cm; simply sliding downhill. But on the next sol it was able to gain grip on firmer ground and by the end of October it was free. Controllers then designed a drive on a 25-degree slope to reach layers at the base of the Burns Cliff, but this excursion had to be called off when a sand field was discovered and the route was revised.[316,317] By Sol 285 Opportunity achieved its most easterly position inside the crater, but it was suffering so much slippage that its drive had to be interrupted again. On finally reaching the near (western) edge of the Burns Cliff in November it paused for several sols to provide a mosaic of the cliff wall that comprised 985 megabits of data and showed the cliff to consist of at least three rock units totaling 7 meters of thickness, recording the history of water in Meridiani Planum. As to drive further along the base of the cliff would involve either crossing a field of fine sand or negotiating a slope that exceeded 30 degrees, it was unsafe to proceed any further. This meant that the cliff could be investigated only by using the mast-mounted cameras. Ten sols were spent taking panoramas and infrared scans looking up at the wall of the cliff. Unfortunately, it was not possible to investigate in detail where layers intersected at a point on the eastern edge of the cliff.

On Sol 295 Opportunity began its drive out of Endurance. There were two exits available. One offered a shortcut, but in places the slope reached 30 degrees. The other was the Karatepe slope, near where the rover had entered the crater 6 months earlier. As this path had shallower slopes and few obstacles, the team decided in favor of it. Just before exiting, Opportunity brushed and analyzed two final targets on the transition between light and dark rock layers. After spending 181 sols inside Endurance, the rover re-emerged onto the plain on Sol 315 (12 December). It spent several sols taking remote-sensing measurements and driving to examine the wheel tracks that it had made prior to entering the crater. To reach the heat shield involved driving south over level ground. By Sol 324 the rover was just 30 meters from its heat shield. On impacting at in excess of 150 km per hour, the shield had flipped itself inside out and broken into two parts. The shallow 2.8-meter-diameter crater which it produced was partially filled with fine charred material. In order to preclude contaminating the rover, it surveyed the site from a distance and then, on 30 December, in making its final drive of the year, it drew up alongside the debris of the heat shield.

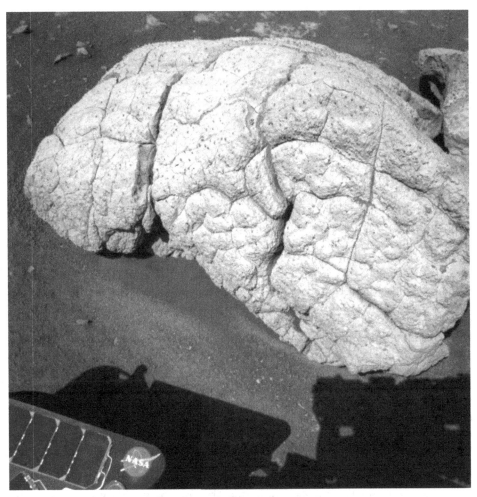

The bizarre 'brain-like' rock Wopmay. (NASA/JPL/Caltech)

A mosaic of 46 images of the Burns Cliff inside Endurance, showing extensive layering. The mosaic was taken from Sol 287 to 294. (NASA/JPL/Caltech)

The impact site of Opportunity's heat shield. The shield itself, at left, had been turned inside out by the impact. (NASA/JPL/Caltech)

Opportunity ended 2004 having driven a total distance of 2,051 meters. During their first terrestrial year on Mars, which ended on the respective anniversaries of their landings, the two rovers had returned in excess of 62,000 pictures in addition to 86 Gbits of other data.

Spirit began 2005 by resuming its drive to Larry's Lookout, a vista point on the Columbia Hills where Mars Global Surveyor imagery suggested that bedrock was exposed. The team had to plan the route very carefully due to the need to keep the solar panels facing the Sun on such a rough terrain. Moreover, a great deal of dust was accumulating on the panels and greatly reducing the power that they generated.[318,319] The rover experienced very high slippage in driving uphill. On the way, it paused for 10 sols to examine a rock named Peace. Because this appeared to be part of the original bedrock, its context was known. Although layered at the centimeter-scale, the microscope showed it to be very coarsely grained at the millimeter-scale. The spectrometers showed the grains to be volcanic basalt inglobated and cemented in a matrix of magnesium and calcium sulfate salts. The next target was a similar rock named Alligator, located only a few meters from Peace. The basalt in both of these rocks was rich in olivine, magnetite and magnesium salts. Altogether these measurements suggested either the alteration of basaltic grains by acid sulfate or evaporation of water. On Sol 388 Spirit resumed its drive to Larry's Lookout. As it did so, scientists noted that the wheels were unearthing bright-colored material and so on Sol 400 the rover paused to analyze this site, which was named Paso Robles. It proved to have a percentage of sulfate salts even higher than in the material at Meridiani Planum, with smaller quantities of phosphates and bromine salts. Spirit finally reached Larry's Lookout on Sol 407, from where it took panoramas of the valley below. It then examined a finely layered rock named Watchtower using its spectrometers and microscope. Rich in phosphates, this had a similar composition to Wishstone but it was more weathered. In early March, the rover's energy output was doubled to 800 W-h, apparently when its solar panels were cleaned by wind or dust devils. Images of the calibration target on the top deck of the rover confirmed it. During this period, moreover, Spirit's camera documented dust devils sweeping across the floor of Gusev below. The increase in power would make traverses less conservative. As the rover's Principal Investigator put it, "we are no longer slaves to the Sun position". On finishing at

A 40-cm wide detail view of bright material churned up by Spirit's wheels in Paso Robles, photographed on Sol 400, or 16 February 2005. The terrain was shown by the alpha-ray spectrometer to contain large percentages of sulfate salts. (NASA/JPL/Caltech)

A panorama of Husband Hill (at the center) and of the floor of Gusev taken by Spirit between Sol 410 and 413, from Larry's Lookout. The rover drove to this point coming from the right and going to the summit of Husband Hill, 200 meters away and 45 meters higher in elevation. The area where the rover tracks make a zig-zag closest to the camera is Paso Robles. (NASA/JPL/Caltech)

This image from Spirit, taken on Sol 532, shows a dust devil scouring the floor of Gusev. The dust devil is the bright column seen right of center just below the horizon. (NASA/JPL/Caltech)

Watchtower, Spirit drove back to an already surveyed site for the first time on its mission; to further investigate Paso Robles to better ascertain the nature of the salts. (The analysis of the first measurements had only just become available.) The thermal emission spectrometer even found signs of water, probably ice captured in the crystalline lattice as hydrated salt. Based on the relatively high concentration of sulfur, phosphorus and bromine, this material appeared to be an evaporite similar to those found by Opportunity in Eagle crater, with chemicals being carried by water and then concentrated by evaporation.[320]

The team next tried to send the rover to the top of Husband Hill, but the route proved too steep. Spirit also tried zigzagging, but the 15-degree slope littered with large rocks interspersed with sand was just too difficult to negotiate. In the end an easier but longer course would have to be found. After many sols trying to drive, during which imaging and trenching were the only scientific activities, Spirit spent some time examining the Methuselah outcrop. After studying the thinly-layered rock Keystone, Spirit devoted several sols to the Jibsheet outcrop, including a 21-hour Mossbauer integration. It then drove to Larry's Outcrop, where it conducted a detailed study of Paros, including a 46-hour Mossbauer integration. Some of these rocks were found to contain ilmenite, an iron oxide with titanium which formed by the solidification of lava quite different in composition to that found elsewhere in Gusev. The rover reached Sol 500 while examining outcrops and soils at Larry's Outcrop. Three sols later it resumed the drive to the summit of Husband Hill, now less than 200 meters away, by going around its perimeter in search of a viable way up the slope. During this drive it analyzed the Backstay rock, which proved to be an unaltered basalt similar to those on the floor of Gusev, albeit with significant differences, and then it stopped to take panoramic views to assist its controllers in route planning.[321] After successfully driving almost 50 meters in two separate sols, Spirit examined the layered rock Independence around the 4th July 2005 American holiday and over the ensuing week. As the teeth of the abrasion tool were wearing out, the rover also used its front-left wheel to scuff the surface of the rock in order to gain deeper access. While examining Independence on Sol 532, Spirit was able to document 9 minutes of the progress of a dust devil crossing the floor of Gusev. As summer approached, dust devils in this area became more common, with their direction changing as the seasons changed. The largest examples were estimated to be 100 meters across and 300 meters tall, lifting 1 kg of dust every second. Based on Spirit's observations, scientists estimated that in Gusev alone should there must be about 90,000 dust devils per sol![322]

As of July 2005, the two rovers had returned in excess of 135 Gbits of data, the majority of which (97 per cent) was relayed by the orbiters; 92 per cent of this via Mars Odyssey and the remainder via Mars Global Surveyor. A relay through Mars Express was also tested as a backup. The direct link to Earth was usually used by a rover only to receive commands at the start of a sol's work. Transmitting directly to Earth took so much power that a rover could overheat. As the mission continued and more dust settled on the solar panels, the power available declined and the use of the direct link to return data would become impractical. But with the dust devils cleaning the solar panels, power was not as critical an issue as had been expected. The real threat was thermal stress and mechanical wear. Nevertheless, except for Spirit's right-front wheel and several minor anomalies both of the MER vehicles were remarkably healthy.[323]

Having spent more than two weeks examining Independence, Spirit resumed its drive for the summit of Husband Hill. Starting on Sol 551 it examined Descartes, Burgeoisie, Hausmann and Assemblée. Containing rounded pebbles centimeters in size embedded in a finer matrix, Hausmann showed signs of a complex evolution. Despite driving mostly on alternate sols in order to preclude draining its batteries during the evening Mars Odyssey passes, Spirit made good progress. On Sol 581 (23 August 2005) it finally achieved the summit. By this point it had driven a total of 4,810 meters. The top of the 100-meter-tall hill was a plateau with an occasional outcrop, a few rocks, and a small summit mound. During the ensuing sols, Spirit assembled a full-color 360-degree panorama and also aimed the camera downward onto its top deck to obtain a self-portrait. The walls of Gusev, 80 km away, were visible through the haze, as was the degraded rim of the smaller Thira whose near rim was 15 km away. Closer in, the panorama showed a number of possible future scientific targets within the South Basin valley on the far side of Husband Hill. In addition to a number of bedrock outcrops, there was the enigmatic bright and flat circular feature named Home Plate. First spotted in Mars Global Surveyor images, this was about 90 meters across and stood several meters high. Seen from Husband Hill, it showed signs of being extensively layered, representing perhaps the largest exposure of layered bedrock yet encountered by either rover. This made it a high-priority target. The area around Home Plate appeared to have suffered substantial erosion, as it was less cratered than was typical for the floor of Gusev crater. The images were also scrutinized for suitable north-facing slopes on which Spirit could be parked during the next winter, if it managed to survive that long.[324] While these activities were being undertaken by the parked rover, the Mossbauer spectrometer integrated an analysis of an undisturbed patch of soil.

Over the next few sols Spirit analyzed targets at Irvine, thought to be a volcanic dike, and Cliffhanger. On Sol 620 it reached the true summit of Husband Hill, 106

A view of the valley south of Husband Hill. The flat, circular Home Plate, where Spirit would spend most of its remaining mission, is visible at center. Right of it is the mound named von Braun and the hollow named the Goddard. The hill visible left of center is the summit of Ramon Hill. (NASA/JPL/Caltech)

meters above the floor of Gusev. Returning to its previous location and downhill to the South Basin, Spirit halted from Sol 625 to 634 to examine the Hillary outcrop, named after Edmund Hillary, one of the first men to reach the summit of Mount Everest. On panoramic images the outcrop clearly appeared heavily layered, but it proved difficult to reach using the arm-mounted instruments because the rover had to be parked in an awkward position.[325] Analysis of panoramic imagery identified a safe route across ridge lines to the South Basin that would not tilt the rover more than 20 degrees. These ridge lines were informally called the Haskin Upper Ridge and Haskin East Ridge. Leaving Hillary on Sol 635, Spirit reached its 5,000-meter distance and then, heading towards the valley, it stopped at the outcrop Kansas to examine a rocky target named Kestrel. Quite long traverses were still being made. For example, on Sol 655 Spirit drove 94.5 meters including both an uphill and a downhill leg, suffering only slight slippage on the slopes. While parked at Larry's Bench several sols later it made a number of night-time observations. Starting in late August 2005 the remarkably clean solar panels were producing enough power during the day to enable the batteries to sustain the computers and the cameras and their heaters at night.

Given the fact that many members of the rover team were either professional or amateur astronomers it was decided to attempt to perform night-time astronomical observations. In particular, by timing transits and eclipses as Phobos and Deimos entered or left the planet's shadow it would be possible to refine the orbits of the moonlets. And observing the sky around the time that the planet was predicted to cross the orbits of known periodic comets would provide new information on the flux of meteoroids, if meteor trails could be detected. A transit of both Phobos and Deimos, with the latter a little more than a moving star, was studied on 30 August. Eclipses of both moonlets were recorded in November and December. Night-time imaging when Mars crossed the orbits of Halley's comet and comet LONEOS was also performed to place constraints on the flux of largish meteors in deep space, as the only available data (from dust detectors on spacecraft such as Ulysses, Galileo and Cassini) was only for very small particles. A total of 353 images were taken, representing a total of over 2.5 hours of observations, but no *bona fide* meteors were seen. Several meteor-like streaks were recorded, but these were more likely caused by cosmic rays striking Spirit's CCDs. In retrospect, the 2004 observation of something which was interpreted at that time as either a meteor or the Viking 2 orbiter was more likely to have been a cosmic-ray strike.[326] On 16 December, the camera was aimed at the Large Magellanic Cloud and obtained the first image of another galaxy from the surface of another planet. On the morning of 29 December it captured Earth and Jupiter raising over the eastern horizon.[327]

Meanwhile, on 21 November 2005 Spirit marked the completion of a local year on the surface of Mars. It was then busy examining the Seminole outcrop prior to driving to study the Algonquin and Comanche outcrops. Seen by the microscope, Comanche showed a granular surface texture. It was reported years later (after the team finally gained time to analyze the scientific data collected by the Mossbauer spectrometer) that Comanche had concentrations of magnesium iron carbonate that closely matched the composition of carbonate granules in the Martian meteorite

A mosaic of panoramic camera pictures of the dune field El Dorado. (NASA/JPL/ Caltech)

ALH84001. Carbonates constituted up to about 25 per cent of the rock's volume. This was therefore a long-overdue identification of Martian carbonates. Moreover, although a volcanic origin could not be excluded, the spectra were well reproduced by carbonates from a non-acidic hydrothermal environment potentially conducive for life.[328] By the end of 2005, Spirit was driving across the El Dorado dune field where its wheels scuffed some of the sand just deep enough for the instruments to inspect the trench. The sand appeared to consist of rounded particles that had a similar composition to a typical Gusev soil, albeit somewhat richer in magnesium and poorer in silicon and chlorine. By this point, Spirit had driven a total distance of 5,829 meters.[329]

On the first day of 2005 Opportunity used its microscope to examine the debris of its heat shield. Engineers at JPL wanted a close look to see how much of the ablative material had been burned off. This insight could assist in designing lighter heat shields in the future. A proposal to abrade part of it was canceled when it was realized that this would produce a shower of very fine powder. Microscope images showed charred ablative material as well as crushed supporting honeycomb. To the surprise of the engineers, some of the plastic thermal blankets which had protected the heat shield in interplanetary space were still in place; they had been expected to burn off early in the entry. The aerodynamic torques created by the pieces that remained could explain attitude disturbances recorded in the descent.[330] In fact, the rover paid a total of three visits to its heat shield, backing up and then closing in on it again from different angles. While the microscope inspected the heat shield, the alpha spectrometer was repeatedly placed over the magnets to integrate a spectrum of typical windblown dust, establishing there to be a resemblance between the dust and the planet's bright albedo features. The presence of olivine implied that water was not involved in the processes which formed the airborne dust.[331]

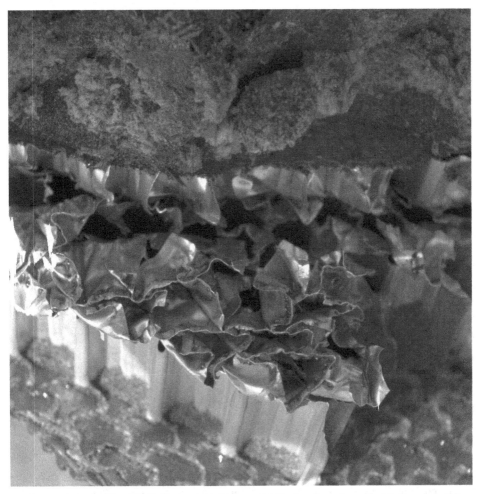

A microscope close-up of Opportunity's heat shield showing charred ablative material (at top) as well as crushed aluminum honeycomb cells. (NASA/JPL/Caltech)

After more than two weeks of inspecting the remains of the spacecraft that had delivered it to Mars, Opportunity drove 10 meters north towards a heavily pitted 25-cm rock which bore a striking resemblance to a metallic meteorite. The thermal emission spectrometer on the mast, and the spectrometers on the arm, confirmed that this Heat Shield Rock was indeed a meteorite, and specifically an iron-nickel kamacite. The fall appeared to be sufficiently ancient that any impression it made on the surface had been weathered away. This meteorite (like others found later by this rover) indirectly showed that the planet's atmosphere was much denser in the past, since air as thin as that today would barely have braked the meteorite, leaving it to shatter on impact; either that or the meteorites fell billions of years ago, were buried, and were then exhumed. Moreover, pits similar to those on Heat Shield Rock are evident in iron meteorites recovered on Earth. These pits are believed to have

The Heat Shield Rock was one of many metallic meteorites identified by Opportunity in Meridiani. (NASA/JPL/Caltech)

originated when iron sulfides contained in the meteorite came in contact with water and the resulting sulfuric acid ate away some of the iron. If this process was at work on Mars, then it would have required the presence of water.

After a brief stop back at the heat shield Opportunity set off in late January on a 1.5-km drive south to a 60-meter circular feature dubbed Vostok. In Mars Global Surveyor imagery Vostok displayed attributes which could be attributed to either a crater or some sort of a depression. On the way the rover was to examine a few small craters, starting with Argo, 300 meters away. Afterwards it was to continue south to investigate a patch of 'etched' terrain. If all went well, Opportunity would then try to cross the etched terrain to reach the 800-meter Victoria crater in which as much as 50 meters of stratigraphy was exposed. This billion-year old impact crater stood on a plateau some 30 meters higher than Endurance. It was informally named after the only ship of Ferdinand Magellan's fleet to complete the circumnavigation of Earth. However, Victoria was 6 km south of Eagle and Endurance, which was a long way even on level terrain. As if to prove its mettle, Opportunity broke its distance record twice during the drive to Argo; driving 154.65 meters on Sol 360 and 156.55 meters on Sol 362. On Sol 365 it observed Strange Rock, then went to dig a trench in Ripple Crest, a spot entirely covered with 'berries' that it studied in detail. Meanwhile, software upgrades gave it new mobility capabilities. In the next few sols this new software allowed Opportunity to exceed the 3,000-meter mark. It then stopped to examine a rock named Russet and several small craters, including Naturaliste. On reaching the rim of Vostok on Sol 399, Opportunity revealed it to be an impact crater filled with dust. It studied a soil target named Laika and a rock named Gagarin. After spending just 6 sols at Vostok the rover continued south to Viking and Voyager, two small craters about 1 km further south. Along the way it established a new series of driving records, culminating with 220 meters on Sol 410; this still stands as the longest distance driven on Mars during a single sol. It reached Viking on Sol 422 and Voyager on Sol 424. In each case it paused only to obtain panoramic images before resuming its drive south for the crater Erebus, some 4 km south of the landing site.

In early April Opportunity entered the etched terrain spied from orbit, where the sand ripples became larger and more spaced. The fact that some small craters were present within the ripples meant the particles had stuck together to form a crust, as otherwise the craters would have been filled in by the wind. It was speculated that the dunes of Meridiani were more static than elsewhere because the wind could not move "the carapace of blueberries" that shielded the finer dust. On Sol 433, in making a turn at the end of a 151-meter drive, the steering actuator of the right-forward wheel jammed, probably as its gear became stuck. Fortunately this wheel seized angled only slightly inward, and the other three steering wheels were easily able to compensate; albeit with reduced precision in positioning. This was not the only hardware problem on Opportunity, because by this time the thermal emission spectrometer often sent back only fragments of data.[332] However, another mishap was imminent that would interrupt the drive south for over a month. On Sol 446 (25 April) Opportunity was some 400 meters from Erebus and on the boundary of the etched terrain. It had been told to execute a 90-meter drive which cut obliquely across ripples several tens of centimeters high and more widely-spaced dunes. The rover was driving backwards as part of the technique to keep a good lubrication on the wheel gears. On driving over a dusty dune that stood 30 cm tall and was a mere 2.5 meters wide, it became trapped in material as fine as flour located beneath the thin crust. After the wheels had spun for the specified 40 meters, Opportunity was to turn and drive forward, but it performed its remaining motions with the wheels spinning uselessly in place, causing them to sink in to about three-quarters of their diameters, which was deeper than they had ever excavated previously. Opportunity 'learned' of the problem only when it took navigation images to verify its heading. It then ceased any additional motion and awaited instructions from Earth. Pictures showed the tracks made as the wheels churned deep into increasingly fine material until eventually the wheels had simply spun in place. The hazard cameras showed that all six wheels had dug into the sand of what was promptly named the Purgatory Dune.[333] Before any action was taken, extensive tests were carried out in JPL's Mars Yard filled with 2 tonnes of sand. In the end, however, the technique chosen was simply to spin the wheels and attempt to back out. An early test of this technique on Sol 463 spun the wheels for two and a half rotations, but the rover moved only a few centimeters. Unfortunately, it was found that Earth simulations could not fully replicate the characteristics of Purgatory, taking only half as many revolutions to move than on Mars. Finally on Sol 484 (4 June), after 38 sols in place, Opportunity regained good traction and was free. Its wheels had spun for 177 meters but it moved only 90 cm and ended up some 2 meters from where it had been trapped.

Opportunity's front hazard camera took this image of the rover's tracks as it got stuck in the Purgatory Dune, driving backwards. (NASA/JPL/Caltech)

At that point it turned to face forward and was carefully sent back to Purgatory in order to study the dune in more detail as a science objective and to ensure that a sand trap like that could be recognized and avoided in the future. It was not clear at that time what had actually halted Opportunity: was it the fine sand, the slope or the height of the dune? A simple algorithm was created and uploaded over the next sols to enable the rover to determine whether it was making progress. It compared two front camera picture taken one-wheel-turn apart. If the view had changed, the rover had moved; if it was identical, no progress had been made and the rover was to halt. Other rules monitored tilt, roll and pitch limits.[334,335] Remarkably, during this hiatus the thermal emission spectrometer appeared to have come back to life. After spending June inspecting the Purgatory Dune, Opportunity took panoramic images that identified a path to the east which would cross only very small ripples on the way to a broad trough in which Opportunity could resume rolling south to Erebus in early July. It would drive along ripple troughs, with images being used to determine where it would be safe to drive over a crest to get from one trough to the next. Driving along this Erebus Highway, the rover began to encounter small rocks and pebbles after having spent many sols crossing only dunes and ripples on the exposed bedrock. By early August 2005, several tens of meters from the crater, the rover entered a cobble field, some of which it paused to abrade and examine. It took up position to examine the Fruit Basket outcrop but after several sols of this work it suffered a software reset on Sol 563. Opportunity awakened autonomously just after the reset, but controllers kept it essentially idle for several sols while they were troubleshooting the problem. In particular, the thermal emission spectrometer was not used because this was suspected of malfunctioning and causing the reset.

Opportunity finally resumed rolling down the Erebus Highway on Sol 576. By Sol 590 pictures showed the interior of Erebus. The plan had been for the rover to enter the crater if worthy targets were identified within it. However, on Sol 596, while the rover was studying some rocks along the way, its computer rebooted and entered safe mode, missing communication windows with Mars Odyssey. It was recovered within two sols and driven nearer the rim of Erebus along a zigzagging route that circumnavigated counter-clockwise the western rim. By the beginning of October 2005 the rover had passed the 6,000-meter mark. Another software reset occurred on Sol 610 but was soon recovered. At one point the software uploaded after the Purgatory Dune was invoked when wheel slippage exceeded 40 per cent and a different way had to be plotted. The rover continued driving along the rim of Erebus in an area of sand dunes until Sol 628 when it was obliged to enter safe mode because a major dust storm had reached Meridiani Planum and the opacity of the atmosphere increased to such an extent that the solar power available fell to the level which invoked the self-protection software. Opportunity spent many sols examining outcrops along the northwestern rim of Erebus. But then on Sol 654 the shoulder joint motor of the arm stalled as it tried to deploy. It took two weeks to diagnose the fault, which was a broken wire in the motor winding, and to devise a workaround. To prevent a recurrence it was decided that henceforth Opportunity would be driven with the arm's elbow joint projecting out in front instead of being completely stowed. While the glitch was being investigated, the static vehicle built up a color panorama and made atmospheric observations. On 12 December it too completed one local year on the surface of the planet. By the end of 2005 it had driven a total distance of 6.5 km plus a few meters.[336,337]

Spirit began 2006 by wrapping up its study of the El Dorado dunes and making slow progress towards Home Plate despite excessive wheel slippage on sandy soil and brake anomalies on two of its steering actuators. Along the way it paused to examine Arad, where it found white soil of a powdery appearance that contained iron-bearing sulfate and other salts. With a composition similar to Paso Robles, it provided another salty clue to the ancient presence of water on the planet. On Sol 735 a diagnostic test showed that the brake problems could safely be ignored. The rover then encountered a strangely jagged rock and used its microscope to study it in detail. Although this rock, named Gong Gong, looked remarkably like the Heat Shield Rock meteorite at Meridiani Planum, it was probably a frothy gas-rich lava that had been eroded by wind and dust. On Sol 744 (5 February) Spirit reached the northwestern edge of Home Plate, 94 sols after leaving Husband Hill. Home Plate was a thick stack of layers, with the bottom unit, named Barnhill, consisting of alternating coarse and finer layers embedding millimeter-sized pebbles. The upper unit, Rogan, was more finely grained. On the topmost 10 cm, the layering became indistinct and could not be resolved. The upper layers exhibited complex characteristics typical of wind deposition. Although just 3 meters tall, Home Plate showed some of the most extensive, intricate and spectacular layering thus far seen by Spirit, leading scientists to refer to it as the 'Burns Cliff of Gusev'. Microscopic images could not help to decide the nature of grains. Rounded grains in the lower layers could be lapilli, and their infrared spectra implied the presence of volcanic glass.

A number of rocks were hurriedly analyzed before Spirit was sent to spend the winter on a north-facing slope on McCool Hill. Targets Posey and Cool Papa Bell on the upper layers were ground prior to analysis by the alpha spectrometer as the rover climbed up onto the small plateau. It then drove clockwise around the rim of Home Plate, pausing at the small irregular loose rock Fuzzy Smith which could not be abraded. Mossbauer analyses revealed this to have a unique mineralogy which could indicate volcanic acid leaching of the originally basaltic pebble. Home Plate rocks all appeared to be basalts with a high alkali content; not too different from others found elsewhere in Gusev crater, although somewhat richer of some minor elements such as chlorine and bromine. Amongst other things, the mast-mounted thermal emission spectrometer found traces of sulfates. Both it and the Mossbauer spectrometer detected olivine, pyroxene and magnetite. All this implied a volcanic origin for Home Plate. Furthermore, chemical analyses hinted at a hydrothermal explosive mechanism. Perhaps, it formed when alkali-rich basaltic lava came into contact with water, ice or wet sediments. Images of the lower layers of the plateau showed at least one 4-cm basaltic 'sag' where a 'bomb' landed and deformed the layers of ash beneath. It was not possible to state whether Home Plate itself was originally a volcanic vent, but it must have been quite near to one. It was possible that one of the nearby layered hills was the actual vent.[338]

Thirty sols after reaching Home Plate, Spirit started to back away from it, since a

Spirit acquired this mosaic of the edge of Home Plate during February 2006. The layering of the wall, only 2 meters tall, is evident, as is the different texture between the fine grained top and the coarse grained bottom. (NASA/JPL/Caltech)

detailed study would have to await the spring. It headed for McCool Hill, a few hundred meters to the southeast. Having suffered intermittent problems since mid-2004 the actuator of the front-right wheel stalled on Sol 779 (13 March 2006) and would not resume working. While engineers investigated this, the rover was driven on five wheels. However, this limited its mobility typically to less than 10 meters per sol. Next it encountered an impassable sandy patch which blocked its route to McCool Hill. The team had it head back towards Home Plate in search of a north-facing slope of at least 16 degrees that would maximize the insolation on the solar panels during the winter. Spirit was parked from April to November 2006 on the flank of a hill in a complex terrain named McMurdo owing to its resemblance to a landscape in Antarctica, and the rocks and soil targets were named after Antarctic stations and explorers. Spirit had traveled 6,876 meters since landing. From there it built the highest-resolution panorama of Mars yet, consisting of no fewer than 1,449 images taken over a period of 88 sols. It also took images and spectra of the soil targets Progress and Halley which were within reach of the arm, one disturbed by the wheels and the other undisturbed, and observed the atmosphere. Otherwise, it "acted like a lander". It characterized the narrow Mitcheltree Ridge to the east of Home Plate for possible exploration in the spring. It and nearby Low Ridge were volcanic deposits that included lapilli stones and were capped with basaltic rocks. Scientists were particularly interested in collecting images and spectra of Tyrone, a patch of white-yellowish material that was exposed by Spirit's failed wheel on Sol 784, because it looked like a hydrated salt. From its parking position, Spirit also spotted two boulders that looked remarkably similar to Opportunity's Heat Shield Rock; they were probably meteorites and were named Zhang Shan and Alan Hills.

By the southern winter solstice on 8 August 2006, Spirit was surviving on less than one-third of the power available in the summer. It was spending every second sol

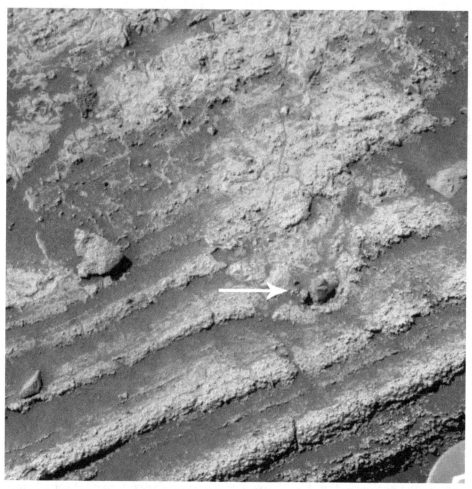

A volcanic 'bomb' embedded in the lower layers of Home Plate. (NASA/JPL/Caltech)

simply recharging its batteries. The fact that the battery heaters were switching on automatically meant that the temperature of the interior of the body had fallen below −20°C. While Spirit was sitting out the winter, a new software release was uploaded to both rovers. Part of the motivation was to assess some of the features planned for the 2009 Mars Science Laboratory rover, including image recognition, map-building hazard-recognition systems, and selecting targets for examination by instruments without prompting from Earth. Moreover, it would sift through images of the sky to determine which ones showed clouds or dust devils and mark them for sending to Earth. Communications with both rovers were intermittent from 16 October to 10 November as the planet passed through solar conjunction. Although the rovers did not receive commands, they performed some pre-programmed tasks and collected an average of 15 Mbits of data daily. On 26 October, Spirit marked 1,000 sols of operation.[339]

On Sol 779, the actuator of Spirit's front-right wheel stalled and the rover had to drive dragging along the wheel, resulting in the track visible at right in this image, taken two days later. (NASA/JPL/Caltech)

Spirit restarted rolling on Sol 1,010 (5 November) with a 33-degree turn and a drive of 0.71 meters in order to reach some interesting material that its wheels had disturbed over 200 sols earlier. When it resumed longer drives in early December it was sent to a nearby rock named King George Island. This showed some of the most rounded grains yet seen on Mars, possibly formed from particles ground by wind or water. This examination provided an opportunity to test the new software. Spirit then moved on, very slowly due to the failed wheel, towards a rock outcrop named Esperanza. On Sol 1,061 (27 December) Gusev was suddenly engulfed in a dust storm. The rover suspended scientific activities and drove to a shallow slope in order to better face its solar panels at the dim Sun. At the onset of this storm, the solar panel output had reached the all-time low of 267 W-h.

As 2006 began Opportunity was able to resume using its arm, which it did by

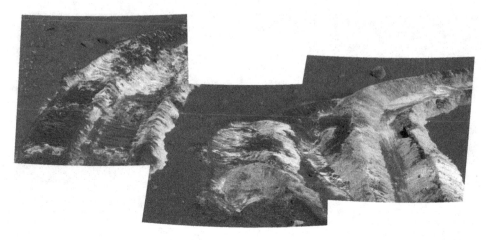

Bright material unearthed by Spirit's wheels in Tyrone. (NASA/JPL/Caltech)

examining the targets Hunt and Ted. It then drove to the rock Overgaard, located in the Olympia outcrop near the rim of Erebus crater. It was the best example yet of centimeter-size festoons, which are rocky sedimentary ripples formed by waves lapping on a beach. Due to a series of glitches with the arm actuators, an extensive study by the microscope and other instruments took longer than expected.[340]

After more than 30 days at Olympia, Opportunity proceeded around the western rim of Erebus to the Payson outcrop less than 100 meters away. Like Overgaard, Payson had some very well preserved ripples. It finally left Erebus on Sol 760 and headed southeast for Victoria crater, trying to avoid crossing a large dune field that was situated due south. Along the way it examined a series of outcrops and rocks including Brookville, Pecos River, Cheyenne and Pueblo. On Sol 833, within 1 km of Victoria and safely in a trough between the crests of two ripples, Opportunity was ordered to make a 24-meter drive, but after traveling just 1.5 meters it rolled into another soft dune. This time the algorithm designed to detect precisely this occurrence successfully halted the rover before its wheels could become buried in the sand. Extraction was relatively easy, and by Sol 841 the vehicle was free. This dune was named Jammerbugt after a bay on the coast of Denmark known for its many shipwrecks, and to honor that nation's contribution to the MER project. The size of the berries had diminished as the rover traveled south of Vostok, but their number increased and their shapes became irregular. But on approaching Victoria the largish spherules reappeared. The berries were probably ejecta from the crater but there was no other ejecta because the soft material would have long-since been eroded by the wind. Where blueberries were lacking, sand ripples became evident, and vice versa. Reaching the edge of the ejecta blanket, it detoured to the 35-meter Beagle crater to inspect some outcrops. After examining several rocks Opportunity reached Beagle in early August 2006. There it used its abrasion tool for the first time in 200 sols. It spent several sols standing on the rim, obtaining full-color panoramas that showed the interior to be extensively layered. After two weeks, Opportunity departed Beagle and resumed its drive for Victoria. It paused for several sols after suffering another

Fossilized shallow water ripples and mud cracks in Payson. These suggested that the rock formed from sediments on a beach with lapping waves. (NASA/JPL/Caltech)

arm stall, and then again to inspect the small crater Emma Dean which was located 50 meters from the rim of Victoria. It reached the rim of Victoria on Sol 951 (27 September 2006), shortly prior to the October solar conjunction. It had driven a total of 9,279 meters since landing.

As already observed in orbital images, Victoria had a complex jagged rim made of alternating 'promontories' and 'gulfs', which were named after the capes and bays discovered by Ferdinand Magellan during the first circumnavigation of Earth. The promontory nearest to Opportunity was named Cape Verde, and it provided an excellent vista point for taking a panorama across Victoria. The cavity was about 750 meters across and 75 meters deep, and had steep, almost vertical promontories along the rim separating rounded, gently sloping bays. The promontories exposed the stratigraphic record for tens of meters. There was intact bedrock near the base of

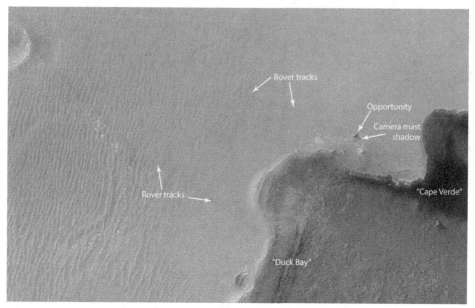

Opportunity on the rim of crater Victoria seen by the high-resolution camera on the Mars Reconnaissance Orbiter on 3 October 2006, five days after it arrived at the crater. (NASA/JPL/University of Arizona)

A mosaic of images of the Cape Verde promontory taken from Duck Bay. This layered promontory was the target of Opportunity's excursion into the crater during 2007. (NASA/JPL/Caltech)

the exposure, with the rock above fractured by the impact shock. On the floor of the crater was a field of sand dunes. The walls appeared to have been widened by wind erosion. In fact, the original crater was estimated to have been no more than 600 meters across.[341] Plainly visible to the south of the rover was the extensively layered 15-meter-tall Cabo Frio promontory and a small crater close to the rim that was named Sputnik.

On 3 October, while Opportunity was assembling a full panorama of the crater, both it and its tracks were imaged at an amazingly high resolution by the recently arrived Mars Reconnaissance Orbiter from an altitude of 297 km. By combining the panorama taken by the rover with the overhead images and the 3-dimensional model derived from them, the rover's controllers would better be able to command its operations. Scientists wanted to send Opportunity inside Victoria, as they had at Endurance. It was decided to survey interesting features on the northern rim of the crater while searching for a suitable route in. Even if a route offering a reasonable chance of escaping from the crater was not found, the team would probably send the rover in any way since such massive exposures would provide context for the Burns Cliff and, as such, were too good to miss.[342,343] But first Opportunity had to stand down for the solar conjunction. It remained in place on Cape Verde, with its Mossbauer spectrometer positioned to collect data on the target Cha.

On Sol 992, after contact was re-established, Opportunity started its clockwise tour of the northern rim by heading to Cape St. Mary, from which it would be able to image the northeast-facing cliff of Cape Verde. On the way there, it took stereo images every about 10 meters for a complete 3-dimensional reconstruction of that part of the crater. Images of Cape Verde clearly showed the demarcation between bedrock and rocks shattered by the impact. In contrast, Cape St. Mary revealed the stratification of 'petrified' sand dunes. On 16 November Opportunity reached its 1,000th day on the surface of Mars. After it finished imaging at Cape St. Mary, the rover drove to the next gulf, named the Bottomless Bay, which was judged to be a good entry point into the crater. Over the holiday season it examined the rock Rio de Janeiro, and by the end of the year was crossing the promontory named Cabo Anonimo heading for the rock named Santa Catarina.

The beginning of 2007 found Spirit still engulfed in the dust storm and short of power. Nevertheless, it monitored the storm by measuring the transparency of the atmosphere. The sol after the third (terrestrial) anniversary of its landing, the rover was finally able to move to the layered outcrop named Troll. Three layers of Troll were analyzed using all of the available instruments: near the bottom (Montalva), the middle (Riquelme) and the top (Zucchelli). While parked at this position, the rover documented a number of eclipses of Phobos. It also tried to image the bright comet McNaught in the predawn sunlight but this proved impossible. Spirit was next sent to analyze the bright terrain that its wheels had exposed at Tyrone. To the thermal emission spectrometer this appeared to contain deposits of hydrated ferric sulfate. However, to preclude the risk of getting stuck in soft soil the team did not send the five-wheeled rover directly to its objective. With the onset of local spring in early February Spirit was able to resume exploring Home Plate, which was then some 50 meters away. The plan was to make the preliminary assessment of Home Plate from

An amazing detail of the wall of the Cape St. Mary promontory seen from Cape Verde. Cape St. Mary showed the stratification of an ancient, fossilized dune field. (NASA/ JPL/Caltech)

a distance and then drive along the narrow 'valley' to the east and study the relationship between Home Plate and Mitcheltree Ridge immediately beyond, in particular to determine whether they were part of the same feature. The floor of the valley contained several flat outcrops of a different color to their surroundings that could be ash deposits. For reasons that will soon be explained, this site gained the moniker Silicon Valley. Along the way, the rover stopped at the rocky outcrop Bellingshausen. During March it studied a number of rocks on Mitcheltree Ridge. The sampling targets were named after the 'pulp fiction' Barsoom saga by Edgar Rice Burroughs, which was set on an inhabited Mars. One, Torquas, was abraded and analyzed by both of the arm-mounted spectrometers. John Carter was analyzed only by the alpha-particle spectrometer. Analyses of the rocks King George, Troll and Torquas revealed a high percentage of potassium and the microscope spotted small basaltic spheres. While scientists were busy examining Mitcheltree Ridge, the rover's controllers looked for a way to access Home Plate from the southeast. First, however, it made a stop at the pockmarked outcrop Madeline English on the southern edge of Mitcheltree Ridge.

Owing to its failed right-front wheel, Spirit was driving at most several meters per sol. Dragging the stuck wheel proved a blessing in disguise because it scraped a trench several centimeters deep, excavating below the superficial layer of dust. On Sol 1,148 (27 March) it unearthed bright soil. Aiming the thermal emission spectrometer directly at the trench the spectrum of silica clearly stood out, so the scientists asked that the rover be driven back and analyze it using the arm-mounted spectrometers. Before this soil, dubbed Gertrude Weise, could be investigated in detail, Spirit was sent to analyze an outcrop made of light-toned rocky fragments

Opportunity's track during its exploration of crater Victoria between 2006 and 2007, superimposed on a Mars Reconnaissance Orbiter high-resolution view of the crater. (NASA/JPL/University of Arizona)

dubbed Elizabeth Mahon and the Madeline English outcrop. Sample targets in this area were named after players and teams of the All American Girls Professional Baseball League. The interest in these two outcrops derived from the fact that the thermal emission spectrometer saw hints of silica masked by dust and dirt. It took many sols to reach Madeline English because the rover required to make a number of maneuvers while dragging its stuck wheel, but it finally arrived on Sol 1,166. In fact, both of these outcrops proved to be rich in silica: up to 72 per cent in the case of Elizabeth Mahon. On the way to Gertrude Weise, Spirit stopped to examine the knobby rock named Good Question. By Sol 1,187 it was finally able to place first the Mossbauer spectrometer and then the alpha-particle spectrometer on the target Kenosha Comets in the Gertrude Weise material. Many hours of integration built up

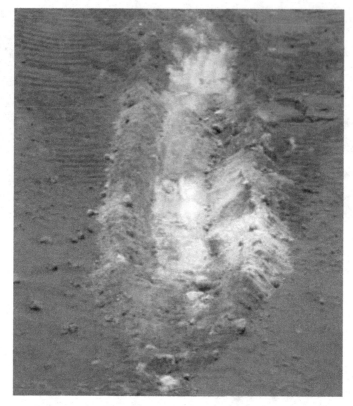

Bright, hydrated silica-rich terrain unearthed by Spirit's stuck wheel at Gertrude Weise. (NASA/JPL/Caltech)

over several sols showed that this bright material consisted of almost pure fine-grained hydrated silica.

Scientists then decided to investigate a number of targets in the Silicon Valley between Home Plate and Mitcheltree Ridge. Operations were aided by dust devils which once again cleaned dust off Spirit's solar panels, restoring the power supply. Whilst continuing to study the stratigraphy of Home Plate several meters away, it examined the rock outcrops Betty Wagoner, Elizabeth Emery and Nancy Warren. It even 'stepped' on some outcrops in an effort to break loose fragments, but was unable to do this. When it unintentionally crushed a small rock this was named the Innocent Bystander. The surfaces of freshly exposed fragments were analyzed and found to be rich of silica, demonstrating that this was present in rocks to a depth of at least several centimeters. The extensive silica deposits and the close association with volcanism made it very likely that Home Plate was formed in hydrothermal conditions by the leaching of basaltic materials by acid. The presence of titanium imposed a constraint on the acidity of the water in which the silica formed: it must have had a relatively low pH. Of course, what was particularly interesting about this discovery was that hydrothermal conditions such as those found in fumaroles and hot springs on Earth constitute favorable environments for microbial life. In fact, on Earth, silica would be a good indicator of habitable conditions and it might even preserve traces of fossils and biological chemistry.[344]

In late June a large dust storm erupted with several local billowing cells, some extremely vigorous. As a result the opacity of the atmosphere increased, reducing the solar power available first to Opportunity and then to Spirit. The storm raged until late August. At its worst, only 1 per cent of the sunlight reached the ground. Not only was there less light, dust accumulated on the solar panels of both rovers. The temperatures plunged by 20°C during the day and even more at night-time.[345] In order to survive, a rover required its solar panels to provide at least 60 W-h, and 128 W-h was available. The storm also impacted the scientific productivity of the single European and two US orbiters. On Sol 1,294 (24 August) Spirit exceeded Viking 2's operational life on the surface of the planet. Finally, at the end of August the storm began to clear. For Spirit the only problem encountered afterward was dust on the optics of the cameras. As the dust had removed most of the rover's tracks, one of its first tasks after the storm abated was to re-dig a trench at Gertrude Weise. By scraping away the recently deposited dust to re-expose the pure silica which had been previously examined by the thermal emission spectrometer, it would be possible to determine the degree to which the optics had been coated by dust during the storm, so that this signature could be subtracted from future measurements.

Spirit resumed driving to Home Plate fully 18 months after leaving it, and by Sol 1,306 it was ascending a ramp essentially due east of the circular plateau. The plan was then to drive more or less clockwise around its rim in the coming months, simultaneously conducting in-situ investigations and looking for a suitable parking place for the third Martian winter. On Sol 1,321 Spirit completed the longest dash yet on five wheels, heading to a platy rock surface on the southern edge of Home Plate that had been named Texas Chili, where it was to spend several sols making an in-depth analysis. Next it drove to a nearby field of boulders that showed some spectral similarity to rocks found on Husband Hill. While surveying these rocks, which were individually named after mountains in Colorado, the rover passed the milestone of 2 local years on the surface of Mars. While brushing Humboldt Peak the abrasion tool suffered some glitches, which in turn reduced the effectiveness of the arm-mounted instruments. A new five-wheel driving record was established on Sol 1,363 as Spirit reached what was described as the 9 o'clock point on Home Plate, where it investigated the rocky target Pecan Pie. Then, as Spirit was driving to the northern side of the plateau, its dragging wheel snagged on a rock causing the rover to pivot and enter a small crater named Tartarus that was filled with very fine dust – and hence was 'dangerous ground'. It took from Sol 1,380 to 1,387 for the vehicle to struggle out, which, ironically, it did by exploiting the drag of the stuck wheel. By mid-December it was safely parked on a 20-degrees slope on the north side of Home Plate, ready for winter.

At the beginning of 2007 Opportunity was still at Cabo Anonimo, one of the promontories on the rim of Victoria. On the very first day of the year it analyzed Santa Catarina, which, at 14 cm across, was the largest of a dozen bright rocks that

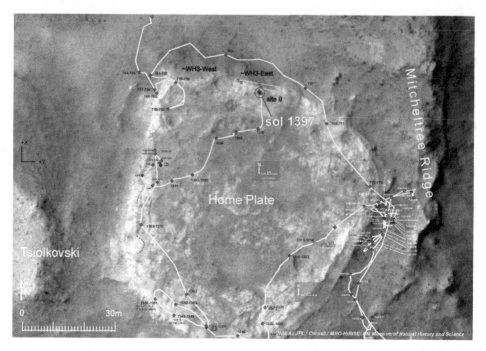

A map of the circuitous route of Spirit's exploration of Home Plate up to the end of 2007, superimposed on a Mars Reconnaissance Orbiter high-resolution view. (NASA/ JPL/University of Arizona)

all had similar spectral characteristics. It was unusual for Meridiani by being rich in troilite, an iron sulfide, but was compatible with a kamacite meteorite. This type of meteorite is relatively rare, and the fact that Santa Catarina at Victoria and Heat Shield Rock at Endurance were found within several kilometers of each other was an indication that they were part of the same 'fall'. Next Opportunity headed to the Bay of Toil and Cape Desire, the next gulf/promontory around the northern rim. It reached Cape Desire in early February, then moved on to Bahia Blanca and Cabo Corrientes to the east. On Sol 1,080, while performing a 50-meter drive between Cape Desire, Bahia Blanca and Cabo Corrientes, the odometer hit the 10-km mark. The rover then spent a week on the rim of Cape of Good Hope prior to driving to the Valley Without Peril. On the way it took detailed 3-dimensional panoramas in order to assess entry points into the crater. Like Cape St. Mary, Cape St. Vincent incorporated extensive ancient dune stratification. In addition to studying targets from a distance using the thermal emission spectrometer and viewing some rocks using the microscope, near the Valley Without Peril the rover performed a number of spectroscopic analyses of the prominent dark rays that could be seen in orbital images and appeared to originate from the rim. It then moved to Tierra del Fuego, the next promontory around the rim, and took stereoscopic images looking back at Cape St Vincent. By this time the rover had traveled about 90 degrees clockwise around Victoria and it was evident that there were only a few points into the crater which sloped less than 25 degrees. The most promising was Duck Bay, near where Opportunity made its initial approach to the crater, between Cape Verde and Cabo Frio. It was therefore decided to return to Duck Bay. Meanwhile, on Sol 1,157 (26 April 2007) Opportunity surpassed the distance traveled by Lunokhod 1 on the Moon. The drive back to Duck Bay began on Sol 1,160 and took a route that kept well away from the rim of the crater, apart from when it made an detour onto the Cape of Good Hope where it exercised its abrader, long unused, on a rocky target named Viva La Rata in preparation for an analysis using the alpha-particle spectrometer. The rover also examined Cercedilla, a rock several tens of centimeters in size near the Cape of Good Hope in which there were large blueberries. After being abraded, this was examined by the microscope and alpha spectrometer. It was found to have a similar chemistry to rocks from the deepest portions of Endurance, implying that it represented material excavated by the impact which made Victoria. The occasion was also taken during the drive back to Duck Bay to test a number of new driving algorithms and techniques. Images were taken of the drill bit of the abrader to estimate its remaining cutting ability.

Opportunity returned to Duck Bay on Sol 1,215 and the plan was to enter the crater on 9 July. The first objective was to be a 1-meter-thick whitish rocky layer that cut through the wall of the crater "like a bathtub ring". It was 12 meters from the rim, and appeared to represent the surface at the time of the impact. Analysis of this material would hopefully reveal the state of the surface and the atmosphere at that time. The idea was to drive as straight as possible to it without taking any detour or stopping for other investigations unless there was a compelling reason to do so. However, just like Spirit on the other side of the planet, Opportunity was hit by the dust storm of that summer. At first it seemed that the rover would be able to attempt its descent in mid-to-late July, after the passage of the storm. The power output was slow to stabilize, as sol-to-sol dustiness changed both significantly and unpredictably. Although at times the storm showed signs of abating, it continued to obscure the Sun. Power margins started to increase in the second half of July, and by the 23rd of that month the batteries were recharged to almost their maximum capacity.[346,347,348] It was not until September that conditions were again favorable for driving. But first the instruments were checked for possible infiltration by dust, and recalibrated. The cover of the microscope was opened and closed as the arm was shaken, in order to clear dust away from its optics. The microscope then took images of the mirror of the thermal emission spectrometer, showing this to have been heavily contaminated by dust.

Finally ready, on Sol 1,291 (11 September) Opportunity made a 'toe dip' test to assess wheel slippage on the slope at Duck Bay. Two sols later it drove into the crater heading for Steno, the uppermost of three 'bathtub ring' layers visible in the wall. The lower layers were Smith and Lyell. Slowly driving on the steep slope it reached Steno on Sol 1,305, carefully unstowed its arm, brushed the rock and built up spectra and took other data over the next several sols. The team took every care not to unbalance the tilted rover while moving the arm and the instruments that it carried. They also had to contend with a malfunctioning arm shoulder joint motor. After more than two weeks spent examining this first location on Steno, the rover moved

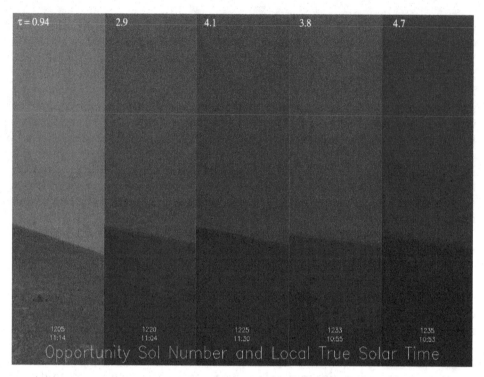

A sequence of images taken by Opportunity during the dust storm of mid-2007. The 'tau' coefficient is a measure of atmospheric opacity. (NASA/JPL/Caltech)

to a second. Fully one month after entering Victoria it reached Smith. But before it could study Smith, engineers had to solve the problem of how to operate the brush and rock abrader after the loss of two of the encoders used to control the motors. It was found that simply monitoring the current drawn by the motors gave an indication of whether the cutting bits were in contact with rock or not. Another engineering activity carried out during this visit to Victoria was to test the multiple relay 'constellation' envisaged for the Phoenix lander that was to arrive at Mars in May 2008. For this exercise, the orbits of the prime relay (Mars Odyssey) and of the secondary relays (Mars Express and Mars Reconnaissance Orbiter) had been synchronized to pass over Victoria. While the problematic drill and arm azimuth joint were being studied, the rover took panoramas looking up at the Cape Verde promontory, and images of Cabo Frio and other promontories. In mid-December Opportunity wrapped up its analyses of targets in Smith and drove some 7 meters further to reach Lyell, the lowest layer. It was still collecting spectra there at the end of the year.[349]

The fall and winter of 2008 were very difficult for Spirit: with its solar panels producing as little as 225 W-h, one-third of that at the start of the mission; with a series of minor mishaps; and the need to preserve such heat as was available. In order to save power, data was stored onboard until the memory was nearly full and then uplinked to Mars Odyssey in one session. The rover moved only a few meters during this time, onto a better winter parking spot that sloped 30 degrees to the north. However, it took several panoramas, in particular of Tuskegee, so-named after the 'Tuskegee airmen', a blacks-only squadron during the Second World War, and of Bonestell, after Chesley Bonestell the creator of many classic 'space paintings' of the 1950s. Long-integration-time spectra of the rocks Chanute and Wendell Pruitt were obtained, as well as of the patch of soil Arthur C. Harmon. Then the arm was placed in a hibernation position in which it would not hinder the later movement of the rover in the event that the articulation mechanism failed during the winter. On 22 March the rover reached the 1,500-sol milestone. Before winter solstice, which occurred on 25 June, it suffered a curious mishap which put it into safe mode for a few sols. On Sol 1,547 it did not receive commands uplinked from Earth because the transmission was blocked by the Moon crossing the line of sight between Earth and Mars. The situation started to improve only in October, when operations were slowly ramped up. In particular, seasonal changes of argon in the atmosphere were monitored by the alpha-particle spectrometer. Before the rover could start to drive towards von Braun, which was a small but steep mesa south of Home Plate, and a 'bowl' named Robert Goddard on the flank of McCool Hill, it would have to pass through another solar conjunction in December. Scientists hoped to discover more evidence of water and volcanism around von Braun and Robert Goddard, the latter resembling a volcanic explosion pit.

The von Braun mesa south of Home Plate that was to be Spirit's next target, but the rover never managed to reach it. (NASA/JPL/Caltech)

Congressional cuts and the shift of NASA priorities towards human spaceflight cast doubt on extra funding for the rovers. The NASA Administrator decided that in spite of overruns by many projects, the budget for science programs would have to remain flat in order to pay for the early work on developing the successor to the Space Shuttle. When a $4 million cut in the Mars Exploration Rover budget was posited in order to cover a small part of a $200 million overrun of the 2009 Mars Science Laboratory rover, JPL said that this would mean terminating either Spirit or Opportunity. This outraged the public and Congress, and resulted in the head of NASA's Scientific Mission Directorate, Alan Stern, resigning after just one year in his post.[350,351]

On Sol 1,709 (23 October) Spirit resumed driving, inching back uphill onto the plateau of Home Plate. Then just as the team prepared for the conjunction, another regional dust storm swept the site with the atmosphere becoming so opaque that on Sol 1,725 the energy production fell to a new low of just 89 W-h. All activity was suspended. To increase the chances of survival a number of heaters were switched off, including that of the thermal emission spectrometer. On regaining contact after the storm and conjunction Spirit was found to be in a very good shape, although more dust had accumulated on its solar panels. Owing to the difficulties of driving up onto Home Plate, where scientists would have liked to continue the exploration of the boulder fields at its southern edge, it was decided to drive downhill and pass around the eastern side of Home Plate to reach Goddard and von Braun, and this is how Spirit ended the year.[352]

At the beginning of 2008 Opportunity was still examining Lyell, the lowermost of the three 'bathtub rings' in the wall of Victoria crater. All three layers proved to be similar in composition, but had different textures and grain sizes. The top layer, Steno, was fine-to-medium grained and finely layered with abundant blueberries. The middle layer, Smith, was smoother and lighter-toned with even finer layering. Lyell was darker, very likely because it contained almost-black basaltic sand.[353] The sulfur-bearing compounds in all three layers implied that the water must have had a very high salinity; higher than any known terrestrial micro-organism is able to withstand. On finishing its study of Lyell around the fourth anniversary of its landing on Mars the rover progressed deeper into Victoria, first to the Buckland outcrop and then to another layer, the lowest visible in Duck Bay, named Gilbert, which it reached in late February. In the microscope Gilbert resembled Lyell, but without layering. However, the Mossbauer analysis showed it to be richer in iron in the form of hematite than any Mars rock yet measured. Minerals in the lowest layers of Victoria therefore appeared to have formed in less acidic conditions. As data from the Mars Express orbiter had suggested, possibly the water had become ever more acidic as volcanism injected sulfur into the atmosphere. If life had ever developed on the planet, the increasing acidity of the water may well have killed it off.

In late March Opportunity started to drive toward the Cape Verde promontory, some 30 meters away on sandy terrain. Just after passing the 1,500-sol milestone, the actuator of the shoulder joint of its arm failed. A new 'stowed' position for the arm was developed in which it was held mostly unstowed in order to minimize the stress on this joint whilst also providing a clear field of view for the cameras and sufficient ground clearance. While this issue was being sorted out, the rover took some time to make panoramas and to study cobbles and the trenches which it had scraped. After 45 sols of troubleshooting it resumed driving on Sol 1,547. But the route to Cape Verde required the rover to drive over soft ground where its wheels tended to dig in. Finally, on Sol 1,557, it again had all its wheels on a rocky ledge. Other routes were devised to approach Cape Verde. Care had to be taken not only to avoid soft ground but also to avoid the shadow of the promontory falling on the solar panels for a significant fraction of the time. In early July the rover started to drive upslope to a flat rock named Nevada, which it was to try to reach by using a zigzagging path through the sand. But it never made it because the motor of the left-front wheel started to malfunction on Sol 1,600 and engineers reasoned that if it seized, as it had on Spirit, then Opportunity would probably become stuck inside Victoria. It was decided instead to exit the crater. On Sol 1,634 (28 August), after negotiating sandy and steep terrain, the rover climbed out of Victoria having spent 340 sols inside it.

The team decided to attempt an amazing excursion that would require a drive of 12 km over the period of a full Martian year, in the process more than doubling Opportunity's odometry. Its target was Endeavour, a 22-km-diameter crater to the southeast whose 300-meter depth promised to expose layers that would highlight even more of Meridiani's history. As seen by the high-resolution camera of Mars Reconnaissance Orbiter the northwestern rim of Endeavour, which Opportunity would approach, showed layering with fracture patterns up to several meters thick. In these areas infrared spectra showed that the rover would probably find patches of smectite iron-bearing and magnesium-bearing clays produced by the interaction of olivine with water. Unlike sulfates, clays would have formed under less acidic conditions that would have been more conducive to the development of life.[354] The thermal emission spectrometer could be particularly useful in identifying clays, but it had been blinded by dust adhering to its mirror during the 2007 storms and was still unusable. Opportunity set off with a 10-meter drive on Sol 1,659. On the way it observed the promontories Cape Victory, Cape Pillar and Cape Agulhas on the western side of Victoria and the small crater Sputnik on its rim. Abundant solar power and the flat terrain enabled it to rapidly leave Victoria behind. On Sol 1,691 it drove 216 meters in a single sol; its second-longest trek.[355] It stopped only once prior to standing down for the solar conjunction, and that was to examine a cobble named Santorini. There was a radio hiatus from 30 November to 13 December, and then, on finishing with Santorini, the rover drove to the Crete outcrop.

Spirit started 2009 by finally driving downslope on Sol 1,782 (6 January), thus departing Home Plate for good. It then wrapped up its scientific observations for several sols prior to heading south to Goddard and von Braun on the eastern side of Home Plate. On Sol 1,800, however, just as Spirit outlived its planned mission duration by a factor of twenty, it refused to move, did not write that sol's activities in its non-volatile memory, and had difficulty in locating the Sun. Although it was subsequently able to resume driving, several sols of diagnostics failed to reveal the source of the glitch. Shortly thereafter, the planned road proved to be blocked by loose soil and Spirit was directed to another route that bypassed Home Plate to the east. An attempt to drive up-slope to avoid obstacles on the northeastern corner of the plateau was frustrated, and Spirit gained a mere 15 meters over a period of two weeks. The team reluctantly decided to switch from a clockwise to a much longer counterclockwise route around the plateau, around its unexplored western side. It paused between Sol 1,861 and 1,866 to analyze freshly unearthed silica deposits. Meanwhile, on Sol 1,856 it exceeded the record for the distance traveled on five wheels by driving 25.82 meters. A regional dust storm made the sky hazier during March but weather reports from the color camera of Mars Reconnaissance Orbiter were enabling rover planners to make better predictions of the atmospheric opacity and to better plan the daily sorties. In April 2009 a repeat of the problem suffered in January caused Spirit's computer to reboot several times for no apparent reason. After troubleshooting this, the team told the rover to resume driving.

When Spirit reached an area named Troy its wheels broke through a thin crust which masked a layer of very fine, slippery sand and it was able to gain only a few centimeters per sol. Fortunately, thanks to the dust devils which were cleaning its solar panels, Spirit, for the first time in years, had more power available than its sibling, Opportunity. On Sol 1,899 (6 May) the motor of the left-middle wheel stalled and Spirit halted with its wheels buried up to their hubs in the fine sand. While the team studied means of escaping this area, the rover obtained a panorama of Home Plate from Troy. Prior to moving Spirit, JPL tested the extraction plan by reproducing the situation in its 'sandbox', as it had in 2005 when Opportunity became bogged down in the Purgatory Dune. But Spirit would have to rely on only five working wheels. Furthermore, to assess the situation the arm positioned the microscope to view directly beneath the belly of the vehicle – the fuzzy and unfocused images seemed to show that Spirit was almost resting on a pointed rock. While engineers assessed the situation, the rover continued to undertake scientific observations. Rocks within reach of the arm were brushed and analyzed, and night-time observations were scheduled to exploit the abundant power available. The sand churned up by the spinning wheels had more surprises in store. Spirit proved to be sitting on the rim of an ancient crater (named Scamander) that was 8 meters across and masked by a thick blanket of fine dust. This dust was made of at least three layers of differently-colored soil covered by a darker layer of sand. Where the wheels had dug into this soil, microscope images showed it to be quite cohesive and able to retain a steep slope without collapsing. The layers appeared to consist of basaltic, sulfur-rich and silicon-rich sand. In fact, the sulfur-rich sand in Troy had the highest sulfur content ever recorded on Mars. This stratification hinted at a process in which surface water, probably in the form of snow, dissolved iron sulfates to leave a crust of calcium sulfates. Intriguingly, this activity seemed to have occurred in geologically recent times.

On 18 August Spirit broke the 2,000-sol mark. But then a small regional dust storm drastically reduced the available output. A series of brake faults in the drive mechanism of the high-gain antenna required almost a month to resolve. Finally, on Sol 2,088 in mid-November, Spirit was able to begin maneuvering to attempt to extricate itself from Troy. The first sequence called for it to spin its five working wheels forward for six revolutions in the hope of driving upslope relative to the Scamander crater, then pause to enable the results to be assessed. On the very first

Engineers on Earth tried to assess ways to extract Spirit from its dust-trap on the rim of Scamander using a full-scale rover mockup and a 'Martian sandbox', but to no avail. (NASA/JPL/Caltech)

A heavily out of focus microscope view of the underside of Spirit, trapped in Troy. (NASA/JPL/Caltech)

While stuck at Troy, scientists used the arm-mounted instruments to investigate Spirit's final resting place. (NASA/JPL/Caltech)

sol, the activity was interrupted by the tilt-limitation algorithm even before it could start. NASA acknowledged that if the extraction process failed, Spirit would likely be stuck for good. Although scientific observations would probably be permitted to continue with the vehicle immobilized, it was not in a very good orientation to survive the coming winter. On Sol 2,092, after it had moved several centimeters, the right-rear wheel abruptly stalled. A number of explanations were offered, none of which were encouraging for the effort to drive free. These included a gear jam, a failed motor, and a small fragment of rock wedged in the wheel. In an effort to improve the situation, the team tried to use the long-failed right-front wheel, which proved able to move on an intermittent basis. However, the rover's wheels merely dug themselves deeper into the sand. At one point the priority would have to shift from attempting to extricate the vehicle to trying to improve its tilt to generate the power required to sustain itself in winter, because if it could not achieve a better parking position it was unlikely to remain operational beyond May 2010.[356]

Opportunity began 2009 by examining several targets on the Crete outcrop. It then suffered a new mechanical glitch when an encoder involved in the movement of the rock abrader failed. This required a modification of the manner in which this tool was used. Undaunted, the rover continued its drive southward, pausing only to reconnoiter small craters like Ranger. On these long drives, the right-front wheel motor began to draw excessive current. This was the wheel whose steering motor had failed in 2005. Opportunity performed diagnostic maneuvers to investigate the malfunction and then, to alleviate the situation, drove backwards for several sols in order to distribute the lubrication evenly in the wheel motor's gearbox. While this problem was under investigation, scientists used all of the available instruments to inspect a cluster of about fifty craters, named the Resolution cluster, scattered over an area of 100 × 140 meters located some 2.5 km south of Victoria, only four of which exceeded 5 meters in diameter. They appeared to be the youngest craters yet encountered by Opportunity (excepting, of course, the one dug by its heat shield). Based on the appearance of dust ripples, the cluster was estimated to be no older than 100,000 years.[357] Images taken from Resolution provided the first glimpse of the rim of Endeavour, over 10 km away. It resumed driving and by Sol 1,870 was 3 km south of Victoria. Unfortunately, the wheel motor was requiring the rover to pause to rest ever more frequently. Nevertheless, in May it managed to reach the 16-km mark. Along the way, it examined the soil and rocks in search of trends in their composition which might correlate with the topography. On Sol 1,915, while allowing its wheel motors to rest, the rover used its accelerometers to undertake a seismometry experiment.

At the end of June, Opportunity halted for a few sols to investigate the outcrop Absecon. On Sol 1,950, while skirting a dune field, the team spotted a suspiciously dark rock off to the right and ordered the rover to make a U-turn to inspect it. It took 9 sols to reach this rock, named Block Island. At 70-cm across it was the size of a watermelon and its mass was estimated at 240 kg. In August the arm-mounted spectrometers integrated data. The results confirmed the suspicion that this was an iron-nickel meteorite like the Heat Shield Rock. It had been decided not to abrade meteorites such as this, to preclude excessive wear of the tool which had long ago exceeded the intended number of brushings. The presence a pedestal beneath the rock and the berries in its cavities were hints that Block Island had fallen on Mars billions years ago. As with the Heat Shield Rock, its intact condition argued for a denser atmosphere. Its pristine surface implied that the climate had been dry since. Meanwhile, communications were reduced because Mars Reconnaissance Orbiter, the most capable of NASA's two remaining orbital relays, entered safe mode and several months of troubleshooting were required to coax it back into full operation. On finishing with Block Island, Opportunity resumed its drive south to Endurance traveling on average 50 meters per sol, but it was obliged to take detours to avoid fields of dunes. Several hundred meters beyond Block Island, it encountered two smaller meteorites, dubbed Shelter Island and Mackinac. As they shared the same appearance, they were presumed to be part of the same fall.

On Sol 2,058 in November, Opportunity detoured to examine an isolated 30-cm rock named Marquette Island that stood out from its surroundings. Since its solar panels were relatively clean, the rover had sufficient power available to abrade this rock. It was the 38th target for this particular tool, and one of the hardest ever. The brushed part of the rock possessed a brightly glinting crystalline surface, together with dark and light patches. The low percentage of nickel indicated that it was not another meteorite, as initially thought. However, it showed no trace of sulfur or the usual sedimentary processes found in almost every Meridiani rock with the notable exception of Bounce at Eagle crater. This was confirmed by the Mossbauer results which showed water-incompatible olivine and pyroxene. Marquette was probably ejecta, possibly even from the same impact as Bounce, although there were significant differences between the two rocks. Unable to agree what kind of rock Marquette was, the team split into two groups. To some, it was a basaltic, coarse-grained rock, rich in large crystals which implied that it cooled slowly after formation. To others, it seemed to be a collection of crystals embedded in a matrix which could include carbonates and clays.[358,359]

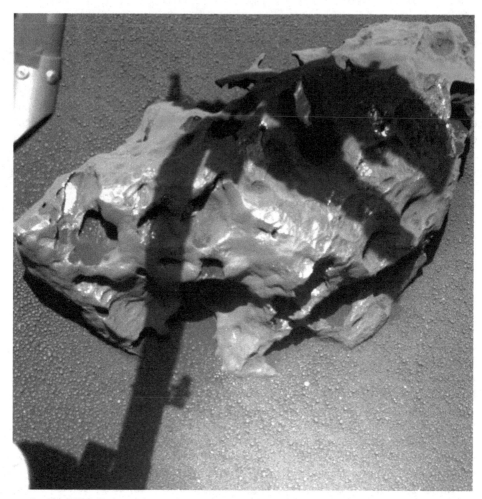

On Sol 2,029 Opportunity took a sequence of funny pictures of the shadow of its robot arm projecting the outline of a dinosaur's head onto the Shelter Island meteorite! (NASA/JPL/Caltech)

As 2010 began, Spirit was making little progress in extricating itself from the fine sand of Troy. Various methods were employed, including turning the steering wheels sideways prior to driving in order to let material fall into the trenches they were excavating, but only a few centimeters were gained. Although it had traveled only about 39 cm since November, engineers reckoned several more weeks should enable it to exit the sand trap. However, at the end of January, with the solstice due in May, NASA switched the priority to having the rover try to tilt its solar panels northward for the winter. After that, if it survived, it would continue its mission as a stationary lander because the agency directed the team not to make any further attempts at extrication. Spirit appeared to be sitting on a small hill and backing up slightly would probably suffice to face the solar panels to the Sun. Alternatively, it could use the

Signs of abrasion on Marquette Island. (NASA/JPL/Caltech)

wheels on one side to dig itself deeper into the sand and achieve the desired tilt in that manner. If the situation could not be improved then the rover would draw increasingly on its batteries and when these dropped below a certain charge the onboard computer would fall into a pre-programmed 'low-power fault' hibernation mode in which it would switch off all systems apart from its clock and would not attempt to contact Earth until the power was restored, which would not occur for several months.[360] Spirit made one last 'drive' on Sol 2,169 (8 February) in an attempt to tilt its solar panels northward, but was unsuccessful and had to be left with a distinctly unfavorable 9-degree tilt to the south.

Meanwhile, plans were drawn up to operate Spirit as a static lander. One task it could usefully perform was to precisely determine the direction of the planet's spin axis to measure the precession angle since the Viking landers and Mars Pathfinder. In fact, the X-Band radio system on Spirit was better suited to such an experiment than was the S-Band system of the Vikings. Although Mars Pathfinder was able to perform this experiment only briefly, the tracking obtained in 3 months provided a measurement of the spin axis about as accurate as that from 2 years of tracking the Vikings. Tracking Spirit for 6 months would pinpoint the location of the rover and 'subtract' the effects of the relative motions of Earth and Mars to reveal the tiny signal sought. A precise determination of the precession of the axis of Mars would impose constraints on the distribution of mass in its interior and the size of its core, and provide insight into whether it still possesses a liquid core.[361]

The preparations included configuring the arm for atmospheric measurements, positioning the high-gain antenna and camera mast to minimize their shadows on the solar panels, and informing Spirit of the scheduled communications windows for contacting Mars Odyssey and Earth for the remainder of 2010 and early 2011. On an optimistic note, a panorama was obtained for comparison with one after the winter. Finished, the team could only observe as the power available to the rover dwindled from one sol to the next. Finally, on Sol 2,218 (30 March) Mars Odyssey received no telemetry as it flew over Gusev. Efforts to re-establish contact started on 26 July by transmitting commands asking Spirit to 'beep' at Earth. Although it was estimated that by then the rover would be receiving just sufficient power to awaken from hibernation, there was no response.

Opportunity began 2010 by leaving Marquette Island on Sol 2,122 to resume its marathon drive for Endeavour. On Sol 2,138 it paused to inspect a young-looking crater named Concepción that was 10 meters across. At about the same time Mars Reconnaissance Orbiter took a picture of Opportunity. The rim of Concepción was marked by blocky ejecta that appeared to lie on top of the sand dunes; evidence of its extreme youth, possibly as young as 1,000 years. The interior of the crater was filled with dust, and an unusual grayish material coated the rocks and filled in their fractures. The material laying atop the square rock Chocolate Hills was analyzed and found to be the same hematite as the blueberries, giving it the appearance of "a blueberry sandwich". It is possible that this resulted from berries being melted by the impact which made Concepción, but it could also predate the formation of the crater – in which case the material could result from berries liberated by water and packed along fractures in the rock. Before leaving Concepción, Opportunity made a complete circumnavigation of the crater. The rim of Endeavour was particularly evident from this position. Moreover, pictures showed hills corresponding to the rim of Iazu beyond, and even the raised rim of crater Bopolu, some 65 km away on the southwestern horizon.

New capabilities were implemented with a software upgrade that was uploaded in early March. These included the ability to recognize rocks that met criteria such as a given shape or color. On Sol 2,191 (24 March) Opportunity passed the 20-km mark. In late March it inspected San Antonio, a pair of craters several meters apart that appeared to be recent and may have been created by a meteorite that broke in two just prior to reaching the surface. Opportunity maneuvered to get a good view of the interior of one of the craters but made no observations apart from imaging, and after a few sols it headed southeast toward Endeavour. By May it had reached a position from which a gentle southward slope allowed a good view of the rim of Endeavour,

A close-up of the unusual coating of the rock Chocolate Hill. The coating appeared to be molten blueberries. (NASA/JPL/Caltech)

then 13 km away. Rim features were named after places and people on James Cook's epic voyages. The plan was for Opportunity to reach the rim in the vicinity of Cape York, a small 'island' on the rim of Endeavour. A problem later that month implied the azimuth actuator of the panoramic camera mast had failed. If this were so, then it would greatly impair the mission. However, tests showed that the actuator was functional. Although the thermal emission spectrometer should have remained off after becoming unusable in the 2007 dust storms, this appeared to have generated the glitch. Having finally cleared the most dangerous-looking dune fields, Opportunity turned due east for a straight approach to Endeavour. Around the winter solstice in May, each drive was designed to end with the vehicle oriented to favorably tilt its solar panels sunward to recharge its batteries. On Sol 2,253 (20 May) it beat the Viking 1 lander as the longest operating spacecraft on the surface of Mars. During

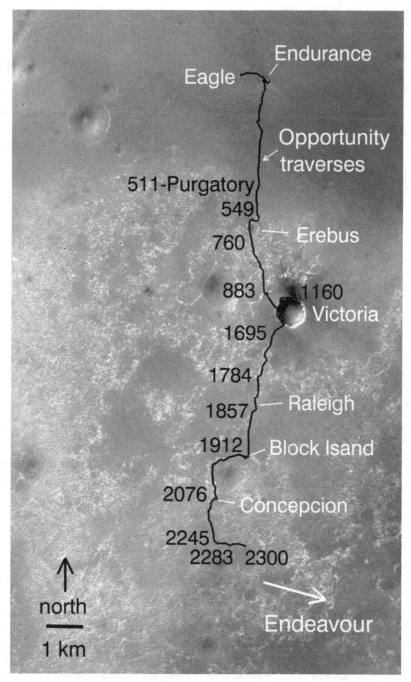

Opportunity's route from landing up to Sol 2,300, superimposed on a Mars Reconnaissance Orbiter context camera view. (MSSS)

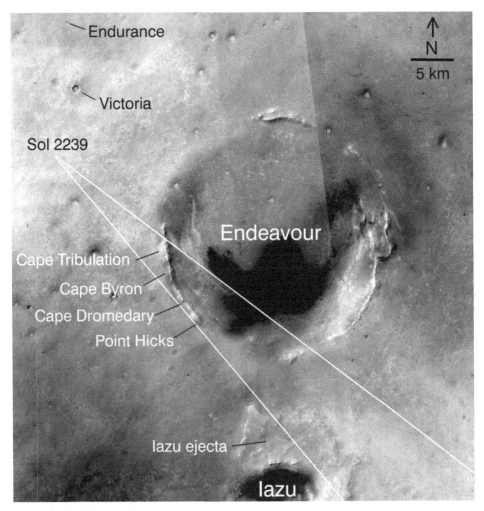

A Mars Reconnaissance Orbiter context camera view of Endeavour crater. The white lines show Opportunity's panoramic camera coverage of the rim shown in the next image. (MSSS)

the ensuing months, it tested its autonomous capabilities by selecting several targets for study. It paused in late August at the Cambridge Bay outcrop which seemed to mark the 'contact' between two different geological units.

In September, Opportunity stopped to investigate yet another meteorite dubbed Oileán Ruaidh. It had by then reached rocky patches of outcrop and accelerated its eastward trek. Meanwhile, the iron-detecting Mossbauer spectrometer showed the first signs of malfunctioning. In early November, Opportunity passed a number of small, shallow craters, including the double crater Paramore, heading for Intrepid the largest of the cluster. Some 17 meters across, Intrepid had exposed bedrock in its walls. It and nearby Yankee Clipper were named respectively for the Apollo 12 lunar

Opportunity's view of the rim of Endeavour on Sol 2,239. (NASA/JPL/Caltech)

Opportunity's view of sand dunes on the floor of Santa Maria, the last largish crater encountered on the road to Endeavour. (NASA/JPL/Caltech)

and command modules. On 14 November (Sol 2,420), on its way to a fresh-looking crater named Santa Maria that seemed to have penetrated the bedrock, the rover passed the 25 km mark. It was possible that Santa Maria, a polygonal crater 80 to 90 meters across with asymmetric ejecta rays some 6 km from Cape York, could sport hydrated rocks.

Opportunity reached Santa Maria on Sol 2,451 (16 December 2010), arriving at a promontory dubbed Palos after the Spanish port from which Columbus' three ships, Santa Maria, Nina and Pinta, set sail for the Indies. The rim of Santa Maria was strewn with big boulders, and a vast dune field occupied its floor. There was a shallow area, possibly a secondary crater on the southeastern rim, named Yuma, where the Mars Reconnaissance Orbiter had spied the presence of water-bearing sulfates. Engineers and scientists planned to investigate the area over the February 2011 conjunction, but there were no plans to enter the crater, unless a good reason was found to do so. The rover ended 2010 driving counterclockwise to a location at the southern edge of the crater dubbed Wanamani.

Attempts to re-establish contacts with Spirit stepped up in January 2011, mid-Spring at Gusev crater, just in time for the seventh anniversary of the landing. It was considered likely that during the low-power periods the rover's clock had drifted significantly, causing it to attempt communications at unpredictable times, so controllers expanded the contact windows. However, the possibility also existed that the winter chill had changed the receiving or transmitting frequency of Spirit's radio. But as the silence continued, it became more likely that the intense cold had caused mission-ending damage. At Spirit's position south of the equator, and with its solar panel tilt, the insolation would reach a maximum some 30 days before the 9 April solstice. However, none of the communication strategies produced a result, not even those for improbable multiple failures. The final communication attempt was made on 25 May, by which time it was estimated that the likelihood of hearing from the rover, after fourteen months of silence, were nil. Although the Deep Space Network would continue to listen for signals whenever its schedule permitted, the mission was effectively over.

In addition to taking a total of 124,000 images, Spirit had ground fifteen targets and brushed 92 others. It eventually traveled a total of 7,730 meters from its landing point.

Opportunity began 2011 on the rim of Santa Maria, analyzing rocks named after the crew of Christopher Columbus. It drove eastward to the bright rock which had meteorite-like cavities, dubbed Luis de Torres, located at the edge of Yuma. It then made a preliminary analysis of the undisturbed surface of the rock before brushing it using its tool for the first time in months, and integrating a Mossbauer spectrum. The radioactive cobalt source had so decayed that it required several days during the late-January solar conjunction to obtain a good quality spectrum. Atmospheric observations were also made during the conjunction. A dust devil possibly swept by during the period Opportunity was out of contact, boosting the power output of its solar panels. The rover then moved northeastward and counterclockwise to a blue-colored rock dubbed Ruiz Garcia. On 1 March, as it was taking microscopic images of this rock, the Mars Reconnaissance Orbiter overhead took a picture of it on the

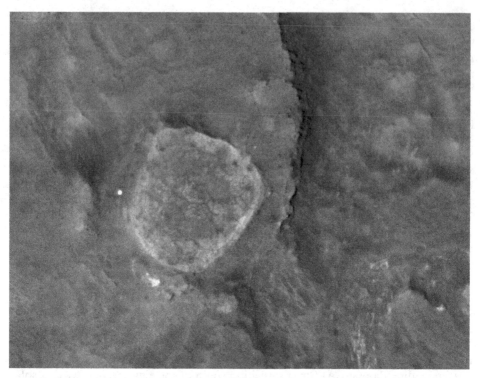

A Mars Reconnaissance Orbiter high-resolution view of the bright, silent Spirit perched on the rim of Home Plate on 31 March 2011. (NASA/JPL/University of Arizona)

rim of Santa Maria. Opportunity finally drove a few tens of meters to take 3-dimensional images of the interior of Santa Maria and resumed heading for Cape York. Not many stops would be taken along the final dash of the route. It crossed the 28-km distance traveled mark on 19 April driving over a terrain scattered with ridges of fractured bedrock and a week later it made the longest backward drive ever, of 152.18 meters. In doing so, it surpassed the distance driven by the Apollo 15 astronauts on the Moon. Opportunity drove toward a group of several small craters named after the US Mercury spacecraft to commemorate the 50th anniversary of the suborbital flight of Alan Shepard. Freedom 7, the largest of the cluster, dedicated to his spacecraft, was about 25 meters across and was estimated to be about 200,000 years old. Meanwhile, diagnostic tests of the thermal emission spectrometer indicated that the instrument exhibited "more anomalous behavior", even failing to draw power as if it were not turning on. Other small craters that the rover drove by in May were named after historical manned spacecraft, including Skylab and Gumdrop, the command module of Apollo 9. On 1 June, Opportunity passed the 30 km mark during a 147-meter drive, and by the end of July it was within a few hundred meters of Cape York.

On Sol 2,681 (9 August) Opportunity made its 'landfall' at Spirit Point, on the southern tip of Cape York, close to the small crater Odyssey. It had driven a total of 33.49 km, and approximately 21 km since leaving Victoria. The edge of the cape was a bench that gave the appearance of being sedimentary rock with brighter veins of material carried by water. The far rim of Endeavour, as well as its floor, were visible in the distance. The rover then started to explore the area by taking a peek at some of the rocks scattered on the rim of Odyssey. A flat, crumbly rock dubbed Tisdale was the first target to be examined in detail. This was a breccia, an agglomerate of fragments fused together by the shock of the impact which created Odyssey. Many of the Apollo lunar samples were breccias. It had an anomalous concentration of zinc and bromine, a possible clue of ancient hydrothermal activity in the area. Next, it inspected, abraded and "alpha-rayed" Chester Lake, an outcrop of bedrock on the inboard side of Cape York. After it had finished at Chester Lake in late-October, Opportunity drove along Cape York heading northeast to Shoemaker Ridge, a few tens of meters distant, where there were indications of clays. There, it found veins of bright bedrock emerging from beneath the dirt. One of these, Homestake, a few centimeters wide and tens of centimeters long, was photographed by the microscope and analyzed by the alpha-ray spectrometer. The instrument identified calcium and sulfur, suggesting relatively pure calcium sulfate possibly in the form of gypsum. Thus, Homestake was probably a vein of gypsum deposited when sulfur-rich water flowed through fractures in the soil.

The rover was to spend the winter on a northward slope. By virtue of being closer to the equator, Opportunity, unlike Spirit, did not need to be parked for the winter but it was a precaution as there had been no cleaning events recently and the solar panels were accumulating dust and producing less and less power. The following spring or summer, it was to travel south to Cape Tribulation. Two suitable candidate parking sites within 20 meters of each other were identified that had northward tilts of 10 to 20 degrees. The southernmost of these was dubbed Turkey Haven because the vehicle was parked there during the Thanksgiving holiday. There, it analyzed an impact breccia named Transvaal. It was then parked at Greeley Haven at the end of 2011, where it would remain until mid-2012. During this time, it would collect long-integration Mossbauer spectra of a target named Amboy. And because Opportunity would remain stationary for a few months, scientists planned to perform the same kind of Doppler radio tracking experiment that they had planned for Spirit after it was immobilized at Scamander.

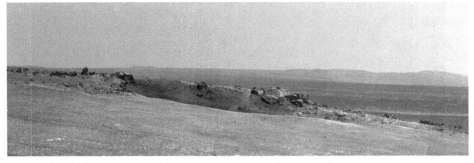

A panoramic camera mosaic of Spirit Point on the rim of Endeavour, showing the small Odyssey crater as well as the floor of Endeavour in the background. (NASA/JPL/Caltech)

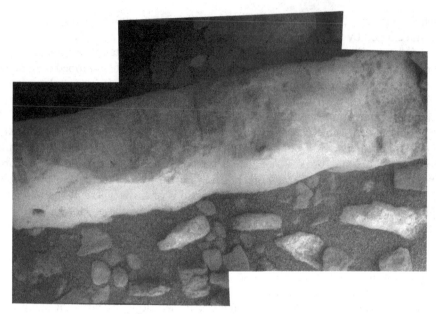

A mosaic of microscope images of Homestake, a vein of bright material possibly made of gypsum. (NASA/JPL/Caltech)

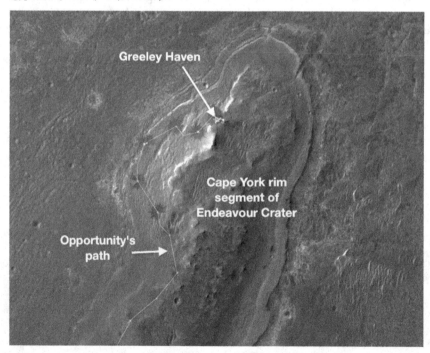

Opportunity's current (early 2012) resting place at 'Greeley Haven', on the northern tip of Cape York. (NASA/JPL-Caltech/UA)

By the end of 2011, Opportunity, at 34,361 meters, is the third most traveled planetary rover, after Lunokhod 2, which traveled 37 km on the Moon and the 'Moon buggy' of Apollo 17, which traveled 35.89 km. Whether it can beat these records remains to be seen. It cannot be predicted how long Opportunity will continue to function. What is certain, however, is that the two Mars Exploration Rovers greatly exceeded their primary missions. Opportunity could well reach clay deposits just before the next NASA rover arrives at Mars in August 2012 to investigate a similar ancient watery environment in Gale crater.

## REFERENCES

1  For the Martian magnetic field see Part 2, pages 398–401
2  Isakowitz-2000
3  Oya-1998
4  Okada-1998
5  Mukai-1998
6  Yamamoto-1998
7  Taguchi-2000
8  Sasaki-1999
9  AWST-2000a
10  Matsuura-2005
11  Abe-1991
12  Kawaguchi-2003b
13  Kimura-1999
14  Yamakawa-1999
15  Tokadoro-1999
16  Nakagawa-2008
17  Sasaki-2005
18  Miyasaka-2005
19  Hidaka-2001
20  Yoshikawa-2005
21  Nakatani-2004
22  Ryné-2004
23  Taylor-1999
24  Malin-1999
25  NASA-1999d
26  NASA-1994
27  Covault-1994
28  Smith-1994
29  Kulikov-1996
30  Cowley-1999
31  Smith-2001a
32  Bonitz-1998
33  Bonitz-2001
34  Boynton-2001
35  For microphones on Veneras 11 to 14 see Part 1, pages 272, 277 and 288
36  NASA-1998b
37  Mitcheltree-1997
38  Lorenz-2000
39  Smrekar-1999
40  Westwick-2007
41  Euler-2001
42  NASA-1999e
43  McCurdy-2005a
44  Westwick-2007
45  Smrekar-1999
46  JPL-2000
47  NASA-1998b
48  JPL-2000
49  Warwick-2000
50  Westwick-2007
51  Rieke-2006b
52  Malin-2005
53  Tytell-2006
54  For the 'rocker-bogie' see Part 2, pages 443–444
55  Arvidson-2000
56  Volpe-2000
57  Shirley-1999
58  Squyres-2005a
59  JPL-1997
60  Bonitz-2000
61  Matousek-1998
62  Desai-1998
63  Bille-2004
64  Wilcox-2000b
65  Reichhardt-1999
66  Kminek-2000
67  ASI-1999

68  Magnani-2004
69  Battistelli-2004
70  Angrilli-2004
71  Colangeli-2000
72  Cordier-2000
73  Golombek-2000
74  JPL-1999
75  Price-2000
76  O'Neill-1999
77  For Mars airplanes see Part 1, pages
    258–259
78  Morton-2000
79  MSSS-1998
80  Furniss-1999
81  Guynn-2003
82  Colozza-2003
83  Zubrin-1993
84  Cesarone-1999
85  Edwards-1999
86  Hastrup-1999
87  Taverna-1999a
88  Jordan-1999
89  Morton-1999
90  AWST-2000b
91  Mecham-2000
92  Covault-2001
93  Covault-2001
94  Smith-2001b
95  Keating-2003
96  Crowley-2007
97  Mazarico-2007
98  Smith-2002a
99  Saunders-2004
100 Mase-2005
101 Smith-2002b
102 Bell-2002
103 Feldman-2002
104 Mitrofanov-2002
105 Boynton-2002
106 Mitrofanov-2003
107 Litvak-2006
108 Head-2003
109 Baker-2003
110 Levrard-2004
111 Mangold-2004
112 Titus-2003
113 Bandfield-2007
114 Baker-2003
115 Christensen-2003a

116 Christensen-2003b
117 Christensen-2005
118 Osterloo-2008
119 Cushing-2007
120 Barbieri-2003
121 Gunnarsdottir-2007
122 Sprague-2004
123 Forget-2004
124 Zuber-2003
125 Litvak-2006
126 Hurley-2005
127 Kieffer-2006b
128 Baratoux-2011
129 Kremer-2011
130 For Mars 96 see Part 2, pages 434–439
131 Bibring-2009a
132 For the Kepler mission see Part 2, page
    136
133 Taverna-1997
134 Schmidt-1998
135 For a description of the camera see Part
    2, page 430
136 Bertaux-2004
137 Barabash-2006
138 Schmidt-1999
139 Chicarro-2003
140 Griffiths-2003
141 Coué-2007
142 Pullan-2004
143 Pillinger-2003
144 Towner-2000
145 Chicarro-2003
146 Fallon-2003
147 Taverna-1999b
148 Butler-2003
149 Parkinson-2004
150 Powell-2003
151 Bibring-2009b
152 ESA-2003
153 Taverna-2000a
154 ESA-2004
155 Nelson-2004
156 Sims-2004a
157 Sims-2004b
158 Peplow-2004
159 House of Commons-2004
160 Abdel-jawad-2008
161 Pillinger-2005
162 Pillinger-2007

163 Bibring-2009c
164 Taverna-2004
165 For 'Swiss cheese' terrain see Part 2, pages 411–412
166 Bibring-2004
167 Borde-2004
168 Denis-2006
169 Adams-2005
170 Oberst-2006
171 Formisano-2004
172 Encrenaz-2008a
173 Wolkenberg-2008
174 Bertaux-2005a
175 Soobiah-2005
176 Bertaux-2005b
177 Perrier-2005a
178 Fast-2009
179 Perrier-2005b
180 Murray-2005b
181 Neukum-2004
182 Hauber-2005
183 Levrard-2004
184 Head-2005
185 Neukum-2005
186 Hand-2008
187 Poulet-2008
188 Bibring-2006
189 Poulet-2005
190 Langevin-2006
191 Schmidt-2010
192 Bibring-2005
193 Langevin-2005a
194 Langevin-2005b
195 Gendrin-2005
196 Arvidson-2005
197 Mustard-2005
198 Douté-2007
199 Bibring-2009d
200 Naeye-2005
201 Chicarro-2006
202 Encrenaz-2008b
203 Lundin-2004
204 Holmström-2007
205 Denis-2006
206 Adams-2005
207 Picardi-2005
208 Watters-2006
209 Plaut-2007
210 Barabash-2007
211 Denis-2009
212 Watters-2007
213 Schultz-2007
214 Willner-2008
215 Basilevsky-2008
216 Giuranna-2010
217 Maltagliati-2011
218 Heavens-2011
219 Jansen-2012
220 For MARSNET and INTERMARSNET see Part 2, pages 461–462
221 Taverna-2000b
222 Smith-2000
223 Taverna-2000c
224 Marsal-1999
225 Marsal-2002
226 Nadalini-2004
227 CNES-2002
228 Crabb-2003
229 Squyres-2005b
230 Squyres-2005c
231 Smith-2000
232 Squyres-2005d
233 Taylor-2005
234 Lindemann-2005
235 Squyres-2005c
236 Baumgertner-2005
237 Bell-2003
238 Smith-2004
239 Wolff-2005
240 Mitcheltree-2004
241 Squyres-2005e
242 Powell-2003
243 Dornheim-2003a
244 Dornheim-2003b
245 Golombek-2005
246 Covault-2003a
247 Covault-2003b
248 Covault-2003c
249 Erickson-2004
250 Covault-2003d
251 Withers-2006
252 Desai-2008d
253 Dornheim-2004c
254 Dornheim-2004d
255 Reeves-2005
256 Squyres-2005f
257 Desai-2008d

258  Covault-2004c
259  Covault-2004d
260  Dornheim-2004e
261  Dornheim-2004f
262  Squyres-2005g
263  Dornheim-2004g
264  Morring-2004a
265  Chan-2004
266  Bibring-2007
267  Covault-2004e
268  Morring-2004b
269  Fairén-2004
270  Bell-2005
271  Covault-2004f
272  Sullivan-2005
273  Squyres-2004a
274  Bell-2004a
275  Squyres-2004b
276  Soderblom-2004c
277  Herkenhoff-2004a
278  Arvidson-2004a
279  Christensen-2004a
280  Klingelhöfer-2004
281  Rieder-2004
282  Smith-2004
283  Dornheim-2004h
284  Dornheim-2004i
285  Dornheim-2004j
286  Reeves-2005
287  Dornheim-2004k
288  Dornheim-2004l
289  Selsis-2005
290  Dornheim-2004m
291  Haskin-2005
292  Squyres-2004c
293  Bell-2004b
294  Grant-2004
295  Arvidson-2004b
296  Herkenhoff-2004b
297  Gellert-2004
298  Morris-2004
299  Christensen-2004b
300  Lemmon-2004
301  Squyres-2005h
302  Covault-2004g
303  Dornheim-2004n
304  Haskin-2005
305  Squyres-2005i
306  Covault-2004h

307  Covault-2004i
308  Dornheim-2004o
309  Corneille-2005a
310  Dornheim-2004n
311  Squyres-2005j
312  Dornheim-2004n
313  Corneille-2004
314  Covault-2004h
315  Covault-2004i
316  Dornheim-2004o
317  Corneille-2005a
318  Dornheim-2005a
319  Dornheim-2005b
320  Dornheim-2005c
321  Corneille-2005b
322  Dornheim-2005d
323  Edwards-2006
324  Dornheim-2005d
325  Covault-2005
326  Domokos-2007
327  Bell-2006
328  Morris-2010
329  Corneille-2006
330  Desai-2008d
331  Goetz-2005
332  Dornheim-2005c
333  Dornheim-2005e
334  Dornheim-2005f
335  Corneille-2005b
336  Corneille-2006
337  Squyres-2006
338  Squyres-2007
339  Corneille-2007
340  Squyres-2006
341  Corneille-2007
342  Squyres-2009
343  Cull-2007
344  Squyres-2008
345  Covault-2007a
346  Covault-2007b
347  Covault-2007c
348  Covault-2007d
349  Corneille-2008
350  Covault-2008
351  Lawler-2008
352  Corneille-2009
353  Squyres-2009
354  Wray-2009
355  Corneille-2009

356  Corneille-2010

357  Golombek-2010

358  Mittlefehldt-2010

359  Corneille-2010

360  Norris-2010

361  Banerdt-2010

# Glossary

ACE: Advanced Composition Explorer

Aerobraking: A maneuver where a spacecraft's orbit is changed by reducing its energy through repeated passages in a planet's atmosphere.

Aerocapture: A maneuver where a spacecraft is inserted in orbit around a planet by slowing it down through a passage in the planet's atmosphere.

Aerogel: A silicon-based foam in which the liquid component of a gel has been replaced with gas or, for use in space, effectively with vacuum, resulting in a very low density solid.

Albedo: in first approximation a measure of the reflecting power of a surface.

Aphelion: The point of maximum distance from the Sun of a solar orbit. Its contrary is the perihelion.

APL: Applied Physics Laboratory

Apoapsis: The point of maximum distance from the central body of any elliptical orbit. This word has been used to avoid complicating the nomenclature, but a term tailored to the central body is often used. The only exceptions used herein owing to their importance were for Earth (apogee) and the Sun (aphelion). The contrary of apoapsis is periapsis.

Apogee: The point of maximum distance from the Earth of a satellite orbit. Its contrary is the perigee.

ASI: Agenzia Spaziale Italiana (Italian Space Agency)

ASPERA: Automatic Space Plasma Experiment with a Rotating Analyzer

Astronomical Unit: To a first approximation the average distance between the Earth and the Sun is 149,597,870,691 ($\pm$30) meters.

AU: Astronomical Unit

AXAF: Advanced X-ray Astrophysical Facility

BMDO: Ballistic Missile Defense Organization

BNSC: British National Space Council

Booster: Auxiliary rockets used to boost the lift-off thrust of a launch vehicle.

Bus: A structural part common to several spacecraft.

CAESAR: Comet Atmosphere Encounter and Sample Return or Comet Atmosphere and Earth Sample Return

CCD: Charge Coupled Device

CHON: Carbon, Hydrogen, Oxygen, Nitrogen rich molecules

CISR: Comet Intercept and Sample Return

CMOS: Complementary Metal–Oxide Semiconductor

CNES: Centre National d'Etudes Spatiales (the French National Space Studies Center)

Conjunction: The time when a solar system object appears close to the Sun as seen by an observer. A conjunction where the Sun is between the observer and the object is called 'superior conjunction'. A conjunction where the object is between the observer and the Sun is called 'inferior conjunction'. See also opposition.

CONTOUR: Comet Nucleus Tour

Cosmic velocities: Three characteristic velocities of spaceflight:

First cosmic velocity: Minimum velocity to put a satellite in a low Earth orbit. This amounts to some 8 km/s.

Second cosmic velocity: The velocity required to exit the terrestrial sphere of attraction for good. Starting from the ground, this amounts to some 11 km/s. It is also called 'escape' speed.

Third cosmic velocity: The velocity required to exit the Solar System for good.

CRAF: Comet Rendezvous/Asteroid Flyby

Cryogenic propellants: These can be stored in their liquid state under atmospheric pressure at very low temperature; e.g. oxygen is a liquid below −183°C.

DASH: Demonstrator of Atmospheric reentry System with Hyperbolic velocity

Deep Space Network: A global network built by NASA to provide round-the-clock communications with robotic missions in deep space.

DeeDri: Deep Driller

Direct ascent: A trajectory on which a deep-space probe is launched directly from the Earth's surface to another celestial body without entering parking orbit.

DS: Deep Space

DSN: Deep Space Network

Ecliptic: The plane of the Earth's orbit around the Sun.

EELV: Evolved Expandable Launch Vehicle

Ejecta: Material from a volcanic eruption or a cratering impact that is deposited all around the source.

ESA: European Space Agency

Escape speed: See Cosmic velocities

FIDO: Field Integrated, Design, and Operations

Flyby: A high relative speed and short duration close encounter between a spacecraft and a celestial body.

GPS: Global Positioning System

GRB: Gamma-Ray Bursts

GRO: Gamma-Ray Observatory

GSFC: Goddard Space Flight Center

HER: Halley Earth Return

HST: Hubble Space Telescope

Hypergolic propellants: Two liquid propellants that ignite spontaneously on coming into contact, without requiring an ignition system. Typical hypergolics are hydrazine and nitrogen tetroxide.

IBEX: Interstellar Boundary Explorer

ICE: International Cometary Explorer

IRAS: InfraRed Astronomical Satellite

ISAS: Institute of Space and Astronautical Sciences

ISO: Infrared Space Observatory

ISS: Cassini's Imaging Science Subsystem

ISS: International Space Station

ITAR: International Traffics in Arms Regulations

IUS: Inertial Upper Stage (previously: Interim Upper Stage)

JAXA: Japanese Aerospace Exploration Agency

JPL: Jet Propulsion Laboratory (a Caltech laboratory under contract to NASA)

Lagrangian Points: Five equilibrium points for a gravitational system comprising two large bodies (e.g. the Sun and a planet) and a third body of negligible mass.

Lander: A spacecraft designed to land on another celestial body.

LaRC: Langley Research Center

Launch window: A time interval during which it is possible to launch a spacecraft to ensure that it attains the desired trajectory.

Lidar: laser radar

LINEAR: Lincoln Near Earth Asteroid Research

Lyman-alpha: The emission line corresponding to the first energy level transition of an electron in a hydrogen atom.

MAGE: Mars Airborne Geophysical Explorer

MAV: Mars Ascent Vehicle

MCO: Mars Climate Orbiter

MER: Mars Exploration Rovers

MESSENGER: Mercury Surface, Space Environment, Geochemistry and Ranging

MGS: Mars Global Surveyor

MINERVA: Micro/Nano Experimental Robot Vehicle for Asteroid

MIT: Massachusetts Institute of Technology

MPF: Mars Pathfinder

MPL: Mars Polar Lander

MRO: Mars Reconnaissance Orbiter

MUSES: MU [rocket] Space Engineering Satellite

NAS: National Academy of Sciences

NASA: National Aeronautics and Space Administration

NASDA: National Space Development Agency

NEAR: Near-Earth Asteroid Rendezvous

NEAT: Near-Earth Asteroid Tracking program

NEP: Nuclear Electric Propulsion

NExT: New Exploration of Tempel 1

NOTSNIK: Naval Ordnance Test Station "Sputnik"

NSTAR: NASA Solar Electric Propulsion Technology Application Readiness

Occultation: When one object passes in front of and occults another, at least from the point of view of the observer.

OMEGA: Observatoire pour la Minéralogie, l'Eau, les Glaces et l'Activité, observatory for mineralogy, water, ices and activity

OMV: Orbital Maneuvering Vehicle

Orbit: The trajectory on which a celestial body or spacecraft is traveling with respect to its central body. There are three possible cases:

Elliptical orbit: A closed orbit where the body passes from minimum distance to maximum distance from its central body every semiperiod. This is the orbit of natural and artificial satellites around planets and of planets around the Sun.

Parabolic orbit: An open orbit where the body passes through minimum distance from its central body and reaches infinity at zero velocity in infinite time. This is a pure abstraction, but the orbits of many comets around the Sun can be described adequately this way.

Hyperbolic orbit: An open orbit where the body passes through minimum distance from its central body and reaches infinity at non-zero speed. This describes adequately the trajectory of spacecraft with respect to planets during flyby maneuvers.

Opposition: The time when a solar system object appears opposite to the Sun as seen by an observer.

Orbiter: A spacecraft designed to orbit a celestial body.

OSCAR: Orbiting Sample Capture and Return

Parking orbit: A low Earth orbit used by deep-space probes before heading to their targets. This relaxes the constraints on launch windows and eliminates launch vehicle trajectory errors. Its contrary is direct ascent.

PAW: Position Adjustable Workbench

Periapsis: The minimum distance point from the central body of any orbit. See also apoapsis.

PEPE: Plasma Experiment for Planetary Exploration

Perigee: The minimum distance point from the Earth of a satellite. Its contrary is apogee.

Perihelion: The minimum distance point from the Sun of a solar orbit. Its contrary is the aphelion.

PFF: Pluto Fast Flyby

PICA: Phenolic Impregnated Carbon Ablator

PKE: Pluto Kuiper Express

PLUTO: Planetary Underground Tool

PREMIER: Programme de Retour d'Echantillons Martiens et Installation d'Expériences en Reseau, Mars sample return and network experiment establishment program

'Push-broom' camera: A digital camera consisting of a single row of pixels, with the second dimension created by the motion of the camera itself.

RAT: Rock Abrasion Tool

Rendezvous: A low relative speed encounter between two spacecraft or celestial bodies.

REP: Radioisotope Electric Propulsion

Retrorocket: A rocket whose thrust is directed opposite to the motion of a spacecraft in order to brake it.

Rj: Jupiter radii (approximately 71,200 km)

Rover: A mobile spacecraft to explore the surface of another celestial body.

Rs: Saturn radii (approximately 60,330 km)

RTG: Radioisotope Thermal Generator

RTH: Radioisotope Thermal Heater

SEDSat: Students for the Exploration and Development of Space Satellite

SERT: Space Electric Rocket Test

SEP: Solar Electric Propulsion

SIRTF: Shuttle (or Space) Infrared Telescope Facility

SMART: Small Missions for Advanced Research in Technology

SOCCER: Sample of Comet Coma Earth Return

Sol: A Martian solar day, lasting 24 Terrestrial hours, 39 minutes, and 35.244 seconds

Solar flare: A solar chromospheric explosion creating a powerful source of high energy particles.

Space probe: A spacecraft designed to investigate other celestial bodies from a short range.

Spectrometer: An instrument to measure the energy of radiation as a function of wavelengths in a portion of the electromagnetic spectrum. Depending on the wavelength the instrument is called, e.g. ultraviolet, infrared, gamma-ray spectrometer etc.

Spin stabilization: A spacecraft stabilization system where the attitude is maintained by spinning the spacecraft around one of its main inertia axes.

Synodic period: The period of time between two consecutive superior or inferior conjunctions or oppositions of a solar system body.

TEGA: Thermal and Evolved Gas Analyzer

Telemetry: Transmission by a spacecraft via a radio system of engineering and scientific data.

THEMIS: Thermal-Emission Imaging System

3-axis stabilization: A spacecraft stabilization system where the axes of the spacecraft are kept in a fixed attitude with respect to the stars and other references (the Sun, the Earth, a target planet etc.)

UTC: Universal Time Coordinated (essentially Greenwich Mean Time)

UTTR: Utah Test and Training Range

VESAT: Venus Environmental Satellite

Vidicon: A television system based on resistance changes of some substances when exposed to light. It has been replaced by the CCD.

VLBI: Very Long Baseline Interferometry

WIRE: Wide-field Infrared Explorer

WSB: Weak Stability Boundaries

# Appendices

## CHRONOLOGY OF SOLAR SYSTEM EXPLORATION 1997–2003

| Date | Event |
|------|-------|
| 23 September 1999 | Mars Climate Orbiter is lost as it crashes on Mars |
| 3 December 1999 | Mars Polar Lander is lost as it crashes on Mars |
| 22 September 2001 | Deep Space 1 flies by comet Borrelly |
| 24 October 2001 | Mars Odyssey enters orbit around Mars |
| 25 December 2003 | Mars Express enters orbit around Mars while the Beagle 2 lander is lost |

| Related milestones | |
|------|-------|
| 2 January 2004 | Stardust flies by comet Wild 2 |
| 4 January 2004 | Spirit lands on crater Gusev, Mars |
| 25 January 2004 | Opportunity lands in Meridiani Planum, Mars |
| 1 July 2004 | Cassini enters orbit around Saturn |
| 8 September 2004 | Genesis crashes to Earth, returning samples of the solar wind |
| 15 January 2005 | Huygens lands on Titan |
| 12 September 2005 | Hayabusa rendezvous with asteroid Itokawa |
| 15 January 2006 | Stardust returns samples of comet Wild 2 to Earth |
| 13 June 2010 | Hayabusa returns samples of asteroid Itokawa to Earth |
| 15 February 2011 | Stardust flies by comet Tempel 1 |

## PLANETARY LAUNCHES 1997–2003

| Launch Date | Name | Main Target | Launcher | Nation |
|---|---|---|---|---|
| 15 October 1997 | Cassini | Saturn | Titan IVB | USA/Italy |
| | Huygens | Titan | | ESA |
| 3 July 1998 | (Nozomi) | Mars | M-V | Japan |
| 24 October 1998 | Deep Space 1 | Asteroid + Comet | Delta 7326 | USA |
| 11 December 1998 | (MCO) | Mars | Delta 7425 | USA |
| 3 January 1999 | (MPL) | Mars | Delta 7425 | USA |
| 7 February 1999 | Stardust | Comet | Delta 7426 | USA |
| 7 April 2001 | Mars Odyssey | Mars | Delta 7925 | USA |
| 30 June 2001 | WMAP | L2 | Delta 7425-10 | USA |
| 8 August 2001 | (Genesis) | Sun probe | Delta 7326 | USA |
| 3 July 2002 | (CONTOUR) | Comet | Delta 7425 | USA |
| 3 May 2003 | Hayabusa | Asteroid | M-V | Japan |
| 2 June 2003 | Mars Express | Mars | Soyuz-FG | ESA |
| | (Beagle 2) | Mars | | UK/ESA |
| 10 June 2003 | Spirit | Mars | Delta 7925 | USA |
| 8 July 2003 | Opportunity | Mars | Delta 7925H | USA |
| 25 August 2003 | Spitzer | Solar orbit | Delta 7920H | USA |

Missions in parentheses are missions that failed, but the status of Genesis is disputed. Despite crashing to Earth instead of landing the sample-return mission looks set to achieve its objectives.

## CASSINI TARGETED ENCOUNTERS

| Date | Satellite | Distance |
|---|---|---|
| 11 June 2004 | Phoebe | 2,068 km |
| 26 October 2004 | Titan (Ta) | 1,174 km |
| 13 December 2004 | Titan (Tb) | 1,192 km |
| 14 January 2005 | Titan (Tc) | 60,003 km |
| 15 February 2005 | Titan (T3) | 1,579 km |
| 17 February 2005 | Enceladus (E0) | 1,261 km |
| 9 March 2005 | Enceladus (E1) | 497 km |
| 31 March 2005 | Titan (T4) | 2,404 km |
| 16 April 2005 | Titan (T5) | 1,027 km |
| 14 July 2005 | Enceladus (E2) | 166 km |
| 22 August 2005 | Titan (T6) | 3,660 km |
| 7 September 2005 | Titan (T7) | 1,075 km |
| 24 September 2005 | Tethys | 1,495 km |
| 26 September 2005 | Hyperion | 479 km |
| 11 October 2005 | Dione | 499 km |
| 28 October 2005 | Titan (T8) | 1,353 km |
| 26 November 2005 | Rhea | 504 km |

| | | |
|---|---|---|
| 26 December 2005 | Titan (T9) | 10,411 km |
| 15 January 2006 | Titan (T10) | 2,043 km |
| 27 February 2006 | Titan (T11) | 1,812 km |
| 19 March 2006 | Titan (T12) | 1,949 km |
| 30 April 2006 | Titan (T13) | 1,856 km |
| 20 May 2006 | Titan (T14) | 1,879 km |
| 2 July 2006 | Titan (T15) | 1,906 km |
| 22 July 2006 | Titan (T16) | 950 km |
| 7 September 2006 | Titan (T17) | 1,000 km |
| 23 September 2006 | Titan (T18) | 960 km |
| 9 October 2006 | Titan (T19) | 980 km |
| 25 October 2006 | Titan (T20) | 1,030 km |
| 12 December 2006 | Titan (T21) | 1,000 km |
| 28 December 2006 | Titan (T22) | 1,297 km |
| 13 January 2007 | Titan (T23) | 1,000 km |
| 29 January 2007 | Titan (T24) | 2,631 km |
| 22 February 2007 | Titan (T25) | 1,000 km |
| 10 March 2007 | Titan (T26) | 981 km |
| 26 March 2007 | Titan (T27) | 1,010 km |
| 10 April 2007 | Titan (T28) | 991 km |
| 26 April 2007 | Titan (T29) | 981 km |
| 12 May 2007 | Titan (T30) | 959 km |
| 28 May 2007 | Titan (T31) | 2,299 km |
| 13 June 2007 | Titan (T32) | 965 km |
| 29 June 2007 | Titan (T33) | 1,933 km |
| 19 July 2007 | Titan (T34) | 1,332 km |
| 31 August 2007 | Titan (T35) | 3,324 km |
| 10 September 2007 | Iapetus | 1,622 km |
| 2 October 2007 | Titan (T36) | 973 km |
| 19 November 2007 | Titan (T37) | 999 km |
| 5 December 2007 | Titan (T38) | 1,298 km |
| 20 December 2007 | Titan (T39) | 970 km |
| 5 January 2008 | Titan (T40) | 1,014 km |
| 22 February 2008 | Titan (T41) | 1,000 km |
| 12 March 2008 | Enceladus (E3) | 48 km |
| 25 March 2008 | Titan (T42) | 999 km |
| 12 May 2008 | Titan (T43) | 1,001 km |
| 28 May 2008 | Titan (T44) | 1,400 km |
| 31 July 2008 | Titan (T45) | 1,614 km |
| 11 August 2008 | Enceladus (E4) | 49 km |
| 9 October 2008 | Enceladus (E5) | 25 km |
| 31 October 2008 | Enceladus (E6) | 169 km |
| 3 November 2008 | Titan (T46) | 1,105 km |
| 19 November 2008 | Titan (T47) | 1,023 km |
| 5 December 2008 | Titan (T48) | 961 km |
| 21 December 2008 | Titan (T49) | 971 km |
| 7 February 2009 | Titan (T50) | 967 km |
| 27 March 2009 | Titan (T51) | 963 km |
| 4 April 2009 | Titan (T52) | 4,147 km |

| | | |
|---|---|---|
| 20 April 2009 | Titan (T53) | 3,599 km |
| 5 May 2009 | Titan (T54) | 3,242 km |
| 21 May 2009 | Titan (T55) | 966 km |
| 6 June 2009 | Titan (T56) | 968 km |
| 22 June 2009 | Titan (T57) | 955 km |
| 8 July 2009 | Titan (T58) | 966 km |
| 24 July 2009 | Titan (T59) | 956 km |
| 9 August 2009 | Titan (T60) | 971 km |
| 25 August 2009 | Titan (T61) | 970 km |
| 12 October 2009 | Titan (T62) | 1,300 km |
| 2 November 2009 | Enceladus (E7) | 99 km |
| 21 November 2009 | Enceladus (E8) | 1,603 km |
| 12 December 2009 | Titan (T63) | 4,850 km |
| 28 December 2009 | Titan (T64) | 955 km |
| 12 January 2010 | Titan (T65) | 1,073 km |
| 28 January 2010 | Titan (T66) | 7,490 km |
| 2 March 2010 | Rhea | 101 km |
| 5 April 2010 | Titan (T67) | 7,462 km |
| 7 April 2010 | Dione | 503 km |
| 28 April 2010 | Enceladus (E9) | 99 km |
| 18 May 2010 | Enceladus (E10) | 435 km |
| 20 May 2010 | Titan (T68) | 1,400 km |
| 5 June 2010 | Titan (T69) | 2,044 km |
| 21 June 2010 | Titan (T70) | 880 km |
| 7 July 2010 | Titan (T71) | 1,005 km |
| 13 August 2010 | Enceladus (E11) | 2,550 km |
| 24 September 2010 | Titan (T72) | 8,175 km |
| 11 November 2010 | Titan (T73) | 7,921 km |
| 30 November 2010 | Enceladus (E12) | 48 km |
| 21 December 2010 | Enceladus (E13) | 48 km |
| 11 January 2011 | Rhea | 76 km |
| 18 February 2011 | Titan (T74) | 3,651 km |
| 19 April 2011 | Titan (T75) | 10,053 km |
| 8 May 2011 | Titan (T76) | 1,873 km |
| 20 June 2011 | Titan (T77) | 1,359 km |
| 12 September 2011 | Titan (T78) | 5,821 km |
| 1 October 2011 | Enceladus (E14) | 99 km |
| 19 October 2011 | Enceladus (E15) | 1,231 km |
| 6 November 2011 | Enceladus (E16) | 496 km |
| 12 December 2011 | Dione | 99 km |
| 13 December 2011 | Titan (T79) | 3,586 km |
| 2 January 2012 | Titan (T80) | 29,415 km |
| 30 January 2012 | Titan (T81) | 31,131 km |
| 19 February 2012 | Titan (T82) | 3,803 km |
| 27 March 2012 | Enceladus (E17) | 74 km |
| 14 April 2012 | Enceladus (E18) | 74 km |
| 2 May 2012 | Enceladus (E19) | 74 km |
| 22 May 2012 | Titan (T83) | 955 km |
| 7 June 2012 | Titan (T84) | 959 km |

| | | |
|---|---|---|
| 24 July 2012 | Titan (T85) | 1,012 km |
| 26 September 2012 | Titan (T86) | 956 km |
| 13 November 2012 | Titan (T87) | 973 km |
| 29 November 2012 | Titan (T88) | 1,014 km |
| 17 February 2013 | Titan (T89) | 1,978 km |
| 9 March 2013 | Rhea | 997 km |
| 5 April 2013 | Titan (T90) | 1,400 km |
| 23 May 2013 | Titan (T91) | 970 km |
| 10 July 2013 | Titan (T92) | 964 km |
| 26 July 2013 | Titan (T93) | 1,400 km |
| 12 September 2013 | Titan (T94) | 1,400 km |
| 14 October 2013 | Titan (T95) | 961 km |
| 1 December 2013 | Titan (T96) | 1,400 km |
| 1 January 2014 | Titan (T97) | 1,400 km |
| 2 February 2014 | Titan (T98) | 1,236 km |
| 6 March 2014 | Titan (T99) | 1,500 km |
| 7 April 2014 | Titan (T100) | 963 km |
| 17 May 2014 | Titan (T101) | 2,994 km |
| 18 June 2014 | Titan (T102) | 3,659 km |
| 20 July 2014 | Titan (T103) | 5,103 km |
| 21 August 2014 | Titan (T104) | 964 km |
| 22 September 2014 | Titan (T105) | 1,400 km |
| 24 October 2014 | Titan (T106) | 1,013 km |
| 10 December 2014 | Titan (T107) | 980 km |
| 11 January 2015 | Titan (T108) | 970 km |
| 12 February 2015 | Titan (T109) | 1,200 km |
| 16 March 2015 | Titan (T110) | 2,275 km |
| 7 May 2015 | Titan (T111) | 2,722 km |
| 16 June 2015 | Dione | 516 km |
| 7 July 2015 | Titan (T112) | 10,953 km |
| 17 August 2015 | Dione | 474 km |
| 28 September 2015 | Titan (T113) | 1,036 km |
| 14 October 2015 | Enceladus (E20) | 1,839 km |
| 28 October 2015 | Enceladus (E21) | 49 km |
| 13 November 2015 | Titan (T114) | 11,920 km |
| 19 December 2015 | Enceladus (E22) | 4,999 km |
| 16 January 2016 | Titan (T115) | 3,817 km |
| 1 February 2016 | Titan (T116) | 1,400 km |
| 16 February 2016 | Titan (T117) | 1,018 km |
| 4 April 2016 | Titan (T118) | 990 km |
| 6 May 2016 | Titan (T119) | 971 km |
| 7 June 2016 | Titan (T120) | 975 km |
| 25 July 2016 | Titan (T121) | 976 km |
| 10 August 2016 | Titan (T122) | 1,599 km |
| 27 September 2016 | Titan (T123) | 1,737 km |
| 14 November 2016 | Titan (T124) | 1,582 km |
| 29 November 2016 | Titan (T125) | 3,223 km |
| 22 April 2017 | Titan (T126) | 979 km |

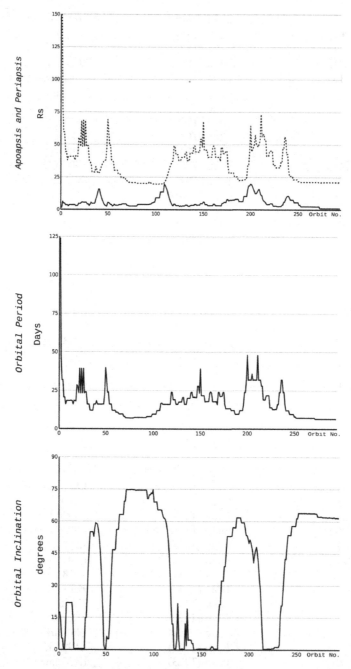

The orbital evolution of Cassini during its 13-year exploration mission. Orbit modifications were due in large part to Titan gravity-assists. Note the periods spent in almost circular orbits with periapsis and apoapsis coincident, as well as the different high and low inclination portions of the mission.

# Chapter references

[Abdel-jawad-2008] Abdel-jawad, M.M., Goldsworthy, M.J., Macrossan, M.N., "Stability Analysis of Beagle 2 in Free Molecular and Transition Regimes", Journal of Spacecraft and Rockets, 45, 2008, 1207-1212

[Abe-1991] Abe, T., Kawaguchi, J., Suzuki, K., "Feasibility Study of Mars Exploration by Using Aerocapture", paper dated 1991

[Abe-2006a] Abe, M., et al., "Near-Infrared Spectral Results of Asteroid Itokawa from the Hayabusa Spacecraft", Science, 312, 2006, 1334-1338

[Abe-2006b] Abe, S., et al., "Mass and Local Topography Measurements of Itokawa by Hayabusa", Science, 312, 2006, 1344-1347

[Abe-2011] Abe, S., et al., "Near-Ultraviolet and Visible Spectroscopy of HAYABUSA Spacecraft Re-entry", arXiv astro-ph/1108.5982 preprint

[Adams-2005] Adams, D., Sabahi, D., Mobrem, M., "MARSIS Antenna Deployment Testing and Analysis", presentation at the Spacecraft and Launch Vehicle Dynamic Environments Workshop, El Segundo, June 2005

[Allton-2005] Allton, J.H., Stansbery, E.K., McNamara, K.M., "Size Distribution of Genesis Solar Wind Array Collector Fragments Recovered", paper presented at the XXXVI Lunar and Planetary Science Conference, Houston, 2005

[Anderson-2007] Anderson, J.D., et al., "Saturn's Gravitational Field, Internal Rotation, and Interior Structure", Science 317, 2007, 1384-1386

[Angrilli-2004] Angrilli, F., et al., "IPSE: The Italian Package for Scientific Experiments on Mars", Planetary and Space Science, 52, 2004, 41-45

[Arvidson-2000] Arvidson, R.E., et al., "FIDO Field Trials in Preparation for Mars Rover Exploration and Discovery and Sample Return Missions", paper presented at the Workshop on Concepts and Approaches for Mars Exploration, July 18-20, 2000, Houston, Texas

[Arvidson-2004a] Arvidson, R.E., et al., "Localization and Physical Property Experiments Conducted by Opportunity at Meridiani Planum", Science, 306, 2004, 1730-1733

[Arvidson-2004b] Arvidson, R.E., et al., "Localization and Physical Properties Experiments Conducted by Spirit at Gusev Crater", Science, 305, 2004, 821-824

[Arvidson-2005] Arvidson, R.E., et al., "Spectral Reflectance and Morphologic Correlations in Eastern Terra Meridiani, Mars", Science, 307, 2005, 1591-1594

[ASI-1999] "Deep Drill System (DeeDri) for Mars Surveyor Program 2003 Subsystem Proposal Information Package (S_PIP)", Rome, ASI and Tecnospazio, 1999

[Atreya-2007] Atreya, S., "Titan's Organic Factory", Science, 316, 2007, 843-844

[AWST-1989] "Cassini to Provide Detailed, Extended Views of Saturn", Aviation Week and Space Technology, 9 October 1989, 109-110

[AWST-2000a] "Nozomi on Target for Mars", Aviation Week & Space Technology, 11 December 2000, 84

[AWST-2000b] Aviation Week & Space Technology, 30 October 2000, 24

[Bagenal-2005] Bagenal, F., "Saturn's Mixed Magnetosphere", Nature, 433, 2005, 695-696

[Bagenal-2007] Bagenal, F., "A New Spin on Saturn's Rotation", Science, 316, 2007, 380-381

[Baines-1995] Baines, K.H., et al., "VESAT: The Venus Environmental Satellite Discovery Mission", Acta Astronautica, 35, 1995, 417-425

[Baines-2005] Baines, K.H., et al., "The Atmospheres of Saturn and Titan in the Near-Infrared: First Results of Cassini/VIMS", Earth, Moon, and Planets 96, 2005, 119–147

[Baker-2003] Baker, V.R., "Icy Martian Mysteries", Nature, 426, 2003, 779-780

[Baland-2011] Baland, R.M., et ql., "Titan's Obliquity as Evidence for a Subsurface Ocean?", arXiv astro-ph/1104.2741 preprint

[Bandfield-2007] Bandfield, J.L., "High-Resolution Subsurface Water-Ice Distributions on Mars", Nature, 447, 2007, 64-67

[Banerdt-2010] Banderdt, W.B., "Mars Exploration Rovers Science Results from 6 ¼ Years on Mars", presentation at the Planet Mars III Workshop, Les Houches, April 2010

[Barabash-2006] Barabash, S., et al., "The Analyzer of Space Plasmas and Energetic Atoms (ASPERA-3) for the Mars Express Mission", Space Science Reviews, 126, 2006, 113–164

[Barabash-2007] Barabash, S., et al., "Martian Atmospheric Erosion Rates", Science, 315, 2007, 501-503

[Baratoux-2011] Baratoux, D., et al., "Thermal History of Mars Inferred from Orbital Geochemistry of Volcanic Provinces", Nature, 472, 2011, 338-341

[Barbieri-2003] Barbieri, A., et al., "Specular Reflection of Odyssey's UHF Beacon from the Northern Latitudes of Mars", paper presented at the Fall 2003 Meeting of the American Geophysical Union

[Barnes-2005] Barnes, J.W., et al., "A 5-Micron-Bright Spot on Titan: Evidence for Surface Diversity", Science, 310, 2005, 92-95

[Barnes-2008] Barnes, J., "Titan: Earth in Deep Freeze", Sky & Telescope, December 2008, 26-32

[Barnes-2009a] Barnes, J.W., et al., "VIMS Spectral Mapping Observations of Titan during the Cassini Prime Mission", Planetary and Space Science, 2009, 57, 1950–1962

[Barnes-2009b] Barnes, J.W., et al., "Shoreline features of Titan's Ontario Lacus from Cassini/VIMS observations", Icarus, 201, 2009, 217–225

[Barnes-2011] Barnes, J.W., et al., "Wave Constraints for Titan's Jingpo Lacus and Kraken Mare from VIMS Specular Reflection Lightcurves", Icarus, 211, 2011, 722-731

[Barraclough-2003] Barraclough B.L., et al., "The plasma ion and electron instruments for the Genesis Mission", Space Science Review, 105, 2003, 627-660

[Barraclough-2004] Barraclough, B.L., et al., "The Genesis Mission Solar Wind Collection: Solar Wind Statistics over the Period of Collection", paper presented at the XXXV Lunar and Planetary Science Conference, Houston, 2004

[Basilevsky-2008] Basilevsky, A.T., et al., "New MEX HRSC/SRC Images of Phobos and the Fobos-Grunt Landing Sites", paper presented at the 48th Vernadsky/Brown Micro-symposium on Comparative Planetology, Moscow, October 2008

[Battistelli-2004] Battistelli, E., et al., "Scientific Instruments Studied by Galileo Avionica for Mars Surface Exploration", Planetary and Space Science, 52, 2004, 47-53

[Baumgertner-2005] Baumgertner, E.T., et al., "The Mars Exploration Rover Instrument Positioning System", paper presented at the 2005 IEEE Aerospace Conference

[Beckes-2005] Beckes, H., et al., "Titan's Magnetic Field Signature During the First Cassini Encounter", Science, 308, 2005, 992-995

[Beckman-1986] Beckman, J., Scoon, G.E.N., "Project Cassini – A Potential Collaborative ESA/NASA Saturn Orbiter and Titan Probe Mission", Acta Astronautica, 14, 1986, 185-194

[Bell-2002] Bell, J., "Tip of the Martian Iceberg?", Science, 297, 2002, 60-61

[Bell-2003] Bell, J.F. III, et al., "The Panoramic Camera (Pancam) Investigation on the NASA 2003 Mars Exploration Rover Mission", paper presented at the XXXIV Lunar and Planetary Science Conference, Houston, 2003

[Bell-2004a] Bell, J.F. III, et al., "Pancam Multispectral Imaging Results from the Opportunity Rover at Meridiani Planum", Science, 306, 2004, 1703-1709

[Bell-2004b] Bell, J.F. III, et al., "Pancam Multispectral Imaging Results from the Spirit Rover at Gusev Crater", Science, 305, 2004, 800-806

[Bell-2005] Bell, J.F. III, et al., "Solar Eclipses of Phobos and Deimos Observed from the Surface of Mars", Nature, 436, 2005, 55-57

[Bell-2006] Bell, J., "Backyard Astronomy from Mars", Sky & Telescope, August 2006, 41-44

[Bertaux-2004] Bertaux, J.-L., et al., "SPICAM: Studying the Global Structure and Composition of the Martian Atmosphere", in: "Mars Express: The Scientific Payload", Noordwijk, ESA SP-1240, August 2004, 95-120

[Bertaux-2005a] Bertaux, J.-L., et al, "Discovery of an Aurora on Mars", Nature, 435, 2005, 790-794

[Bertaux-2005b] Bertaux, J.-L., et al., "Nightglow in the Upper Atmosphere of Mars and Implications for Atmospheric Transport", Science, 307, 2005, 566-569

[Bertotti-2003] Bertotti, B., Iess, L., Tortora, P., "A Test of General Relativity Using Radio Links with the Cassini Spacecraft", Nature, 425, 2003, 374-376

[Bertucci-2008] Bertucci, C., et al., "The Magnetic Memory of Titan's Ionized Atmosphere", Science, 321, 2008, 1475-1478

[Bibring-2004] Bibring, J.-P., et al., "Perennial Water Ice Identified in the South Polar Cap of Mars", Nature, 428, 2004, 627-630

[Bibring-2005] Bibring, J.-P., et al., "Mars Surface Diversity as Revealed by the OMEGA/ Mars Express Observations", Science, 307, 2005, 1576-1581

[Bibring-2006] Bibring, J.-P., et al., "Global Mineralogical and Aqueous Mars History Derived from OMEGA/Mars Express Data", Science, 312, 2006, 400-404

[Bibring-2007] Bibring, J.-P., et al., "Coupled Ferric Oxides and Sulfates on the Martian Surface", Science, 317, 2007, 1206-1210

[Bibring-2009a] Bibring, J.-P., "Mars: Planète Bleue?" (Mars, Blue Planet?), Odile Jacob, 2009, 133-135 (in French)

[Bibring-2009b] ibid., 141-142

[Bibring-2009c] ibid., 142

[Bibring-2009d] ibid.,145-170

[Bille-2004] Bille, M., Lishock, E., "The First Space Race", Texas A&M University Press, 2004, 140-150

[Bird-2005] Bird, M.K., et al., "The Vertical Profile of Winds on Titan", Nature, 438, 2005, 800-802

[Boehnardt-1999] Boehnardt, H., et al., "The Nuclei of Comets 26P/Grigg-Skjellerup and 73P/ Schwassmann-Wachmann 3", Astronomy and Astrophysics, 341, 1999, 912-917

[Bolton-2002] Bolton, S.J., et al., "Ultra-Relativistic Electrons in Jupiter's Radiation Belts", Nature, 415, 2002, 987-991

[Bonitz-1998] Bonitz, R.G., "Mars Surveyor '98 Lander MVACS Robotic Arm Control

Systen Design Concepts", in: Proceedings of the IEEE International Conference on Robotics and Automation, 1997, 2465-2470

[Bonitz-2000] Bonitz, R.G., Nguyen, T.T., Kim, W.S., "The Mars Surveyor '01 Rover and Robotic Arm", paper presented at the IEEE Aerospace Conference, March 2000

[Bonitz-2001] Bonitz, R.G., et al., "MVACS Robotic Arm", Journal of Geophysical Research, 106, 2001, 17623-17634

[Borde-2004] Borde, J., Poinsignon, V., Schmidt, R., "Mars Express Mission Outcome: Scientific and Technological Return of the First European Satellite Around the Red Planet", paper presented at the 55th International Astronautical Congress, Vancouver, 2004

[Borovicka-2011] Borovicka, J., et al., "Photographic Observations of the Hayabusa Re-entry", arXiv astro-ph/1108.6006 preprint

[Boynton-2001] Boynton, W.V., et al., "Thermal and Evolved Gas Analyzer: Part of the Mars Volatile and Climate Surveyor Integrated Payload", Journal of Geophysical Research, 106, 2001, 17683-17698

[Boynton-2002] Boynton, W.V., et al., "Distribution of Hydrogen in the Near Surface of Mars: Evidence for Subsurface Ice Deposits", Science, 297, 2002, 81-85

[Brad Dalton-2005] Brad Dalton, J., "Saturn's Retrograde Renegade", Nature, 433, 2005, 695-696

[Britt-2004] Britt, D.T., et al., "The Morphology and Surface Processes of Comet 19/P Borrelly", Icarus, 167, 2004, 45-53

[Brown-2006] Brown, R.H.., et al., "Composition and Physical Properties of Enceladus' Surface", Science, 311, 2006, 1425-1428

[Brown-2008a] Brown, M.E., et al., "Discovery of Lake-Effect Clouds on Titan", arXiv astro-ph/0809.1841 preprint

[Brown-2008b] Brown, R.H., et al., "The Identification of Liquid Ethane in Titan's Ontario Lacus", Nature, 454, 2008, 607-610

[Brownlee-1996] Brownlee, D.E., et al., "Stardust: Finessing Expensive Cometary Sample Returns", Acta Astronautica, 39, 1996, 51-60

[Brownlee-2003] Brownlee, D.E., et al., "Stardust: Comet and Interstellar Dust Sample Return Mission", Journal of Geophysical Research, 108, 2003

[Brownlee-2004] Brownlee, D.E., et al., "Surface of Young Jupiter Family Comet 81P/Wild 2: View From the Stardust Spacecraft", Science, 304, 2004, 1764-1769

[Brownlee-2006] Brownlee, D.E., et al., "Comet 81P/Wild 2 Under a Microscope", Science, 314, 2006, 1711-1716

[Buratti-2004a] Buratti, B.J., et al., "9969 Braille: Deep Space 1 Infrared Spectroscopy, Geometric Albedo, and Classification", Icarus, 167, 2004, 129-135

[Buratti-2004b] Buratti, B.J., et al., "Deep Space 1 Photometry of the Nucleus of Comet 19P/ Borrelly", Icarus, 167, 2004, 16-29

[Buratti-2009] Buratti, B.J., Faulk, S.P., "A Search for Plume Activity on Mimas, Tethys, and Dione with Cassini VIMS High Solar Phase Angle Observations", NASA Undergraduate Student Research Program paper dated August 2009

[Burch-2007] Burch, J.L., et al., "Tethys and Dione as Sources of Outward-Flowing Plasma in Saturn's Magnetosphere", Nature, 447, 2007, 833-835

[Burnett-2003] Burnett D.S., et al., "The Genesis Discovery Mission: Return of solar matter to Earth", Space Science Review, 105, 2003, 509-534

[Burnett-2005] Burnett, D.B., et al., "Molecular Contamination on Anodized Aluminium Components of the Genesis Science Canister", paper presented at the XXXVI Lunar and Planetary Science Conference, Houston, 2005

[Burnett-2006a] Burnett, D.S., "NASA Returns Rocks from a Comet", Science, 314, 2006, 1709-1710

[Burnett-2006b] Burnett, D.B., and the Genesis Science Team, "Genesis Mission: Overview and Status", paper presented at the XXXVII Lunar and Planetary Science Conference, Houston, 2006

[Burnett-2011] Burnett, D.S., and Genesis Science Team, "Solar composition from the Genesis Discovery Mission", Proceedings of the National Academy of Sciences of the United States of America, 108, 2011, 19147-19151

[Burns-2010] Burns, J.A., "The Birth of Saturn's Baby Moons", Nature, 465, 2010, 701-702

[Butler-2003] Butler, D., "Are You on Board?", Nature, 423, 2003, 476

[Calcutt-1992] Calcutt, S., et al., "The Composite Infrared Spectrometer", Journal of the British Interplanetary Society, 45, 1992, 381-386

[Canup-2010] Canup, R.M., "Origin of Saturn's Rings and Inner Moons by Mass Removal from a Lost Titan-Sized Satellite", Nature, 468, 2010, 943-946

[Carroll-1993] Carroll, M.W., "Cheap Shots", Astronomy, August 1993, 38-47

[Carroll-1995] Carroll, M., "New Discoveries on the Horizon: NASA's Next Missions", Astronomy, November 1995, 36-43

[Carroll-1997] Carroll, M., "Europa: Distant Ocean, Hidden Life?", Sky & Telescope, December 1997, 50-55

[Carusi-1985] Carusi, A., et al., "Long-Term Evolution of Short-Period Comets", Bristol, Adam Hilger, 1985

[Casani-1996] Casani, E.K., Stocky, J.F., Rayman, M.D., "Solar Electric Propulsion", paper presented at the First IAA Symposium on Realistic Near-Term Advanced Scientific Space Missions, Aosta, 25–27 June 1996

[Cesarone-1999] Cesarone, R.J., et al., "Architectural Design for a Mars Communications & Navigation Orbital Infrastructure", paper AAS 99-300

[Chan-2004] Chan, M.A., et al., "A Possible Terrestrial Analogue for Haematite Concretions on Mars", Nature, 429, 2004, 731-734

[Charnoz-2005] Charnoz, S., et al., "Cassini Discovers a Kinematic Spiral Ring around Saturn", Science, 310, 2005, 1300-1304

[Charnoz-2007] Charnoz, S. et al., "The Equatorial Ridges of Pan and Atlas: Terminal Accretionary Ornaments?", Science, 318, 2007, 1622-1624

[Charnoz-2009] Charnoz, S., "Physical Collisions of Moonlets and Clumps with the Saturn's F-Ring Core", arXiv astro-ph/0901.0482 preprint

[Charnoz-2010] Charnoz, S., Salmon, J., Crida, A., "The Recent Formation of Saturn's Moonlets from Viscous Spreading of the Main Rings", Nature, 465, 2010,752-754

[Cheng-2000] Cheng, A.F., Barnouin-Jha, O.S., Pieters, C.M., "Aladdin: Sample Collection from the Moons of Mars", paper presented at the Workshop on Concepts and Approaches for Mars Exploration, July 18-20, 2000, Houston, Texas

[Chicarro-2003] Chicarro, M., Martin, P., Troutenet, R., "Mars Express – Unravelling the Scientific Mysteries of Mars", ESA Bulletin, 115, 2003, 18-25

[Chicarro-2006] Chicarro, M., "One Martian Year in Orbit – The Science from Mars Express", ESA Bulletin, 125, 2006, 17-19

[Christensen-2003a] Christensen, P.R., et al., "Formation of Recent Martian Gullies through Melting of Extensive Water-Rich Snow Deposits", Nature, 422, 2003, 45-48

[Christensen-2003b] Christensen, P.R., et al., "Morphology and Composition of the Surface of Mars: Mars Odyssey THEMIS Results", Science, 300, 2003, 2056-2061

[Christensen-2004a] Christensen, P.R., et al., "Mineralogy at Meridiani Planum from the Mini-TES Experiment on the Opportunity Rover", Science, 306, 2004, 1733-1739

[Christensen-2004b] Christensen, P.R., et al., "Initial Results from the Mini-TES Experiment in Gusev Crater from the Spirit Rover", Science, 305, 2004, 837-842

[Christensen-2005] Christensen, P.R., et al., "Evidence for Magmatic Evolution and Diversity on Mars from Infrared Observations", Nature, 436, 2005, 504-509

[Clark-2005a] Clark, R.N., et al., "Compositional Maps of Saturn's Moon Phoebe from Imaging Spectroscopy", Nature, 435, 2005, 66-69

[Clark-2005b] Clark, B.E., Grant, K.B., "Japan's Asteroid Archaeologist", Sky & Telescope, June 2005, 34-37

[Clark-2009] Clark, R.N., "Detection of Adsorbed Water and Hydroxyl on the Moon", Science, 2009, 326, 562-564

[Clarke-2005] Clarke, J.T., et al., "Morphological Differences between Saturn's Ultraviolet Aurorae and Those of Earth and Jupiter", Nature, 433, 2005, 717-719

[Clayton-2011] Clayton, R.N., "The Earth and the Sun", Science, 332, 2011, 1509-1510

[CNES-2002] "2007 Orbiter Mission Specification", document MARS-TS-00-001-CNES dated 25 January 2002

[Coates-1992] Coates, A.J., et al., "The Electron Spectrometer for the Cassini Spacecraft", Journal of the British Interplanetary Society, 45, 1992, 387-392

[Cochran-2002] Cochran, A., et al., "The COmet Nucleus TOUR (CONTOUR), A NASA Discovery Mission", Earth Moon and Planets, 89, 2002, 289-300

[Colangeli-2000] Colangeli, L., "The Martian Atmospheric Grain Observer (MAGO) for In-Situ Dust Analysis", paper presented at the Workshop on Concepts and Approaches for Mars Exploration, July 18-20, 2000, Houston, Texas

[Colozza-2003] Colozza, A., Landis, G., Lyons, V., "Overview of Innovative Aircraft Power and Propulsion Systems and Their Applications for Planetary Exploration", NASA TM-2003-212459

[Conway-2007] Conway, E.M., Flores, M., "Deep Space 1: A Revolution in Space Exploration", Quest, 14, No. 2, 2007, 41-51

[Cordier-2000] Cordier, B., et al., "MA-FLUX: The X-Ray Fluorescence Experiment inside the IPSE Laboratory", paper presented at the Workshop on Concepts and Approaches for Mars Exploration, July 18-20, 2000, Houston, Texas

[Corneille-2004] Corneille, P., "High Life on Mars", Spaceflight, October 2004, 389-392

[Corneille-2005a] Corneille, P., "Roving on the Red Planet", Spaceflight, March 2005, 102-106

[Corneille-2005b] Corneille, P., "Extended Mission for MER Twins", Spaceflight, September 2005, 339-343

[Corneille-2006] Corneille, P., "Two Years of MER Operations on Mars", Spaceflight, April 2006, 140-144

[Corneille-2007] Corneille, P., "Three Years of MER Operations", Spaceflight, January 2007, 16-19

[Corneille-2008] Corneille, P., "Four Years of MER Operations on Mars", Spaceflight, January 2008, 22-25

[Corneille-2009] Corneille, P., "Five Years of MER on the Red Planet", Spaceflight, February 2009, 59-63

[Corneille-2010] Corneille, P., "Six Years of MER Operations on Mars", Spaceflight, January 2010, 20-25

[Coué-2007] Coué, P., "La Chine Veut la Lune" (China Wants the Moon), Paris, A2C Medias, 2007, 150 (in French)

[Covault-1994] Covault, C., "U.S., Russia Plan New Mars Mission", Aviation Week & Space Technology, 6 June 1994, 24-25

[Covault-1997] Covault, C., "Saturn's Mysteries Beckon Cassini", Aviation Week & Space Technology, 20 October 1997, 22-24

[Covault-2001] Covault, C., "U.S. Poised for Return to Mars", Aviation Week & Space Technology, 2 April 2001, 36-38

[Covault-2002] Covault, C., "Boeing Delta Rockets Contour toward Comets", Aviation Week & Space Technology, 8 July 2002, 25

[Covault-2003a] Covault, C., "Mars Beckons", Aviation Week & Space Technology, 16 June 2003, 61-62

[Covault-2003b] Covault, C., "Limited Opportunity", Aviation Week & Space Technology, 7 July 2003, 34-35

[Covault-2003c] Covault, C., "Fast Action for Rover", Aviation Week & Space Technology, 14 July 2003, 34-35

[Covault-2003d] Covault, C., "Taking Mars", Aviation Week & Space Technology, 30 November 2003, page unknown

[Covault-2004a] Covault, C., "Titan Revealed", Aviation Week & Space Technology, 1 November 2004, 42-43

[Covault-2004b] Covault, C., "Stardust's Adventure", Aviation Week & Space Technology, 12 January 2004, 29

[Covault-2004c] Covault, C., "Bedrock and Pay Dirt", Aviation Week & Space Technology, 2 February 2004, 32-36

[Covault-2004d] Covault, C., "Probing Martian Mysteries", Aviation Week & Space Technology, 9 February 2004, 31-32

[Covault-2004e] Covault, C., "Men from Earth", Aviation Week & Space Technology, 31 May 2004, 62-68

[Covault-2004f] Covault, C., "Sailing a Martian Sea", Aviation Week & Space Technology, 19 July 2004, 180-184

[Covault-2004g] Covault, C., "Roll On, Rovers!" Aviation Week & Space Technology, 12 April 2004, 53

[Covault-2004h] Covault, C., "Martian Mountaineering", Aviation Week & Space Technology, 23 August 2004, 40-42

[Covault-2004i] Covault, C., "The Real Martians", Aviation Week & Space Technology, 1 November 2004, 30-37

[Covault-2005] Covault, C., "Martian Bonus", Aviation Week & Space Technology, 25 July 2005, 27

[Covault-2007a] Covault, C., "Over the Edge", Aviation Week & Space Technology, 9 July 2007, 22

[Covault-2007b] Covault, C., "Surviving the Storm", Aviation Week & Space Technology, 16 July 2007, 31

[Covault-2007c] Covault, C., "Opportunity's Knocks", Aviation Week & Space Technology, 30 July 2007, 34

[Covault-2007d] Covault, C., "Victorian Image", Aviation Week & Space Technology, 8 October 2007, 39

[Covault-2008] Covault, C., "The Outer Limit", Aviation Week & Space Technology, 14 April 2008, 30-32

[Cowley-1999] Cowley, L.T., Schroeder, M., "Forecasting Martian Halos", Sky & Telescope, December 1999, 60-64

[Crabb-2003] Crabb, C., "NASA Bails Out of French-Led Mars Mission", Science, 300, 2003, 719

[Crary-2005] Crary, F.J., et al., "Solar Wind Dynamic Pressure and Electric Field as the Main Factors Controlling Saturn's Aurorae", Nature, 433, 2005, 720-722

[Crida-2010] Crida, A., Charnoz, S., "Recipe for Making Saturn's Rings", Nature, 468, 2010, 904-905

[Crovisier-1996] Crovisier, J., et al., "What Happened to Comet 73P/Schwassmann-Wachmann 3?", Astronomy and Astrophysics, 310, 1996, L17-L20

[Crowley-2007] Crowley, G., Tolson, R.H., "Mars Thermospheric Winds from Mars Global Surveyor and Mars Odyssey Accelerometers", Journal of Spacecraft and Rockets, 44, 2007, 1188-1194

[Cruikshank-1972] Cruikshank, D.P., Morrison, D., "Titan and Its Atmosphere", Sky & Telescope, August 1972, 83-85

[Cruikshank-2007] Cruikshank, D.P., et al., "Surface composition of Hyperion", Nature, 448, 2007, 54-57

[Cull-2007] Cull, S., "Martian Photo Opportunity", Sky & Telescope, January 2007, 30-32

[Cushing-2007] Cushing, G.E., et al., "THEMIS Observes Possible Cave Skylights on Mars", paper presented at the XXXVIII Lunar and Planetary Science Conference, Houston, 2007

[Cuzzi-2010] Cuzzi, J.N., "An Evolving View of Saturn's Dynamic Rings", Science, 327, 2010, 1470-1475

[Cyranoski-2010] Cyranoski, D., "Space Capsule Probed for Asteroid Dust", Nature, 466, 2010, 16-17

[Dandouras-2009] Dandouras, I., et al., "Titan's Exosphere and its Interaction with Saturn's Magnetosphere", Philosophical Transactions of the Royal Society A: Mathematical, Physical and Engineering Sciences, 367, 2009, 743-752

[Davies-1988] Davies, J.K., "Satellite Astronomy: The Principles and Practice of Astronomy from Space", Chichester, Ellis Horwood, 1988, 156-158

[Davis-2007] Davis, D. C., Patterson, C., Howell, K. C., "Solar Gravity Perturbations to Facilitate Long-Term Orbits: Application to Cassini" Paper AAS 07-275

[Déau-2009] Déau, E., et al., "The Opposition Effect in Saturn's Rings Seen by Cassini/ISS: I. Morphology of Phase Curves", arXiv astro-ph/0901.0289 preprint

[Demura-2006] Demura, H., et al., "Pole and Global Shape of 25143 Itokawa", Science, 312, 2006, 1347-1349

[Denis-2006] Denis, M., et al., "Deployment of the MARSIS Radar Antennas On-Board Mars Express", paper presented at the Space Operations Conference, Rome, Italy, June 2006

[Denis-2009] Denis, M., et al., "Ordinary Camera, Extraordinary Places – Visual Monitoring Cameras in the ESA Fleet", ESA Bulletin, 139, 2009, 29-33

[Denk-2010] Denk, T., et al., "Iapetus: Unique Surface Properties and a Global Color Dichotomy from Cassini Imaging", Science, 327, 2010, 435-439

[Desai-1998] Desai, P.N., "Mars Ascent Vehicle Flight Analysis", Paper AIAA-98-2850

[Desai-2000] Desai, S., et al., "Autonomous Optical Navigation (AutoNav) Technology Validation Report". In: "Deep Space 1 Technology Validation Reports", Washington, NASA, 2000

[Desai-2008a] Desai, P.N., et al., "Entry, Descent, and Landing Operations Analysis for the Stardust Entry Capsule", Journal of Spacecraft and Rockets, 45, 2008, 1262-1268

[Desai-2008b] Desai, P.N., Qualls, G.D., "Stardust Entry Reconstruction", paper AIAA-2008-1198

[Desai-2008c] Desai, P.N., Qualls, G.D., Schoenberger, M., "Reconstruction of the Genesis Entry", Journal of Spacecraft and Rockets, 45, 2008, 33-38

[Desai-2008d] Desai, P.N., Knocke, P.C., "Mars Exploration Rovers Entry, Descent and Landing Trajectory Analysis", Journal of the Astronautical Sciences, 55, 2008

[Deutsch-2002] Deutsch, L.J., "Resolving the Cassini/Huygens Relay Radio Anomaly", paper presented at the 2002 IEEE Aerospace Conference, Big Sky, Montana

[Dombard-2007] Dombard, A.J., "Cracks under Stress", Nature, 447, 2007, 276-277

[Domokos-2007] Domokos, A., et al., "Measurement of the meteoroid flux at Mars", Icarus, 191, 2007, 141-150

[Domokos-2009] Domokos, G., et al., "Formation of Sharp Edged and Planar Areas of Asteroids by Polyhedral Abrasion", arXiv astro-ph/0904.4423 preprint

[Dornheim-1996] Dornheim, M.A., "Cassini Mission to Saturn Caps Era of Grand Spacecraft", Aviation Week & Space Technology, 9 December 1996, 71-75

[Dornheim-1998] Dornheim, M.A., "Cassini Gets First Boost from Venus", Aviation Week & Space Technology, 4 May 1998, 41

[Dornheim-2001a] Dornheim, M.A., "ESA Board Suggests Procedures for Huygens Telemetry", Aviation Week & Space Technology, 8 January 2001, 21

[Dornheim-2001b] Dornheim, M.A., "Cassini and Galileo Spacecraft Jointly Observe Jupiter", Aviation Week & Space Technology, 8 January 2001, 20

[Dornheim-2003a] Dornheim, M.A., "Red Rover, Red Rover", Aviation Week & Space Technology, 26 May 2003, 54-56

[Dornheim-2003b] Dornheim, M.A., "Can $$$ Buy Time?", Aviation Week & Space Technology, 26 May 2003, 56-58

[Dornheim-2004a] Dornheim, M.A., "Discoveries Already", Aviation Week & Space Technology, 5 July 2004, 22-24

[Dornheim-2004b] Dornheim, M.A., "Nabbing Titan", Aviation Week & Space Technology, 12 July 2004, 27-29

[Dornheim-2004c] Dornheim, M.A., "Spirit Itches to Move", Aviation Week & Space Technology, 12 January 2004, 24-28

[Dornheim-2004d] Dornheim, M.A., "Rover Crunch Time", Aviation Week & Space Technology, 26 January 2004, 27-28

[Dornheim-2004e] Dornheim, M.A., "Martian Pearls", Aviation Week & Space Technology, 16 February 2004, 32-34

[Dornheim-2004f] Dornheim, M.A., "In the Trenches", Aviation Week & Space Technology, 23 February 2004, 35-36

[Dornheim-2004g] Dornheim, M.A., "Berries are Hematite", Aviation Week & Space Technology, 22 March 2004, 31

[Dornheim-2004h] Dornheim, M.A., "Extended Mission", Aviation Week & Space Technology, 5 April 2004, 33-34

[Dornheim-2004i] Dornheim, M.A., "Separated at Birth?", Aviation Week & Space Technology, 19 April 2004, 61-63

[Dornheim-2004j] Dornheim, M.A., "To a Richer Outcrop", Aviation Week & Space Technology, 3 May 2004, 40

[Dornheim-2004k] Dornheim, M.A., "Spirited Revival", Aviation Week & Space Technology, 2 February 2004, 36-38

[Dornheim-2004l] Dornheim, M.A., "Ready to Roll", Aviation Week & Space Technology, 9 February 2004, 33-34

[Dornheim-2004m] Dornheim, M.A., "Spirit at Big Crater", Aviation Week & Space Technology, 15 March 2004, 32-33

[Dornheim-2004n] Dornheim, M.A., "Time Machine", Aviation Week & Space Technology, 10 May 2004, 25-28

[Dornheim-2004o] Dornheim, M.A., "Martian Sand Traps", Aviation Week & Space Technology, 15 November 2004, 75-76

[Dornheim-2005a] Dornheim, M.A., "Mineral Water", Aviation Week & Space Technology, 10 January 2005, 37

[Dornheim-2005b] Dornheim, M.A., "Mission Rejuvenation", Aviation Week & Space Technology, 31 January 2005, 24-25

[Dornheim-2005c] Dornheim, M.A., "Mars Mission Grows", Aviation Week & Space Technology, 25 April 2005, 68-70

[Dornheim-2005d] Dornheim, M.A., "King of the Hill", Aviation Week & Space Technology, 12 September 2005, 60-61

[Dornheim-2005e] Dornheim, M.A., "Freeing Opportunity", Aviation Week & Space Technology, 9 May 2005, 33-35

[Dornheim-2005f] Dornheim, M.A., "Free at Last", Aviation Week & Space Technology, 20 June 2005, 64

[Dougherty-2005] Dougherty, M.K., et al., "Cassini Magnetometer Observations During Saturn Orbit Insertion", Science, 307, 2005, 1266-1270

[Dougherty-2006] Dougherty, M.K., et al., "Identification of a Dynamic Atmosphere at Enceladus with the Cassini Magnetometer", Science, 311, 2006, 1406-1409

[Douté-2007] Douté, S., et al., "South Pole of Mars: Nature and Composition of the icy Terrains from Mars Express OMEGA Observations", Planetary and Space Science, 55, 2007, 113–133

[Dunham-2004] Dunham, D.W., Farquhar, R.W., "Background and Applications of Astrodynamics for Space Missions of the Johns Hopkins Applied Physics Laboratory". In: Belbruno, E., Folta, D., Gurfil, P., "Astrodynamics, Space Missions and Chaos", Annals of the New York Academy of Sciences, 1017, 2004

[Duxbury-2004] Duxbury, T.C., et al., "Asteroid 5535 Annefrank Size, Shape, and Orientation: Stardust First Results", Journal of Geophysical Research, 109, 2004, E02002.1-E02002.5

[Dyudina-2008] Dyudina, U.A., et al., "Dynamics of Saturn's South Polar Vortex", Science, 319, 2008, 1801

[Ebihara-2011] Ebihara, M., et al., "Neutron Activation Analysis of a Particle Returned from Asteroid Itokawa", Science, 333, 2008, 1116-1121

[Edwards-1999] Edwards, C., "Mars Network: First Stop on the Interplanetary Internet", presentation of the Mars Network Project Office Telecommunications and Mission Operations Directorate, 5 October 1999

[Edwards-2006] Edwards, C.D. Jr., et al., "Relay Communications Strategies for Mars Exploration Through 2020", Acta Astronautica, 59, 2006, 310-318

[Elachi-2005] Elachi, C., et al., "Cassini Radar Views the Surface of Titan", Science, 308, 2005, 970-974

[Elachi-2006] Elachi, C., et al., "Titan Radar Mapper Observations from Cassini's T3 Fly-by", Nature, 441, 2006, 709-713

[Encrenaz-2008a] Encrenaz, T., "Search for Methane on Mars: Observations, Interpretation and Future Work", Advances in Space Research, 42, 2008, 1-5

[Encrenaz-2008b] Encrenaz, T., et al., "A Study of the Martian Water Vapor over Hellas using OMEGA and PFS Aboard Mars Express", Astronomy & Astrophysics, 484, 2008, 547-553

[Erickson-2004] Erickson, J.K., Manning, R., Adler, M., "Mars Exploration Rover: Launch, Cruise, Entry, Descent and Landing", paper presented at the 55th International Astronautical Congress, Vancouver, October 2004

[ESA-2003] Mars Express Project Team, "Mars Express – Closing in on the Red Planet", ESA Bulletin, 115, 2003, 10-17

[ESA-2004] "Beagle 2: ESA/UK Commission of Inquiry", report dated 5 April 2004

[ESF-1998] European Science Foundation, National Research Council, "U.S.-European

Collaboration in Space Science", Washington, National Academy press, 1998, 60-64 and 152-153

[Esposito-2005] Esposito, L.W., et al., "Ultraviolet Imaging Spectroscopy Shows an Active Saturnian System", Science, 307, 2005, 1251-1255

[Esposito-2008] Esposito, L.W., et al., "Moonlets and Clumps in Saturn's F ring", Icarus, 194, 2008, 278-289

[Euler-2001] Euler, E.E:, Jolly, S.D., Curtis, H.H., "The Failures of the Mars Climate Orbiter and Mars Polar Lander: A Perspective from the People Involved", paper AAS 01-074

[Fairén-2004] Fairén, A.G., et al., "Inhibition of Carbonate Synthesis in Acidic Oceans on Early Mars", Nature, 431, 2004, 423-426

[Fallon-2003] Fallon, E.J. II, Sinclair, R., "Design and Development of the Main Parachute for the Beagle 2 Mars Lander", paper AIAA-2003-2153

[Farnham-2002] Farnham, T.L., Cochran, A.L., "A McDonald Observatory Study of Comet 19P/Borrelly: Placing the Deep Space 1 Observations into a Broader Context", arXiv astro-ph/0208445 preprint

[Farquhar-1999] Farquhar, R.W., "The use of Earth-return trajectories for missions to comets", Acta Astronautica, 44, 1999, 607-623

[Farquhar-2011] Farquhar, R.W., "Fifty Years on the Space Frontiers: Halo Orbits, Comets, Asteroids, and More", Denver, Outskirt Press, 2011, 209-223

[Fast-2009] Fast, K.E., et al., "Comparison of HIPWAC and Mars Express SPICAM Observations of Ozone on Mars 2006-2008 and Variation from 1993 IRHS Observations", Icarus, 203, 2009, 20-27

[Feldman-2002] Feldman, W.C., et al., "Global Distribution of Neutrons from Mars: Results from Mars Odyssey", Science, 297, 2002, 75-78

[Fink-1976] Fink, D.E., "JPL Shapes Broad Planetary Program", Aviation Week & Space Technology, 9 August 1976, 37–43

[Fischer-2006] Fischer, G., et al., "Discrimination between Jovian Radio Emissions and Saturn Electrostatic Discharges", Geophysical Research Letters, 33, 2006, L21201

[Fischer-2007] Fischer, G., et al., "Nondetection of Titan Lightning Radio Emissions with Cassini/RPWS after 35 Close Titan Flybys", Geophysical Research Letters, 34, 2007, L22104

[Fischer-2008] Fischer, G., et al., "Atmospheric Electricity at Saturn", Space Sci Review, 137, 2008, 271–285

[Fischer-2011] Fischer, G., et al., "A Giant Thunderstorm on Saturn", Nature, 475, 2011, 75-77

[Flamini-1998] Flamini, E., Somma, R., "Italian Participation to Interplanetary Exploration: The Cassini-Huygens Mission", paper presented at the Second IAA Symposium on Realistic Near-Term Advanced Scientific Space Missions, Aosta, Italy, June 29-July 1, 1998

[Flasar-2005a] Flasar, F.M., et al., "Temperatures, Winds, and Composition in the Saturnian System", Science, 307, 2005, 1247-1251

[Flasar-2005b] Flasar, F.M., et al., "Titan's Atmospheric Temperatures, Winds, and Composition", Science, 307, 2005, 975-978

[Fletcher-2008] Fletcher, L.N., et al., "Temperature and Composition of Saturn's Polar Hot Spots and Hexagon", Science, 319, 2008, 79-81

[Fletcher-2011] Fletcher, L.N., et al., "Thermal Structure and Dynamics of Saturn's Northern Springtime Disturbance", Science, 332, 2011, 1413-1417

[Flight-2000] "NASA Nano-Rover Axed after Cost and Weight Problems", Flight International, 14 November 2000, 40

[Flynn-2006] Flynn, G.J., et al., "Elemental Compositions of Comet 81P/Wild 2 Samples Collected by Stardust", Science, 314, 2006, 1731-1735

[Forget-2004] Forget, F., "Alien Weather at the Poles of Mars", Science, 306, 2004, 1298-1299

[Formisano-2004] Formisano, V., et al., "Detection of Methane in the Atmosphere of Mars", Science, 306, 2004, 1758-1761

[Francillon-1970] Francillon, R.J., "Japanese Aircraft of the Pacific War", London, Putnam, 1970, 206-214

[Fujiwara-2006] Fujiwara, A., et al., "The Rubble-Pile Asteroid Itokawa as Observed by Hayabusa", Science, 312, 2006, 1330-1334

[Fulchignoni-2005] Fulchignoni, M., et al., "In Situ Measurements of the Physical Characteristics of Titan's Environment", Nature, 438, 2005, 785-791

[Furniss-1999] Furniss, T., "Martian Gliders", Flight International, 10 March 1999, 56

[Garnier-2007] Garnier, P., et al., "The Exosphere of Titan and its Interaction with the Kronian Magnetosphere: MIMI Observations and Modeling", Planetary and Space Science, 55, 2007, 165-173

[Gellert-2004] Gellert, R., et al., "Chemistry of Rocks and Soils in Gusev Crater from the Alpha Particle X-ray Spectrometer", Science, 305, 2004, 829-832

[Gendrin-2005] Gendrin, A., et al., "Sulfates in Martian Layered Terrains: The OMEGA/Mars Express View", Science, 307, 2005, 1587-1591

[Giampieri-2006] Giampieri, G., et al., "A Regular Period for Saturn's Magnetic Field that May Track its Internal Rotation", Nature, 441, 2006, 62-63

[Giuranna-2010] Giuranna, M., et al., "Compositional Interpretation of PFS/MEx and TES/MGS Thermal Infrared Spectra of Phobos", paper presented at the European Planetary Science Congress 2010, Rome

[Gladstone-2002] Gladstone, G.R., et al., "A Pulsating Auroral X-ray Hot Spot on Jupiter", Nature, 415, 2002, 1000-1002

[Goetz-2005] Goetz, W., et al., "Indication of Drier Periods on Mars from the Chemistry and Mineralogy of Atmospheric Dust", Nature, 436, 2005, 62-65

[Golombek-2000] Golombek, M., et al., "Preliminary Evaluation of Engineering Constrains of Mars Sample Return Landing Sites", paper presented at the XXXI Lunar and Planetary Science Conference, Houston, 2000

[Golombek-2005] Golombek, M.P., et al., "Assessment of Mars Exploration Rover Landing Site Predictions", Nature, 436, 2005, 44-48

[Golombek-2010] Golombek, M., et al., "Constraints on Ripple Migration at Meridiani Planum from Observations of Fresh Craters by Opportunity And HiRISE", paper presented at the XLI Lunar and Planetary Science Conference, Houston, 2010

[Gombosi-2005] Gombosi, T.I., Hansen, K.C., "Saturn's Variable Magnetosphere", Science, 307, 2005, 1224-1226

[Grant-2004] Grant, J.A., et al., "Surficial Deposits at Gusev Crater Along Spirit Rover Traverses", Science, 305, 2004, 807-810

[Griffith-2003] Griffith, C.A., et al., "Evidence for the Exposure of Water Ice on Titan's Surface", Science, 300, 2003, 628-630

[Griffith-2005] Griffith, C.A., et al., "The Evolution of Titan's Mid-Latitude Clouds", Science, 310, 2005, 474-477

[Griffith-2006a] Griffith, C.A., "Titan's Exotic Weather", Nature, 442, 2006, 362-363

[Griffith-2006b] Griffith, C.A., et al., "Evidence for a Polar Ethane Cloud on Titan", Science, 313, 2006, 1620-1622

[Griffiths-2003] Griffiths, A.D., et al., "The Scientific Objectives of the Beagle 2 Stereo Camera System", paper presented at the XXXIV Lunar and Planetary Science Conference, Houston, 2003

[Grimberg-2006] Grimberg. A., et al., "Solar Wind Neon from Genesis: Implications for the Lunar Noble Gas Record", Science, 314, 2006, 1133-1135

[Gunnarsdottir-2007] Gunnarsdottir, H.M., et al., "Martian Surface Roughness Using 75-cm Bistatic Surface Echoes Received by Mars Odyssey", paper presented at the XXXVIII Lunar and Planetary Science Conference, Houston, 2007

[Gurnett-2001] Gurnett, D.A., et al., "Non-Detection at Venus of High-Frequency Radio Signals Characteristic of Terrestrial Lightning", Nature, 409, 2001, 313-315

[Gurnett-2002] Gurnett, D.A., et al., "Control of Jupiter's Radio Emission and Aurorae by the Solar Wind", Nature, 415, 2002, 985-987

[Gurnett-2005] Gurnett, D.A., et al., "Radio and Plasma Wave Observations at Saturn from Cassini's Approach and First Orbit", Science, 307, 2005, 1255-1259

[Gurnett-2007] Gurnett, D.A., et al., "The Variable Rotation Period of the Inner Region of Saturn's Plasma Disk", Science, 316, 2007, 442-445

[Gurnett-2009] Gurnett, D.A., et al., "Discovery of a North-South Asymmetry in Saturn's Radio Rotation Period", Geophysical Research Letters, 36, 2009, L16102

[Guynn-2003] Guynn, M.D., et al., "Evolution of a Mars Airplane Concept for the ARES Mars Scout Mission", paper AIAA 2003-6578

[Hand-2008] Hand, E., "When Water Gushed on Mars", Nature, 453, 2008, 1153

[Hansen-2006] Hansen, K.C., et al., "Enceladus' Water Vapor Plume", Science, 311, 2006, 1422-1425

[Hansen-2008] Hansen, C.J., et al., "Water Vapour Jets inside the Plume of Gas Leaving Enceladus", Nature, 456, 2008, 477-479

[Hartogh-2011] Hartogh, P., et al., "Direct Detection of the Enceladus Water Torus with Herschel", Astronomy & Astrophysics, 532, 2011, L2

[Harvey-2000] Harvey, B., "The Japanese and Indian Space Programmes", Chichester, Springer-Praxis, 2000, 4-6

[Haskin-2005] Haskin, L.A., et al., "Water Alteration of Rocks and Soils on Mars at the Spirit Rover Site in Gusev Crater", Nature, 436, 2005, 66-69

[Hassan-1997] Hassan, H., Jones, J.C., "The Huygens Probe", ESA Bulletin, 92, 1997, 33-43

[Hastrup-1999] Hastrup, R.C., et al., "Mars Comm/Nav MicroSat Network", paper SSC99-VII-5

[Hauber-2005] Hauber, H, et al., "Discovery of a Flank Caldera and Very Young Glacial Activity at Hecates Tholus, Mars", Nature, 434, 2005, 356-361

[Hayes-2011] Hayes, A.G., "Transient Surface Liquid in Titan's Polar Regions from Cassini", Icarus, 211, 2011, 655-671

[Head-2003] Head, J.W., et al., "Recent Ice Ages on Mars", Nature, 426, 2003, 797-802

[Head-2005] Head, J.W., et al., "Tropical to Mid-Latitude Snow and Ice Accumulation, Flow and Glaciation on Mars", Nature, 434, 2005, 346-351

[Heavens-2011] Heavens, N.G., "Sunshine on a Cloudy Forecast", Science, 333, 2011, 1832-1833

[Heber-2007] Heber, V.S., et al., "Helium and Neon Isotopic and Elemental Composition in Different Solar Wind Regime Targets from the Genesis Mission", paper presented at the XXXVIII Lunar and Planetary Science Conference, Houston, 2007

[Heber-2009] Heber, V.S., et al., "Fractionation Processes in the Solar Wind Revealed by Noble Gases Collected by Genesis Regime Targets", paper presented at the XL Lunar and Planetary Science Conference, Houston, 2009

[Hedman-2007] Hedman, M.M., et al., "The Source of Saturn's G Ring", Science, 317, 2007, 653-656

[Hedman-2009a] Hedman, M.M., et al., "Three Tenuous Rings/Arcs for Three Tiny Moons", Icarus, 199, 2009, 378-386

[Hedman-2009b] Hedman, M.M., et al., "Aegeon (Saturn LIII), a G-Ring Object", arXiv astro-ph/0911.0171 preprint
[Hedman-2011a] Hedman, M.M., et al., "Saturn's Curiously Corrugated C Ring", Science, 332, 2011, 708-711
[Hedman-2011b] Hedman, M.M., et al., "Physical properties of the small moon Aegaeon (Saturn LIII)", paper presented at the European Planetary Science Congress, Nantes, 2011
[Helled-2011] Helled, R., "Constraining Saturn's Core Properties by a Measurement of Its Moment of Inertia - Implications to the Solstice Mission", arXiv astro-ph/1105.5068 preprint
[Herkenhoff-2004a] Herkenhoff, K.E., et al., "Evidence from Opportunity's Microscopic Imager for Water on Meridiani Planum", Science, 306, 2004, 1727-1730
[Herkenhoff-2004b] Herkenhoff, K.E., et al., "Textures of the Soils and Rocks at Gusev Crater from Spirit's Microscopic Imager", Science, 305, 2004, 824-826
[Hidaka-2001] Hidaka, T., et al., "Trajectory Plan and Earth Swingby Operation of NOZOMI". Paper dated 2001
[Hill-2002] Hill, T.W., "Magnetic Moments at Jupiter", Nature, 415, 2002, 965-966
[Hirst-1999] Hirst, E.A., Yen, C.-W. L., "Stardust Mission Plan", JPL Document SD-75000-100, 1 February 1999
[Hittle-2006] Hittle, J.D., et al., "Genesis Spacecraft Science Canister Preliminary Inspection and Cleaning", paper presented at the XXXVII Lunar and Planetary Science Conference, Houston, 2006
[Hohenberg-2006] Hohenberg, C.M., et al., "Light Noble Gases from Solar Wind Regimes Measured in Genesis Collectors from Different Arrays", paper presented at the XXXVII Lunar and Planetary Science Conference, Houston, 2006
[Holmström-2007] Holmström, M., et al., "Mars Express/ASPERA-3/NPI and IMAGE/ LENA Observations of Energetic Neutral Atoms in Earth and Mars Orbit", arXiv astro-ph/0711.1678 preprint
[Hong-2002] Hong, P., Carlisle, G., Smith, N., "Look Ma, no HANS!", paper at the 2002 Aerospace Conference, Big Sky, March 2002
[Hörz-2006] Hörz, F., et al., "Impact Features on Stardust: Implications for Comet 81P/Wild 2 Dust", Science, 314, 2006, 1716-1719
[House of Commons-2004] "Government Support for Beagle 2", ordered by The House of Commons, London, The Stationery Office Limited, 2 November 2004
[Hsu-2009] Hsu, H.W., et al., "Stream Particles Observation during the Cassini-Huygens Flyby of Jupiter", paper presented at the European Planetary Science Congress, Potsdam, 2009
[Hueso-2006] Hueso, R., Sanchez-Lavega, A., "Methane Storms on Saturn's Moon Titan", Nature, 442, 2006, 428-431
[Hurford-2007] Hurford, T.A., et al., "Eruptions Arising from Tidally Controlled Periodic Openings of Rifts on Enceladus", Nature, 447, 2007, 292-294
[Hurley-2005] Hurley, K., et al., "An Exceptionally Bright Flare from SGR 1806-20 and the Origins of Short-Duration gamma-Ray Bursts", Nature, 434, 2005, 1098-1103
[IAUC-8389] "International Astronomical Union Circular No. 8389", 16 August 2004
[IAUC-8401] "International Astronomical Union Circular No. 8401", 9 September 2004
[IAUC-8432] "International Astronomical Union Circular No. 8432", 8 November 2004
[IAUC-8524] "International Astronomical Union Circular No. 8524", 6 May 2005
[IAUC-8759] "International Astronomical Union Circular No. 8759", 11 October 2006
[IAUC-8773] "International Astronomical Union Circular No. 8773", 14 November 2006
[IAUC-8857] "International Astronomical Union Circular No. 8857", 18 July 2007

[IAUC-8970] "International Astronomical Union Circular No. 8970", 5 September 2008

[IAUC-9023] "International Astronomical Union Circular No. 9023", 3 March 2009

[IAUC-9091] "International Astronomical Union Circular No. 9091", 2 November 2009

[Iess-2010] Iess, L., et al., "Gravity Field, Shape, and Moment of Inertia of Titan", Science, 327, 1367-1369

[Iorio-2008] Iorio, L., "On the Recently Determined Anomalous Perihelion Precession of Saturn", ArXiv gr-qc/0811.0756v2 preprint

[Isakowitz-2000] Isakowitz, S.J., Hopkins, J.P. Jr., Hopkins, J.B., "International Reference Guide to Space Launch Systems", 3rd Edition, Reston, AIAA, 2000, 245-256

[Ishii-2003] Ishii, N., et al., "System Description and Reentry Operation Scenario of MUSES-C Reentry Capsule", The Institute of Space and Astronautical Science Report SP No. 17, March 2003

[Ishii-2008] Ishii, H.A., et al., "Comparison of Comet 81P/Wild 2 Dust with Interplanetary Dust from Comets", Science, 319, 2008, 447-450

[Jacobson-2006] Jacobson, R.A., et al., "The Gravity Field of the Saturnian System from Satellite Observations and Spacecraft Tracking Data", The Astronomical Journal, 132, 2006, 2520-2526

[Jaffe-1997] Jaffe, L.D., Herrell, L.M., "Cassini/Huygens Science Instruments, Spacecraft, and Mission", Journal of Spacecraft and Rockets, 34, 1997, 509-521

[Jäkel-1996] Jäkel, E., et al., "Drop Testing the Huygens Probe", ESA Bulletin, 85, 1996, 51-54

[Jansen-2012] Jansen, F.A., "Mars Express Status", presentation to the Mars Exploration Program Analysis Group (MEPAG) meeting #25, Washington, DC, February 2012

[Jewitt-2007] Jewitt, D., Haghighipour, N., "Irregular Satellites of the Planets: Products of Capture in the Early Solar System". In: "Annual Review of Astronomy and Astrophysics", 45, 2007, 261-297

[Johnson-2005] Johnson, T.V., Lunine, J.I., "Saturn's Moon Phoebe as a Captured Body from the Outer Solar System", Nature, 435, 2005, 69-71

[Jones-1997] Jones, J.C., Giovagnoli, F., "The Huygens Probe System Design". In: "Huygens: Science, Payload and Mission", Noordwijk, ESA, 1997, 25-45

[Jones-2006] Jones, G.H., et al., "Enceladus' Varying Imprint on the Magnetosphere of Saturn", Science, 311, 2006, 1412-1415

[Jones-2008] Jones, G.H., et al., "The Dust Halo of Saturn's Largest Icy Moon, Rhea", Science, 319, 2008, 1380-1384

[Jordan-1999] Jordan, J.F., Miller, S.L., "The Mars Surveyor Program Architecture", Journal of Space Mission Architecture, 1, 1999, 1-10

[JPL-1997] "Mars Surveyor Program Announcement of Opportunity – 2001 Lander Mission Proposal Information Package", Pasadena, JPL, 1997

[JPL-1999] "Mars Sample Return Mission Lander Additional Payload (AP) Proposal Information Package", Pasadena, JPL, 1999

[JPL-2000] JPL Special Review Board, "Report of the Loss of the Mars Polar Lander and Deep Space 2 Missions", Pasadena, JPL, 22 March 2000

[Jurewicz-2003] Jurewicz A.J.G., et al., "The Genesis solar-wind collector materials", Space Science Review, 105, 2003, 535-560

[JWG-1986] Joint Working Group on Cooperation in Planetary Exploration, "United States and Western Europe Cooperation in Planetary Exploration", Washington, National Academy Press, 1986, 88-128

[Kargel-2006] Kargel, J.S., "Enceladus: Cosmic Gymnast, Volatile Miniworld", Science, 311, 2006, 1389-1391

[Karkoschka-2007] Karkoschka, E., et al., "DISR Imaging and the Geometry of the Descent of the Huygens Probe within Titan's Atmosphere", Planetary and Space Science, 55, 2007, 1896-1935

[Kawaguchi-1995] Kawaguchi, J., et al., "On the Low Cost Sample and Return Mission to Near Earth Asteroid Nereus via Electric Propulsion", Acta Astronautica, 35 supplement, 1995, 193-200

[Kawaguchi-1996] Kawaguchi, J., et al., "The MUSES-C, World's First Sample and Return Mission from Near Earth Asteroid: Nereus", Acta Astronautica, 39, 1996, 15-23

[Kawaguchi-1999] Kawaguchi, J., et al., "The MUSES-C, Mission Description and its Status", Acta Astronautica, 45, 1999, 397-405

[Kawaguchi-2003a] Kaweguchi, J., Uesugi, K., Fujiwara, A., "The MUSES-C Mission for the Sample and Return – Its Technology Development Status and Readiness", Acta Astronautica, 52, 2003, 117-123

[Kawaguchi-2003b] Kaweguchi, J., et al., "Synthesis of an Alternative Flight Trajectory for Mars Explorer, Nozomi", Acta Astronautica, 52, 2003, 189-195

[Kawaguchi-2004] Kawaguchi, J., Fujiwara, A., Uesugi, T.K., "The Ion Engine Cruise Operation and the Earth Swingby of 'Hayabusa' (MUSES-C)", paper presented at the 55th International Astronautical Congress, Vancouver, October 2004

[Kazeminejad-2007] Kazeminejad, B., et al., "Huygens' Entry and Descent through Titan's Atmosphere – Methodology and Results of the Trajectory Reconstruction", Planetary and Space Science, 55, 2007, 1845-1876

[Keating-2003] Keating, G.M., et al., "Global Measurements of the Mars Upper Atmosphere: Insitu Accelerometer Measurements from Mars Odyssey 2001 and Mars Global Surveyor", paper presented at the XXXIV Lunar and Planetary Science Conference, Houston, 2003

[Keller-2006] Keller, L.P., et al., "Infrared Spectroscopy of Comet 81P/Wild 2 Samples Returned by Stardust", Science, 314, 2006, 1728-1731

[Kelly Beatty-2004] Kelly Beatty, J., "Saturn's Phoebe: Small Moon, Grand Debut", Sky & Telescope, September 2004, 30-32

[Kempf-2005a] Kempf, S., et al., "High-Velocity Streams of Dust Originating from Saturn", Nature, 433, 2005, 289-291

[Kempf-2005b] Kempf, S., et al., "Composition of Saturnian Stream Particles", Science, 307, 2005, 1274-1276

[Kennedy-2004] Kennedy, B.M., Carranza, E., Williams, K., "1-AU Calibration Activities for Stardust Earth Return", paper AAS 04-134

[Kerr-1999] Kerr, R.A., "Deep Space 1 Traces Braille Back to Vesta", Science, 285, 1999, 993-994

[Kerr-2006] Kerr, R.A., "A Dry View of Enceladus Puts a Damper on Chances for Life There", Science, 314, 2006, 1668

[Kerr-2008] Kerr, R.A., "Electron Shadow Hints at Invisible Rings Around a Moon", Science, 319, 2008, 1325

[Kerr-2011a] Kerr, R.A., "Prime Science Achieved at Asteroid", Science, 332, 2011, 302

[Kerr-2011b] Kerr, R.A., "Hayabusa Gets to the Bottom of Deceptive Asteroid Cloaking", Science, 333, 2008, 1081

[Kieffer-2006a] Kieffer, S.W., et al., "A Clathrate Reservoir Hypothesis for Enceladus' South Polar Plume", Science, 314, 2006, 1764-1766

[Kieffer-2006b] Kieffer, H.H., Christensen, P.R., Titus, T.N., "$CO_2$ Jets Formed by Sublimation Beneath Translucent Slab Ice in Mars' Seasonal South Polar Ice Cap", Nature, 442, 2006, 793-796

[Kieffer-2008] Kieffer, S.W., Jakosky, B.M., "Enceladus-Oasis or Ice Ball?", Science, 320, 2008, 1432-1433

[Kimura-1999] Kimura, M., et al., "PLANET-B ('NOZOMI') Orbital Plan". In: Proceedings of the 8th Workshop on Astrodynamics and Flight Mechanics, ISAS, 1999

[Kissel-2004] Kissel, J., et al., "The Cometary and Interstellar Dust Analyzer at Comet 81P/ Wild 2", Science, 304, 2004, 1774-1776

[Kivelson-2006] Kivelson, M.G., "Does Enceladus Govern Magnetospheric Dynamics at Saturn?", Science, 311, 2006, 1391-1392

[Kivelson-2007] Kivelson, M.G., "A Twist on Periodicity at Saturn", Nature, 450, 2007, 178-179

[Klingelhöfer-2004] Klingelhöfer, G., et al., "Jarosite and Hematite at Meridiani Planum from Opportunity's Mossbauer Spectrometer", Science, 306, 2004, 1740-1745

[Kloster-2009] Kloster, K.W., Tam, C.H., Longuski, J.M., "Saturn Escape Options for Cassini Encore Missions", Journal of Spacecraft and Rockets, 46, 2009, 874-882

[Kminek-2000] Kminek, G., et al., "MOD-An In-Situ Organic Detector for the MSR 2003 Mission", paper presented at the XXXI Lunar and Planetary Science Conference, Houston, 2000

[Kohlhase-1997] Kohlhase, C., Peterson, C.E., "The Cassini Mission to Saturn and Titan", ESA Bulletin, 92, 1997, 5562

[Koon-1999] Koon, W.S., Lo, M.W., Marsden, J.E., "The Genesis Trajectory and Heteroclinic Connections", paper presented at the AAS/AIAA Astrodynamic Specialist Conference , August 16-19, 1999

[Kremer-2011] Kremer, K., "Record-Breaking Martian Odyssey", Spaceflight, March 2011, 96-99

[Krimigis-2002] Krimigis, S.M., et al., "A Nebula of Gases from Io Surrounding Jupiter", Nature, 415, 2002, 994-996

[Krimigis-2005] Krimigis, S.M., et al., "Dynamics of Saturn's Magnetosphere from MIMI During Cassini's Orbital Insertion", Science, 307, 2005, 1270-1273

[Krimigis-2009] Krimigis, S.M., et al., "Imaging the Interaction of the Heliosphere with the Interstellar Medium from Saturn with Cassini", Science, 326, 2009, 971-973

[Królikowska-2006] Królikowska, M., Szutowicz, S., "Non-Gravitational Motion of the Jupiter-Family Comet 81P/Wild 2. I. The Dynamical Evolution", Astronomy & Astrophysics, 448, 2006, 401-409

[Kronk-1984a] Kronk, G.W., "Comets: A Descriptive Catalog", Hillside, Henslow, 1984, 319-320

[Kronk-1984b] ibid., 148

[Kronk-1984c] ibid., 225-226

[Kronk-1999a] Kronk, G.W., "Cometography – A Catalog of Comets. Volume 1: Ancient-1799", Cambridge University Press, 1999, 481-482

[Kronk-1999b] ibid., 367-368

[Krot-2011] Krot, A.N., "Bringing Part of an Asteroid Back Home", Science, 333, 2008, 1098-1099

[Krüger-2000] Krüger, F.R., Kissel, J., "Erste Direkte Chemische Analyse Interstellarer Staubteilchen" (First Direct Chemical Analysis of Interstellar Particles), Sterne und Weltraum, 39, 2000, 326-329 (in German)

[Kubota-2003] Kubota, T., et al., "An Autonomous Navigational and Guidance System for MUSES-C Asteroid Landing", Acta Astronautica, 52, 2003, 125-131

[Kubota-2005] Kubota, T., Yoshimitsu, T., "Asteroid Exploration Rover", presentation at the 2005 IEEE ICRA Planetary Rover Workshop

[Kulikov-1996] Kulikov, S., "Top-Priority Space Projects", Aerospace Journal, November 1996, page unknown

[Kurth-2002] Kurth, W.S., et al., "The Dusk Flank of Jupiter's Magnetosphere", Nature, 415, 2002, 991-994

[Kurth-2005] Kurth, W.S., et al., "An Earth-Like Correspondence between Saturn's Auroral Features and Radio Emissions", Nature, 433, 2005, 722-724

[Kurth-2008] Kurth, W.S., et al., "An Update to a Saturnian Longitude System Based on Kilometric Radio Emissions", Journal of Geophysical Research, 113, 2008, A05222

[Lakdawalla-2009] Lakdawalla, E., "Ice Worlds of the Ringed Planet", Sky & Telescope, June 2009, 26-34

[Lämmerzahl-2006] Lämmerzahl, C., Preuss, O., Dittus, H., "Is the Physics Within the Solar System Really Understood?" arXiv gr-qc/0604052 preprint

[Lamy-1998] Lamy, P.L., Toth, I., Weaver, H.A., "Hubble Space Telescope Observations of the Nucleus and Inner Coma of Comet 19P/1904 Y2 (Borrelly)" Astronomy & Astrophusics, 337, 1998, 945-954

[Lamy-2011a] Lamy, L., et al., "Properties of Saturn Kilometric Radiation Measured within its Source Region", arXiv astro-ph/1101.3842 preprint

[Lamy-2011b] Lamy, L., "Variability of Southern and Northern SKR Periodicities", arXiv astro-ph/1102.3099 preprint

[Lancaster-2006] Lancaster, N., "Linear Dunes on Titan", Science, 312, 2006, 702-703

[Langevin-2005a] Langevin, Y., et al., "Summer Evolution of the North Polar Cap of Mars as Observed by OMEGA/Mars Express", Science, 307, 2005, 1581-1584

[Langevin-2005b] Langevin, Y., et al., "Sulfates in the North Polar Region of Mars Detected by OMEGA/Mars Express", Science, 307, 2005, 1584-1586

[Langevin-2006] Langevin, Y., et al., "No Signature of Clear CO2 Ice from the "Cryptic" Regions in Mars' South Seasonal Polar Cap", Nature, 442, 2006, 790-792

[Lauer-2005] Lauer, H.V., et al., "Genesis: Removing Contamination from Sample Collectors", paper presented at the XXXVI Lunar and Planetary Science Conference, Houston, 2005

[Lawler-2008] Lawler, A., "NASA's Stern Quits Over Mars Exploration Plans", Science, 320, 2008, 31

[Lazzarin-2001] Lazzarin, M., et al., "Groundbased Investigation of Asteroid 9969 Braille, Target of the Spacecraft Mission Deep Space 1", Astronomy & Astrophysics, 375, 2001, 281-284

[Lebreton-1988] Lebreton, J.-P., Scoon, G., "Cassini – A Mission to Saturn and Titan", ESA Bulletin, 55, 1988, 24-30

[Lebreton-1997] Lebreton, J.-P., Matson, D.L:, "The Huygens Probe: Science, Payload and Mission Overview". In: "Huygens: Science, Payload and Mission", Noordwijk, ESA, 1997, 5-24

[Lebreton-2005] Lebreton, J.-P., et al., "An Overview of the Descent and Landing of the Huygens Probe on Titan", Nature, 438, 2005, 758-764

[Lemmon-2004] Lemmon, M.T., et al., "Atmospheric Imaging Results from the Mars Exploration Rovers: Spirit and Opportunity", Science, 306, 2004, 1753-1756

[Levison-2011] Levison, H.F., et al., "Ridge Formation and De-Spinning of Iapetus via an Impact-Generated Satellite", arXiv astro-ph/1105.1685

[Levrard-2004] Levrard, B., et al., "Recent Ice-Rich Deposits Formed at High Latitudes on Mars by Sublimation of Unstable Equatorial Ice During Low Obliquity", Nature, 431, 2004, 1072-1075

[Leyrat-2008] Leyrat, C., et al., "Infrared Observations of Saturn's Rings by Cassini CIRS : Phase Angle and Local Time Dependence", Planetary and Space Science, 56, 2008, 117-133

[Lindemann-2005] Lindemann, R.A., Voorhees, C.J., "Mars Exploration Rover Mobility

Assembly Design, Test and Performance", paper presented at the 2005 International Conference on Systems, Man, and Cybernetics, Hawaii, 10-12 October 2005

[Litvak-2006] Litvak, M.L., et al., "Comparison between Polar Regions of Mars from HEND/ Odyssey Data", Icarus, 180, 2006, 23-37

[Lopes-2010] Lopes,R.M.C., et al., "Distribution and Interplay of Geologic Processes on Titan from Cassini Radar Data", Icarus, 205, 2010, 540-558

[Lorenz-1994] Lorenz, R.D., "Huygens Probe Impact Dynamics", ESA Journal, 18, 1994, 93-117

[Lorenz-1997] Lorenz, R.D., "Lightning and Triboelectric Charging Hazard Assessment for the Huygens Probe". In: "Huygens: Science, Payload and Mission", Noordwijk, ESA, 1997, 265-270

[Lorenz-2000] Lorenz, R.D., et al., "Penetration Tests on the DS-2 Mars Microprobes: Penetration Depth and Impact Accelerometer", Planetary and Space Science, 48, 2000, 419-436

[Lorenz-2006] Lorenz, R.D., et al., "The Sand Seas of Titan: Cassini RADAR Observations of Longitudinal Dunes", Science, 312, 2006, 724-727

[Lorenz-2008a] Lorenz, R. Mitton, J., "Titan Unveiled: Saturn's Mysterious Moon Explored", Princeton University Press, 2008, 158-159

[Lorenz-2008b] ibid., 204-209

[Lorenz-2008b] Lorenz, R.D., et al., "Titan's Rotation Reveals an Internal Ocean and Changing Zonal Winds", Science, 319, 2008, 1649-1651

[Lorenz-2008c] Lorenz, R.D., West, R.D., Johnson, W.T.K., "Cassini RADAR Constraint on Titan's Winter Polar Precipitation", Icarus, 195, 2008, 812-816

[Lorenz-2009] Lorenz, R.D., "Titan Mission Studies – A Historical Review", Journal of the British Interplanetary Society, 62, 2009, 162-174

[Lorenz-2010] Lorenz, R.D., "Winds of Change on Titan", Science, 329, 2010, 519-520

[Lundin-2004] Lundin, R., et al., "Solar Wind–Induced Atmospheric Erosion at Mars: First Results from ASPERA-3 on Mars Express", Science 305, 2004, 1933-1936

[Lunine-2008] Lunine, J.I., et al., "Cassini Radar's Third and Fourth Looks at Titan", Icarus, 195, 2008, 415-433

[Magnani-2004] Magnani, P.G., et al., "Deep Drill (DeeDri) for Mars Application", Planetary and Space Science, 52, 2004, 79-82

[Malin-1999] Malin, M.C., et al., "The Mars Color Imager (MARCI) Investigation on the Mars Climate Orbiter Mission", paper presented at the XXX Lunar and Planetary Science Conference, Houston, 1999

[Malin-2005] Malin, M.C., "Hidden in Plain Sight: Finding Martian Landers", Sky & Telescope, July 2007, 42-44

[Maltagliati-2011] Maltagliati, L., et al., "Evidence of Water Vapor in Excess of Saturation in the Atmosphere of Mars", Science, 333, 2011, 1868-1871

[Mangold-2004] Mangold, N., et al., "Evidence for Precipitation on Mars from Dendritic Valleys in the Valles Marineris Area", Science, 305, 2004, 78-81

[Marsal-1999] Marsal, O., et al., "NetLander: The First Scientific Lander Network on the Surface of Mars", paper presented at the L Congress of the International Astronautical Federation, Amsterdam, 1999

[Marsal-2002] Marsal, O., et al., "The NetLander Geophysical Network on the Surface of Mars: General Mission Description and Technical Design Status", Acta Astronautica, 51, 2002, 379-386

[Martin Marietta-1976] "A Titan Exploration Study – Science, Technology, and Mission Planning Options", NASA CR-137846, June 1976

[Martínex-Fríaz-2007] Martínex-Fríaz, J., Nna-Mvondo, D., Rodríguez-Losada, J.A., "Stardust's Hydrazine (N2H4) Fuel: A Potential Contaminant for the Formation of Titanium Nitride (Osbornite)", Energy & Fuels, 21, 2007, 1822-1823

[Marty-2011] , Marty, B., et al., "A 15N-Poor Isotopic Composition for the Solar System As Shown by Genesis Solar Wind Samples", Science, 332, 2011, 1533-1536

[Mase-2005] Mase, R.A., et al., "Mars Odyssey Navigation Experience", Journal of Spacecraft and Rockets, 42, 2005, 386-393

[Matousek-1998] Matousek, S., et al., "A Few Good Rocks: The Mars Sample Return Mission Architecture", Paper AIAA-98-4282

[Matsuura-2005] Matsuura, S., "Osorubeki Tabiji - Kasei Tansaki 'Nozomi' no Tadotta 12 nen" (The Terrifying Journey - 12 Years of the Mars Probe Nozomi), Tokio, Opendoors, 2005 (in Japanese)

[Matzel-2010] Matzel, J.E.P., et al., "Constraints on the Formation Age of Cometary Material from the NASA Stardust Mission", Science, 328, 2010, 483-486

[Mauk-2003] Mauk, B.H., et al., "Energetic Neutral Atoms from a Trans-Europa Gas Torus at Jupiter", Nature, 421, 2003, 920-922

[Mazarico-2007] Mazarico, E., et al., "Atmospheric Density During the Aerobraking of Mars Odyssey from Radio Tracking Data", Journal of Spacecraft and Rockets, 44, 2007, 1165-1171

[McCarthy-1996] McCarthy, C., Hassan, H., "Lightning Susceptibility of the Huygens Probe", ESA Bulletin, 85, 1996, 55-57

[McCurdy-2005a] McCurdy, H.E., "Low-Cost Innovation in Spaceflight: the Near Earth Asteroid Rendezvous (NEAR) Shoemaker Mission", Washington, NASA, 2005, 38

[McFarling-2002] McFarling, U.L., "Missions to Pluto, Europa Canceled", Los Angeles Times, 13 April 2002, A14

[McKeegan-2006] McKeegan, K.D., et al., "Isotopic Compositions of Cometary Matter Returned by Stardust", Science, 314, 2006, 1724-1728

[McKeegan-2011] McKeegan, K.D., et al., "The Oxygen Isotopic Composition of the Sun Inferred from Captured Solar Wind", Science, 332, 2011, 1528-1532

[McNamara-2005] McNamara, K.M., and the Genesis Contingency Team, "Genesis Field Recovery", paper presented at the XXXVI Lunar and Planetary Science Conference, Houston, 2005

[McNeil Cheatwood-2000] McNeil Cheatwood, F., et al., "Dynamic Stability Testing of the Genesis Sample Return Capsule", Paper AIAA 2000-1009

[Mecham-2000] Mecham, M., "Red Team Preps Odyssey to Mars", Aviation Week & Space Technology, 11 December 2000, 78-79

[Mecham-2006] Mecham, M., "Stardust's Return Points NASA Toward More Deep Space Missions", Aviation Week & Space Technology, 23 January 2006, 20-21

[Meech-2004] Meech, K.J., Hainaut, O.R., Marsden, B.G., "Comet Nucleus Size Distributions from HST and Keck Telescopes", Icarus, 170, 2004, 463-491

[Michel-2006] Michel, P., Yoshikawa, M., "Dynamical Origin of the Asteroid (25143) Itokawa: the Target of the Sample-Return Hayabusa Space Mission", Astronomy & Astrophysics, 449, 2006, 817-820

[Mitchell-2000] Mitchell, R.T., "The Cassini/Huygens Mission to Saturn and Titan", paper presented at the 51th International Astronautical Congress, Rio de Janeiro, October 2000

[Mitchell-2002] Mitchell, R.T., "The Cassini/Huygens Mission to Saturn and Titan", paper presented at The World Space Congress 2002, Houston

[Mitchell-2003] Mitchell, R.T., "Cassini/Huygens at Saturn and Titan", paper presented at the 56th International Astronautical Congress, Bremen, October 2003

[Mitchell-2004] Mitchell, R.T., "Cassini/Huygens Arrives at Saturn", paper presented at the 55th International Astronautical Congress, Vancouver, October 2004

[Mitchell-2005] Mitchell, R.T., "The Cassini/Huygens Mission to Saturn", paper presented at the 54th International Astronautical Congress, Fukuoka, October 2005

[Mitchell-2006] Mitchell, C.J., et al., "Saturn's Spokes: Lost and Found", Science, 311, 2006, 1587-1589

[Mitchell-2007] Mitchell, R.T., "The Cassini Mission at Saturn", Acta Astronautica, 61, 2007, 37-43

[Mitcheltree-1997] Mitcheltree, R.A., et al., "Aerodynamics of the Mars Microprobe Entry Vehicles", AIAA paper 97-3658

[Mitcheltree-1999] Mitcheltree, R.A., et al., "Aerodynamics of Stardust Sample Return Capsule" Journal of Spacecraft and Rockets, 36, 1999, 429-435

[Mitcheltree-2004] Mitcheltree, R.A., et al., "Mars Exploration Rover Mission: Entry, Decent, and Landing System Validation", paper presented at the 55th International Astronautical Congress, Vancouver, 2004

[Mitrofanov-2002] Mitrofanov, I., et al., "Maps of Subsurface Hydrogen from the High Energy Neutron Detector, Mars Odyssey", Science, 297, 2002, 75-81

[Mitrofanov-2003] Mitrofanov, I.G., et al., "CO2 Snow Depth and Subsurface Water-Ice Abundance in the Northern Hemisphere of Mars", Science, 300, 2003, 2081-2084

[Mittlefehldt-2010] Mittlefehldt, D.W., et al., "Marquette Island: A Distinct Mafic Lithology Discovered by Opportunity", paper presented at the XLI Lunar and Planetary Science Conference, Houston, 2010

[Miyasaka-2005] Miyasaka, H., et al., "ACE/NOZOMI Multispacecraft Observations of Solar Energetic Particles", paper presented at the 29th International Cosmic Ray Conference, Pune, 2005

[Morita-2003] Morita, Y., et al., "Demonstrator of Atmospheric Reentry with Hyperbolic Velocity – DASH", Acta Astronautica, 52, 2003, 29-39

[Morring-2002a] Morring, F. Jr, "Hopes Are Fading Fast for Lost Comet Probe", Aviation Week & Space Technology, 26 August 2002, 68-69

[Morring-2002b] Morring, F. Jr, "Contour Team Wants to Build a New Probe", Aviation Week & Space Technology, 2 September 2002, 40-41

[Morring-2004a] Morring, F. Jr., "Splashdown", Aviation Week & Space Technology, 29 March 2004, 32-34

[Morring-2004b] Morring, F. Jr., Dornheim, M.A., "Follow the Water", Aviation Week & Space Technology, 8 March 2004, 26-27

[Morring-2005a] Morring, F. Jr, Taverna, M.A., "Following Up", Aviation Week & Space Technology, 31 January 2005, 22-23

[Morring-2005b] Morring, F. Jr, Taverna, M.A., Dornheim, M.A., "Rover Territory", Aviation Week & Space Technology, 24 January 2005, 24-26

[Morring-2006] Morring, F. Jr., "NASA Applies Genesis-Failure Recommendations to Its Human Exploration Vehicle Developments", Aviation Week & Space Technology 19 June 2006, 36

[Morring-2007] Morring, F. Jr., "Plume Plunge", Aviation Week & Space Technology, 6 August 2007, 38

[Morring-2008] Morring, F., Jr., "Hot Stuff", Aviation Week & Space Technology, 14 January 2008, 13

[Morris-2004] Morris, R.V., et al., "Mineralogy at Gusev Crater from the Mossbauer Spectrometer on the Spirit Rover" Science, 305, 2004, 833-836

[Morris-2010] Morris, R.V., et al., "Identification of Carbonate-Rich Outcrops on Mars by the Spirit Rover", Science, 329, 2010, 421-424

[Morton-1999] Morton, O., "To Mars, En Masse", Science, 283, 2009, 1103-1104

[Morton-2000] Morton, O., "Mars Air: How to Built the First Extraterrestrial Airplane", Air & Space, December 1999–January 2000, 34-42

[MSSS-1998] "Airplane Proposed for Mars Survey on Centennial of Wright Brothers First Flight", Malin Space Science Systems press release, 20 July 1998

[Mukai-1998] Mukai, T., et al., "Observations of Mars and Its Satellites by the Mars Imaging Camera (MIC) on Planet-B", Earth Planets Space, 50, 1998, 183-188

[Murray-1992] Murray, C.D., "The Cassini Imaging Science Experiment", Journal of the British Interplanetary Society, 45, 1992, 359-364

[Murray-2005a] Murray, C.D., et al., "How Prometheus Creates Structure in Saturn's F Ring", Nature, 437, 2005, 1326-1329

[Murray-2005b] Murray, J.B., et al., "Evidence from the Mars Express High Resolution Stereo Camera for a Frozen Sea Close to Mars' Equator", Nature, 434, 2005, 352-356

[Murray-2008] Murray, C.D., et al., "The Determination of the Structure of Saturn's F Ring by Nearby Moonlets", Nature, 453, 2008, 739-744

[Mustard-1999] Mustard, J.F., et al., "Mapping the Mineralogy of Environments of Mars with Aladdin", paper presented at the XXX Lunar and Planetary Science Conference, Houston, 1999

[Mustard-2005] Mustard, J.F., et al., "Olivine and Pyroxene Diversity in the Crust of Mars", Science, 307, 2005, 1594-1597

[Nadalini-2004] Nadalini, R., Bodendieck, F., "The Thermal Control System for a Network Mission on Mars: The Experience of the NetLander Mission", paper presented at the LV Congress of the International Astronautical Federation, Vancouver, 2004

[Naeye-2005] Naeye, R., "Europe's Eye on Mars", Sky & Telescope, December 2005, 30-36

[Nagao-2011] Nagao, K., et al., "Irradiation History of Itokawa Regolith Material Deduced from Noble Gases in the Hayabusa Samples", Science, 333, 2008, 1128-1131

[Nakagawa-2008] Nakagawa, H., et al., "UV Optical Measurements of the Nozomi Spacecraft Interpreted with a Two-Component LIC-Flow Model", Astronomy & Astrophysics, 491, 29-41

[Nakamura-2001] Nakamura, T., et al., "Multi-Band Imaging Camera and its Sciences for the Japanese Near-Earth Asteroid Mission MUSES-C", Earth Planets Space, 53, 2001, 1047-1063

[Nakamura-2008] Nakamura, T., et al., "Chondrulelike Objects in Short-Period Comet 81P/ Wild 2", Science, 321, 2008, 1664-1667

[Nakamura-2011] Nakamura, T., et al., "Itokawa Dust Particles: A Direct Link Between S-Type Asteroids and Ordinary Chondrites", Science, 333, 2008, 1113-1116

[Nakatani-2004] Nakatani, I., "Nozomi Exploration Operations Challenges", presentation at the 8th International Conference on Space Operations, Montreal, May 2004

[NASA-1994] Joint U.S./Russian Technical Working Groups, "Mars Together and Fire & Ice", NASA CR-19884, October 1994, 17-31

[NASA-1998a] "Deep Space 1 Launch Press Kit", Washington, NASA, October 1998

[NASA-1998b] "1998 Mars Missions Press Kit", Washington, NASA, December 1998

[NASA-1999a] "Stardust Launch Press Kit", Washington, NASA, February 1999

[NASA-1999b] "Europa Orbiter: Mission and Project Information", NASA Announcement of Opportunity 99-OSS-04, Appendix C, 1999

[NASA-1999c] "Solar Probe: Mission and Project Information", NASA Announcement of Opportunity 99-OSS-04, Appendix G, 1999

[NASA-1999d] "Mars Climate Orbiter Arrival Press Kit", Washington, NASA, September 1999

[NASA-1999e] "Mars Climate Orbiter Mishap Investigation Board Phase I Report", Washington, NASA, 10 November 1999

[NASA-2001] "Genesis Launch Press Kit", NASA, July 2001

[NASA-2002] "CONTOUR-Comet Nucleus Tour Launch Press Kit", NASA, July 2002

[NASA-2003] "Comet Nucleus Tour-CONTOUR Mishaps Investigation Board Report", NASA, 31 May 2003

[NASA-2004] "Genesis Sample Return Press Kit", NASA, September 2004

[NASA-2005] "Genesis Mishaps Investigation Board Report Volume 1", NASA, July 2005

[Nelson-2004] Nelson, L., "Project Structure Blamed for Beagle 2 Loss", Nature, 429, 2004, 330

[Neugebauer-2003] Neugebauer M., et al., "Genesis on-board determination of the solar wind flow regime", Space Science Review, 105, 2003, 661-679

[Neukum-2004] Neukum, G., et al., "Recent and Episodic Volcanic and Glacial Activity on Mars Revealed by the High Resolution Stereo Camera", Nature, 432, 2004, 971-979

[Neukum-2005] Neukum, G., et al., "High Resolution Stereo Camera on Mars Express", presentation to the 1st Mars Express Science Conference, Noordwijk, 2005

[Newburn-2003] Newburn, R.L. Jr., et al., "Phase Curve and Albedo of Asteroid 5535 Annefrank", Journal of Geophysical Research, 108, 2003

[Niemann-2005] Niemann, H.B., et al., "The Abundances of Constituents of Titan's Atmosphere from the GCMS Instrument on the Huygens Probe", Nature, 438, 2005, 779-784

[Nimmo-2006] Nimmo, F., Pappalardo, R.T., "Diapir-Induced Reorientation of Saturn's Moon Enceladus", Nature, 441, 2006, 614-615

[Nimmo-2007] Nimmo, F., et al., "Shear Heating as the Origin of the Plumes and Heat Flux on Enceladus", Nature, 447, 2007, 289-291

[Nixon-2010] Nixon, C.A., "Abundances of Jupiter's Trace Hydrocarbons from Voyager and Cassini", Planetary and Space Science, 58, 2010, 1667-1680

[Noguchi-2011] Noguchi, T., et al., "Incipient Space Weathering Observed on the Surface of Itokawa Dust Particles", Science, 333, 2008, 1121-1125

[Nordholt-2003] Nordholt J.E., "The Genesis solar wind concentrator", Space Science Review, 2003 105, 561-599

[Norris-2010] Norris, G., "Survival Spirit", Aviation Week & Space Technology, 1 February 2010, 36-38

[Oberst-2006] Oberst, J., et al., "Astrometric Observations of Phobos and Deimos with the SRC on Mars Express", Astronomy & Astrophysics, 447, 2006, 1145–1151

[Ohtsuka-2007] Ohtsuka, K., et al., "Are There Meteors Originated from Near Earth Asteroid (25143) Itokawa?", arXiv astro-ph/0808.2671 preprint

[Okada-1998] Okada, T., Ono, T., "Application of Altimeter Experiments of Planet-B Orbiter to the Exploration of Martian Surface and Subsurface Layers", Earth Planets Space, 50, 1998, 235-240

[Okada-2006] Okada, T., et al., "X-ray Fluorescence Spectrometry of Asteroid Itokawa by Hayabusa", Science, 312, 2006, 1338-1341

[O'Neill-1999] O'Neill, W., Casaux, C., "The Mars Sample Return Project", Paper IAF-99-Q.3.02

[Osterloo-2008] Osterloo, M.M., et al., "Chloride-Bearing Materials in the Southern Highlands of Mars", Science, 319, 2008, 1651-1654

[Ostro-2004] Ostro, S.J., et al., "Radar observations of asteroid 25143 Itokawa (1998 SF36)", Meteoritics & Planetary Science, 39, 2004, 1-18

[Ostro-2006] Ostro, S.J., et al., "Cassini RADAR Observations of Enceladus, Tethys, Dione, Rhea, Iapetus, Hyperion, and Phoebe", Icarus, 183, 2006, 479-490

[Owen-1986] Owen, T., "The Cassini Mission", In: NASA Goddard Institute for Space Studies, "The Jovian Atmospheres", 1986, 231-237

[Owen-1999] Owen, T., "Titan". In: Kelly Beatty, J., Petersen, C.C., Chaikin, A. (eds.), "The New Solar System", Cambridge University Press, 4th edition, 1999, 277-284

[Owen-2005] Owen, T., "Huygens Rediscovers Titan", Nature, 438, 2005, 756-757

[Oya-1998] Oya, H., Ono, T., "A New Altimeter for Mars Land Shape Observations Utilizing the Ionospheric Sounder System Onboard the Planet-B Spacecraft", Earth Planets Space, 50, 1998, 229-234

[Paganelli-2007] Paganelli, F., et al., "Titan's Surface from Cassini RADAR SAR and High Resolution Radiometry Data of the First Five Flybys", Icarus, 191, 2007, 211-222

[Parkinson-2004] Parkinson, B., Kemble, S., "A Micromission for Mars Sample Return", Journal of the British Interplanetary Society, 57, 2004, 256-261

[Peplow-2004] Peplow, M., "Beagle Team Hounds Space Agency over Lost Lander", Nature, 430, 2004, 954

[Perrier-2005a] Perrier, S., et al., "Ozone Retrieval from SPICAM UV and Near IR Measurements: A First Global View of Ozone on Mars", presentation to the 1st Mars Express Science Conference, Noordwijk, 2005

[Perrier-2005b] Perrier, S., et al., "Spatially Resolved UV Albedo of Phobos with SPICAM on Mars Express", presentation to the 1st Mars Express Science Conference, Noordwijk, 2005

[Picardi-2005] Picardi, G., et al., "Radar Soundings of the Subsurface of Mars", Science, 310, 2005, 1925-1928

[Pieters-2000] Pieters, C.M., et al. "Aladdin: Exploration and Sample Return from the Moons of Mars", paper presented at the Workshop on Concepts and Approaches for Mars Exploration, July 18-20, 2000, Houston, Texas

[Pillinger-2003] Pillinger, C., "Beagle: From Sailing Ship to Mars Spacecraft", London, Faber and Faber, 2003

[Pillinger-2005] Pillinger, C., et al., "A Combined Exobiology and Geophysics Mission to Mars 2009", presentation to the 1st Mars Express Science Conference, Noordwijk, 2005

[Pillinger-2007] Pillinger, C., "Space is a Funny Place", London, Barnstorm, 2007, 191-197

[Plaut-2007] Plaut, J.J., et al., "Subsurface Radar Sounding of the South Polar Layered Deposits of Mars", Science, 316, 2007, 92-95

[Podolak-2007] Podolak, M., "The Case of Saturn's Spin", Science, 317, 2007, 1330

[Porco-2003] Porco, C.C., et al., "Cassini Imaging of Jupiter's Atmosphere, Satellites, and Rings", Science, 299, 2003, 1541-1547

[Porco-2005a] Porco, C.C., et al., "Cassini Imaging Science: Initial Results on Phoebe and Iapetus", Science, 307, 2005, 1237-1242

[Porco-2005b] Porco, C.C., et al., "Cassini Imaging Science: Initial Results on Saturn's Rings and Small Satellites", Science, 307, 2005, 1226-1236

[Porco-2005c] Porco, C.C., et al., "Imaging of Titan from the Cassini Spacecraft", Nature, 434, 2005, 159-168

[Porco-2005d] Porco, C.C., et al., "Cassini Imaging Science: Initial Results on Saturn's Atmosphere", Science, 307, 2005, 1243-1247

[Porco-2006] Porco, C.C., et al., "Cassini Observes the Active South Pole of Enceladus", Science, 311? 2006, 1393-1401

[Porco-2007] Porco, C.C., et al. "Saturn's Small Inner Satellites: Clues to Their Origins", Science, 318, 2007, 1602-1607

[Porco-2008] Porco, C., "The Restless World of Enceladus", Scientific American, December 2008, 52-63

[Postberg-2008] Postberg, F., et al., "The E-Ring in the Vicinity of Enceladus II. Probing the Moon's Interior — The Composition of E-Ring Particles", Icarus, 193, 2008, 438-454

[Postberg-2009] Postberg, F., et al., "Sodium Salts in E-Ring Ice Grains from an Ocean below the Surface of Enceladus", Nature, 459, 2009, 1098-1101

[Postberg-2011] Postberg, F., et al., "A Salt-Water Reservoir as the Source of a Compositionally Stratified Plume on Enceladus", Nature, 474, 2011, 620-622

[Poulet-2005] Poulet, F., et al., "Phyllosilicates on Mars and Implications for Early Martian Climate", Nature, 438, 2005, 623-627

[Poulet-2008] Poulet, F., et al., "Abundance of Minerals in the Phyllosilicate-Rich Units on Mars", Astronomy & Astrophysics, 487, 2008, L41-L44

[Powell-2003] Powell, J.W., "Symbolic Mars", Spaceflight, September 2003, 364-365

[Price-2000] Price, H., et al., "Mars Sample Return Spacecraft Systems Architecture", paper presented at the 2000 IEEE Aerospace Conference, March 18-25, 2000, Big Sky, Montana

[Pryor-2008] Pryor, W., et al., "Radiation Transport of Heliospheric Lyman-a from Combined Cassini and Voyager Data Sets", Astronomy & Astrophysics, 491, 2008, 21–28

[Pryor-2011] Pryor, W.R., et al., "The Auroral Footprint of Enceladus on Saturn", Nature, 472, 2011, 331-333

[Pullan-2004] Pullan, D., et al., "Beagle 2: the Exobiological Lander of Mars Express", Noordwijk, ESA SP-1240, August 2004, 165-204

[Racca-1998] Racca, G.D., Whitcomb, G.P., Foing, B.H., "The SMART-1 Mission", ESA Bulletin, 95, 1998, 72-81

[Rapp-1996] Rapp, D., et al., "The Suess-Urey Mission (Return of Solar Matter to Earth)", Acta Astronautica, Vol. 39, 1996, 229-238

[Ratcliff-1992] Ratcliff, P.R., et al., "The Cosmic Dust Analyser", Journal of the British Interplanetary Society, 45, 1992, 375-380

[Rayman-1999] Rayman, M.D., et al., "Results from the Deep Space 1 Technology Validation Mission", paper presented at the L Congress of the International Astronautical Federation, Amsterdam, 1999

[Rayman-2000] Rayman, M.D., Varghese, P., "The Deep Space 1 Extended Mission", paper presented at the LI Congress of the International Astronautical Federation, Rio de Janeiro, 2000

[Rayman-2002a] Rayman, M.D., "The Deep Space 1 Extended Mission: Challenges in Preparing for an Encounter with Comet Borrelly", Acta Astronautica, 51, 2002, 507-516

[Rayman-2002b] Rayman, M.D., "The Successful Conclusion of the Deep Space 1 Mission: Important Results without a Flashy Title", paper presented at the LIII Congress of the International Astronautical Federation, Houston, 2002

[Read-2009] Read, P.L., Dowling, T.E., Schubert, G., "Saturns Rotation Period from its Atmospheric Planetary-Wave Configuration", Nature, 460, 2009, 608-610

[Read-2011] Read, P., "Storm-Clouds Brooding on Towering Heights", Nature, 475, 2011, 44-45

[Reeves-2005] Reeves, G., Neilson, T., "The Mars Rover Spirit FLASH Anomaly", paper presented at the 2005 IEEE Aerospace Conference

[Reichhardt-1999] Reichhardt, T., "The One-Pound Problem", Air & Space Smithsonian, October/November 1999, 50-57

[Reichhardt-2000] Reichhardt, T., "Doubts and Uncertainties Slow NASA's Schedule", Nature, 405, 2000, 4

[Reisenfeld-2005] Reisenfeld, D.B., et al., "The Genesis Mission Solar Wind Samples:

Collection Times, Estimated Fluences and Solar-Wind Conditions", paper presented at the XXXVI Lunar and Planetary Science Conference, Houston, 2005

[Reynolds-2001] Reynolds, E., et al., "The Use of Hibernation Modes for Deep Space Missions as a Method to Lower Mission Operation Costs", paper presented at the 15th annual AIAA/USU Conference on Small Satellites, August 2001

[Richter-2001] Richter, I., et al., "First Direct Magnetic Field Measurements of an Asteroidal Magnetic Field: DS1 at Braille", Geophysical Research Letters, 28, 2001, 1913-1916

[Riedel-2000] Riedel, J.E., et al., "Using Autonomous Navigation for Interplanetary Missions: The Validation of Deep Space 1 AutoNav", paper presented at the IV IAA International Conference on Low-Cost Planetary Missions, Laurel, 2-5 May 2000

[Rieder-2004] Rieder, R., et al., "Chemistry of Rocks and Soils at Meridiani Planum from the Alpha Particle X-ray Spectrometer", Science, 306, 2004, 1746-1749

[Rieke-2006a] Rieke, G.H., "The Last of the Great Observatories: Spitzer and the Era of Faster, Better, Cheaper at NASA", Tucson, University of Arizona Press, 2006

[Rieke-2006b] ibid., 119

[Rodriguez-2009a] Rodriguez, S., et al., "Cassini/VIMS hyperspectral observations of the HUYGENS landing site on Titan" arXiv astro-ph/0906.5476 preprint

[Rodriguez-2009b] Rodriguez, S., et al., "Global Circulation as the Main Source of Cloud Activity on Titan", Nature, 459, 2009, 678-682

[Ryne-2004] Ryne, M., Nandi, S., "Nozomi Earth Swingby Orbit Determination", Paper AAS 04-131

[Saito-2006] Saito, J., et al., "Detailed Images of Asteroid 25143 Itokawa from Hayabusa", Science, 312, 2006, 1341-1344

[Sanchez-Lavega-1989] Sanchez-Lavega, A., "Saturn's Great White Spots", Sky & Telescope, August 1989, 141-143

[Sanchez-Lavega-2005] Sanchez-Lavega, A., "How Long Is the Day on Saturn?", Science, 307, 2005, 1223-1224

[Sanchez-Lavega-2011] Sanchez-Lavega, A., et al., "Deep Winds beneath Saturn's Upper Clouds from a Seasonal Long-Lived Planetary-Scale Storm", Nature, 475, 2011, 71-77

[Sandford-2006] Sandford, S.A., et al., "Organics Captured from Comet 81P/Wild 2 by the Stardust Spacecraft", Science, 314, 2006, 1720-1724

[Sanford-1992] Sanford, M.C.W., "The Cassini/Huygens Mission and the Scientific Involvement of the United Kingdom", Journal of the British Interplanetary Society, 45, 1992, 355-358

[Sasaki-1999] Sasaki, S., et al., "Initial Results of Mars Dust Counter (MDC) on Board Nozomi: Leonids Encounter", paper presented at the XXX Lunar and Planetary Science Conference, Houston, 1999

[Sasaki-2005] Sasaki, S., et al., "Summary of Observation of Interplanetary and Interstellar Dust by Mars Dust Counter on Board Nozomi", paper presented at the Workshop on Dust in Planetary Systems, Kaua'i, September 2005

[Saunders-2004] Saunders, R.S., et al., "2001 Mars Odyssey Mission Summary", Space Science Reviews, 110, 2004, 1-36

[Schaller-2006] Schaller, E.L., et al, "A Large Cloud Outburst at Titan's South" Pole", Icarus, 182, 2006, 224-229

[Schaller-2009] Schaller, E.L., "Storms in the Tropics of Titan", Nature, 460, 2009, 873-875

[Schipper-2006] Schipper, A.M., Lebreton, J.-P., "The Huygens Probe – Space History in Many Ways", Acta Astronautica, 59, 2006, 319-334

[Schmidt-1998] Schmidt, R., et al., "The Mars Express Mission Concept – A New Management Approach", ESA Bulletin, 95-1998, 66-71

[Schmidt-1999] Schmidt, R., et al., "ESA's Mars Express Mission – Europe on Its Way to Mars", ESA Bulletin, 98, 1999, 56-66

[Schmidt-2010] Schmidt, F., et al., "Sublimation of the Martian CO2 Seasonal South Polar Cap", arXiv preprint astro-ph/1003.4453v1

[Schneider-2012] Schneider, T., et al., "Polar Methane Accumulation and Rainstorms on Titan from Simulations of the Methane Cycle", Nature, 481, 2012, 58-61

[Schultz-2007] Schultz, P.H., "Hidden Mars", Science, 318, 2007, 1080-1081

[Schwochert-1997] Schwochert, M., "Stardust Navigation Camera Instrument Description Document", JPL Document SD-74000-100, 30 June 1997

[Science-2008] "Cooking up the Solar System from the Right Ingredients", Science, 319, 2008, 1756

[Scotti-1998] "Fleeting Expectations: The Tale of an Asteroid", Sky & Telescope, July 1998, 30-34

[Sears-2004] Sears, D., et al., "The Hera Mission: Multiple Near-Earth Asteroid Sample Return", Advances in Space Research, 34, 2004, 2270-2275

[Sekanina-2004] Sekanina, Z., et al., "Modeling the Nucleus and Jets of Comet 81P/Wild 2 Based on the Stardust Encounter Data", Science, 304, 2004, 1769-1774

[Sekigawa-2003] Sekigawa, E., Mecham, M., "Pick up Bits", Aviation Week & Space Technology, 19 May 2003, 40-41

[Selsis-2005] Selsis, F., et al., "A Martian Meteor and its Parent Comet", Nature, 435, 2005, 581

[Shemansky-2005] Shemansky, D.E., et al., "The Cassini UVIS Stellar Probe of the Titan Atmosphere", Science, 308, 2005, 978-982

[Shiibashi-2010] Shiibashi, K., "Back to Base", Aviation Week & Space Technology, 28 June 2010, 31-32

[Shirley-1999] Shirley, D.L., "Touching Mars: 1998 status of the Mars Robotic Exploration Program", Acta Astronautica, 45, 1999, 249-265

[Showalter-2005] Showalter, M.R., "Saturn's Strangest Ring Becomes Curiouser and Curiouser", Science, 310, 2005, 1287-1288

[Showman-2009] Showman, A.P., "Windy clues to Saturn's spin", Nature, 460, 2009, 582

[Sims-2004a] Sims, M.R. (ed.), "Beagle 2 Mars: Lessons Learned", University of Leicester, 2004

[Sims-2004b] Sims, M.R. (ed.), "Beagle 2 Mars: Mission Report", University of Leicester, 2004

[Smith-1994] Smith, B.A., "U.S./Russian Flights to Planets Discussed", Aviation Week & Space Technology, 20 June 1994, 60

[Smith-1997a] Smith, B.A., "Cassini Readied for Marathon Mission", Aviation Week & Space Technology, 5 May 1997, 42-45

[Smith-1997b] Smith, B.A., "Cassini Team Refines Science Sequences", Aviation Week & Space Technology, 5 May 1997, 45-47

[Smith-1998] Smith, J.C., "Description of Three Candidate Cassini Satellite Tours", paper AAS 98-106

[Smith-2000] Smith, B.A., "NASA Weighs Mission Options", Aviation Week & Space Technology, 11 December 2000, 54-59

[Smith-2001a] Smith, P.H., et al., "The MVACS Surface Stereo Imager on Mars Polar Lander", Journal of Geophysical Research, 106, 2001, 17589-17607

[Smith-2001b] Smith, B.A., "Odyssey Goes On Despite Glitch", Aviation Week & Space Technology, 27 August 2001, 31

[Smith-2002a] Smith, J.C., Bell, J.L., "2001 Mars Odyssey Aerobraking", paper AIAA 2002-4532

[Smith-2002b] Smith, B.A., "Mars Odyssey Poised to Deploy GRS Boom", Aviation Week & Space Technology, 20 May 2002, 65

[Smith-2003] Smith, N.G., et al., "Genesis - The Middle Years", paper presented at the 2003 IEEE Aerospace Conference , March 8-15, 2003

[Smith-2004] Smith, M.D., et al., "First Atmospheric Science Results from the Mars Exploration Rovers Mini-TES", Science, 306, 2004, 1750-1753

[Smrekar-1999] Smrekar, S., et al., "Deep Space 2: the Mars Microprobe Mission", Journal of Geophysical Research, 104, 1999, 27,013-27,030

[Soderblom-2000a] Soderblom, L.A., et al. "Miniature Integrated Camera Spectrometer (MICAS) Validation Report", paper presented at the Deep Space 1 Technology Validation Symposium, Pasadena, 2000

[Soderblom-2000b] Soderblom, L.A., "New Short-Wavelength Infrared Spectra of Mars (1.3 to 2.5 mm) from the Miniature Integrated Camera Spectrometer (MICAS) on Deep Space 1", paper presented at the XXXI Lunar and Planetary Science Conference, Houston, 2000

[Soderblom-2001] Soderblom, L.A., Yelle, R.V., "Near-Infrared Reflectance Spectroscopy of Mars (1.4 to 2.6 mm) from the Deep Space 1 Miniature Integrated Camera Spectrometer (MICAS)", paper presented at the XXXII Lunar and Planetary Science Conference, Houston, 2001

[Soderblom-2002] Soderblom, L.A., et al., "Observations of Comet 19P/Borrelly by the Miniature Integrated Camera and Spectrometer Aboard Deep Space 1", Science, 296, 2002, 1087-1091

[Soderblom-2004a] Soderblom, L.A., et al., "Imaging Borrelly", Icarus, 167, 2004, 4-15

[Soderblom-2004b] Soderblom, L.A., et al., "Short Wavelength Infrared (1.3-2.6 mm) Observations of the Nucleus of Comet 19P/Borrelly", Icarus, 167, 2004, 100-112

[Soderblom-2004c] Soderblom, L.A., et al., "Soils of Eagle Crater and Meridiani Planum at the Opportunity Rover Landing Site", Science, 306, 2004, 1723-1726

[Soderblom-2007a] Soderblom, L.A., et al., "Topography and Geomorphology of the Huygens Landing Site on Titan", Planetary and Space Science, 55, 2007, 2015-2024

[Soderblom-2007b] Soderblom, L.A., et al., "Correlations between Cassini VIMS Spectra and RADAR SAR Images: Implications for Titan's Surface Composition and the Character of the Huygens Probe Landing Site", Planetary and Space Science, 55, 2007, 2025-2036

[Soderblom-2009] Soderblom, L.A., et al., "The Geology of Hotei Regio, Titan: Correlation of Cassini VIMS and RADAR", Icarus, 204, 2009, 610-618

[Sohl-2010] Sohl, F., "Revealing Titan's Interior", Science, 327, 2010, 1338-1339

[Somma-2008] Somma, R., "Some Recent Results from the Cassini Titan Radar Mapper", Journal of the British Interplanetary Society, 61, 2008, 295-299

[Soobiah-2005] Perrier, S., et al., "Observations of Magnetic Anomaly Signatures in Mars Express ASPERA-ELS Data", presentation to the 1st Mars Express Science Conference, Noordwijk, 2005

[Sotin-2005] Sotin, C., et al., "Release of Volatiles from a Possible Cryovolcano from Near-Infrared Imaging of Titan", Nature, 435, 2005, 786-789

[Sotin-2007] Sotin, C., "Titan's Lost Seas Found", Nature, 445, 2007, 29-30

[Sotin-2008] Sotin, C., Tobie, G., "Titan's Hidden Ocean", Science, 319, 2008, 1629-1630

[Southwood-1992] Southwood, D.J., Balogh, A., Smith, E.J., "Dual Technique Magnetometer Experiment for the Cassini Orbiter Spacecraft", Journal of the British Interplanetary Society, 45, 1992, 371-374

[Spaceflight-1992] "A Minor Planet with a Tail", Spaceflight, October 1992, 315

[Spahn-2006a] Spahn, F., Schmidt, J., "Saturn's Bared Mini-Moons", Nature, 440, 2006, 614-615

[Spahn-2006b] Spahn, F., et al., "Cassini Dust Measurements at Enceladus and Implications for the Origin of the E Ring", Science, 311, 2006, 1416-1417

[Sparaco-1996] Sparaco, P., "Huygens Planetary Probe Set for Titan Landing in 2004", Aviation Week & Space Technology, 9 December 1996, 77-80

[Spencer-2006] Spencer, J.R., et al., "Cassini Encounters Enceladus: Background and the Discovery of a South Polar Hot Spot", Science, 311, 2006, 1401-1405

[Spencer-2010] Spencer, J.R., et al., "Formation of Iapetus' Extreme Albedo Dichotomy by Exogenically Triggered Thermal Ice Migration", Science, 327, 2010, 432-435

[Spilker-2010] Spilker, L., "Cassini-Huygens Solstice Mission", White paper for the Solar System Decadal Survey 2013- 2023

[Spitale-2006] Spitale, J.N., et al., "The Orbits of Saturn's Small Satellites Derived from Combined Historic and Cassini Imaging Observations", The Astronomical Journal, 132, 2006, 692-710

[Spitale-2007] Spitale, J.N., Porco, C.C., "Association of the Jets of Enceladus with the Warmest Regions on its South-Polar Fractures", Nature, 449, 2007, 695-697

[Spitale-2009] Spitale, J.N., Porco, C.C., "Detection of Free Unstable Modes and Massive Bodies in Saturn's Outer B Ring", arXiv astro-ph/0912.3489 preprint

[Sprague-2004] Sprague, A.L., et al., "Mars' South Polar Ar Enhancement: A Tracer for South Polar Seasonal Meridional Mixing", Science, 306, 2004, 1364-1367

[Squyres-2004a] Squyres, S.W., et al., "The Opportunity Rover's Athena Science Investigation at Meridiani Planum", Science, 306, 2004, 1698-1703

[Squyres-2004b] Squyres, S.W., et al., "In Situ Evidence for an Ancient Aqueous Environment at Meridiani Planum, Mars", Science, 306, 2004, 1709-1714

[Squyres-2004c] Squyres, S.W., et al., "The Spirit Rover's Athena Science Investigation at Gusev Crater, Mars" Science, 305, 2004, 794-799

[Squyres-2005a] Squyres, S., "Roving Mars", New York, Hyperion, 42-53

[Squyres-2005b] ibid., 71-72

[Squyres-2005c] ibid., 73-85

[Squyres-2005d] ibid., 86-93

[Squyres-2005e] ibid., 120-141

[Squyres-2005f] ibid., 237-287

[Squyres-2005g] ibid., 288-321

[Squyres-2005h] ibid., 322-329

[Squyres-2005i] ibid., 350-378

[Squyres-2005j] ibid., 329-349

[Squyres-2006] Squyres, S.W., et al., "Two Years at Meridiani Planum: Results from the Opportunity Rover" Science, 313, 2006, 1403-1407

[Squyres-2007] Squyres, S.W., et al., "Pyroclastic Activity at Home Plate in Gusev Crater, Mars", Science, 316, 2007, 738-742

[Squyres-2008] Squyres, S.W., et al., "Detection of Silica-Rich Deposits on Mars", Science, 320, 2008, 1063-1067

[Squyres-2009] Squyres, S.W., et al., "Exploration of Victoria Crater by the Mars Rover Opportunity", Science, 324, 2009, 2058-2061

[Sremcevic-2007] Sremcevic, M., et al., "A Belt of Moonlets in Saturn's A Ring", Nature, 449, 2007, 1019-1021

[ST-1996] "Comet Schwassmann-Wachmann 3 Breaks Up", Sky & Telescope, March 1996, 11

[ST-1999] "Getting a Feel for Braille", Sky & Telescope, October 1999, 28

[Staehle-1999] Staehle, R.L., et al., "Ice & Fire: Missions to the Most Difficult Solar System Destinations... on a Budget", Acta Astronautica, 45, 1999, 423-439

[Stallard-2008] Stallard, T., et al., "Complex Structure within Saturn's Infrared Aurora", Nature, 456, 2008, 214-217

[Stansbery-2001] Stansbery E.K., et al., "Genesis Discovery mission: Science canister processing at JSC", paper presented at the Lunar and Planetary Science Conference XXXII, Houston, 2001

[Stansbery-2005] Stansbery, E.K., and Genesis Recovery Processing Team, "Genesis Recovery Processing", paper presented at the XXXVI Lunar and Planetary Science Conference, Houston, 2005

[Steinberg-2003] Steinberg, J.T., et al., "Science Rationale for Observations from a Spacecraft in a Distant Retrograde Orbit: Case Study Using Genesis", Los Alamos National Laboratory paper LA-UR-03-6205

[Stephan-2008] Stephan, T., Leitner, J., van der Bogert, C.H., "Comparing Wild 2 Matter with Halley's Dust and Interplanetary Dust Particles", paper presented at the European Planetary Science Congress, Münster, 2008

[Stephan-2010] Stephan, K., et al., "Specular Reflection on Titan: Liquids in Kraken Mare", Geophysical Research Letters, 37, 2010, L07104

[Stiles-2008] Stiles, B.W., et al., "Determining Titan's Spin State from Cassini RADAR Images", The Astronomical Journal, 135, 2008, 1669-1680

[Stofan-2007] Stofan, E.R., et al., "The Lakes of Titan", Nature, 445, 2007, 61-64

[Stofan-2008] Stofan, E.R., "Varied Geologic Terrains at Titan's South Pole: First Results from T39", paper presented at the Lunar and Planetary Science Conference XXXIX, Houston, 2008

[Strange-2002] Strange, N. J., Goodson, T. D., Hahn, Y., "Cassini Tour Redesign for the Huygens Mission", paper presented at the AIAA Astrodynamics Specialist Conference, Monterey, August 2002

[Strange-2002] Strange, N. J., Goodson, T. D., Hahn, Y., "Cassini Tour Redesign for the Huygens Mission", paper presented at the AIAA Astrodynamics Specialist Conference, Monterey, August 2002

[Suess-1956] Suess H.E., Urey H.C., "Abundances of the elements", Reviews of Modern Physics, 28, 1956, 53–74

[Sullivan-2005] Sullivan, R., et al., "Aeolian Processes at the Mars Exploration Rover Meridiani Planum Landing Site", Nature, 436, 2005, 58-61

[Taguchi-2000] Taguchi, M., et al., "Ultraviolet Imaging Spectrometer (UVS) Experiment on Board the NOZOMI Spacecraft: Instrumentation and Initial Results", Earth Planets Space, 52, 2000, 49-60

[Taverna-1997] Taverna, M.A., Anselmo, J.C., "New Cooperative Spirit Bodes Well for Mars Exploration", Aviation Week & Space Technology, 13 October 1997, 24-25

[Taverna-1999a] Taverna, M.A., "Microsats to Back Up Sample Return Missions", Aviation Week & Space Technology, 15 February 1999, 22-23

[Taverna-1999b] Taverna, M.A., "U.K. Funding Boost Mars Lander Project", Aviation Week & Space Technology, 16 August 1999, 22-23

[Taverna-2000a] Taverna, M.A., "Europe Targets 2003 Mars Touchdown", Aviation Week & Space Technology, 11 December 2000, 71-75

[Taverna-2000b] Taverna, M.A., "ESA Firms up Mars Lander Project, Explores Role in 2005 Mission", Aviation Week & Space Technology, 27 November 2000, 45

[Taverna-2000c] Taverna, M.A., "Europe to Have Major Sample Return Role", Aviation Week & Space Technology, 11 December 2000, 60-63

[Taverna-2004] Taverna, M.A., "Mars in 3D" Aviation Week and Space Technology, 2 February 2004, 38-39

[Taylor-1999] Taylor, F.W., Calcutt, S.B., Vellacott, T., "An Experimental Investigation into the Present-Day Climate of Mars: the PMIRR Experiment on the Mars Climate Orbiter". In: Hiscox, J.H. (ed.), "The Search for life on Mars", London, British Interplanetary society, 1999, 89-92

[Taylor-2005] Taylor, J., et al., "Mars Exploration Rover Telecommunications", JPL DESCANSO Design and Performance Summary Series, Article 10, October 2005

[Teolis-2010] Teolis, B.D., et al., "Cassini Finds an Oxygen-Carbon Dioxide Atmosphere at Saturn's Icy Moon Rhea", Science, 330, 2010, 1813-1815

[Thomas-2007] Thomas, P.C., et al., "Hyperion's sponge-like appearance", Nature, 448, 2007, 50-53

[Tiscareno-2006] Tiscareno, M.S. et al., "100-Metre-Diameter Moonlets in Saturn's A Ring from Observations of 'Propeller' Structures", Nature, 440, 2006, 648-650

[Tiscareno-2007] Tiscareno, M.S., "Ringworld Revelations", Sky & Telescope, February 2007, 32-39

[Tiscareno-2010] Tiscareno, M.S., et al., "Physical Characteristics and Non-Keplerian Orbital Motion of 'Propeller' Moons Embedded in Saturn's Rings", Astrophysical Journal Letters, 718, 2010, L92-L96

[Titus-2003] Titus, T.N., et al., "Exposed Water Ice Discovered Near the South Pole of Mars", Science, 299, 2003, 1048-1051

[Tokadoro-1999] Tokadoro, H., et al., "Nozomi Transmars Orbit". In: Proceedings of the 8th Workshop on Astrodynamics and Flight Mechanics, ISAS, 1999

[Tokano-2006] Tokano, T., et al., "Methane Drizzle on Titan", Nature, 442, 2006, 432-435

[Tokar-2006] Tokar, R.L., et al., "The Interaction of the Atmosphere of Enceladus with Saturn's Plasma", Science, 311, 2006, 1409-1412

[Tokar-2012] Tokar, R.L., et al., "Detection of exospheric $O2+$ at Saturn's moon Dione", Geophysical Research Letters, 39, 2012, L03105

[Tomasko-1997] Tomasko, M.G., et al., "The Descent Imager/Spectral Radiometer (DISR) Aboard Huygens". In: "Huygens: Science, Payload and Mission", Noordwijk, ESA, 1997, 109-138

[Tomasko-2005] Tomasko, M.G., et al., "Rain, Winds and Haze During the Huygens Probe's Descent to Titan's Surface", Nature, 438, 2005, 765-778

[Tortora-2004] Tortora, P., et al., "Precise Cassini Navigation During Solar Conjunctions Through Multifrequency Plasma Calibrations", Journal of Guidance, Control, and Dynamics, 27, 2004, 251-257

[Towner-2000] Towner, M.C:, et al., "The Beagle 2 Environmental Sensors: Instrument Measurements and Capabilities", paper presented at the XXXI Lunar and Planetary Science Conference, Houston, 2000

[Tsou-1993] Tsou, P., Brownlee, D.E., Albee, A.L., "Intact Capture of Hypervelocity Particles on Shuttle", paper presented at the XXIV Lunar and Planetary Science Conference, Houston, 1993

[Tsou-2004] Tsou, P., et al., "Stardust Encounters Comet 81P/Wild 2", Journal of Geophysical Research, 109, 2004

[Tsuchiyama-2011] Tsuchiyama, A., et al., "Three-Dimensional Structure of Hayabusa Samples: Origin and Evolution of Itokawa Regolith", Science, 333, 2008, 1125-1128

[Tsurutani-2004] Tsurutani, B.T., et al., "Plasma Clouds Associated with Comet P/Borrelly Dust Impacts", Icarus, 167, 2004, 89-99

[Turtle-2009] Turtle, E.P., et al., "Cassini Imaging of Titan's High-Latitude Lakes, Clouds, and South-Polar Surface Changes", Geophysical Research Letters, 36, 2009, L02204

[Turtle-2011] Turtle, E.P., et al., "Rapid and Extensive Surface Changes Near Titan's Equator: Evidence of April Showers", Science, 331, 2011, 1414-1417

[Tuzzolino-2004] Tuzzolino, A.J., et al., "Dust Measurements in the Coma of Comet 81P/Wild 2 by the Dust Flux Monitor Instrument", Science, 304, 2004, 1776-1780

[Tytell-2004] Tytell, D., "NASA's Ringmaster", Sky & Telescope, November 2004, 38-42

[Tytell-2006] Tytell, D., "Mars Polar Lander Still Missing", Sky & Telescope, January 2006, 22

[Veazey-2004] Veazey, G.R., "Recovery System Explained", Aviation Week & Space Technology, 8 November 2004, 6

[Verbiscer-2009] Verbiscer, A.J., Skrutskie, M.F., Hamilton, D.P., "Saturn's Lagest Ring", Nature, 461, 2009, 1098-1100

[Veverka-1999] Veverka, J., Yeomans, D.K., "Comet Nucleus Tour (CONTOUR) A Mission to Study the Diversity of Comet Nuclei", paper presented at the Torino Impact Workshop, June 1999

[Veverka-2011] Veverka, J., et al., "Return to Comet Tempel 1: Results from Stardust-NExT", paper presented at the European Planetary Science Congress, Nantes, 2011

[Volpe-2000] Volpe, R., et al., "Technology Development and Testing for Enhanced Mars Rover Sample Return Operations", paper presented at the 2000 IEEE Aerospace Conference, March 18-25, 2000, Big Sky, Montana

[Waite-2005a] Waite, J.H. Jr., et al., "Oxygen Ions Observed Near Saturn's A Ring", Science, 307, 2005, 1260-1262

[Waite-2005b] Waite, J.H. Jr., et al., "Ion Neutral Mass Spectrometer Results from the First Flyby of Titan", Science, 308, 2005, 982-986

[Waite-2006] Waite, J.H. Jr, et al., "Cassini Ion and Neutral Mass Spectrometer: Enceladus Plume Composition and Structure", Science, 311, 2006, 1419-1422

[Waite-2007] Waite, J.H. Jr., et al., "The Process of Tholin Formation in Titan's Upper Atmosphere", Science, 316, 2007, 870-875

[Waite-2009] Waite, J.H. Jr., et al., "Liquid Water on Enceladus from Observations of Ammonia and 40Ar in the Plume", Nature, 460, 2009, 487-490

[Wald-2009] Wald, C., "In Dune Map, Titan's Winds Seem to Blow Backward", Science, 323, 2009, 1418

[Waller-2003] Waller, W.H., "NASA's Space Infrared Telescope Facility: Seeking Warmth in the Cosmos", Sky & Telescope, February 2003, 42-48

[Wang-2000] Wang, J., et al., "Deep Space One Investigations of Ion Propulsion Plasma Environment", Journal of Spacecraft and Rockets, 37, 2000, 545-555

[Warwick-2000] Warwick, G., "Overload Caused Mars Failures", Flight International, 11 April 2000, 41

[Watters-2006] Watters, T.R., et al., "MARSIS Radar Sounder Evidence of Buried Basins in the Northern Lowlands of Mars", Nature, 444, 2006, 905-908

[Watters-2007] Watters, T.R., et al., "Radar Sounding of the Medusae Fossae Formation Mars: Equatorial Ice or Dry, Low-Density Deposits?", Science, 318, 2007, 1125-1128

[West-2005] West, R.A., et al., "No Oceans on Titan from the Absence of a Near-Infrared Specular Reflection", 436, 2005, 670-672

[Westphal-2010] Westphal, A.J., et al., "Analysis of 'Midnight' Tracks in the Stardust Interstellar Dust Collector: Possible Discovery of a Contemporary Interstellar Dust Grain", paper presented at the XXXXI Lunar and Planetary Science Conference, Houston, 2010

[Westwick-2007] Westwick, P.J., "Into the Black: JPL and the American Space Program 1976-2004", New Haven, Yale University Press, 2007, 276-279

[Whipple-1950] Whipple, F.L., "A Comet Model. I. The Acceleration of Comet Encke", the Astrophysical Journal, 111, 1950, 375-394

[Wilcox-2000a] Wilcox, B.H., Jones, R.M. "The MUSES-CN Nanorover Mission and Related Technology", paper presented at the 2000 IEEE Aerospace Conference

[Wilcox-2000b] Wilcox, B.H., "A Miniature Mars Ascent Vehicle", paper presented at the Workshop on Concepts and Approaches for Mars Exploration, July 18-20, 2000, Houston, Texas

[Willner-2008] Willner, K., et al., "New Astrometric Observations of Phobos with the SRC on Mars Express", Astronomy & Astrophysics, 488, 2008, 361–364

[Wilson-2002] Wilson, R., "Genesis: Mission Design and Operations", paper presented at the conference on Libration Points and Applications , June 10-14, 2002

[Wilson-2004] Wilson, R.S., Barden, B.T., Chung, M.-K.J., "Trajectory Design for the Genesis Backup Orbit and Proposed Extended Mission", paper AAS 04-227

[Withcomb-1988] Withcomb, G.P., Corradini, M., Volonté, S., "The 1998 Scientific-Programme Selection", ESA Bulletin, 55, 1988, 10-11

[Withers-2006] Withers, P., Smith, M.D., "Atmospheric entry profiles from the Mars Exploration Rovers Spirit and Opportunity", Icarus, 185, 2006, 133-142

[Woerner-1998] Woerner, D.F., "Revolutionary Systems and Technologies for Missions to the Outer Planets", paper presented at the Second IAA Symposium on Realistic Near-Term Advanced Scientific Space Missions, Aosta, 29 June-1 July 1998

[Wolf-1998] Wolf, A.A., "Incorporating Icy Satellite Flybys in the Cassini Orbital Tour", paper AAS 98-107

[Wolf-2007] Wolf, A., et al., "Stardust New Exploration of Tempel-1 (NExT)", paper presented at the Seventh IAA International Conference on Low-Cost Planetary Missions, Pasadena, 12-14 September 2007

[Wolff-2005] Wolff, M.J., Smith, M.D., "Things Are Looking Up", Sky & Telescope, March 2005, 44-45

[Wolkenberg-2008] Wolkenberg, P., et al., "Simultaneous observations of Martian atmosphere by PFS-MEX and Mini-TES-MER", paper presented at the European Planetary Science Congress, Münster, 200

[Wray-2009] Wray, J.J., et al., "Phyllosilicates and Sulfates at Endeavour Crater, Meridiani Planum, Mars", Geophysical Research Letters, 36, 2009, L21201

[Yam-2007] Yam, C.H., et al., "Saturn Impact Trajectories for Cassini End-of-Life", Paper AAS 07-257

[Yamakawa-1999] Yamakawa, H., "PLANET-B Orbit Around Mars (2004 Arrival)". In: Proceedings of the 8th Workshop on Astrodynamics and Flight Mechanics, ISAS, 1999

[Yamamoto-1998] Yamamoto, Y., Tsuruda, K., "The PLANET-B Mission", Earth Planets Space, 50, 1998, 175-181

[Yano-2006] Yano, H., et al., "Touchdown of the Hayabusa Spacecraft at the Muses Sea on Itokawa", Science, 312, 2006, 1350-1353

[Yoshikawa-2005] Yoshikawa, M., et al., "Summary of the Orbit Determination of NOZOMI Spacecraft for all the Mission Period", Acta Astronautica, 57, 2005, 510-519

[Yoshimitsu-2006] Yoshimitsu, T., Kubota, T., Nakatani, I., "The opearation [sic] and scientific data of MINERVA rover in Hayabusa mission", Paper COSPAR 2006-A-02987

[Young-2004] Young, D.T., et al., "Solar Wind Interactions with Comet 19P/Borrelly", Icarus, 167, 2004, 80-88

[Young-2005] Young, D.T., et al., "Composition and Dynamics of Plasma in Saturn's Magnetosphere", Science, 307, 2005, 1262-1266

[Zarka-2007] Zarka, P., et al., "Modulation of Saturn's Radio Clock by Solar Wind Speed", Nature, 450, 2007, 265-267

[Zarka-2008] Zarka, P., et al., "Ground-Based and Space-Based Radio Observations of Planetary Lightning", Space Science Reviews, 137, 2008, 257-269

[Zarnecki-1992] Zarnecki, J.C., "Surface Science Package for the Huygens Titan Probe", Journal of the British Interplanetary Society, 45, 1992, 365-370

[Zarnecki-2005] Zarnecki, J.C., et al., "A Soft Solid Surface on Titan as Revealed by the Huygens Surface Science Package", Nature, 438, 2005, 792-795

[Zebker-2009] Zebker, H.A., et al., "Size and Shape of Saturn's Moon Titan", Science, 324, 2009, 921-923

[Zimmerman-2001] Zimmerman, W., Bonitz, R., Feldman, J., "Cryobot: An Ice Penetrating Robotic Vehicle for Mars and Europa.", paper presented at the 2001 IEEE Aerospace Conference, Big Sky

[Zolensky-2006] Zolensky, M.E., et al., "Mineralogy and Petrology of Comet 81P/Wild 2 Nucleus Samples", Science, 314, 2006, 1735-1739

[Zuber-2003] Zuber, M.T., "Learning to Think Like Martians", Science, 302, 2003, 1694-1695

[Zubrin-1993] Zubrin, R., et al., "A New MAP for Mars", Aerospace America, September 1993, 20-24

# Further reading

## BOOKS

Godwin, R., (editor), "Deep Space: The NASA Mission Reports", Burlington, Apogee, 2005
Godwin, R., (editor), "Mars: The NASA Mission Reports Volume 2", Burlington, Apogee, 2004
Kelly Beatty, J., Collins Petersen, C., Chaikin, A. (editors), "The New Solar System", 4th edition, Cambridge University Press, 1999
Shirley, J.H., Fairbridge, R.W., "Encyclopedia of Planetary Sciences", Dordrecht, Kluwer Academic Publishers, 1997

## MAGAZINES

Aerospace America
l'Astronomia (in Italian)
Aviation Week & Space Technology
Espace Magazine (in French)
Flight International
Novosti Kosmonavtiki (in Russian)
Science
Scientific American
Sky & Telescope
Spaceflight

## INTERNET SITES

ESA (www.esa.int)
Jonathan's Space Home Page (planet4589.org/space/space.html)
JPL (www.jpl.nasa.gov)
Malin Space Science Systems (www.msss.com)
NASA NSSDC (nssdc.gsfc.nasa.gov)
Novosti Kosmonavtiki (www.novosti-kosmonavtiki.ru)

Space Daily (www.spacedaily.com)
Spaceflight Now (www.spaceflightnow.com)
The Planetary Society (planetary.org)

# Previous volumes in this series

**Part 1: The golden age 1957–1982**

**Part 2: Hiatus and renewal  1983–1996**

# Index